淀粉-琼脂糖凝胶释放电泳
——发现和研究

Starch-agarose Gel Release Electrophoresis
——Discovery and Research

秦文斌 著
Author　Qin Wenbin

科学出版社
北 京

内 容 简 介

本书的主要内容：一是在基础理论方面，发现红细胞内各种蛋白质的两种存在状态：①红细胞内 HbA_2 与 HbA_1 结合存在，还有 PRX_2（过氧化物还原酶2）参与；②红细胞内 HbA_1 与 CA_1（碳酸酐酶1）结合存在。二是在临床实践方面，发现各种疾病释放情况不同：①溶血性疾病时全血 HbA_2 现象消失；②地中海贫血时，全血再释放增强；③球形红细胞增多症时，释放现象消失，而且红细胞内没有 HbA_3；④糖尿病时，血糖浓度与全血的多带再释放相关；⑤阻塞性黄疸时，红细胞和全血再释放都增强；⑥溶血性黄疸时，红细胞和全血再释放都减弱；⑦肝细胞性黄疸时，位于二者之间，红细胞再释放减弱、全血再释放增强。

本书可作为生物专业及医学相关专业研究人员的参考用书。

图书在版编目(CIP)数据

淀粉-琼脂糖凝胶释放电泳：发现和研究 / 秦文斌著. —北京：科学出版社，2018.8
　ISBN 978-7-03-058338-3

　Ⅰ. ①淀⋯　Ⅱ. ①秦⋯　Ⅲ. ①电泳–应用–琼脂–淀粉制品–研究
Ⅳ. ①TQ151.7

中国版本图书馆 CIP 数据核字（2018）第 165284 号

责任编辑：周　园 / 责任校对：郭瑞芝
责任印制：张欣秀 / 封面设计：陈　敬

科学出版社 出版
北京东黄城根北街 16 号
邮政编码：100717
http://www.sciencep.com

北京虎彩文化传播有限公司 印刷
科学出版社发行　各地新华书店经销

*

2018 年 8 月第 一 版　　开本：787×1092　1/16
2018 年 8 月第一次印刷　　印张：28 3/4
字数：720 000
定价：298.00 元
（如有印装质量问题，我社负责调换）

著者简介

秦文斌，男，1928年生，沈阳人，1953年毕业于中国医科大学研究生班，后留校任助教，1956年支援边疆来到内蒙古，扎根边疆。目前虽已离休，但仍在做实验研究，继续写书和发表文章。著者长期从事血红蛋白研究，1978年参加第一届全国科学大会，获先进个人奖（相当于全国劳动模范待遇）。1984年出版《血红蛋白病》；2015年出版专著《红细胞内血红蛋白的电泳释放——发现和研究》；2016年出版专著《基因诊断多重PCR和通用引物PCR》；2017年再出版专著《无氧条件下红细胞内血红蛋白的电泳释放——发现和研究》。此次出版的《淀粉-琼脂糖凝胶释放电泳——发现和研究》，是著者的终身之作。

1981年著者开始淀粉-琼脂糖凝胶释放电泳研究，时至今日已经过去30多年。在这漫长的岁月里，著者做过成千上万次电泳实验，积累下许多经验和教训，发现这种凝胶电泳具备很多优点。第一，电泳释放作用，将完整红细胞或全血加入凝胶进行电泳，发现从红细胞释放出来一些"特殊的血红蛋白"，即初释放血红蛋白类和再释放血红蛋白类。第二，这种电泳还发现血红蛋白的相互作用，包括交叉互作和混合互作。第三，它能由红细胞分离和准备各种血红蛋白等，用于进一步分析。第四，进一步分析中，最明显的例子就是用来做质谱分析，从而进入著者提出的"电泳释放蛋白质组学"，它与经典蛋白质组学不同，是对由细胞释放出来的蛋白质进行的蛋白质组学研究。

总之，著者创建的淀粉-琼脂糖凝胶释放电泳，在基础理论和临床应用两方面都有许多新发现。

第一，在基础理论方面，发现红细胞内各种血红蛋白等的不同存在状态：①红细胞内HbA_2与HbA_1结合存在，还有PRX_2（过氧化物还原酶2）参与，它又与红细胞膜疏松结合；②红细胞内HbA_1与CA_1（碳酸酐酶1）结合存在，它们又与红细胞膜牢固结合。第二，在临床应用方面发现：①全血HbA_2现象，有助于了解溶血性疾病的存在；②地中海贫血时，全血再释放增强；③球形红细胞增多症时，其红细胞内快泳血红蛋白消失；④糖尿病时，血糖浓度与全血的多带再释放相关；⑤阻塞性黄疸时，红细胞和全血再释放都增强；⑥溶血性黄疸时，红细胞和全血再释放都减弱；⑦肝细胞性黄疸时，位于二者之间，红细胞再释放减弱、全血再释放增强。

著者认为，科研无止境，永远在路上。

主要参与者

多年来和我一起做过实验的人们（包括家人和好友）(按姓氏汉语拼音排列)

白桂兰	白利平	白秀梅	宝勿仁	曹国栋	车桂花	常建萍	常江	陈宁	陈德喜
陈启明	陈晓东	陈言东	崔丽霞	崔珊娜	党彤	丁国平	丁海麦	丁海涛	丁慧荣
丁晓岭	董乐乐	董艳丽	窦君	杜茂林	额尔登	冯慧琼	付建刚	高敏	高阳
高桂英	高丽君	高雅琼	高永生	高长青	葛华	郭俊	郭春林	郭晓玲	韩丽红
韩丽莎	韩明友	郝吉林	郝亚胜	郝艳梅	何海英	何培生	和姬苓	和彦苓	贺其图
洪高明	侯安国	胡凤英	华淋	华伟	黄颖	霍建新	霍秀丽	贾粉	贾恭
贾璐	贾春梅	贾存德	贾国荣	贾尼娅	贾瑞平	贾彦彬	姜慧荣	姜素媛	焦健霞
焦玲君	焦勇钢	金树琦	经鑫爱	居红格	睢天林	康耀霞	孔凡青	寇晓丽	李斌
李贵	李静	李莉	李琴	李薇	李鑫	李喆	李成义	李豪侠	李嘉欣
李金萍	李俊峰	李丕宇	李晓红	李晓晶	李月春	李增艳	梁彩云	梁友珍	林爱卿
刘芬	刘佳	刘健	刘丽	刘睿	刘丽萍	刘素梅	刘文学	龙桂芳	卢艳
卢文英	吕莲英	马强	马登峰	马宏杰	孟俊	孟祥军	孟云清	孟云霞	那日苏
南蕾	潘桂兰	裴娟慧	奇那顺	乔姝	秦浩	秦然	秦良光	秦良伟	秦良谊
秦佩媛	秦艳晶	秦玉珍	尚忠义	邵国	申玲	沈靖	沈木生	石继海	石瑞丽
史平	斯琴	宋芳	宋瑞琪	宋玉娥	苏燕	苏丽娅	孙刚	孙桥	孙洪英
孙丽蓉	孙小荣	折志刚	腾喻	田伟	田春燕	田慧芳	田园青	田志华	王程
王辉	王琪	王燕	王英	王步云	王彩丽	王翠峰	王大光	王海龙	王建勋
王媚媚	王秋凤	王树平	王小利	王晓明	王晓平	王兴业	王秀娜	王迎新	王颖慧
王宇晗	王玉珍	王云丹	王占黎	魏枫	魏春华	乌兰	吴涤	吴刚	吴丽娥
武莎莎	席海燕	谢基明	辛佳音	邢娟	邢少姬	邢晓雁	熊睿	徐春秋	徐春忠
徐秀菊	闫斌	闫宇	闫春华	闫国珍	闫巧梅	闫少春	闫秀兰	杨静	杨森
杨颖	杨国安	杨文杰	杨艳红	杨占君	姚莉萍	尹卫东	于玲	于慧	于昌连
袁桂梅	袁晓俊	岳秀兰	张坤	张敏	张园	张爱萍	张春阳	张宏旺	张宏伟
张建国	张晶晶	张巨峰	张利荣	张茂林	张素英	张向今	张秀兰	张学明	张永红
张永强	张咏梅	张友良	赵敏	赵明清	赵淑梅	赵喜君	郑玉云	周成江	周家瑛
周俊红	周立社	朱王亮	朱秀珍	卓纳					

特此致谢：睢天林、岳秀兰和周立社等在早期实验中贡献明显；高丽君、韩丽红、高雅琼、宝勿仁等在后来实验中贡献明显。其中，高丽君的贡献是多方面的，除了在实验中，在制图、协助写书、排版、校对等方面亦贡献明显；苏燕在发表 SCI 文章方面贡献明显，高雅琼也逐渐开始发表 SCI 文章；周立社在出版和发放书籍方面贡献明显；邵国将分子进化概念引入血红蛋白研究；王占黎、于慧从生物信息学角度加强了血红蛋白研究。还有国内的折志刚、邢晓雁、白利平、王玉珍、王程、李琴、王大光、丁海涛等，国外的曹国栋、陈德喜、李金萍、武莎莎（医生）也都对本书做出了明显贡献。

关于序言的说明

本书是前两本书的集成和发展,故将它们的序言也都放在这里,衷心感谢两位序言作者,并留做永久纪念。第一本书:《红细胞内血红蛋白的电泳释放——发现和研究》2015 年由科学出版社出版,由曾溢滔院士作序,有中文和英文两种文字。第二本书:《无氧条件下红细胞内血红蛋白的电泳释放——发现和研究》2017 年由科学出版社出版,由广西医科大学龙桂芳教授作序。具体见后。

序　一

秦文斌教授和我是同行，共同研究血红蛋白多年，还一起发表过文章。我国20世纪60年代曾有一个"血红蛋白协作组"，组长是广西医科大学的梁徐教授（已过世），副组长是我和秦文斌教授。秦教授于1984年出版的《血红蛋白病》（人民卫生出版社），我在2003年出版的《人类血红蛋白》（科学出版社），医科院基础所张俊武教授和广西医科大学龙桂芳教授于2003年出版的《血红蛋白与血红蛋白病》（广西科学技术出版社），这些书全面介绍了国内外血红蛋白研究的新进展。

以上三本书，还有国外出版的许多有关血红蛋白的书籍，其中提到的血红蛋白都是来自红细胞裂解后的"溶血液"。过去认为，红细胞里边的血红蛋白与溶血液中血红蛋白相同，从1981年开始，秦教授用完整的红细胞做凝胶电泳，先后发现了血红蛋白的"初释放"和"再释放"现象。"初释放"是指红细胞电泳通电一次由红细胞释放出来的血红蛋白，"再释放"是指通电两次（停电—再通电）由红细胞释放出来的血红蛋白，多次停电—再通电也属于"再释放"。通过一系列实验，秦教授发现红细胞里血红蛋白存在下述许多特殊现象。

秦教授在"初释放"中发现，由红细胞释放出来的血红蛋白与溶血液的血红蛋白不同。在红细胞里的血红蛋白 A_2 是与血红蛋白 A_1 结合存在，刚释放出来时还在一起，随着电泳的进行，血红蛋白 A_1 带着血红蛋白 A_2 往前移动，慢慢分开，出现血红蛋白 A_2 与溶血液血红蛋白 A_2 电泳位置的差异，这就是秦教授所提出的"血红蛋白 A_2 现象"（1981年）。这一发现，揭示出红细胞的一个奥秘，那就是红细胞内的血红蛋白不是孤立存在，而是彼此依存，相互作用，共同完成运氧等生理功能。

秦教授在"再释放"中观察到，又有血红蛋白由红细胞再释放出来，它不是血红蛋白 A_2 与 A_1 的复合物，而是血红蛋白 A_1 与碳酸酐酶2（CA_2）结合存在，这是秦教授发现的红细胞的又一个奥秘。"再释放"血红蛋白与临床关系密切，不同形态的红细胞（如球形红细胞与靶形红细胞）"再释放"结果互异，中药也能影响"再释放"等，上述现象在秦教授的书里都有详细记载；在 Biji T. Kurien 和 R. Hal Scfield 教授主编的 *Protein Electrophoresis*(Humana Press,2012)里也有一章专门介绍秦教授的上述原创工作。

秦教授常说自己是中国本土科技工作者，我知道，他毕业于中国医科大学（沈阳），支援边疆来到内蒙古包头医学院，就在那里完成了他毕生的血红蛋白研究事业。今年他已经86岁，想出此书留给后人，可敬可贺，我祝福他身体健康、安度晚年。

中国工程院院士　曾溢滔
2014年6月于上海

Foreword

Professor Wenbin Qin and I used to work in the same field of hemoglobin research. We studied hemoglobin(Hb) for many years, and published some articles as co-authors. In 1960's, there was a "Hemoglobin Research Group"in our country. The leader of this group was Professor Liang Xu (died) of Guangxi Medical University,the vice leaders of this group were Professor Wenbin Qin and me. In 1984,Professor Qin completed a book Hemoglobin Disease(published by People's Medical Publishing House). In 2003,my book Human Hemoglobin(Science Press),and then Professors Junwu Zhang and Guifang Long's book Hemoglobin and Hemoglobin Disease(Guangxi Science and Technology Press)were also published. In these books the progresses of hemoglobin researches were introduced comprehensively.

Hemoglobin referred in the above three books and the other related books published in other countries are all isolated from hemolysate. In the past. hemoglobin isolated from red blood cells(RBCs)was thought to be identical to that from hemolysate. From 1981,Professor Qin began to utilize RBCs gel electrophoresis, and found the"initial release"and"re-release"phenomenon of hemoglobin. The"initial release"refers to the release of Hb from RBC by continuous electrophoresis without power-off of electricity. The"re-release"refers to the release of Hb by uncontinuous electrophoresis with more than two times of power off electricity. Through a series of experiments,Professor Qin discovered the following findings.

Professor Qin observed that during"initial release",hemoglobin released from RBCs was different from that released from hemolysate. In RBCs,HbA_2 was combined with HbA_1. These two Hbs were firstly released together from the RBCs and then moved forward. However,with the electrophoresis,HbA_2 was separated with HbA_1 gradually,and the band position of RBC HbA_2 was found to be proceeding different from that of hemolysate HbA_2. This was the"Hemoglobin A_2 phenomenon"proposed by Professor Qin in 1981. This finding reveals was that hemoglobin does not exist separately in RBCs,they could interact with each other to function physiologically in oxygen transportation.

Professor Qin also found that hemoglobin could be further released from RBCs during"rerelease electrophoresis". The re-released Hb is not HbA_2 and HbA_1, but a complex of HbA_1 and carbonic anhydrase$_2$(CA_2). This was another important finding by Professor Qin in RBCs. Re-released hemoglobin is closely associated with clinical diseases. Different morphological RBCs(such as spherocytes and target RBCs)have difierent"re-release"results. Chinese medicine can also affect their"re-release". These above phenomena were wrote in detail in this book,and the original work of Professor Qin was also described in Protein Electrophoresis(Humana Press,2012)edited by Professors Biji T. Kurien and R. Hal Scfield.

Professor Qin always says that he is a native Chinese scientist. I know that he was graduated from Chinese Medical University(Shenyang),and came to Baotou Medical College,Inner Mongolia for intelligent support. He completed his lifelong hemoglobin research career there. Now,he is eighty-six years old,wrote this book for young researchers. I am honored to write this preface and wish him a healthy and happy life.

Yitao Zeng

Academician of Chinese Academy of Engineering

June 2014, Shanghai

序　　二

秦文斌教授是我的同行，共同研究过血红蛋白，寄过血红蛋白病患者血液标本，以配合秦教授的血红蛋白释放研究。我曾是博导，评审博士生毕业论文时，多次请秦教授为评委，帮助完成博士生的论文答辩。

20世纪60年代，中国研究血红蛋白的人很多，比较有影响的带头人是：中国医学科学院的梁植权教授（院士）（已过世），上海儿童医院的曾溢滔教授，包头医学院血红蛋白研究室的秦文斌教授，广西医科大学儿科梁徐教授，（已过世）和我本人。那时候全国有一个血红蛋白科研协作组，组长是梁徐教授，副组长是秦文斌教授和曾溢滔教授。这几位都出过书，秦教授于1984年出版《血红蛋白病》（人民卫生出版社），曾溢滔教授（院士）于2003年出版《人类血红蛋白》（科学出版社），同年，我和医科院基础所的张俊武教授共同出版《血红蛋白与血红蛋白病》（广西科学技术出版社）。2015年，秦文斌教授又出版《红细胞内血红蛋白的电泳释放——发现和研究》（科学出版社），曾溢滔院士作序。

前边那几本书里研究的"血红蛋白"，都是使红细胞溶血，游离出来血红蛋白，研究它的结构及与疾病的关系。秦文斌教授2015年这本书，是用电泳的方法使血红蛋白离开红细胞，看看电泳释放出来的血红蛋白与来自溶血液的血红蛋白有无差异。实验结果显示，红细胞释放出来的血红蛋白与来自溶血液者不尽相同，其中包括"初释放现象"（也称"血红蛋白A_2现象"）和"再释放现象"。A_2现象的核心内容是，红细胞内HbA_2与HbA_1结合存在，而溶血液里二者是彼此分开的。"再释放现象"显示，红细胞内有一部分血红蛋白与细胞膜结合牢固，第二次通电才能释放出来，与多种疾病相关。

以上说的是秦教授2015年出版的书，在本书《无氧条件下红细胞内血红蛋白的电泳释放——发现和研究》中，秦教授想研究镰状细胞贫血患者红细胞的电泳释放，此病多见于美国黑人，在国内未能找到标本。此时，他想到梅花鹿的红细胞也是镰状，决定用它研究。实验过程中，普通电泳看不出梅花鹿红细胞的特点，后来，在凝胶里加入偏重亚硫酸钠，发现多带释放明显增强。将这套办法应用于人体红细胞，又发现一系列新情况。偏重亚硫酸钠还原能力很强，此时血红蛋白处于还原状态和无氧状态，所以称之为"无氧条件下红细胞内血红蛋白的电泳释放"。无氧条件下，人体红细胞的血红蛋白A_2现象消失，红细胞内的血红蛋白的相互作用消失。无氧条件下，红细胞外的血红蛋白之间的交叉互作也消失，当HbA_1穿过HbA_2时，后者不出现应有的区带变形。电泳过程中，无氧条件时血红蛋白区带拐弯，而有氧条件时不拐弯，说明无氧血红蛋白(Hb)与有氧血红蛋白 $(Hb-4O_2)$的电泳行为明显不同。这样，就可以推测出，肺侧（有氧）和组织侧（无氧）红细胞内血红蛋白的存在状态。

科研无止境，永远在路上。秦文斌教授今年已经八十九岁，还在实验研究，还在专心写作，我钦佩他的治学精神，预祝他身体健康，安度晚年。

2017年4月于广西

前　言

今年我已经 89 岁，虽已离休，但还在做实验研究，继续写书和发表文章。

1984 年我出版了第一本书《血红蛋白病》。离休后 2015 年出版第二本书《红细胞内血红蛋白的电泳释放——发现和研究》。2016 年出版第三本书《基因诊断多重 PCR 和通用引物 PCR》。2017 年出版第四本书《无氧条件下红细胞内血红蛋白的电泳释放——发现和研究》。

现在是第五本书《淀粉-琼脂糖凝胶释放电泳——发现和研究》，也是最后一本，是本人一生的"总结性"著作*。

从 1981 年开始淀粉-琼脂糖凝胶释放电泳研究，时至今日，已经过去 30 多年。在这漫长的岁月里，我做过成千上万次电泳实验，积累下许多经验和教训，发现这种凝胶电泳具备很多优点。

1. 它的电泳释放效果很好，将完整红细胞加入凝胶进行电泳，就能从红细胞释放出来一些"特殊的血红蛋白"，即初释放血红蛋白（也称"血红蛋白 A 型血现象"）和再释放血红蛋白。将全血加入凝胶进行电泳，除"血红蛋白 A_2 现象"外，还能发现"纤维蛋白原现象"。

2. 利用这种电泳可以研究蛋白质的相互作用，包括交叉互作和非交叉互作（混合互作）。交叉互作是在凝胶中一种血红蛋白穿过另一种血红蛋白时，后者区带变形，说明发生交叉互作，区带不变形为没有发生交叉互作。混合互作是两种成分相继加入凝胶的同一位置，通电后发生互作。

3. 它可由红细胞分离和制备各种血红蛋白，用于进一步分析。进一步分析中，最明显的例子就是与质谱连接，从而进入蛋白质组学研究。但是，这里的蛋白质组学与众不同，它是由我们提出来的，称之为"电泳释放蛋白质组学"，是对由细胞释放出来的蛋白质进行蛋白质组学研究。

归纳起来，创建"淀粉-琼脂糖凝胶释放电泳（简称淀琼电泳）"后的主要收获如下。

1. 在基础理论方面，利用它发现了红细胞内各种血红蛋白的存在状态：①HbA_2 与 HbA_1 结合存在，还有 PRX_2（过氧化物还原酶 2）参与；②HbA_1 与 CA_1（碳酸酐酶 1）结合存在；③这些都是"电泳释放蛋白质组学"的独特成果。

2. 在临床实践方面，利用它发现：①全血 HbA_2 现象，有助于鉴别溶血性疾病是否存在；②地中海贫血时，全血再释放增强；③球形红细胞增多症时，红细胞内快泳血红蛋白消失；④糖尿病时，血糖浓度与全血的多带再释放相关；⑤阻塞性黄疸时，红细胞和全血再释放都增强；⑥溶血性黄疸时，红细胞和全血再释放都减弱；⑦肝细胞性黄疸时，位于二者之间，红细胞再释放减弱、全血再释放增强。

科研无止境，永远在路上。

秦文斌
2017 年 12 月于鹿城包头

---

* 本书部分内容为作者在不同时期的论文和作品，有的已公开发表，在编辑形成本书时，为与原文保持一致，部分未进行统一处理，因此存在个别单位、体例等方面的不一致。

目 录

第一篇 总 论

前言 ··· 1
第一章 淀粉-琼脂糖混合凝胶电泳 ··· 5
第一节 基本操作 ··· 5
第二节 操作注释 ··· 7
第三节 强调"丽春红-联苯胺染色法" ·· 8
第二章 淀粉-琼脂糖凝胶双向对角线释放电泳 ··· 10
第三章 初释放电泳 举例 ·· 18
第四章 再释放电泳 举例 ·· 19
第五章 交叉互作 举例 ·· 21
第六章 混合互作 举例 ·· 22
第七章 其他细胞 举例 ·· 23
第八章 电泳释放蛋白质组学——红细胞初释放电泳释放蛋白质组学 举例 ··· 24
第九章 制备电泳举例——人 HbA_1 与 HbA_2 的交叉互作研究 ····················· 35
第十章 染色法 举例 ·· 36

第二篇 全血释放电泳

第十一章 全血释放电泳概况 ·· 37
第十二章 红细胞和全血再释放电泳类型及临床意义 ······································ 40
第十三章 血浆成分对红细胞释放血红蛋白的影响 ·· 55
第十四章 纤维蛋白原现象的发现和研究 ·· 73

第三篇 初释放电泳

前言 ··· 80
第十五章 血红蛋白 A_2 现象——Ⅰ.A_2 现象的发现及其初步应用 ················ 82
第十六章 血红蛋白 A_2 现象——Ⅱ.α链异常血红蛋白的 A_2 现象及其意义 ··· 87
第十七章 血红蛋白 A_2 现象——Ⅲ.δ链异常血红蛋白的 A_2 现象及其意义 ··· 91
第十八章 红细胞外 HbA_2 与 HbA 间的相互作用 ·· 95

第十九章　血红蛋白 A_2 现象发生机制的研究——"红细胞 HbA_2"为 HbA_2 与 HbA 的结合产物···100

第二十章　血红蛋白 A_2 现象与红细胞膜关系的研究——I.血红蛋白 A_2 现象与溶血的关系···104

第二十一章　血红蛋白 A_2 现象与红细胞膜关系的研究——II.红细胞外血红蛋白与磷脂间的相互作用···108

第二十二章　血红蛋白 A_2 现象异常症(秦氏病)···114

第二十三章　初释放研究的国外引用情况···118

第四篇　再释放电泳

前言···126

第二十四章　血红蛋白释放试验与轻型β-地中海贫血·······································129

第二十五章　不连续通电的红细胞电泳——一种新的血红蛋白释放试验·················133

第二十六章　普通外科患者血红蛋白释放试验的比较研究··································138

第二十七章　血糖浓度和血红蛋白释放试验的比较研究····································143

第二十八章　尿毒症患者低渗血红蛋白释放试验的初步结果································146

第二十九章　血红蛋白释放试验与血液流变学检测结果相关性的研究····················150

第三十章　肝硬化患者血红蛋白释放试验明显异常···153

第三十一章　人红细胞电泳中再释放蛋白质的分子互作·····································158

第三十二章　剖宫产前后血红蛋白释放试验的连续观察····································164

第三十三章　肝内胆管癌与血红蛋白释放试验··168

第三十四章　血红蛋白释放试验鉴别黄疸类型的初步研究··································178

第三十五章　中药穿心莲和当归对小鼠红细胞血红蛋白再释放的影响·····················181

第三十六章　梅花鹿镰状红细胞内血红蛋白的电泳释放····································186

第三十七章　遗传性球形红细胞增多症患者红细胞内快泳血红蛋白缺失?················190

第三十八章　慢阻肺患者红细胞内血红蛋白的电泳再释放——一例报告··················196

第三十九章　球形红细胞与靶形红细胞电泳释放血红蛋白明显不同·······················199

第四十章　α-地贫与球形红细胞增多症血红蛋白电泳释放的比较研究···················205

第四十一章　缺氧对小鼠红细胞再释放血红蛋白的影响·····································222

第四十二章　贲门癌与血红蛋白释放试验···224

第五篇　交叉互作电泳

前言···228

第四十三章　淀粉-琼脂糖混合凝胶交叉互作电泳··231

第四十四章　血红蛋白 F 与血红蛋白 A_2 之间的交叉互作··································233

第四十五章	血红蛋白 A_3 与血红蛋白 A_2 之间没有交叉互作	236
第四十六章	大鼠血红蛋白之间的交叉互作——大鼠红细胞结晶的可能机制	239
第四十七章	羊膜动物与非羊膜动物血红蛋白的交叉互作行为不同——脊椎动物血红蛋白的分子进化	243
第四十八章	几种鱼类血红蛋白不与人血红蛋白相互作用及其进化意义	253
第四十九章	美国锦龟血红蛋白与人血红蛋白相互作用的生物信息学和计算机模拟研究	258
第五十章	生物信息学方法推测脊椎动物血红蛋白相互作用机制	264
第五十一章	血红蛋白的聚丙烯酰胺凝胶交叉互作电泳	271

第六篇　混合互作电泳

前言		277
第五十二章	蒿甲醚抗疟机制的研究——蒿甲醚与红细胞成分的混合互作	278
第五十三章	青蒿素类药物与一些物质的混合互作——表现为凝集反应	286
第五十四章	偏重亚硫酸钠与红细胞相互作用——将偏重亚硫酸钠加入标本　无氧与有氧同在一胶板	298
第五十五章	肝素与 DNA 之间的相互作用	301
第五十六章	肝素与染料之间的相互作用——用凝胶电泳发现肝素与染料之间的相互作用	304
第五十七章	计算机模拟法分析鸡血红蛋白与溴酚蓝特异性互作的结构基础	311

第七篇　其他细胞内成分的电泳释放

前言		319
第五十八章	血小板内成分电泳释放的初步研究	320
第五十九章	粒细胞内蛋白质电泳释放的初步研究	324
第六十章	淋巴细胞内蛋白质电泳释放的初步研究	328
第六十一章	胃癌细胞内蛋白质电泳释放的初步研究	333
第六十二章	小鼠胚胎成纤维细胞内蛋白质电泳释放的初步研究	337
第六十三章	比较几种细胞的释放结果	340

第八篇　电泳释放蛋白质组学

前言		341
第六十四章	红细胞初释放蛋白质组学	342
第六十五章	红细胞再释放蛋白质组学	343
第六十六章	其他细胞电泳释放蛋白质组学展望	344

第一节	血小板电泳释放蛋白质组学展望	344
第二节	粒细胞电泳释放蛋白质组学展望	345
第三节	淋巴细胞电泳释放蛋白质组学展望	346
第四节	胃癌细胞电泳释放蛋白质组学展望	346
第五节	NIH 3T3 电泳释放蛋白质组学展望	347

第九篇　制　备　电　泳

前言 ································· 348

第六十七章	制备电泳通则	350
第六十八章	制备人 HbA_1 与 HbA_2——用于交叉互作研究	351
第六十九章	制备大鼠四种血红蛋白——用于交叉互作研究	352
第七十章	制备 HbA_2 与 HbA_1 之间成分——用于血红蛋白 A_2 现象机制的研究	353
第七十一章	制备定时再释放区带——用于再释放机制的研究	359
第七十二章	制备电泳拾零	363

第十篇　无氧条件下红细胞内血红蛋白的电泳释放

前言 ································· 365

第七十三章	无氧条件下的血红蛋白电泳释放——偏重亚硫酸钠淀琼胶电泳	366
第七十四章	比较有氧和无氧电泳结果的三种方法	371
第七十五章	无氧条件下正常分娩者红细胞内血红蛋白的电泳释放——将偏重亚硫酸钠加入标本　无氧与有氧同在一胶板	372
第七十六章	无氧条件下脑出血患者红细胞内血红蛋白的电泳释放——将偏重亚硫酸钠加入标本　无氧与有氧同在一胶板	375
第七十七章	无氧条件下乳腺癌患者红细胞内血红蛋白的电泳释放——将偏重亚硫酸钠加入标本　无氧与有氧同在一胶板	378
第七十八章	无氧条件下胃癌患者红细胞内血红蛋白的电泳释放——将偏重亚硫酸钠加入标本　无氧与有氧同在一胶板	381
第七十九章	四种情况(正常分娩、脑出血、胃癌、乳腺癌)的比较和分析——将偏重亚硫酸钠加入标本　无氧与有氧同在一胶板	384
第八十章	无氧条件下血红蛋白之间不能交叉互作——一个电泳槽里两种胶板	386
第八十一章	有氧和无氧条件下糖尿病患者的红细胞指纹图	390
第八十二章	无氧胶中血红蛋白区带泳动时的拐弯现象	393
第八十三章	无氧释放总结	395

第十一篇 红细胞内蛋白成分的存在状态及其电泳释放图解

前言 ·· 396
第八十四章 红细胞内的游离蛋白成分及其电泳释放 ·· 397
第八十五章 红细胞内血红蛋白 A_2 等的初释放现象 ·· 398
第八十六章 红细胞内血红蛋白 A_1 等的再释放现象 ·· 399
第八十七章 红细胞内初释放现象与再释放现象同时存在状态及其电泳释放 ·············· 400
第八十八章 红细胞内全部成分的存在状态及其电泳释放 ······································ 401
第八十九章 无氧条件下红细胞内蛋白成分的存在状态及其电泳释放 ······················· 402
第九十章 球形红细胞增多症时红细胞内蛋白成分的存在状态及其电泳释放 ············· 403
第九十一章 靶形红细胞增多症时红细胞内蛋白成分的存在状态及其电泳释放 ·········· 404

附　　录

附录一　名词注释 ··· 406
附录二　红细胞释放电泳(溶血液电泳)历年发表的文章 ······································· 423
附录三　红细胞释放电泳 ·· 427
附录四　电泳成分染色法 ·· 429
附录五　电泳图画 ·· 438

第一篇 总 论

前 言

1 淀粉-琼脂糖凝胶释放电泳

(1) 淀粉-琼脂糖凝胶释放电泳的特点或优点，在于它的半流动性、可塑性、事实上的等渗性，活体红细胞加入此凝胶后，可保持完整性，通电后释放出其中成分，从而发现内在规律。

(2) 显示电泳结果的是"丽春红-联苯胺染色法"，此染色法是我们自己创建的，它的特点是红(非血红蛋白)蓝(血红蛋白)并存，鲜艳悦目。

(3) 释放内容：初释放和再释放。

(4) 引发出来的内容：交叉互作和非交叉互作(混合互作)。

(5) 红细胞以外的细胞也能释放。

(6) 把释放与质谱分析联系起来，产生"电泳释放蛋白质组学"。

(7) 淀粉-琼脂糖凝胶，也能分离和制备蛋白质。

(8) 国内外有过这类技术吗？

1) 按"淀粉-琼脂糖凝胶电泳"查文献，未查到。

2) 按"琼脂糖-淀粉凝胶电泳"查文献，有如下两篇：①A chalavardjian. Agarose–starch gel electrophoresis of rat serum lipoproteins. Journal of Lipid Research，1971，12(3)：265。发表于1971年，是研究血清脂蛋白，与血红蛋白无关，与释放无关。②XY Yu，XB Guo. Detection of GLO I phenotypes in blood and blood stains using agarose starch gel electrophoretic analysis. Hunan Yi Ke Da Xue Xue Bao，2000，35(2)：203。发表于2000年，而我们的实验是开始于1981年。

(9) 综上所述，淀粉-琼脂糖凝胶释放电泳，是我们的原创技术。

2 红细胞释放电泳

(1) 在我们创建"淀琼胶释放电泳"之前，没有人用完整红细胞做电泳，那时都是用红细胞溶血液做实验，无法知道红细胞内部的情况。

(2) 红细胞释放电泳发现了"初释放现象"，见下文所述。

(3) 红细胞释放电泳还发现了"再释放现象"，见下文所述。

(4) 在"初释放现象"启发下，我们又发现了"交叉互作"，见下文所述。

3 全血释放电泳

(1) 用全血做电泳，更是史无前例，查阅国内外文献，没有人做过"全血电泳"。做"红细胞电泳"，需要先由全血分离红细胞，比较复杂。全血较简单，用样品就可直接上电泳。

(2) 全血电泳也有"初释放"和"再释放"，只是情况比"红细胞电泳"复杂一些，因为全血除了红细胞还有白蛋白、球蛋白和纤维蛋白原等。正因如此，"全血电泳"更能反映疾病的复杂内容。也正因如此，"全血电泳"还能显示血浆成分与红细胞之间的相互作用。

(3) "全血电泳"发现了"纤维蛋白原现象"，这是"红细胞电泳"所没有的。"全血电泳"能揭示许多疾病的特点，"红细胞电泳"在这方面则相对差一点。

4 初释放现象

(1) 将活体红细胞加入凝胶，第一次通电，有血红蛋白由完整红细胞释放出来，我们把这次释放称作"初释放"。初释放的主要内容是，发现了血红蛋白 A_2 现象。

血红蛋白 A_2 现象是指红细胞与其溶血液并排单向电泳时，溶血液 HbA_2 靠后(阴极侧)，"红细胞 HbA_2"靠前(阳极侧)的现象。实际上，"血红蛋白 A_2"就是红细胞内 HbA_2 与 HbA_1 结合产物，详见下一项。

(2) 将活体红细胞加入凝胶，做双向电泳。第二向电泳时 HbA_2 与 HbA_1 分开，上下对应，都脱离对角线，这说明 A_2 现象是 HbA_2 与 HbA_1 的结合产物。质谱分析结果进一步证明，A_2 现象里还有第三者参加，那就是 Prx(过氧化物还原酶)，红细胞内 HbA_2 的存在状态是 HbA_1-HbA_2-Prx。

5 再释放现象

(1) 将活体红细胞加入凝胶，第一次断电后再通电，又有血红蛋白由完整红细胞释放出来，我们把这次释放叫作"再释放"。再释放电泳也有单向和双向之分。

(2) 单向再释放电泳

1) 单向单带再释放电泳：只释放出一个区带，按通电时间控制它的位置，称"定时释放""定点释放"，简称为"定释"。

2) 单向多带再释放电泳：释放出多个区带，称"多带再释放"。因为多带像梯子，也称"梯带再释放"。

(3) 双向再释放电泳

1) 双向再释放电泳也有单带再释放和多带再释放之分。

2) 单向再释放，弄不清再释放出来的血红蛋白是哪种血红蛋白。

3) 双向再释放，证明它是 HbA_1。

4) 质谱分析，进一步证明它是 HbA_1 与 CA_1(碳酸酐酶 1)的结合产物。

6 交叉互作

(1) 血红蛋白 A_2 现象的机制，是红细胞内 HbA_2 与 HbA_1 之间的相互作用，在红细胞外二者也能相互作用吗？带着这个问题，我们进行了一系列实验，结果发现，溶血液 HbA_2 与

溶血液 HbA_1 可以发生交叉互作,即 HbA_1 穿过 HbA_2 时,后者区带变形。

(2) 交叉互作还有许多例子:①人 HbF 能与 HbA_2 发生交叉互作。②大鼠四种血红蛋白(HbA、HbB、HbC、HbD)之间都能发生交叉互作。

(3) HbA_3 不能与 HbA_2 发生交叉互作,这是有待研究的课题。

(4) 交叉互作主要出现于血红蛋白之间,其他蛋白质尚未发现交叉互作。

7 混合互作

这里说的"混合互作",顾名思义,就是两种物质混到一起后,它们就发生相互作用,并且能用电泳方法检测出来。其实,"混合互作",就是一般的"相互作用",因为有一个"交叉互作",为了区分,才起名叫"混合互作"。混合互作,广泛存在,比交叉互作多见。

(1) 血浆成分与红细胞的相互作用:①血浆白蛋白与红细胞的相互作用;②血浆球蛋白与红细胞的相互作用;③氨基酸与红细胞的相互作用;④血糖与红细胞的相互作用;⑤胰岛素与红细胞的相互作用;⑥维生素与红细胞的相互作用;⑦各种无机盐与红细胞的相互作用。

(2) 外来物质与红细胞的相互作用:①蒿甲醚与红细胞的相互作用;②青蒿素与红细胞的相互作用;③双氢青蒿素与红细胞的相互作用。

(3) 其他物质之间的相互作用:①肝素与 DNA 的相互作用;②DNA 之间的相互作用。

8 其他细胞内成分的电泳释放

前面几项说的都是红细胞,其实其他细胞也存在电泳释放问题。例如:①血小板(来自血站);②淋巴细胞(来自急性淋巴细胞白血病患者);③粒细胞(来自慢性粒细胞白血病患者);④胃癌细胞(来自细胞培养);⑤小鼠胚胎成纤维细胞(NIH 3T3,来自细胞培养)。主要做双向电泳,观察以下几点:①原点沉淀;②原点上升成分;③脱离对角线成分。然后再做质谱分析,这就是"电泳释放蛋白质组学"。

9 电泳释放蛋白质组学

(1) 众所周知,蛋白质组学,是当今研究细胞内蛋白质存在状态的最高水平。但是,经典蛋白质组学,都是先破坏细胞(放出蛋白质),然后再往下分析。但是它的研究结果,不能代表完整细胞内蛋白质存在状态。

(2) 我们提出的"电泳释放蛋白质组学",是以完整细胞为基础。将完整细胞加入淀粉-琼脂糖凝胶,通电,由完整细胞释放出蛋白质,接着研究分析,完成细胞的电泳释放蛋白质组学。我们的"红细胞初释放电泳释放蛋白质组学"发现红细胞内 HbA_1-HbA_2-Prx(过氧化物还原酶)的存在。经典的红细胞蛋白质组学并没有发现这种互作。我们的"红细胞再释放电泳释放蛋白质组学"发现红细胞内 HbA_1-CA_1(碳酸酐酶 1)的存在。经典的红细胞蛋白质组学,也没有发现这种互作。

(3) 以上是红细胞的电泳释放蛋白质组学,其他细胞也做了电泳释放,等待进入蛋白质组学分析,估计会得到超出经典蛋白质组学的结果。

10 制备电泳

各种电泳技术都有分离制备内容,分离出来的蛋白质称为"电泳纯的蛋白质"。淀粉-琼脂糖凝胶分离制备蛋白质的操作比较简单,效果良好。具体操作如下所示。①凝胶电泳后丽春红染色;②根据需要,抠取显色后的蛋白质(连同所在的凝胶);③转入大 EP 管;④–20℃冰箱过夜;⑤到时取出,室温融化;⑥高速离心取上清,这里边就是所要的蛋白质;⑦可直接使用,蛋白浓度太低时,可浓缩处理。

11 染色法

大多数电泳成分,需要染色后,观察蛋白质的存在情况。我们使用的染色,有以下几种:①丽春红染色法;②氨基黑 10B 染色法;③考马斯亮蓝染色法;④丽春红-联苯胺染色法;⑤丽春红-考马斯亮蓝染色法。其中,考马斯亮蓝染色法灵敏度最高,但不能区分血红蛋白与非血红蛋白。丽春红-联苯胺染色法灵敏度也很高,而且能够区分血红蛋白(蓝黑色)和非血红蛋白(红色)。

丽春红-联苯胺染色法,属于我们原创,是本书中最常用的染色法。

12 无氧条件下红细胞内血红蛋白的电泳释放

无氧条件,就是加入"偏重亚硫酸钠",造成还原状态、无氧条件。此时,红细胞内血红蛋白的电泳释放发生明显变化。首先,"初释放现象"里的"血红蛋白 A_2 现象"改变到消失。其次,"交叉互作"失效。"再释放现象"增强或减弱到消失。

第一章 淀粉-琼脂糖混合凝胶电泳

第一节 基本操作

1 一般试剂

试验用水均为普通蒸馏水，试剂均为分析纯，室温保存。废弃物按要求处理。

2 制胶

(1) 制胶缓冲液组分：1×TEB 缓冲液 pH 8.6。先将 200ml 蒸馏水加于 1L 的容量瓶中，再称取三氨基甲烷(Tris)5.10g，EDTA 0.50g，硼酸 0.40g 于该容量瓶中，加蒸馏水至满刻度(注释 1)。室温保存。

(2) 所用北京红星化工厂的马铃薯淀粉批号为 781226。市售的各种淀粉如马铃薯淀粉、玉米淀粉、藕粉等均可用(注释 2)。

(3) 琼脂糖，西班牙进口分装。也用过琼脂粉(注释 3)。

3 电泳缓冲液

硼酸盐缓冲液(BN 缓冲液)，pH 9.3：先加约 200ml 蒸馏水于 1L 的容量瓶中。称 18.55g 硼酸，4.0g 氢氧化钠于该容量瓶中，加蒸馏水近刻度。再用氢氧化钠调 pH 至 9.3 后用水补至刻度线(注释 4)。

4 染色液

(1) 漂洗液：先加 200ml 蒸馏水于 1L 的容量瓶中。用量筒量取 50ml 冰醋酸，20ml 丙三醇(甘油)加于该容量瓶中。加蒸馏水至刻度线(注释 5)。

(2) 丽春红染色液：先加 200ml 漂洗液于 1L 的容量瓶中。称取 1.0g 丽春红于该容量瓶中。加漂洗液至满刻度(注释 6)。

(3) 联苯胺染色液：先加 200ml 漂洗液于 1L 的三角烧瓶中。称取 0.5g 联苯胺于该三角烧瓶中。再加漂洗液至 500ml，将烧瓶置于电磁炉上的水浴锅内，加温至 75℃约 0.5 小时，待联苯胺完全溶解(注释 7)。

(4) 硝普钠(亚硝基铁氰化钠)：配制见注释 8。

(5) 30% H_2O_2 溶液：见注释 9。

(6) 考马斯亮蓝染色液：称取 2.5g 的考马斯亮蓝 G-250，450ml 30%冰醋酸，100ml 甲醇，加水至 1L(注释 10)。

(7) 0.5%的氨基黑 10B 染色液：先加 200ml 漂洗液于 1L 的三角烧瓶中，称取 2.5g 联苯胺于该三角烧瓶中。再加漂洗液至 500ml，充分混合，待氨基黑 10B 完全溶解备用。

5　全血

抗凝血(注释 11)。

6　玻璃板

17cm×10cm(小板)和 17cm×17cm(大板)大小，厚 2～3mm(注释 12)。

7　制胶方法

(1) 小板胶的配制：用量筒量取 pH 8.6 的 1×TEB 缓冲液 55ml 于 250ml 的三角瓶中，加入马铃薯淀粉 1.0g，琼脂糖 0.15g。混匀。

(2) 加热：于电磁炉上的沸水锅内煮沸 8 分钟。室温冷却至 50℃左右(注释 13)。

(3) 倒胶：将事先准备好的 17cm×10cm 玻璃板置于水平台(用水平仪调平)上，缓慢把冷至 50℃的凝胶平铺在玻璃板上，待完全凝固后使用(注释 14)。

8　样品的制备

(1) 全血：可以是各种抗凝全血，最常用的是 EDTA 抗凝血，可以直接加样。

(2) 红细胞的制备：取抗凝全血 100μl 于 1.5ml 的大 EP 管中，加生理盐水 1ml，上下翻转混匀，2000～3000 转/分离心 3 分钟。用吸管吸出上清液弃于废液瓶中，再加生理盐水 1ml，上下翻转混匀，离心同上。重复洗 5～6 次。最后一次保留生理盐水与血球比例为 1∶1 即可备用(注释 15)。

(3) 溶血液的制备

1) 四氯化碳法：取上述洗好的红细胞 100μl，加等量四氯化碳在漩涡混匀器反复震荡约 2 分钟后，12 000 转/分离心 4 分钟。上清液即为溶血液。

2) 冻融法：取上述洗好的红细胞 100μl，于–20℃冰箱中过夜。次日取出，室温融化，即可使用。

9　电泳加样

(1) 准备滤纸条。取 1mm 厚的滤纸，根据需要用剪刀剪成 10mm 长，2mm 宽的滤纸条(注释 16)。

(2) 将滤纸条放于事先准备好的一次性的胶片上。按加样顺序摆好。用加样器取待检样品(全血、红细胞或溶血液等)10μl 加在滤纸上，待完全吸收后用镊子夹住滤纸，按事先排好的顺序分别插入离正极边缘约 3cm 处的胶中(注释 17)。

(3) 用记号笔在胶上做好标记。

10 电泳

10.1 单向电泳

(1) 胶板为 17cm×10cm 的，可以是普泳(注释 18)，也可以是定释(注释 19)或梯带泳(注释 20)等。

(2) 把加好样的胶板按照加样端为阳极置于电泳槽中，两侧分别搭好纱布。

(3) 通电，按照每厘米 4～6V 电压进行电泳。

(4) 电泳结束约 3 小时(注释 18)。

10.2 双向电泳

(1) 胶板一般为 17cm×17cm 的正方形。第一向电泳，同上。

(2) 当第一向泳到头后，胶板转方向 90°，并倒极再电泳，约 1.5 小时(注释 21)。

11 染色

(1) 丽春红染色：将事先配制好的丽春红染液倒于 20cm×30cm 的白瓷盘中，再把电泳好的胶板平放进去，让染液没过胶面。约染 3 小时，或过夜。到时取出胶板，照片留图(注释 22)(见图 1)。

(2) 可以自然晾干，也可以烤干。

(3) 复染联苯胺：染色前现配染液。在白瓷盘内先加入硝普钠约 2g，再倒入刚配好的联苯胺染液 500ml，再加入过氧化氢 1.4ml 混匀后。把已晾干或烤干的胶板放入染液中，约 20 分钟。边染边晃动瓷盘以便染色更均匀(注释 23)。

(4) 考马斯亮蓝染色：见注释 23。

(5) 氨基黑 10B 染色：将事先配制好的氨基黑 10B 染色液倒于 20cm×30cm 的白瓷盘中，再把电泳好的胶板平放进去，让染液没过胶面。过夜。到时取出胶板，照片留图。

第二节 操 作 注 释

注释 1. 先加蒸馏水于容量瓶中，可避免加入固体试剂时直接粘在瓶底部，先加 Tris，溶解后再加 EDTA、硼酸，可增加溶解速度。

注释 2. 新鲜制备的马铃薯淀粉，可以使用。市售的各种淀粉，如马铃薯淀粉、红薯淀粉、藕粉等，基本上都能用，但需要找条件。具体如下。

(1) 倒胶时，感觉太稀、容易流出玻板，需要多加淀粉；感觉太稠、流动太慢，需要减少淀粉。

(2) 染色后，容易脱板，胶容易破，考虑多加淀粉；胶厚染色慢，应当减少淀粉。

(3) 做红细胞 A_2 现象的双向电泳，出现标准图像为合格(参见图 1-1)。

注释 3. 最初发现 A_2 现象，是采用的淀粉琼脂混合凝胶，后来的实验均使用琼脂糖。

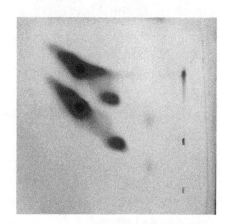

图 1-1 红细胞 A_2 现象双面电泳的标准图像

注释 4. 先加 2.4g 氢氧化钠。加蒸馏水近刻度，再用氢氧化钠调 pH 至 9.3 后，用水补满刻度。共用约 4.0g 氢氧化钠。

注释 5. 甘油(丙三醇)太少胶板干后容易裂，太多胶板不容易干。

注释 6. 丽春红在漂洗液中溶解较慢，提前先配好浓的丽春红染液放至室温。临用前在白瓷盘中做 1∶20 稀释即可使用。

注释 7. 联苯胺染色液必须现用现配。

注释 8. 将固体硝普钠约 2g 先加在白瓷盘中，后将预先配好的联苯胺染液 500ml 倒于白瓷盘中，不停晃动，以加速硝普钠尽快溶解，并均匀分布。

注释 9. 再向此染液中加约 1.4ml 的 30% H_2O_2 溶液。

注释 10. 考马斯亮蓝浓染液事先配好，放置室温保存，临用前白磁盘中按 1∶20 稀释即可。

注释 11. 各种抗凝全血均可，如 EDTA、肝素、枸橼酸钠等。要求不能溶血。最常用的是 EDTA 抗凝。

注释 12. 玻璃板的大小根据电泳槽的大小而定；还可根据自己的实验而定，一般的单相电泳可以是长方形，双向电泳可以是正方形。

注释 13. 加热溶解时边煮边摇，以免淀粉琼脂糖结块。

注释 14. 倒胶时温度不宜过热，也不易过冷，过热胶易流出玻璃板以外，过冷胶不能平铺满玻璃板上。以手摸不烫手为宜。

注释 15. 正常人的全血中血球与血浆的比例大约为 1∶1。

注释 16. 滤纸条的长短与所加的样本数量有关，一般 17cm×10cm 的小板最多可加 8 个样品。

注释 17. 把滤纸垂直插入胶中，并保证胶面一定没有带有样品的滤纸。

注释 18. 普泳，一般为 4~6V/cm 恒电压下，中间不断电，一直泳到电泳结束。

注释 19. 定释，一般为在 4~6V/cm 恒电压下，泳到大约 2.5 小时后，停电 15 分钟，再泳 30 分钟电泳结束。可以用定时器来完成。

注释 20. 梯带，一般为在 4~6V/cm 恒电压下，泳 15 分钟后，停电 15 分钟，再泳 15 分钟后，再停 15 分钟，交替开闭电源，直至电泳结束，需 5~6 个小时。这种交替开闭电源是用定时器(中国慈溪市公牛电器有限公司生产的公牛牌 24 小时定时器)来完成的。

注释 21. 双向电泳可以是双层、三层或多层，根据需要而定。电泳时间也不一样，是根据具体情况而定。

注释 22. 通过丽春红染色，所有蛋白均被染成红色。

注释 23. 通过联苯胺染色，所有血红素蛋白均被染成蓝黑色。

第三节 强调"丽春红-联苯胺染色法"

1 本染色法的原创性

国内外文献中，可以查到"丽春红染色法"，也可以查到"联苯胺染色法"，但查不到"丽春红-联苯胺染色法"，是我们把二者结合起来，该方法红(普通蛋白)蓝(血红蛋白)分明、高效运行，属于"原创行为"。

2 本染色法的特点

本染色法的主要特点：能够用颜色区分血红蛋白呈蓝黑色和非血红蛋白呈红色。染色结果一出来，肉眼就能看出，血红蛋白在哪里，非血红蛋白在哪里。哪些成分增加了，哪些成分减少了，哪些成分消失了，一目了然。

此处专门说明 MHA(高铁血红素白蛋白)：正常情况下白蛋白是红色，如果其中夹杂蓝黑色，表明有了 MHA，说明患者存在溶血性疾病。溶血性贫血时，血红蛋白从红细胞中释放到血浆中，被氧化成 3 价铁的血红素，即高铁血红素，后者再与血浆里的白蛋白结合，形成 MHA。

第二章 淀粉-琼脂糖凝胶双向对角线释放电泳

秦文斌[1*] 高丽君[1#] 韩丽红[1#] 苏 燕[1#] 闫 斌[3] 高雅琼[1]
宝勿仁[1] 龙桂芳[2] 贺其图[3] 王彩丽[3] 魏 枫[3] 和姬苓[3]
葛 华[3] 张咏梅[3] 王翠峰[3] 乔 姝[3]

(1. 包头医学院 血红蛋白电泳释放研究室，包头 014010；2. 广西医科大学 医学科学实验中心，西宁 530021；3. 包头医学院第一附属医院，包头 014010)

摘 要

目的： 报告一种以淀粉-琼脂糖凝胶(淀琼胶)为支持体、以红细胞或非红细胞为样品的释放型双向对角线电泳并探讨其科学意义。

方法： 收取多种疾病的临床血液标本，将红细胞或其他细胞加在淀琼胶的原点处进行双向电泳，电势梯度7V/cm，第一向，电泳3小时后，第二向，电泳2小时，丽春红-联苯胺染色，观察释放出的成分与对角线的关系。

如果在第一向电泳中再加入"断电/再通电"，就会从原点又释放出血红蛋白成分。当"断电/再通电"多次重复时，就会出现多个血红蛋白区带(也称"多带"或"梯带")，梯带强弱与疾病关系密切。

结果： 本文有各种各样的标本和病例(详见正文)，它们的双向对角线电泳结果，涉及蛋白质间相互作用的不同情况。还有血红蛋白梯带释放，各种疾病不尽相同，如梯带最弱的遗传性球形红细胞增多症，梯带最强的地中海贫血等。

结论： 淀粉-琼脂糖凝胶双向对角线细胞电泳，既能发现细胞内蛋白质之间的相互作用，又能通过血红蛋白梯带释放了解疾病情况，应用前景良好。

关键词 淀粉琼脂糖混合凝胶电泳；双向电泳；对角线电泳；细胞电泳；血红蛋白释放试验/HRT；红细胞

1 前言

对角线电泳(diagonal electrophoresis)，是双向电泳的一种特例。试验是将蛋白质样品加在凝胶的一端，第一次电泳后进行某种特殊处理，再将凝胶转90°进行第二次电泳，此双向电泳即为对角线电泳。如果两次电泳间的特殊处理对蛋白质无影响，则电泳图谱中的蛋白点都在对角线上；如果该特殊处理有影响，则电泳图谱中的蛋白点偏离对角线。根据特殊处理的性质，可获得变化蛋白点的相关重要信息。1966年Brown和Hartley首次报告对角线纸上电泳，用于分析牛糜蛋白酶原A的二硫键[1]，后来yano等利用非还原/还原SDS-PAGE(十二烷基硫酸钠聚丙烯酰胺凝胶电泳)来确定硫氧还蛋白的目标蛋白,从而为确定分子间和分子内的二硫键提供一种新的方法[2]。对角线电泳还曾用来鉴定植物中的热休克蛋白和耐寒蛋白[3],Rais 等[4]利用对角线电泳技术在线粒体复合物Ⅰ的 70 多种蛋白质中分离并鉴定了 11 种高疏水性蛋白质。Sánchez 等[5]利用非还原/还原 SDS-PAGE 对角线电泳技术,以天然通道蛋白为研究对象,发现奈瑟球菌属的通道蛋白是由 PorA 和 PorB 组成的异源三聚体,而不是以往报道的由 PorA 和 PorB 组成的同源三聚体。BN-PAGE(blue native polyacrylamide gel-electrophoresis)是 chägger 和 von Jagow[6]为了研究膜整合蛋白而设计的一种改进的 PAGE,BN-PAGE/SDS-PAGE 对角线电泳在线粒体呼吸链复合物[7]、多蛋白复合物[8]和蛋白质包涵体[9]的研究中得到广泛应用。

众所周知,经典的蛋白质组学研究方法包括 IEF/SDS-PAGE 和质谱技术的联用,但由于 IEF 的一些局限性,限制了其更广泛的应用。对角线电泳由于具有不同于 IEF/SDS-PAGE 的特点,在蛋白质组学研究中的应用越来越广泛。从最初的二硫键定位到疏水性膜整合蛋白的研究,再到蛋白质复合物的研究,逐渐发展和成熟。鉴于目前蛋白质复合物的研究越来越受到关注,对角线双向电泳,特别是非还原/还原 SDS-PAGE 对角线电泳也逐步成为蛋白质相互关系研究中的一种新的有效手段。必须指出的是,IEF/SDS-PAGE 与对角线电泳技术各有优缺点,并相互补充,将两者结合起来用于蛋白质组学的研究将会收到更好效果。

如上所述,"对角线电泳"来自双向聚丙烯酰胺凝胶电泳,并在二硫键鉴定、特别是在蛋白互作方面得到广泛应用。本文报告另外一种双向对角线电泳,它的支持体是淀粉琼脂糖混合凝胶,样品为完整的红细胞或其他细胞。我们于 1990 年建立此方法[10],用于解决血红蛋白 A_2 现象[11, 12, 13]的机制问题。后来用于血红蛋白梯带现象(血红蛋白释放试验 HRT)[13-15]的机制研究。本文对角线电泳的特点是两向电泳之间不需任何处理,就能发现有无互作。如果在第一向电泳中加入多次开闭电源,释放出来的血红蛋白梯带,又可反映红细胞内血红蛋白残留和再释放的一些情况,提供重要临床资料,特报道如下。

2 材料和方法

2.1 材料

(1) 血液标本来自包头医学院第一附属医院血液科、检验科等。静脉采血,EDTA 或肝素抗凝。

(2) 电泳仪:DYY-Ⅲ9B 电泳槽(北京六一仪器厂)、梯带定时器(公牛 24 小时定时器,慈溪市公牛电器有限公司)。

(3) 淀粉:马铃薯淀粉或食用藕粉或玉米淀粉,琼脂糖(西班牙产)国内分装。

(4) 缓冲液

1) 制胶缓冲液:1×TEB pH 8.6。先加约 200ml 蒸馏水于 1L 的容量瓶中,再称取 Tris 5.10g,EDTA 0.50g,硼酸 0.40g 于该容量瓶中。加蒸馏水至满刻度,室温保存。

2) 电泳缓冲液:硼酸缓冲液(BN 缓冲液,pH 9.3)。先加约 200ml 蒸馏水于 1L 的容量瓶中,称 18.55g 硼酸,4.0g 氢氧化钠于该容量瓶中。加蒸馏水近刻度,再用氢氧化钠调 pH 至 9.3 后用水补至刻度。

(5) 染色液

1) 丽春红染色液：先加 200ml 漂洗液于 1L 的容量瓶中，称取 1.0g 丽春红于该容量瓶中。加漂洗液至满刻度。

2) 联苯胺染色液：先加 200ml 漂洗液于 1L 的三角烧瓶中，称取 0.5g 联苯胺于该三角烧瓶中。再加漂洗液至 500ml，置于电磁炉上的水浴锅内，加温至 75℃约 0.5 小时，待联苯胺完全溶解(用前加硝普钠，30%过氧化氢)。

3) 考马斯亮蓝染色液：2.5g 的考马斯亮蓝 R-250，450ml 30%冰醋酸溶液，100ml 甲醇，加水至 1L。

(6) 漂洗液： 先加 200ml 蒸馏水于 1L 的容量瓶中。用量筒量取 50ml 冰醋酸，20ml 丙三醇于该容量瓶中。加蒸馏水至满刻度。

2.2 方法

2.2.1 分离红细胞　取抗凝血 100μl 于 1.5ml 的 EP 管中，加生理盐水 1ml，上下翻转混匀，2000～3000 转/分离心 3 分钟。用吸管吸出上清液及红细胞上的白膜，弃于废液瓶中，再加生理盐水 1ml，上下翻转混匀，离心同上。重复洗 5～6 次。最后一次保留生理盐水与血球比例为 1∶1 备用。

2.2.2 淀琼胶双向对角线电泳

(1) 制胶：取 17cm×17cm 的玻璃板一块，称马铃薯淀粉 1.7mg，琼脂糖 0.24mg，加于装有 90ml 1×TEB 缓冲液三角烧瓶内，摇匀后，于沸水锅内煮，边煮边摇 8 分钟。取出后凉至 50℃左右平铺于玻璃板上。待完全凝固即可使用。

(2) 上样：①准备滤纸条。取 1mm 厚的滤纸，根据需要剪成 6mm×2mm 大小。②取待检样品 8μl 加在滤纸上，待完全吸收后用镊子夹住滤纸插入离负极边缘约 3cm 处的胶中。③用记号笔在胶上做好标记。

2.2.3 双向电泳：第一向可以是普泳(普通电泳)或再释放电泳。

(1) 普通电泳：一般为 7V/cm 恒电压下，中间不断电，一直泳到电泳结束，此时称为"初释放"。

(2) 一次再释放电泳：一般为在 7V/cm 恒电压下，泳到大约 2.5 小时后，停电 15 分钟，再泳 30 分钟结束。可以用定时器来完成。

(3) 多次再释放电泳：一般为在 7V/cm 恒电压下，泳 15 分钟后，停电 15 分钟，再泳 15 分钟后，再停 15 分钟，交替开闭电源，直至电泳结束约 6 个小时。这种交替开闭电源用定时器(中国慈溪市公牛电器有限公司生产的公牛牌 24 小时定时器)来完成。第二向一般为普泳。当第一向电泳结束后，胶板转方向 90°，并倒极再电泳，约 1.5 小时，电泳结束。

2.2.4 染色：取下胶板放于丽春红染液中染色，约 1.5 小时。晾干后可以再复染联苯胺或考马斯亮蓝。

3 结果

3.1 正常人的双向双层对角线电泳结果　见图 2-1、图 2-2。

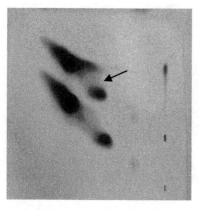 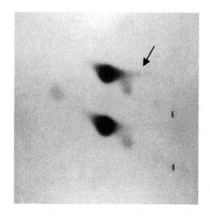

图 2-1　正常人的血红蛋白 A_2 现象
注释：上层为红细胞，下层为溶血液，第一向电泳时没有停电处理(此为"初释放"，没有"再释放")

图 2-2　正常人的血红蛋白多带再释放现象
注释：上层为红细胞，下层为全血，第一向电泳时加入多次停电处理

由图 2-1 可以看出，溶血液中各成分(蓝色的 HbA_3、HbA_1、HbA_2、红色的 CA 碳酸酐酶)都在对角线上，红细胞中各成分(HbA_3、HbA_1、CA)在对角线上，HbA_2 则离开对角线(在对角线的下方)，而在其上方出现一个血红蛋白区带(横向位置相当于 HbA_1)，而且中间有虚带连接(箭头所指)。这说明，在红细胞中，HbA_2 与 HbA_1 有相互作用。

由图 2-2 可以看出，红细胞结果中，除了有血红蛋白 A_2 现象外，又出现了血红蛋白多带(箭头所指)。这说明，正常人红细胞内的血红蛋白，第一次通电时释放出来大部分血红蛋白，但还有少量残留，再次通电时又释放出来，每次释放一点，多次通电形成血红蛋白多带。下层的全血结果中，左上方红带为白蛋白，由于它也含红细胞，也有血红蛋白 A_2 现象和血红蛋白多带现象(减弱)。

3.2　各种临床标本的双向对角线电泳结果　此时，第一向电泳中都加入停电处理，能够看到血红蛋白多带的强弱。

3.2.1　"单层"双向对角线电泳结果　见图 2-3。

图 2-3A 的上样物质为 α 地中海贫血患者的全血，由此图可以看出，此时多带释放增强，在第二向电泳时离开对角线。图 2-3B 是 HbH 患者的全血标本，血红蛋白 H 出现在白蛋白与血红蛋白 A_1 之间(箭头所指)，但它稍微脱离对角线，推测有蛋白互作。图 3C 与图 3B 为

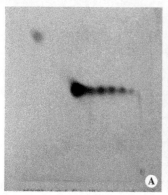 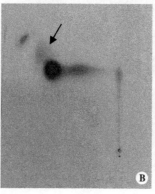 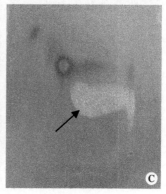

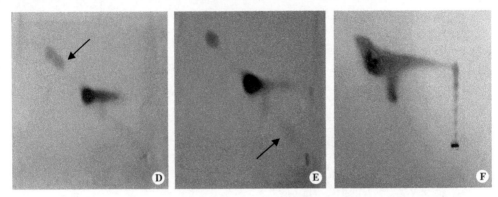

图 2-3 "单层"双向对角线电泳结果

注释：A 为 α 地贫；B、C 为 HbH 病；D 为双白蛋白血症；E 为骨髓瘤；F 为干燥综合征

同一图，C 是来自以红色为背景的照片。此时出现一片类似羊身的白色成分(箭头所指)。这可能是联苯胺反应中加入的 H_2O_2，经过氧化氢酶催化放出 O_2，以气泡形式停留在凝胶中，推测此病同时伴有过氧化氢酶活性增强。图 2-3D 为双白蛋白血症的全血标本，双白蛋白(箭头所指)都在对角线上。图 2-3E 为骨髓瘤患者的全血标本，箭头所指为 M-蛋白，它没离开对角线。图 2-3F 为干燥综合征的红细胞标本，电泳条件为第二向时多带释放，出现似小鸡下蛋的图像。说明第一向电泳后，原点还残留较多血红蛋白，这些多带也不在对角线上。

3.2.2 "双层"双向对角线电泳结果 见图 2-4。

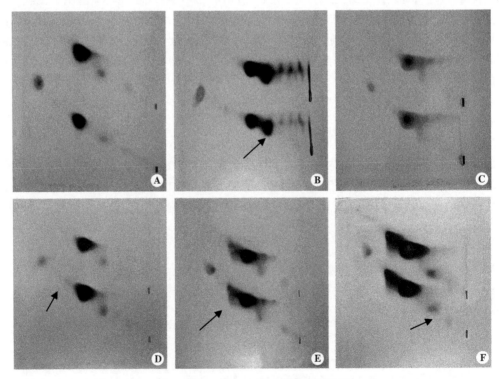

图 2-4 "双层"双向对角线电泳结果

注释：A 为遗传球形红细胞增多症；B 为 HbF 和 HbE；C 为脑梗；D 为剖宫产后；E 为 HbH 病；F 为 CS 型 HbH 病

图 2-4 都是上层为红细胞，下层为全血。图 2-4A 为遗传性球形红细胞增多症的样品。它的特点是红细胞几乎没有多带，全血的血红蛋白 A_2 基本在对角线上，其他标本很少有这

种情况。图 2-4B 为 7 个月婴儿的血液,标本中有大量血红蛋白 F 和 E(箭头所指)。此时,血红蛋白 F 和 E 都在对角线上,红细胞的梯带增强,全血梯带相对稍弱。再释放的梯带显示,上强下弱、上前下后,上边的梯带来自血红蛋白 F,下边梯带来自血红蛋白 E,这是梯带中同时显示两种血红蛋白的例子。图 2-4C 为脑梗标本,红细胞梯带增强,全血梯带也相当明显。图 2-4D 是剖宫产患者术后抽血的实验结果。此时,在全血的对角线上血红蛋白 A_1 之前出现一个次要快泳区带,剖宫产前的血液中没有此成分。图 2-4E 来自又一例血红蛋白 H 病患者,它的特点是,在血红蛋白 H 的下方出现脱离对角线成分(箭头所指),与图 2-3B 脱离对角线情况相反。图 2-4F 是 CS 型血红蛋白 H 病标本,血红蛋白 CS(Constant Spring)(箭头所指)位于对角线上,血红蛋白 H 的下方也出现脱离对角线成分,情况同图 2-4E。

3.2.3 "三层"双向对角线电泳结果　见图 2-5。

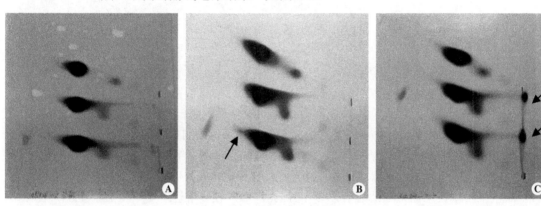

图 2-5　"三层"双向对角线电泳
注释:A 为剖宫产前;B 为剖宫产后;C 为 HbF+HbA$_2$ 增多

图 2-5 都是上层为红细胞溶血液,中层为红细胞,下层为全血。图 2-5A 为剖宫产前的血液标本,图 2-5B 为同一人剖宫产后的血液标本,二者结果差别不大,只是在剖宫产后的全血部分,在血红蛋白 A_1 的前边出现一个"小尖"(箭头所指),它是血红蛋白,相当于血红蛋白 A_3 之前,意义不明。图 2-5C 可能是 β 地贫纯合子的血液标本,它的血红蛋白 F 达到 68.1%,A_2 为 7.1%,原点残留特多(箭头所指)。

3.3 非红细胞的双向对角线电泳结果　见图 2-6。

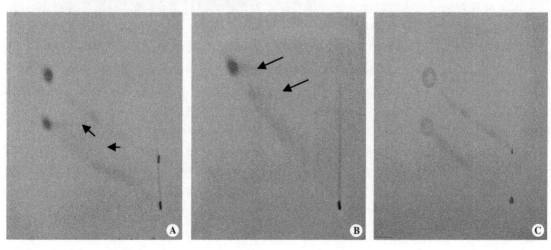

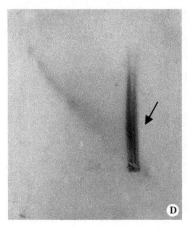 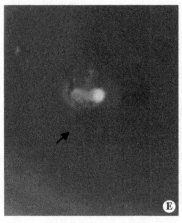 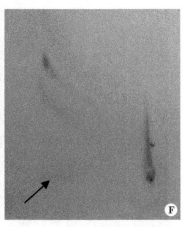

图 2-6 非红细胞的双向对角线电泳

注释：A 为血小板(双层)；B 为血小板(单层)；C 为粒细胞；D 为淋巴细胞；E 为淋巴细胞与纸片移动；F 为胃癌的培养细胞

图 2-6 结果均来自非红细胞双向对角线电泳，图 2-6A 血小板的双层对角线电泳结果，上层为血浆，下层为血小板。血浆成分都在对角线上，血小板则不同，它出现多个脱离对角线的成分(箭头所指)。为了证明这些脱离对角线的成分不是来自血浆，又进行了图 2-6B 的试验。这是血小板本身的单层对角线电泳结果，其脱离对角线的成分与图 2-6A 相同。图 2-6C 为粒细胞(来自慢性粒细胞白血病患者)，上层为血浆、下层为白细胞。血浆成分都在对角线上，粒细胞成分也都在对角线上，此点与血小板不同。原点上升成分也很少，意义不明。图 2-6D 为淋巴细胞(来自淋巴细胞白血病患者)，原点打孔，加入较多淋巴细胞，上边盖上圆片滤纸后开始电泳。第一向电泳时圆片滤纸向阴极移动 5mm，出现图 2-6E 景象，第二向后染氨黑 10B，原点上方出现大量蛋白质(箭头所指)。图 2-6E 图像非常特殊，过去从未见过，推测是淋巴细胞与滤纸一起移动，此淋巴细胞可能带较多正电荷。图 2-6F 为胃癌的培养细胞，上层为细胞培养液，下层为胃癌的培养细胞。与血小板不同，此时脱离对角线的成分出现在左下方(箭头所指)，原点残留也很明显。

4 讨论

本文双向对角线电泳的特殊之处，除支持体为淀粉琼脂糖混合凝胶(不是聚丙胶)外，更重要的是上样物质为细胞(红细胞或非红细胞)，不是细胞裂解液。这里涉及一个概念，细胞成分的电泳释放，即细胞是完整的，其中成分通过电泳而释放出来。细胞成分的电泳释放可分为两大类：红细胞成分的电泳释放和非红细胞成分的电泳释放。红细胞成分的电泳释放再细分为初释放和再释放。所谓初释放(第一次释放)，就是上样后进行普通电泳(打开电源后不再关闭)，此时由红细胞释放出来的血红蛋白 A_2 的电泳位置与来自溶血液的血红蛋白 A_2 不同，我们称之为"血红蛋白 A_2 现象"[11,12]。再释放的概念是开始为普通电泳，后来增加"关闭—再打开电源"过程，此时又有血红蛋白由原点释放出来。如果此过程反复进行，可产生多个血红蛋白区带，形状似竹节或梯子蹬，称"竹节带"或"梯带"，此现象称之为"血红蛋白梯带现象"，属于多次再释放结果[14,15]。在双向细胞电泳中，只有第一向可以涉及初释放和再释放，第二向肯定是再释放，因为它又开闭一次电源。红细胞成分的电泳释放，在电泳过程中就能看到释放出来的红色的血红蛋白，所以也称为"血红蛋白释放试验 HRT"[15]。非红细胞成分的电泳释放，电泳过程中看不见东西，必须染色才能显示结果[16]。

电泳释放也可分单向与双向，本文主要讨论双向电泳，只有双向电泳才能看到对角线成分和脱离对角线的情况。对于一般对角线电泳来说，脱离对角线就是互作成分。对于释放型对角线电泳来说，如果只有初释放，没有再释放，也可以这样理解。如果加入了再释放，脱离对角线成分就出现两种情况，一方面是再释放造成的血红蛋白梯带现象，另一方面是来自初释放的血红蛋白 A_2 现象等。所以，在本文的双向对角线电泳图，能够提供多方面的信息。首先是互作问题，有无互作带，互作带在哪里，互作带多少，各种细胞有何差异，这一点可见非红细胞的双向对角线电泳结果。红细胞的结果是既有互作又有梯带，各种疾病不尽相同，上边各图中均有体现。

淀琼胶双向对角线细胞电泳能发现互作和梯带现象，每个标本的结果都不相同，信息量较大，留下较大的研究空间，它们的许多细节还都需要进一步深入研究。为此，可先取脱离对角线成分做 SDS-PAGE，接着进行质谱分析，最终弄清互作成分。这方面的第一个例子，就是我们用此法发现血红蛋白 A_2 现象来自 HbA_1、HbA_2 与 Prx-2 之间的相互作用[12]。我们认为，这就是"电泳释放蛋白质组学"或者"电泳释放互作蛋白质组学"，它可能给我们带来更多蛋白质之间相互作用方面的新知识。

参 考 文 献

[1] Brown JR, Hartley BS.Location of disulphide bridges by diagonal paper electrophoresis. The disulphide bridges of bovine chymotrypsinogen A. Biochem J, 1966, 101(1): 214-228.
[2] Yano H, Wong J H, Lee Y M, et al. A strategy for the identification of proteins targeted by thioredoxin. Proc natl acade sci usa, 2001, 98(8): 4794-4799.
[3] 史济夫. 对角线电泳的一些应用. 生物学通报, 1999, 34(11): 91.
[4] Rais I, Karas M, Schägger H. Two-dimensional electrophoresis for the isolation of integral membrane proteins and mass spectrometric identification. Proteomics, 2004, 4(9): 2567-2571.
[5] Sánchez S, Arenas J, Abel A, et al. Analysis of outer membrane protein complexes and heat-modifiable proteins in *Neisseria* strains using two-dimensional diagonal electrophoresis. J Proteome Res, 2005, 4(1): 91-95.
[6] Schägger H, von Jagow G. Blue native electrophoresis for isolation of membrane protein complexes in enzymatically active form. Anal Biochem, 1991, 199(2): 223-231.
[7] Nijtmans L G J, Henderson N S, Holt I J. Blue native electrophoresis to study mitochondrial and other protein complexes. Methods, 2002, 26(4): 327-334.
[8] Camacho-Carvajal M M, Wollscheid B, Aebersold R, et al. Two-dimensional blue native/SDS gel electrophoresis of multi-protein complexes from whole cellular lysates. Mol Cell Proteomics, 2004, 3(2): 176-182.
[9] Stegemann J, Ventzki R, Schroel A, et al. Comparative analysis of protein aggregates by blue native electrophoresis and subsequent sodium dodecylsulfate-polyacrylamide gel electrophoresis in a three-dimensional geometry gel. Proteomics, 2005, 5(8): 2002-2009.
[10] 秦文斌. 血红蛋白的 A_2 现象发生机制的研究"红细胞 HbA_2"为 HbA_2 与 HbA 的结合产物. 生物化学与生物物理进展, 1991, 18(4): 286-288.
[11] 秦文斌, 梁友珍. 血红蛋白 A_2 现象Ⅰ. A_2 现象的发现及其初步应用. 生物化学与生物物理学报, 1981, 13(2): 199-205.
[12] Su Y, Gao LJ, Ma Q, et al. Interactions of hemoglobin in live red blood cells measured by the electrophoesis release test. Electrophoresis, 2010, 31: 2913-2920.
[13] 秦文斌. 活体红细胞内血红蛋白的电泳释放.中国科学生命科学, 2011, 41(8): 597-607.
[14] 秦文斌, 高丽君, 苏燕, 等. 血红蛋白释放试验与轻型β地中海贫血. 包头医学院学报, 2007, 23(6): 561-563.
[15] Su Y, Shao G, Gao LJ, et al. RBC electrophoresis with discontinuous power supply – a newly established hemoglobin release test. Electrophoresis, 2009, 30: 3041-3043.
[16] 乔姝, 沈木生, 韩丽红, 等. 血小板成分电泳释放的初步研究. 现代预防医学, 2011, 38(4): 685-688.

第三章 初释放电泳 举例

1 单向初释放电泳

见图 3-1。

1.1 结果 一次通电电泳后，得到如下结果：溶血液 HbA_2 靠后(阴极侧)；红细胞 HbA_2 靠前(阳极侧)。

1.2 讨论 这就是"血红蛋白 A_2 现象"的单向电泳结果，"血红蛋白 A_2 现象"是红细胞初释放的主要内容。

2 双向初释放电泳

见图 3-2。

2.1 结果 下层，各成分，都在对角线上；上层，各成分，有的在对角线上，有的不在对角线上；上下两个箭头(↓↑)所指处，不在对角线上，即脱离对角线；下边箭头(↑)所指处为 HbA_2；上边箭头(↓)所指处为 HbA_1。

2.2 讨论 单向电泳里的"红细胞 HbA_2"，在双向电泳里变成脱离对角线的两个成分，上边是 HbA_1，下边是 HbA_2。这说明，单向电泳里的"红细胞 HbA_2"是 HbA_2 与 HbA_1 的结合产物，在第二向电泳时彼此分开，而且之间还有拖拉成分。同时也说明，红细胞内 HbA_2 是与 HbA_1 结合存在的，单向电泳时它表现为"红细胞 HbA_2"，双向电泳时暴露身份，它们是 HbA_2 和 HbA_1 的结合产物。质谱分析证明，还有 Prx(过氧化物还原酶)参与，结合产物是 HbA_1-HbA_2-Prx。

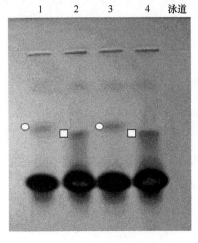

图 3-1 单向电泳 血红蛋白 A_2 现象

注释：单数泳道=溶血液，双数泳道=红细胞；两个〇之间=溶血液 HbA_2，两个□之间=红细胞 HbA_2

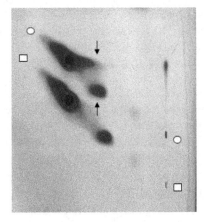

图 3-2 双向电泳 血红蛋白 A_2 现象

注释：上层两个〇之间加样为红细胞；下层两个□之间加样为溶血液；蓝色区带为血红蛋白，红色区带为碳酸酐酶；对角线，两个〇之间或两个□之间划一直线为对角线

这就是初释放电泳结果，它揭露出红细胞内血红蛋白存在状态的奥秘。

第四章 再释放电泳 举例

1 红细胞的单向再释放

1.1 红细胞的单向单带再释放 见图4-1。

1.1.1 结果 单向电泳，停电—再通电，溶血液都没有单带再释放，红细胞都有单带再释放。

1.1.2 讨论 单带再释放，来自红细胞，第一次通电，没有单带再释放，停电—再通电才出来，这说明再释放血红蛋白与红细胞膜结合牢固，不容易释放。

1.2 红细胞的单向多带再释放 见图4-2。

1.2.1 结果 单向电泳，多次停电—多次再通电，溶血液都没有多带再释放，红细胞都有多带再释放。

1.2.2 讨论 多带再释放，来自红细胞，第一次通电，没有单带再释放，停电—再通电才出来，这说明再释放血红蛋白与红细胞膜结合牢固，不容易释放。多次停电—多次再通电都能再释放出来血红蛋白，说明红细胞内存在很多这种血红蛋白。

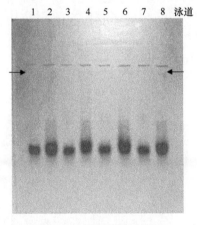

图4-1 单向单带再释放

注释：这是电泳后没染色的情况，红色区带为血红蛋白；泳道1、3、5、7为溶血液；泳道2、4、6、8为红细胞；两个箭头(→←)之间为再释放的单带

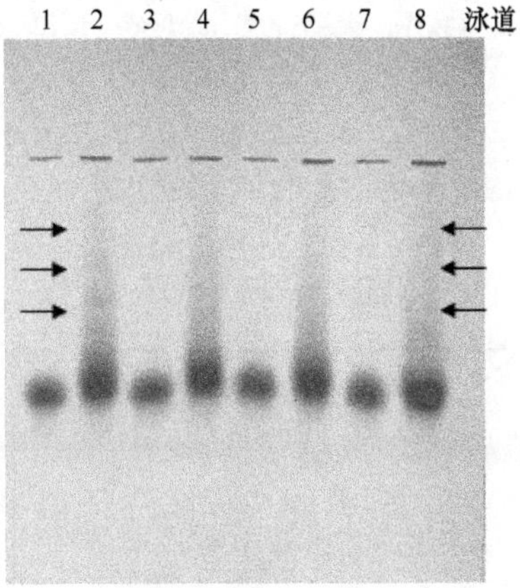

图4-2 单向多带再释放

注释：这是电泳后没染色的情况，红色区带为血红蛋白；泳道1、3、5、7为溶血液；泳道2、4、6、8为红细胞；多个箭头(→←)之间为再释放的多带

1.3 轻型β珠蛋白生成障碍性贫血(β地贫)患者的全血单向多带再释放 见图4-3。

1.3.1 结果 多带来自再释放，患儿和她母亲的结果相同，都是多带再释放明显增强。

1.3.2 讨论 轻型β地贫有多种基因类型，但它们有一个共同的特点，那就是患者的血红蛋白电泳显色 HbA_2 含量增多。本例患者母女都是 HbA_2 增多，遗传关系明确。在此基础上，我们发现轻型β地贫患者的再释放明显增强，从而产生了再释放的诊断意义。

2 双向再释放

见图 4-4。

2.1 结果 HbA_2 减少，见箭头(↑)所指处，多带再释放增强，见多个箭头(↓↓↓)所指处。

2.2 讨论 HbA_2 减少，怀疑为 α 地贫。本例，基因诊断为缺失型 α 地贫。α 地贫也显示多带再释放增强，由此可知，α 地贫和 β 地贫都可表现为再释放增强。

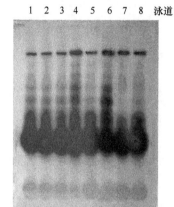

图 4-3　轻型 β 地贫患者全血的单向多带再释放电泳
注释：泳道 4 和 6 为轻型 β 地贫全血标本，其余都是正常人；泳道 4 为患儿(女孩)，泳道 6 为患儿的母亲；上边的黑红色，都是血红蛋白；最下方红色区带为白蛋白

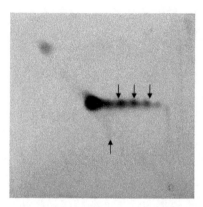

图 4-4　α 地贫全血的双向多带再释放电泳
注释：左上角红色区带为白蛋白；右下角红色区带为碳酸酐酶加纤维蛋白原；箭头(↑)所指处为 HbA_2；多个箭头(↓↓↓)所指处为多带再释放

3 上述再释放结果的模式图

见图 4-5 和图 4-6。

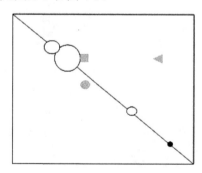

图 4-5　单带再释放型双向对角线电泳
注释：HbA_1／HbA_2 ｝HbA_2 现象　来自初释放
◁ Hb 单带现象　来自再释放

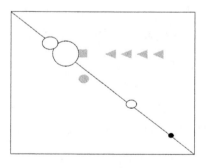

图 4-6　多带再释放型双向对角线电泳
注释：HbA_1／HbA_2 ｝HbA_2 现象　来自初释放
◁◁◁◁ Hb 多带现象　来自再释放

第五章　交叉互作　举例

1　溶血液交叉互作电泳

见图 5-1。

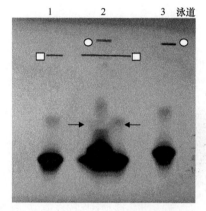

图 5-1　溶血液交叉互作电泳结果
注释：两个○之间是后排的红细胞溶血液；两个□之间是前排的红细胞溶血液；泳道 1 是溶血液对照；泳道 3 是溶血液对照；泳道 2 是交叉互作，溶血液穿过溶血液

1.1　结果　后排溶血液的 HbA_1，穿过前排溶血液的 HbA_2，被穿过的 HbA_2 区带变形，见两个箭头(→←)所指处。

1.2　讨论　被穿过的 HbA_2 区带变形，说明发生了交叉互作。

2　电泳纯血红蛋白交叉互作电泳

见图 5-2。

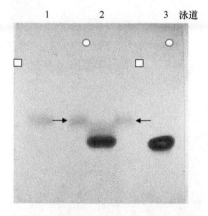

图 5-2　电泳纯血红蛋白交叉互作电泳结果
注释：两个○之间是后排的 HbA_1(抠胶后直接放在这里)；两个□之间是前排的 HbA_2(抠胶后直接放在这里)；泳道 1 是 HbA_2 对照；泳道 3 是 HbA_1 对照；泳道 2 是交叉互作，HbA_1 穿过 HbA_2

2.1　结果　后排的 HbA_1，穿过前排 HbA_2，被穿过的 HbA_2 区带变形，见两个箭头(→←)所指处。

2.2　讨论　被穿过的 HbA_2 区带变形，说明发生了交叉互作。

第六章 混合互作 举例

1 白蛋白对红细胞多带再释放的影响

见图 6-1。

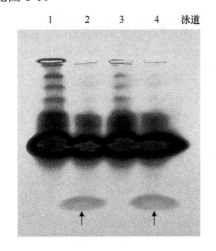

图 6-1 白蛋白对红细胞多带释放的影响电泳
注释：泳道 1、3 为红细胞；泳道 2、4 为红细胞加白蛋白
(见图中箭头↑所指处)

1.1 结果 红细胞存在多带再释放；红细胞加入白蛋白后，多带再释放明显减弱。
1.2 讨论 这就是白蛋白与红细胞的相互作用(混合互作)。

2 丙种球蛋白对红细胞多带再释放的影响

见图 6-2。

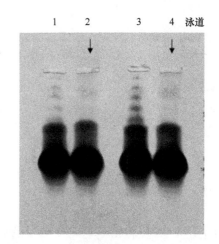

图 6-2 丙种球蛋白对红细胞多带释放的影响电泳
注释：泳道 1、3 为红细胞；泳道 2、4 为红细胞加丙种球蛋白(见图中箭头↓所指处)

2.1 结果 红细胞存在多带再释放，红细胞加入丙种球蛋白后，多带再释放明显减弱。
2.2 讨论 这就是丙种球蛋白与红细胞的相互作用(混合互作)。

第七章 其他细胞 举例

1 血小板与其血浆的双向电泳

见图 7-1。

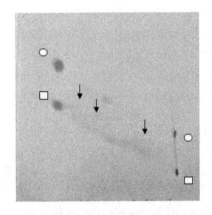

图 7-1 血小板与其血浆的双向电泳结果
注释：上层(两个○之间)为血浆；下层(两个□之间)为血小板

1.1 结果 血浆方面，能看到白蛋白、α_1、α_2、β 球蛋白和纤维蛋白原，它们都在对角线上。血小板方面，除了对角线上成分之外，还有一些脱离对角线成分(箭头↓所指处)。

1.2 讨论 这表明的是血小板成分电泳释放的特点。

2 胃癌细胞培养物的双向电泳

见图 7-2。

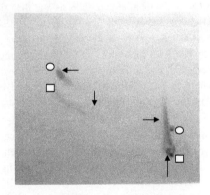

图 7-2 双向电泳比较培养液与胃癌细胞培养物
注释：上层(两个○之间)为细胞培养的培养基；下层(两个□之间)为胃癌细胞的培养物

2.1 结果 培养基里，出现多量快泳成分，箭头←所指处，可能是白蛋白。胃癌细胞的培养物里，上述快泳成分很少，原点沉淀很多(箭头↑所指处)，此沉淀物，在第二向电泳时离开原点，形成上升成分(箭头→所指处)；好像还有脱离对角线成分(箭头↓所指处)，类似血小板结果。

2.2 讨论 这表明的是胃癌培养物电泳释放的特点。

第八章 电泳释放蛋白质组学——红细胞初释放电泳释放蛋白质组学 举例

红细胞电泳释放蛋白质组学与普通的红细胞蛋白质组学有何不同？

以前研究过红细胞的普通蛋白质组学，那是用红细胞溶血液做实验，发现许多斑点，但没有发现红细胞电泳释放时出现各种相互作用。

红细胞电泳释放蛋白质组学，用完整红细胞电泳释放出来的血红蛋白做实验，初释放结果表明红细胞内存在"HbA_1-HbA_2-Prx"式的相互作用，详见下文。

用电泳释放试验检测活体红细胞内血红蛋白之间的相互作用

Interactions of hemoglobin in live red blood cells measured by the electrophoresis release test

Yan Su[1], Lijun Gao[1], Qiang Ma[2], Lishe Zhou[1], Liangyi Qin[3], Lihong Han[1], Wenbin Qin[1]

(1. Laboratory of Hemoglobin, Baotou Medical College, Baotou, China; 2. Analytical Testing Center, Zhejiang Chinese Medical University, Hangzhou, China; 3. Clinical Laboratory, Nanhui Central Hospital, Shanghai, China)

Abstract

To elucidate the protein-protein interactions of hemoglobin(Hb)variants A and A_2, HbA was first shown to bind with HbA_2 in live red blood cells(RBCs)by diagonal electrophoresis and then the interaction between HbA and HbA_2 outside the RBC was shown by cross electrophoresis. The starch-agarose gel electrophoresis of hemolysate, RBCs, freeze-thawed RBCs and the supernatant of freeze-thawed RBCs showed that the interaction between HbA and HbA_2 was affected by membrane integrity. To identify the proteins involved in the interaction, protein components located between HbA and HbA_2 in RBCs(RBC HbA-HbA_2)and hemolysate(hemolysate HbA-HbA_2)were isolated from the starch-agarose gel and separated by 5% ~12% SDS-PAGE. The results show that there was a ≈22 kDa protein band located in the RBC HbA-HbA_2 but not in hemolysate HbA-HbA_2. Sequencing by LC/MS/MS showed that this band was a protein complex that included mainly thioredoxin peroxidase B, alpha globin, delta globin and beta globin. Thus, using our unique in vivo whole blood cell electrophoresis release test(ERT), Hbs were proven for the first time to interact with other proteins in the live RBC.

Corresponding author: Wenbin Qin
Address: Laboratory of Hemoglobin, Baotou Medical College, Baotou, Inner Mongolia, China 014060, E-mail address: qinwenbinbt@sohu.com, Phone number: 86-472-5615442, Fax number: 86-472-5152442
Abbreviations: Hb hemoglobin, RBC red blood cell, MS mass spectrometry, ERT electrophoretic release test, CA carbonic anhydrase, TPx thioredoxin peroxidase, Prx peroxiredoxin, NKEF-B natural killer enhancing factor-B, ROS reactive oxygen species, GPx glutathione peroxidase, TAP tandem affinity purification, GST glutathione-S-transferase

Key words　hemoglobin; red blood cell; interaction; peroxiredoxin

1　Introduction

It is believed that biological processes are carried out through precise protein-protein interactions[1-2]. A better understanding of these interactions is crucial for elucidating the structural/functional relations of proteins, investigating their roles in the development of associated disease and determining potential drug targets for clinical applications[3]. Increasing numbers of interactions are now being identified and the information organized and hosted in many databases with the help of high throughput screening technologies, computational predictions and literature-mining processes[4-8].

The red blood cell(RBC)is the simplest human cell and hemoglobins(Hbs)are iron- containing oxygen-transport metalloproteins located within it[9-10]. The protein structure, function and gene and expression of Hb have been deeply studied during the last century. As we know, Hb variants HbA(over 95%), HbF(<1%)and HbA_2(1.5%～3.5%)all exist in normal adult RBCs. All variants are comprised of two similar types of globin chains that combine to form a tetramer. The subunit compositions of HbA, HbF and HbA_2 are $\alpha_2\beta_2$, $\alpha_2\gamma_2$ and $\alpha_2\delta_2$, respectively. Each subunit is a globin chain with an embedded heme group and each heme group contains an iron atom that can bind one oxygen molecule through ion-induced dipole forces. These subunits are bound to each other by salt bridges, hydrogen bonds and hydrophobic interactions. However, proteins rarely act alone at the biological level[2]. As the most important and abundant proteins in the RBC[11](accounting for more than 90% of the cellular dry weight), the interactions between Hbs become a interesting issue for us to study. During the past decade, with the rapid development of mass spectrometry(MS), increasing work has been done in the field of erythrocyte proteomics and erythrocyte membrane proteomics[11-18]. Increasing insights into the protein constitution of RBC are being found and a recent report showed that 1578 gene products had been identified from the erythrocyte cytosol[11]. These results provided a solid foundation for studying protein-protein interactions inside the RBC. To date, only *in silico* protein-protein interactions in the human erythrocyte have been predicted through bioinformatics[11-12] and few of these have been confirmed by in vivo experiments. Whether or not Hbs form complexes with other proteins, how many proteins are involved in the Hb complex, the nature of the dynamic interaction between these proteins and the biological significance of these remain unresolved.

In 1981, we found that HbA_2 released from RBCs moved faster than that sourced from hemolysate during starch–agarose mixed gel electrophoresis. This phenomenon was named the "HbA_2 phenomenon"[19-21] and an electrophoresis method using live RBCs, the electrophoretic release test(ERT), was developed[21]. Although many people think of this as a simple physical phenomenon, we continue to believe that this is an intrinsic phenomenon that may have some relationship with the interactions between Hbs and other proteins in the RBC. In this paper, we seek to determine the exact molecular mechanism behind the "HbA_2 phenomenon" and try to give some insights into the natural protein-protein interactions of Hbs in live RBCs.

2　Materials and Methods

2.1　Patients and specimens　This study was approved by our local ethics committee. Blood samples from healthy adults were collected randomly from the first affiliated hospital of Baotou Medical College. Before blood samples were collected, all participants in the experiment were asked to sign consent information. Venous blood samples were anti-coagulated with heparin, stored at 4 °C, and generally analyzed within 24 h.

2.2 Preparation of RBCs and hemolysate The anti-coagulated blood was centrifuged at 3 000 rpm for 10 min and the upper plasma was aspirated. Two hundred microliter of the lower RBCs layer was then added to 1 mL saline, mixed gently, centrifuged at 3 000 rpm for 4 min and the supernatant aspirated. This washing operation was repeated four to five times and the RBCs were then used to perform electrophoresis after 1∶1 dilution with saline. Hemolysate was prepared by continuously adding 200 μl water and 100 μl CCl_4 to the RBCs. After turbulent mixing, the sample was centrifuged at 12 000 rpm for 10 min and the upper red hemolysate was pipetted out carefully and stored at 4 °C for later use. Freeze-thawed RBCs were prepared by placing the RBCs in a –80 °C freezer for 20 min, then at room temperature for 20 min. This cycle was repeated at least twice and some of the sample was then aliquoted out as freeze-thawed RBC samples. The remainder of the sample was centrifuged at 12 000 rpm for 10 min and the upper supernatant was pipetted out for later use.

2.3 Diagonal electrophoresis on starch-agarose mixed gel A 2% starch-agarose mixed gel(starch: agarose =4∶1)was prepared with TEB buffer(pH 8.6)as described previously[21]. After adding 8 μl RBC and hemolysate at the indicated position, two vertical direction electrophoresis runs were performed at 5 V/cm and the gel was then sequentially stained with Ponceau Red and Benzidine.

2.4 Cross starch-agarose mixed gel electrophoresis The gel was prepared in the same way as for diagonal electrophoresis. Two rows of sample slots about 1 cm apart were then made on the cathodic side of the gel, about 1.5 cm away from the edge. The front slot was about 3 cm long and was loaded with electrophoretically pure HbA_2. The back slot was about 1 cm long and was loaded with electrophoretically pure HbA. To one side of the middle slots was a corresponding control sample. The electrophoresis and staining steps were the same as those described above.

2.5 Recovery and enrichment of protein from the starch-agarose mixed gel After starch-agarose mixed gel electrophoresis of hemolysate and RBCs, the red band of HbA_2 and the gel located between HbA and HbA_2(HbA-HbA_2)of RBC and hemolysate were cut out separately and frozen at –80°C for at least 30 min. Before use, the gels were taken out and thawed at room temperature. After centrifuging at 12 000 rpm for 10 min, the supernatants were pipetted into new Eppendorf tubes and dried in a vacuum freeze drier(Multi-drier, FROZEN IN TIME, England)for about 6-8 h.

2.6 SDS-PAGE Each of the freeze-dried samples was dissolved in 50～100 μl ultra-pure water. After boiling with 2×loading buffer for 5 min, the samples were loaded onto a 5%～12% SDS-PAGE gel. After electrophoresis, the gel was stained overnight with 0.08% Coomassie Brilliant Blue G250 and then destained with MilliQ water until the background staining was low.

2.7 In-gel digestion The gel band was excised into 1 mm^2 pieces and further destained for 20 min in 100 μl of 100 mM NH_4HCO_3/50% acetonitrile two or three times until completely destained. The gel was then dried in a SpeedVac vacuum airer(Savant Instruments, Holbrook, NY, USA)for 20 min and then allowed to swell in 5～10 μl of 25 mM NH_4HCO_3(pH 8.0)containing 10 ng/μl modified trypsin at 4 °C for 40 min. Finally, it was incubated with another 10 μl of 25 mM NH_4HCO_3 buffer overnight at 37 °C. Digested peptide was extracted with 50～100 μl of 5% TFA solution at 40 °C for 1 h and then extracted with the same volume of 50% ACN/2.5% TFA solution at 30 °C for 1 h. Finally, it was extracted ultrasonically with 50 μl ACN solution. The extracted solution was pooled, dried in the SpeedVac vacuum dryer and resuspended in 3～5 μl 0.1% formic acid prior to analysis.

2.8 MS Digested samples were analyzed by a LC/ MS/MS system, composed of a nano-ACQUITY ultra-performance liquid chromatography system(ACQUITY UPLC ® System, Waters, Milford, MA, USA)and a SYNAPT™ High Definition Mass Spectrometry™ system HDMS™, Waters, Milford, MA, USA)with an electrospray ionization(ESI)source. Twenty microliters of trypsin-digested protein was loaded onto a guard column(180 μm × 20 mm Symmetry C18 column; Waters)and separated by reversed phase chromatography on a nanoACQUITY UPLC BEH C18 column(75 μm × 250 mm; Waters). The flow rate of the mobile phase was set at 200 nL/min and the temperature of the column was 35°C. An aqueous solution containing 0.1% formic acid was used as mobile phase A and acetonitrile containing 0.1% formic acid was used as mobile phase B. The gradient program was as follows: 0～80 min from 1% to 40% of B, 80～90 min from 40% to 80% of B, 90～100 min from 80% B, and 100～120 min from 80 to 1% B. The mass spectrometer was controlled by MassLynx software 4.0 and detected in the positive ion mode. The capillary voltage was set at 2.5 kV; the cone voltage was set at 35 V; the ion source temperature was set at 90°C; the scan ranges of m/z were 350～1600(MS)and 50～2000(MS/MS).

2.9 MS analysis Data searching was performed with the ProteinLynx Global Server(PLGS; Micromass Wythenshawe, UK)software 2.3 by submitting the result to Mascot Search(version 1.9; Matrix Science, London, UK)(http: //www.matrixscience.com)MS/MS Ion Search to match the peptide mass fingerprints[22, 23]. The NCBInr Homosapiens database was used for the search. The search conditions were restricted to trypsin digestion, M oxidation and iodoacetamide alkylation into a variable modification. One missing cleavage was allowed. MS and MS/MS fragment error tolerance≤0.2 Da.

3 Results

3.1 "RBC HbA-HbA$_2$" is a binding product of HbA$_2$ and HbA Diagonal electrophoresis of RBCs and hemolysate were performed using starch-agarose electrophoresis. After two vertical directional electrophoresis runs, the results showed that HbA, HbA$_2$, carbonic anhydrase and the origin were located at one diagonal in the hemolysate sample but the location of HbA and HbA$_2$ deviated from the diagonal in the RBC sample(Fig. 8-1). Thus, "RBC HbA$_2$" and "hemolysate HbA$_2$"

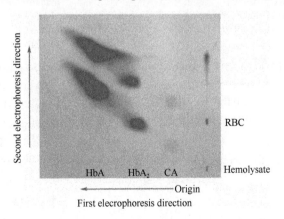

Fig. 8-1 Diagonal electrophoresis of RBCs and hemolysate on a starch-agarose mixed gel
The origins of RBC and hemolysate samples are located at the bottom right corner of the gel. The first electrophoresis direction was from right to left and the second electrophoresis direction was from bottom to top

were not located at the same position in the vertical direction. Furthermore, during the second directional electrophoresis, another HbA was released from "RBC HbA-HbA$_2$" in the first directional electrophoresis but not from the hemolysate sample. This demonstrates that "RBC HbA-HbA$_2$" is a complex of HbA and HbA$_2$.(For simplicity, the speech marks are omitted from the complex names in the remainder of this document.)

3.2 HbA$_2$ could interact with HbA outside the RBC To evaluate whether HbA could bind with HbA$_2$ and form a complex, cross electrophoresis of HbA and HbA$_2$ was performed on a starch-agarose mixed gel. When the fast moving HbA crossed the slow moving HbA$_2$ in the electric field, the middle band of HbA$_2$ was distorted toward the anode(Fig. 8-2). This result demonstrates that HbA$_2$ could interact with HbA transiently outside the RBC.

3.3 RBC membrane mediates the interaction between HbA and HbA$_2$ To clarify whether the interaction between HbA and HbA$_2$ is associated with the RBC membrane, hemolysate, RBCs, freeze-thawed RBCs and the supernatant of freeze-thawed RBCs were subject to electrophoresis on a starch-agarose gel. The results showed that HbA$_2$ in the RBCs group moved the fastest, followed by the freeze-thawed RBCs group, while the hemolysate and the supernatant of the freeze-thawed RBCs group moved the slowest(Fig. 8-3).

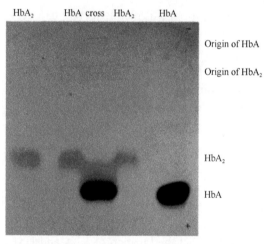

Fig. 8-2 Cross electrophoresis of HbA and HbA$_2$ on a starch-agarose mixed gel

Two rows of sample slots are located at the top of the gel. HbA$_2$ was added in the front slot and HbA was added in the back slots

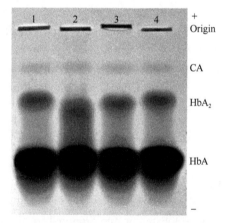

Fig. 8-3 The HbA$_2$ phenomenon was affected by RBC membrane integrity

Lane 1 contained hemolysate, lane 2 contained RBCs, Lane 3 contained the freeze-thawed RBCs and lane 4 contained the supernatant of freeze-thawed RBCs

3.4 A different protein band appeared in RBC HbA-HbA$_2$ during SDS-PAGE To further determine whether other proteins were involved in the Hb complex, hemolysate HbA$_2$, RBC HbA$_2$, hemolysate HbA-HbA$_2$ and RBC HbA-HbA$_2$ extracted from the starch-agarose gel were separated by 5%～12% SDS-PAGE. The results show that no significantly different protein band appeared between the hemolysate HbA$_2$ group and the RBC HbA$_2$ group, although there were content differences in some bands. However, compared with the hemolysate HbA- HbA$_2$ group, a significant protein band(≈22kDa)appeared in the RBC HbA-HbA$_2$ group(Fig. 8-4).

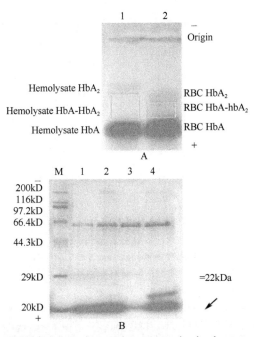

Fig. 8-4 Proteins extracted from the starch-agarose mixed gel separated by SDS-PAGE

A. The excised bands for HbA- HbA$_2$ and HbA$_2$ are indicated in this figure. Lane 1 contained hemolysate and lane 2 contained RBC. B. The result of separation of the extracted proteins by 5-12% SDS- PAGE. Lane M contained protein markers and lanes 1-4 contained hemolysate HbA$_2$, RBC HbA$_2$, hemolysate HbA-HbA$_2$ and RBC HbA- HbA$_2$, respectively

3.5 Identification of the ≈22 kDa band by LC/MS/MS To investigate the composition of the ≈22 kDa protein band, the band was cut out of the gel and sent to the National Center for Biomedical Analysis. After in-gel digestion, peptide sequencing was performed by nanoUPLC-ESI MS/MS. The results show that this band was a protein complex that mainly included thioredoxin peroxidase(TPx)B(gi|9955007)(Fig. 8-5)and the subunits of HbA and HbA$_2$, including α-globin, δ-globin and β-globin(Table 8-1 and the supporting Information).

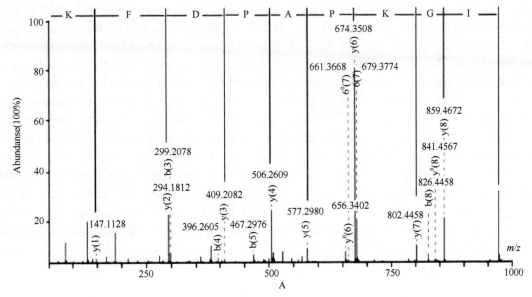

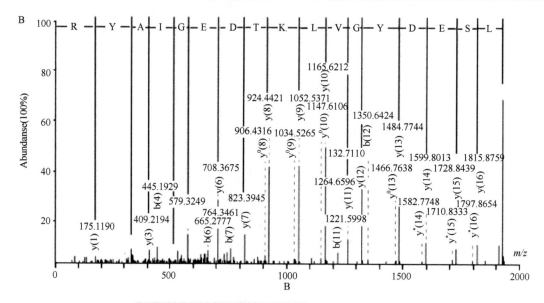

Fig. 8-5 The nanoUPLC-ESI MS/MS spectra for the trypsin digested gel band

(A)The MS spectrum of a peptide with molecular weight 971.5440 and an amino acid sequence IGKPAPDFK is labeled.(B)The MS spectrum of a peptide with molecular weight 1927.9527 and an amino acid sequence LSEDYGVLKTDEGIAYR is labeled.(C)The sequence coverage of thioredoxin peroxidase B is 77%

4 Discussion

The HbA$_2$ phenomenon tells us that Hbs exist in RBC as a complicate complex. In our experiments, diagonal electrophoresis and cross electrophoresis proved that RBC HbA- HbA$_2$ was a combined product of HbA$_2$ and HbA. However, if RBC HbA$_2$ = hemolysate HbA$_2$ + hemolysate HbA, the electrophoretic position of RBC HbA$_2$ should locate at the middle of hemolysate HbA$_2$ and hemolysate HbA. In fact, RBC HbA$_2$ was located near to hemolysate HbA$_2$. Thus, we speculated that there must be some other components(X)involving in this Hb complex. The electrophoresis results for hemolysate, RBCs, freeze-thawed RBCs and the supernatant of freeze-thawed RBCs indicated that integrity of the RBC membrane was necessary to mediate the interaction between HbA and HbA$_2$. When the RBC membrane was removed, the interaction was eliminated. If membrane integrity was damaged, the interaction would obviously be weakened.

To further explore the constitution of the Hb complex, proteins located between HbA and HbA$_2$ were extracted and separated by SDS-PAGE. The results showed a clear ≈22 kDa band in RBC HbA-HbA$_2$ but not in hemolysate HbA-HbA$_2$. LC/MS/MS detection of the digested ≈22 kDa band showed that this band was a peptide mixture mainly composed of TPx B, α-globin, δ-globin and β-globin. Traces of keratin appearing in the result(Supporting Information)was almost certainly due to contamination(Table 8-1).

Table 8-1   LC/MS/MS results for the ≈22 kDa band

NCBI accession no.	Name	Mass	Score	Queries matched
gi|9955007	Chain A, Thioredoxin Peroxidase B from red blood cells	21795	1218	R.SVDEALR.L R.GLFIIDGK.G K.TDEGIAYR.G R.IGKPAPDFK.A K.ATAVVDGAFK.E R.LSEDYGVLK.T R.QITVNDLPVGR.S R.QITVNDLPVGR.S R.GLFIIDGKGVLR.Q K.ATAVVDGAFKEVK.L K.EGGLGPLNIPLLADVTR.R K.EGGLGPLNIPLLADVTR.R R.KEGGLGPLNIPLLADVTR.R K.EGGLGPLNIPLLADVTRR.L K.EGGLGPLNIPLLADVTRR.L R.LSEDYGVLKTDEGIAYR.G R.LSEDYGVLKTDEGIAYR.G K.LGCEVLGVSVDSQFTHLAWINTPR.K + Carbamidomethyl(C) R.KLGCEVLGVSVDSQFTHLAWINTPR.K + Carbamidomethyl(C) R.KLGCEVLGVSVDSQFTHLAWINTPR.K + Carbamidomethyl(C) R.KLGCEVLGVSVDSQFTHLAWINTPR.K + Carbamidomethyl(C) R.LVQAFQYTDEHGEVCPAGWKPGSDTIKPNVDDSK.E + Carbamidomethyl(C) R.LVQAFQYTDEHGEVCPAGWKPGSDTIKPNVDDSK.E + Carbamidomethyl(C) R.LVQAFQYTDEHGEVCPAGWKPGSDTIKPNVDDSKEYFSK.H + Carbamidomethyl(C) R.LVQAFQYTDEHGEVCPAGWKPGSDTIKPNVDDSKEYFSK.H + Carbamidomethyl(C)
gi|4504351	Delta globin	16045	396	K.LHVDPENFR.L K.VNVDAVGGEALGR.L R.LLVVYPWTQR.F K.VVAGVANALAHKYH.- K.EFTPQMQAAYQK.V + Oxidation(M) K.GTFSQLSELHCDK.L + Carbamidomethyl(C) K.VLGAFSDGLAHLDNLK.G R.FFESFGDLSSPDAVMGNPK.V + Oxidation(M)
gi|161760892	Chain D, neutron structure analysis of deoxy human hemoglobin	15869	368	-.RHLTPEEK.S K.LHVDPENFR.L R.LLVVYPWTQR.F K.VNVDEVGGEALGR.L K.EFTPPVQAAYQK.V K.VVAGVANALAHKYH.- K.VLGAFSDGLAHLDNLK.G R.FFESFGDLSTPDAVMGNPK.V + Oxidation(M)
gi|47679341	Hemoglobin beta	11439	256	K.LHVDPENFR.- K.VNVDAVGGEALGR.L R.LLVVYPWTQR.F K.VLGAFSDGLAHLDNLK.G R.FFESFGDLSTPDAVMGNPK.V + Oxidation(M)

Continued

NCBI accession no.	Name	Mass	Score	Queries matched
gi|66473265	Homo sapiens beta globin chain	11480		K.LHVDPENFR.- R.LLVVYPWTKR.F K.VNVDEVGGEALGR.L K.VLGAFSDGLAHLDNLK.G R.FFESFGDLSTPDAVMGNPK.V + Oxidation(M)
gi|229751	Chain A, Structure of hemoglobin in the deoxy quaternary state with ligand bound at the alpha hemes(a2)	15117	249	R.MFLSFPTTK.T + Oxidation(M) K.LRVDPVNFK.L -.VLSPADKTNVK.A K.VGAHAGEYGAEALER.M K.TYFPHFDLSHGSAQVK.G K.TYFPHFDLSHGSAQVK.G K.VADALTNAVAHVDDMPNALSALSDLHAHK.L + Oxidation(M)
gi|179409	Beta-globin	15870	247	K.LHVDPENFR.L K.EFTPPVKAAYQK.V K.VVAGVANALAHKYH.- K.VLGAFSDGLAHLDNLK.G
gi|4929993	Chain A, module-substituted chimera hemoglobin beta-alpha(F133v)	15780	240	K.LRVDPVNFK.L R.LLVVYPWTQR.F K.VNVDEVGGEALGR.L K.VLGAFSDGLAHLDNLK.G R.FFESFGDLSTPDAVMGNPK.V + Oxidation(M)
gi|157838239	Chain A, hemoglobin thionville: an alpha-chain variant with a substitution of a glutamate for valine at na-1 and having an acetylated methionine nh2 terminus(a$_2$)	15278	226	R.MFLSFPTTK.T + Oxidation(M) K.LRVDPVNFK.L K.VGAHAGEYGAEALER.M K.TYFPHFDLSHGSAQVK.G K.TYFPHFDLSHGSAQVK.G K.VADALTNAVAHVDDMPNALSALSDLHAHK.L + Oxidation(M)

TPx B has previously been called torin, calpromotin, thiol-specific antioxidant/protector protein, band-8, natural killer enhancing factor-B and is now named peroxiredoxin 2(Prx 2)[24-25]. The human Prx 2 gene is located at 13q12, coding a 198 amino acid polypeptide with a molecular mass of 22 kDa[24]. There are 6 types of mammalian Prx isoforms(Prx 1-6)[26-28], of which Prx 2 is the third most abundant protein in RBC and has an important peroxidase activity that can protect RBCs against various oxidative stresses[24]. As we know, the RBC contains high levels of O_2 and Hbs and the continual auto-oxidation of Hbs produces many reactive oxygen species(ROS)such as O_2 and H_2O_2. Therefore, compared with other somatic cells, RBCs are exposed to a higher level of oxidative stress[13, 29]. The ROS in RBCs can damage proteins and membrane lipids but, as anucleate cells, RBCs are unable to synthesize new proteins to replace damaged proteins[26, 30]. Thus, RBCs must be well-equipped with many antioxidant proteins, which include catalase, glutathione peroxidase, and the emerging antioxidant enzyme, Prxs. For a long time, it was considered that catalase and glutathione peroxidase constituted the defense against ROS in RBC[14]. Recently, increasing attention has been given to the antioxidant role of Prxs in RBCs. Our experimental results show for the first time that in live RBC, Prx 2 binds with the globin chain of Hb to form a complex. This complex enables Prx 2 to be more effective in protecting Hb from oxidative stress. However, the manner by which Prx 2 interacts with these globins remains unknown. We believe some of the membrane proteins mediate the binding of this Hb complex because the HbA_2 phenomenon appeared only in RBC samples and not in hemolysate or the supernatant of freeze-thawed RBCs in which the RBC membrane was removed.

Furthermore, during our experiments, a unique method, ERT, was established for the study of protein-protein interaction in live cells. An increasing number of approaches, such as yeast two-hybrid systems, tandem affinity purification, protein chip, co-immunoprecipitation and glutathione-S-transferase pull-down methods, have been developed to detect protein-protein interactions[31]. However, functional protein-protein interactions are dynamic processes and many are maintained by non-covalent bonds. Therefore, the detection of interactions in live cells remains difficult. In addition, the loss of internal organelles greatly limits the use of interaction detection approaches such as TAP and GST pull-down in the RBC. With ERT, live RBCs were added directly onto the gel and the electric current perforated the membrane instantaneously. The protein complexes in live RBCs were thus released directly to the electric field and the different electrophoresis behaviors of RBC proteins could be directly compared with hemolysate, in which the protein-protein interactions would have been damaged during preparation, especially interactions mediated by membrane proteins. Thus, *in vivo* interactions in the RBC could be identified by ERT. Furthermore, this method can be used in other live cells. Evidence of *in vivo* protein-protein interactions may be found through finding differences in the electrophoretic behavior of living cells and corresponding cell lysates. However, the low resolving power of starch-agarose gels decreases the detection range and it is therefore mainly limited to the detection of interactions between high-abundance proteins. Interactions involving low-abundance proteins are difficult to detect but we believe that through the use of a high-resolution electrophoresis method, more information on such interactions may be found in the future.

This work was supported by grants from the Major Projects of Higher Education Scientific Research in the Inner Mongolia Autonomous Region(NJ09157)and the Key Science and Technology Research Project of the Ministry of Education. We also especially acknowledge all of the people who donated their blood samples for our research.

The authors have declared no coflict of interest.

References

[1] Kim, K. K. Kim, H. B., *World J. Gastroenterol.* 2009, *15*, 4518-4528.
[2] Bu, D., Zhao, Y., Cai, L., Xue, H., Zhu, X., Lu, H., Zhang, J., Sun, S., Ling, L., Zhang, N., Li, G., Chen, R., *Nucleic Acids Res.* 2003, *31*, 2443-2450.
[3] Hase, T., Tanaka, H., Suzuki, Y., Nakagawa S., Kitano H., *PLoS. Comput. Biol.* 2009, *5*, e1000550.
[4] Frishman, D., Albrecht, M., Blankenburg, H., Bork, P., Harrington, E. D., Hermjakob, H., Jensen, L. J., Juan, D. A., Lengauer, T., Pagel., P., Schachter, V., Valencia, A., in: Frishman, D., Valencia, A.(Eds)*Modern Genome Annotation*, Springer Wien New York, Vienna 2008, pp. 353-410.
[5] Börnke, F., in: Junker B. H., Schreiber F., *Analysis of Biological Networks*, Wiley-Interscience, New York 2008, pp. 207-232.
[6] Jung, S. H., Hyun, B., Jang, W. H., Hur, H. Y., Han D. S., *Bioinformatics* 2010, *26*, 385-391.
[7] Jung, S. H., Jang, W. H., Hur, H. Y., Hyun, B., Han D. S., *Genome Inform.* 2008, *21*, 77-88.
[8] Collura, V., Boissy, G., *Subcell Biochem.* 2007, *43*, 135-183.
[9] Perutz, M. F., Rossmann, M. G., Cullis, A. F., Muirhead, H., Will G., North, A. C., *Nature* 1960, *185*, 416-422.
[10] Perutz, M. F., *Brookhaven Symp. Biol.* 1960, *13*, 165-183.

[11] D'Alessandro, A., Righetti, P. G., Zolla. L., *J. Proteome Res.* 2010, *9*, 144-163.
[12] Goodman, S. R., Kurdia, A., Ammann, L., Kakhniashvili, D., Daescu O., *Exp. Biol. Med.(Maywood)* 2007, *232*, 1391-1408.
[13] Kakhniashvili, D. G., Bulla Jr., L. A., Goodman, S. R., *Mol. Cell Proteomics* 2004, *3*, 501-509.
[14] D'Amici, G. M., Rinalducci, S., Zolla, L., *J. Proteome Res.* 2007, *6*, 3242-3255.
[15] Alvarez-Llamas, G., de la Cuesta, F., Barderas, M. G, . Darde, V. M., Zubiri, I., Caramelo, C., Vivanco, F., *Electrophoresis* 2009, *30*, 4095-4108.
[16] van Gestel, R. A., van Solinge, W. W., van der Toorn, H. W., Rijksen, G., Heck, A. J., van Wijk, R., Slijper, M., *J. Proteomics* 2010, *73*, 456-465.
[17] Zhang, Q., Tang, N., Schepmoes, A. A., Phillips, L. S., Smith, R. D., Metz, T. O., *J. Proteome Res.* 2008, *7*, 2025-2032.
[18] Eleuterio, E., Di Giuseppe, F., Sulpizio, M., di Giacomo, V., Rapino, M., Cataldi, A., Di Ilio, C., Angelucci, S., *Biochim. Biophys. Acta* 2008, *1784*, 611-620.
[19] Qin, W. B., Liang, Y. Z., *Chin. J. Biochem. Biophys.* 1981, *13*, 199-201.
[20] Qin, W. B., *Sheng Wu Hua Xue Yu Sheng Wu Wu Li Jin Zhan* 1991, *18*, 286-289.
[21] Su, Y., Shao, G., Gao, L., Zhou, L., Qin L, Qin W., *Electrophoresis* 2009, *30*, 3041-3043.
[22] Wang, H. X., Jin, B. F., Wang, J., He, K., Yang, S. C., Shen, B. F., Zhang, X. M., *Sheng Wu Hua Xue Yu Sheng Wu Wu Li Xue Bao* 2002, *34*, 630-634.
[23] Xia, Q., Wang, H. X., Wang, J., Liu, B. Y., Hu M. R., Zhang, X. M., Shen, B. F., *Zhongguo Yi Xue Ke Xue Yuan Xue Bao* 2004, *26*, 483-487.
[24] Schröder, E., Littlechild, J. A., Lebedev, A. A., Errington, N., Vagin, A. A., Isupov, M. N., *Structure* 2000, *8*, 605-615.
[25] Wood, Z. A., Schröder, E., Robin Harris, J., Poole, L. B., *Trends Biochem. Sci.* 2003, *28*, 32-40.
[26] Stuhlmeier, K. M., Kao, J. J., Wallbrandt, P., Lindberg, M., Hammarström, B., Broell, H., Paigen, B., *Eur. J. Biochem.* 2003, *270*, 334-341.
[27] Yang, K. S., Kang, S. W., Woo, H. A., Hwang, S. C., Chae, H. Z., Kim, K., Rhee, S. G., *J. Biol. Chem.* 2002, *277*, 38029-38036.
[28] Manta, B., Hugo, M., Ortiz, C., Ferrer-Sueta, G., Trujillo M., Denicola A., *Arch. Biochem. Biophys.* 2009, *484*, 146-154.
[29] Johnson, R. M., Goyette, G. Jr., Ravindranath, Y., Ho, Y. S., *Free Radic. Biol. Med.* 2005, *39*, 1407-1417.
[30] Halliwell, B., Gutteridge, J. M., *Free Radicals in Biology and Medicine*, Oxford University Press Inc., New York 1998.
[31] Drewes, G., Bouwmeester, T., *Curr. Opin. Cell Biol.* 2003, *15*, 199-205.

(*Electrophoresis*, 2010, 31: 2913-2920)

第九章　制备电泳举例——人 HbA_1 与 HbA_2 的交叉互作研究

1 取正常人血液，做单向普通电泳，得到如下结果

　　见图 9-1。

2 抠去 HbA_1 和 HbA_2 后

　　如图 9-2 所示。

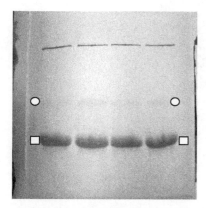

图 9-1　正常人红细胞溶血液的单向电泳结果
注释：两个□之间的区带为 HbA_1；两个○之间的区带为 HbA_2

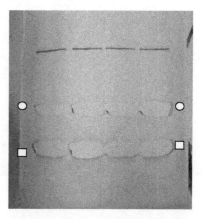

图 9-2　HbA_1 及 HbA_2 的抠胶情况
注释：两个□之间的位置为 HbA_1 抠取后情况；两个○之间的区带为 HbA_2 抠取后情况

3 进行"交叉互作电泳"，HbA_1 穿过 HbA_2，后者区带变形

　　如图 9-3 所示。

4 证明

　　HbA_1 穿过 HbA_2 时，后者区带变形，说明二者发生了相互作用，我们把这种相互作用称之为"交叉互作"。

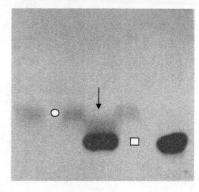

图 9-3　HbA_1 穿过 HbA_2 时，后者区带变形
注释：□两侧区带为 HbA_1，○两侧区带为 HbA_2；箭头(↓)所指处为 HbA_2 区带变形的地方

第十章 染色法 举例

1 血红蛋白 A_2 现象的单向电泳

见图 10-1。

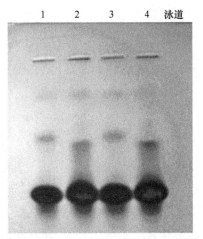

图 10-1 血红蛋白 A_2 现象的单向电泳结果
注释：泳道 1、3=红细胞溶血液；泳道 2、4=红细胞

1.1 结果 蓝色区带（图下方）是血红蛋白；红色区带（图上方）是非血红蛋白，碳酸酐酶。
1.2 讨论 红蓝不同，一目了然，这就是丽春红-联苯胺染色法的特点和优点。

2 血红蛋白 A_2 现象的双向电泳

见图 10-2。

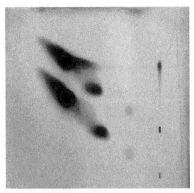

图 10-2 血红蛋白 A_2 现象的双向电泳结果
注释：上层=红细胞，下层=红细胞溶血液

2.1 结果 蓝色区带（图左侧）是血红蛋白；红色区带（图右侧）是非血红蛋白，碳酸酐酶。
2.2 讨论 红蓝不同，一目了然；这就是丽春红-联苯胺染色法的特点和优点。

第二篇　全血释放电泳

第十一章　全血释放电泳概况

1　历史背景

国内外的文献里，没有"全血释放电泳"(whole blood release electrophoresis)这个词。也就是说，没有人用全血做过释放电泳。那我们怎么想起用全血做电泳呢？这要追溯到异常血红蛋白的筛查。20世纪60年代，全世界各国都在大面积筛查异常血红蛋白，此时需要先裂解红细胞释放出来的血红蛋白，然后再用电泳技术筛查异常血红蛋白。由于一次要筛查成千上万个标本，都要先制备成溶血液，工作量很大。于是，我们想，如果制备溶血液太麻烦，能不能直接用全血来筛查异常血红蛋白，就这样，我们用淀琼胶电泳对全血进行了直接筛查，结果在内蒙古自治区发现了大量异常血红蛋白。对此，我们发表了一系列文章，参见本书附录"红细胞释放电泳(溶血液电泳)历年发表的文章"部分。

2　全血释放电泳的特点

全血中除了有红细胞外，还有血浆成分。血浆成分会影响红细胞释放血红蛋白吗？于是我们进行了全血与红细胞并排电泳实验，实验结果如后所述。

3　全血与红细胞的单向并排多带电泳

见图11-1。

由图11-1可以看出，泳道1、2，全血里的多带释放弱于红细胞的多带再释放；泳道3、4，全血里的多带释放弱于红细胞的多带再释放；泳道5、6，全血里的多带释放弱于红细胞的多带再释放；泳道7、8，全血里的多带释放弱于红细胞的多带再释放。这说明，全血里有抑制红细胞多带再释放的物质存在，这只能是全血里的血浆成分在起作用。那么是血浆里的何种成分起作用？情况较为复杂，详见本篇第十二、十三章。

4　全血与红细胞的双向多带电泳

见图11-2。

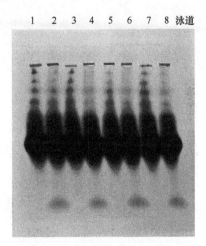

图11-1　全血与红细胞的单向并排多带电泳
注释：这是4份血液标本，单数泳道是红细胞，双数泳道是全血；最下边(阳极侧)的红色区带是血浆白蛋白

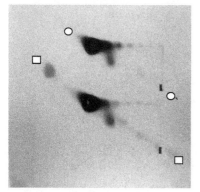

图 11-2 全血与红细胞的双向多带电泳
注释：上层(两个○之间)加红细胞，下层(两个□之间)加全血

由图 11-2 可以看出，红细胞的多带再释放明显强于全血的多带再释放，这说明全血里有抑制红细胞多带再释放的物质存在，这只能是全血里的血浆成分在起作用。那么是血浆里的何种成分起作用？ 情况较为复杂，详见本篇第十三章。

5 全血电泳优于红细胞电泳，能够发现溶血性疾病

见图 11-3、图 11-4。

(1) 由图 11-3 可以看出，A 图全血和红细胞的 HbA_2 都不在对角线上；B 图红细胞的 HbA_2 不在对角线上，全血的 HbA_2 在对角线上，而且，白蛋白里含有 MHA(高铁血红素白蛋白)(箭头↑所指处)。B 图中红细胞的 HbA_2 现象基本正常，B 图中全血的 HbA_2 现象消失，而且出现 MHA，充分证明样品所属者患有溶血性疾病。

(2) 由图 11-4 可以看出，正常人全血的双向电泳，有 A_2 现象，没有 MHA；溶血性疾病患者全血的双向电泳没有 A_2 现象，有 MHA。可见，在判定溶血性疾病方面，全血电泳优于红细胞电泳。

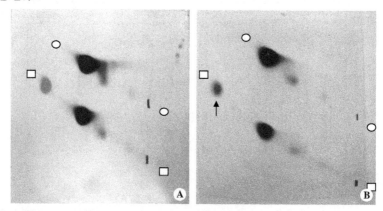

图 11-3 双向电泳，比较正常人和溶血性疾病患者的全血和红细胞
注释：A 图为正常人红细胞和全血的双向电泳，B 图为溶血性疾病患者红细胞和全血的双向电泳；上层(两个○之间)加红细胞，下层(两个□之间)加全血

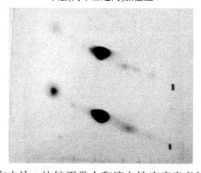

图 11-4 双向电泳，比较正常人和溶血性疾病患者的全血和全血
注释：上层为正常人全血的双向电泳，下层是溶血性疾病患者全血的双向电泳；上层(两个○之间)有 A_2 现象，没有 MHA，下层(两个□之间)没有 A_2 现象，有 MHA

6 全血电泳还有一个"纤维蛋白原现象",这是红细胞电泳没有的

由图 11-5 可以看出,上层各成分都在对角线上,下层各成分都在对角线上,中层除 A_2 现象外,纤维蛋白原(箭头↑所指处)也脱离对角线。全血双向电泳中,纤维蛋白原脱离对角线,这就是"纤维蛋白原现象"。红细胞有 HbA_2 现象,全血既有 HbA_2 现象,又有纤维蛋白原现象。纤维蛋白原现象的深入研究,将会揭露更多生命的奥秘。详见第十四章。

7 全血中红细胞以外有形成分的电泳释放

(1) 全血中红细胞以外有形成分,还有血小板、粒细胞、淋巴细胞。

(2) 其他有形成分的电泳释放,详见本书第七篇。

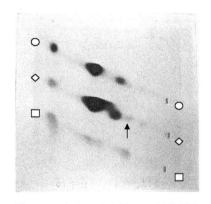

图 11-5　全血双向电泳——纤维蛋白原现象

注释:上层(两个○之间)加全血溶血液,中层(两个◇之间)加全血,下层(两个□之间)加血浆

第十二章 红细胞和全血再释放电泳类型及临床意义

高雅琼[1] 高丽君[2] 任建民[1] 秦文斌[2]※

(1. 包头市第八医院 检验科，包头 014010；2. 包头医学院 血红蛋白研究室，包头 014010)

摘 要

血红蛋白释放试验(HRT)中最常用而且最简单的就是各种血液成分直接电泳，此时红细胞和全血的再释放结果备受关注。红细胞和全血的再释放结果有几种情况，最常见的是红细胞的再释放大于全血的再释放，正常人属于此类，许多疾病也是如此。也有红细胞的再释放小于全血再释放的、红细胞的再释放等于全血再释放的等等，本文将其归纳为五种类型：红细胞再释放大于全血再释放为类型Ⅰ；红细胞再释放等于全血再释放为类型Ⅱ；红细胞再释放小于全血再释放为类型Ⅲ；红细胞再释放大于全血再释放，并且全血再释放大于零为类型Ⅳ；红细胞再释放等于零、全血再释放也等于零为类型Ⅴ。每种类型都有一些对应的疾病，可给临床诊断提供重要参考。

关键词 全血；红细胞；地贫；黄疸；遗传性球形红细胞增多症；再释放电泳

1 前言

临床上用血液做化验的项目很多，常用的项目有血常规或血细胞分析，筛查有无感染、贫血及一些血液病。肝功能化验，辅助肝炎、脂肪肝、胆囊炎等肝功能损伤诊断。肾功能检查，辅助肾炎、肾病综合征等肾功能损伤诊断。血糖测定，用于筛查及辅助诊断糖尿病、低血糖等疾病，糖尿病疗效观察。血沉分析，用于结核、风湿、感染、肿瘤等疾病的辅助诊断。血液流变学提供对心脑血管疾病的风险参考。凝血项目，可提供凝血疾病的辅助诊断和抗凝药效果观察。抗O、类风湿因子，有利于类风湿辅助诊断。血液病的化验更细致，有网织红细胞计数、酸溶血试验(Ham试验)、蔗糖溶血试验、红细胞渗透脆性试验等。

血液的化验还有很多，不再赘述，这里边唯独没有的就是血红蛋白释放试验(HRT)。HRT是我们研发的新领域[1, 2]，它涉及内容很多，本章主要讨论双释放问题。双释放就是电泳观察两种释放——红细胞再释放和全血再释放。在这里将解释双释放的几种类型及它们与疾病的关系。

2 材料及方法

2.1 材料 血液标本来自包头医学院第一附属医院检验科及各临床科室，还有第二附属医院检验科。

2.2 方法 每种类型均做单向及双向释放。

2.2.1 定点释放

※ 通讯作者：秦文斌，电子邮箱：qwb5991309@tom.com

(1) 制胶：①单向的取 10cm×17cm 的玻璃板，双向的取 17cm×17cm 的玻璃板。②单向的 TEB 缓冲液 55ml 于三角瓶中，淀粉 1g，琼脂糖 0.1g，沸水中煮 8 分钟，待 50℃左右倒于玻璃板上，凝固后使用；双向的缓冲液总量为 90ml，其他成分相应增加。

(2) 上样：①取 EDTA 抗凝全血 1 份，同时制备一管带有等量盐水的红细胞。②加红细胞 10μl 于滤纸条上，单向的插入已制备好淀粉-琼脂糖凝胶的左上方，同时加全血 10μl 于另一滤纸条上，插入凝胶的右上方；双向的将红细胞的滤纸条插入胶的右侧中间，全血的插入胶的右侧下方。

(3) 电泳：①电泳方向，从负极向正极泳。②按电势梯度 6V/cm。③单向的先普通泳 2.5 小时，停电 15 分钟，再泳 15 分钟后终止；双向的第一向同单向，第二向普泳 1.5 小时。

(4) 染色：①丽春红染色，将胶板放入丽春红染色液中过夜，第二天取出烤干；②联苯胺染色，将烤干的胶板先预热，再放入联苯胺染液中染色，漂洗晾干。

(5) 拍照留图。

2.2.2 多带释放+后退

(1) 制胶、上样同定点释放。

(2) 电泳：①②同定点释放；③单向的先泳 15 分钟，停电 15 分钟，再泳 15 分钟，再停电 15 分钟，反复进行 6 次，最后一次电泳 30 分钟，倒极使其后退泳 15 分钟，停止电泳。双向的第一向同单向电泳，第二向普泳 1.5 小时。

(3) 染色、拍照留图同定点释放。

3 结果

3.1 双释放类型Ⅰ 全血再释放＜红细胞再释放，包括正常人和多种疾病，参见表 12-1。

3.1.1 单向双释放 见图 12-1。

3.1.2 双向双释放 见图 12-2。

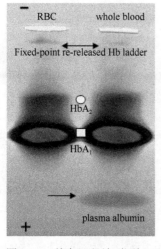

图 12-1 单向双释放(类型Ⅰ)

注释：左侧为红细胞，右侧为全血，□左右两侧为 HbA_1，○左右两侧为 HbA_2，→所指处为血浆白蛋白，以上各项，单向图相同；电泳条件为定点释放，箭头(⟶)所指处为定释带

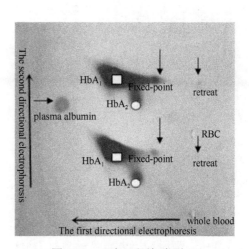

图 12-2 双向双释放(类型Ⅰ)

注释：上层为红细胞，下层为全血，长箭头(↓)所指者为定释带，短箭头(↓)所指者为后退带，□所在处为 HbA_1，○所在处为 HbA_2，→所指处为血浆白蛋白，以上各项，双向图相同；电泳条件为第一向，定释(长箭头)+后退(短箭头)，全血弱于红细胞，第二向，普泳

3.2 双释放类型Ⅱ 全血再释放＝红细胞再释放，二者都增强，包括地贫、阻塞性黄疸

等，参见表 12-1。

 3.2.1 单向双释放 见图 12-3。

 3.2.2 双向双释放 见图 12-4。

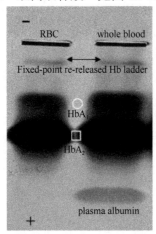
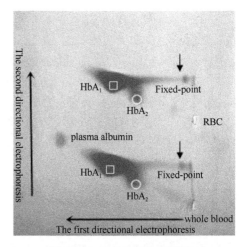

图 12-3 单向双释放(类型Ⅱ)

注释：左侧为红细胞，右侧为全血；电泳条件，定点释放，箭头(←→)所指处为定点带，红细胞与全血强度相同

图 12-4 双向双释放(类型Ⅱ)

注释：上层为红细胞，下层为全血；电泳条件，第一向定释(箭头↓所指处)，红细胞与全血强度相同，第二向普泳

3.3 双释放类型Ⅲ 全血再释放＞红细胞再释放，包括肝细胞性黄疸等，参见表 12-1。

 3.3.1 单向双释放 见图 12-5。

 3.3.2 双向双释放 见图 12-6。

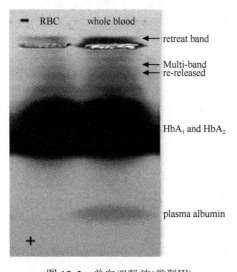
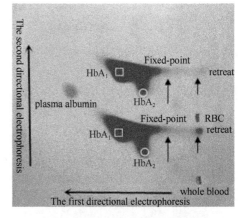

图 12-5 单向双释放(类型Ⅲ)

注释：左侧为红细胞，右侧为全血；电泳条件，多带释放+后退，上边的箭头(←)所指处为后退带，下边两个箭头(←)所指处为多带，全血再释放大于红细胞再释放

图 12-6 双向双释放(类型Ⅲ)

注释：上层为红细胞，下层为全血；电泳条件，第一向多带(长箭头↑所指处)+后退(短箭头↑所指处)，全血强于红细胞，第二向普泳

3.4 双释放类型Ⅳ 全血再释放＜红细胞再释放，而且全血再释放=0，包括溶血性黄疸等，参见表 12-1。

 3.4.1 单向双释放 见图 12-7。

 3.4.2 双向双释放 见图 12-8。

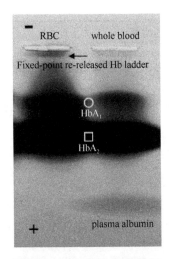

图 12-7 单向双释放(类型Ⅳ)

注释：左侧为红细胞，右侧为全血；电泳条件，定点释放(箭头←所指处)，红细胞有，全血无

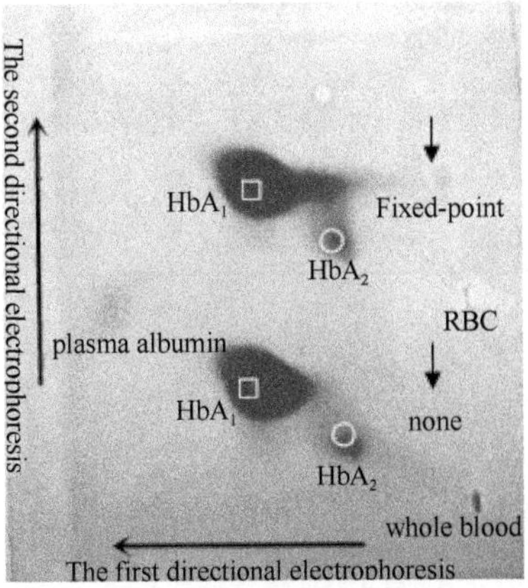

图 12-8 双向双释放(类型Ⅳ)

注释：上层为红细胞，下层为全血；电泳条件，第一向定点(箭头↓所指处)，红细胞有，全血无，第二向普泳

3.5 双释放类型Ⅴ 全血再释放=红细胞再释放，并且二者再释放都是 0，包括遗传球形红细胞增多症，参见表 12-1。

3.5.1 单向双释放 见图 12-9。

3.5.2 双向双释放 见图 12-10。

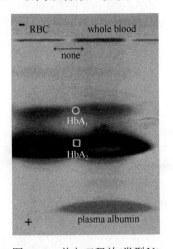

图 12-9 单向双释放(类型Ⅴ)

注释：左侧为红细胞，右侧为全血；电泳条件：定点释放，结果红细胞和全血都没有定释带

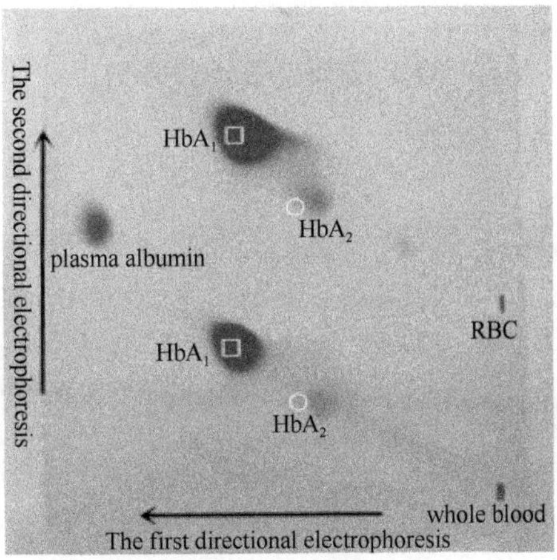

图 12-10 双向双释放(类型Ⅴ)

注释：上层为红细胞，下层为全血；电泳条件，第一向多带，红细胞和全血都没有多带，第二向普泳

3.6 附表

表 12-1 血红蛋白再释放电泳与各疾病的关系

类型	内容	疾病
Ⅰ	全血定释＜红细胞定释	正常人、原发性血小板增多症、真性红细胞增多症、卟啉病、蚕豆病(G-6-PD 缺乏症)、慢性粒细胞白血病、肝豆状核变性、乙型肝炎、肝硬化、IgA 肾病、多发性骨髓瘤、双白蛋白血症、糖尿病

续表

类型	内容	疾病
II	全血定释=红细胞定释	地贫(轻型β地贫、α地贫、HbH病、东南亚型α地贫)、阻塞性黄疸、真性红细胞增多症、腔隙性脑梗死、原发性血小板增多症、干燥综合征
III	全血定释>红细胞定释	肝细胞性黄疸
IV	全血定释=0	溶血性黄疸、自身免疫性溶血性贫血、糖尿病急性间歇性卟啉病
V	全血定释=0，红细胞定释=0	遗传性球形红细胞增多症

4 讨论

红细胞及全血的再释放电泳，可以对比这两种成分的再释放情况，以红细胞为基准比较全血的变化，能够观察出血浆成分对红细胞的影响，这就是所说的"浆胞互作"(血浆与红细胞之间的相互作用)。在这方面明显的例子就是三种黄疸[3-5]，阻塞性黄疸时全血的再释放增强，与红细胞平分秋色(参见类型II)。肝细胞性黄疸时，全血的再释放大于红细胞，显示出自己的特点(参见类型III)。溶血性黄疸则完全相反，此时全血的再释放为零(参见类型IV)。从遗传性疾病角度来看，地中海贫血和遗传性球形红细胞增多症是两个有趣的例子。地中海贫血时红细胞和全血的再释放都增强[6,7](参见类型II)。遗传性球形红细胞增多症则恰恰相反，红细胞和全血的再释放都消失[8,9](参见类型V)。类型II与V的明显差异，也可从红细胞形态角度来理解[10]，轻型β地中海贫血患者红细胞为靶形，遗传性球形红细胞增多症患者红细胞为球形，红细胞形状不同造成电泳释放结果互异，这是一种很有趣的现象，应当扩大范围，研究各种红细胞形态与电泳释放的关系。总之，红细胞内血红蛋白的电泳释放是新开垦的土地，仔细耕耘，应该会有更多的收获。

参 考 文 献

[1] 秦文斌. 活体红细胞内血红蛋白的电泳释放. 中国科学生命科学, 2011, 41(8): 597-607.

[2] 秦文斌. 红细胞内血红蛋白的电泳释放. 北京: 科学出版社, 2015: 6-53.

[3] 王翠峰, 秦文斌, 高丽君, 等. 通过血红蛋白释放试验鉴别黄疸类型的初步研究. 中华医学会第七次全国中青年检验医学学术会议. 2012: 127-128.

[4] 秦文斌. 红细胞内血红蛋白的电泳释放——发现和研究. 北京: 科学出版社, 2015: 196-199.

[5] Su Y, Han LS, Gao LJ, et al. Abnormal increased re-released Hb from RBCs of an intrahepaticbile duct carcinoma patient was detected by electrophoresis release test. Bio-medical Materials and Engineering, 2015, 26: S2049-2054.

[6] 秦文斌, 高丽君, 周立社, 等. 血红蛋白释放实验与轻型β-地中海贫血. 包头医学院学报, 2007, 23(6): 261-263.

[7] Su Y, Shao G, Gao LJ, et al. RBC electrophoresis with discontinuos power supply —a newly established hemoglobin release test. Electrophoresis, 2009, 30: 3041-3043.

[8] 马宏杰, 贾国荣, 高丽君, 等. 遗传性球形红细胞增多症患者红细胞内快泳血红蛋白缺失. 包头医学院学报, 2015, 31(4): 13-15.

[9] Su Y, Ma HJ, Zhang W, et al. Comparative study of the amount of re-released hemoglobin from α-thalassemia and hereditary spherocytosis erythrocytes. Chapter 6 of Book: inheridit hemoglobin disorders, 2016.

[10] 秦文斌. 红细胞内血红蛋白的电泳释放——发现和研究. 北京: 科学出版社, 2015: 224-229.

附 研究的英文翻译

The clinical significance of different re-released electrophoresis types in RBC suspension and whole blood

Yaqiong Gao[1], Lijun Gao[2], Jianmin Ren[1], Wenbin Qin[2]

(1. Department of Clinical Laboratory, the eighth hospital of Baotou, Inner Mongolia, China, Number22, Nanmenwai Street, Donghe District, Baotou, Inner Mongolia; 2. Laboratory of Hemoglobin, Baotou Medical College, Baotou, Inner Mongolia, China, Number31, Jianshe Street, Donghe District, Baotou, Inner Mongolia)

Abstract

The blood in illness persons can be observed by Hemoglobin Release Test HRT [1]. The amount and status of re-released hemoglobin hinds different sorts of diseases. The most common and easiest HRT is electrophoresised directly with all kinds of blood components. Among them [2], we most concerned hemoglobin that re-released from red blood cell(RBC)and whole blood. There are several situations of electrophoresis results. Here, we discussed them into five types: I, The re-released hemoglobin of RBC is more than whole blood, II, The re-released hemoglobin of RBC is equal to whole blood, III, The re-released hemoglobin of RBC is less than whole blood, IV, The re-released of RBC is more than zero, re-released of whole blood is zero, V No re-released hemoglobin in RBC or whole blood. Each type of re-released result has it's matched disease, so as to service clinical diagnosis. Type I is the most common one, normal persons and many diseases are of this kind. HRT is a convenient and novel way to judge the pathological status of hemoglobin in blood, it is a linkage between HRT and clinical practice.

Keywords: whole blood, RBC; thalassemia jaundice; hereditary Spherocytosis; re-released electrophoresis.

1 Introduction

Fire poses one of the most serious disasters to tunnels and concrete structures. The usage of self-consolidating concrete(SCC)further aggravates this situation because it often results in spalling[1, 2], as shown in Fig. 1. Two main mechanisms can simulate and characterize explosive spalling in concrete[3]. One is related to the thermo-mechanical process, which is directly associated with the temperature field in concrete, as illustrated in Fig. 2a. The other one is related to the thermo-hydral process, which is directly associated with the mass transfer of vapor, water and air in the porous network. This thermo-hydral process will result in building up high pore pressures and pressure gradients, as shown in Fig. 1.

The components inside RBC are of abundant proteins, these proteins accumulated for more than 90% of the dry weight of RBC. The studies between interactions of Hbs thus became an interesting issues.

Since the beginning study of the hematology, Hb is a wide range focus for the researchers all over the world. But people are usually pay attention to hemolysates, which is made from RBC and studied the function and further interactions of Hb that fleed from intact RBC. The Hb in hemolysates is a kind of isolate protein, for quite a long time, its function stands for the function

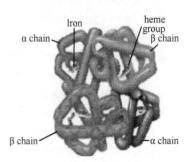

Fig. 1

of Hb within whole blood. Our research take use of intact RBC. The Hb status in intact RBC is more complex than that in hemolysates. The interaction between Hb-RBC membrane and the interaction between HbA_1 and HbA_2 is detected in Hemoglobin electrophoresis[1].In our previous study, we observed "HbA_2 phenomenon" which made us speculate that HbA_2, HbA_1, and perhaps some other membrane components(X)might bind with each other and form "HbA_2-HbA_1-X" complex in RBC[3].

AS a result, "HbA_1-HbA_2-RBC membrane" is a whole for the integrity of RBC biological function. Based on these foundational findings, this experiment makes the comparison of whole blood and RBC suspension in re-released electrophoresis. In one direction and double direction electrophoresis, the variation of amount of re-released Hb in whole blood or RBC suspension is coincidently in certain type of re-released electrophoresis, that is, the fixed re-released Hb in whole blood is higher or lower than that in RBC suspension, or even disappears in both two solutions. We know that plasma is absent in RBC suspension. Therefore, the different performance of fixed re-released Hb is due to interaction between plasma and intact RBC and RBC membrane. For different kinds of diseases, the fixed re-released Hb is varied regularly. The blood test is an important pathway to realize many diseases. For example, blood routine test or blood cell analysis is used to observe anemia, infection or some blood diseases. Liver function test is associated to hepatitis, fatty liver, and cholecystitis. Kidney function test is helpful to diagnose nephritis, nephrotic syndrome, and some injuries related to kidney. Blood glucose assay assist to have a definite diagnosis of hypoglycemia, diabetes, and observation of diabetes curative effect. Erythrocyte sedimentation rate is a way of auxiliary diagnosis to tuberculosis, rheumatism, inflammation, tumor, and so on. Cellular hemorheology can imply the adventure of cardiovascular and cerebrovascular diseases. The coagulation defect is related to disorders of blood coagulation and anticoagulants effect observation. Hemopathy test goes further of reticulocyte count, Ham test, sucrose lysis test, osmotic fragility test of RBC, and so on[4-8].

HRT is absent among these blood tests. HRT is one of our frontier field[8], many detailed phenomenons are involved. Here we mainly discuss the phenomenon of Double Re-released-re-released from RBC and whole blood, and we will reveal the relation between Double Re-released types and diseases.

Other researches around the world are devoted to detect the mutation of hemoglobin chain, interferon or LPS effect on Hb. The molecular substance and bacteria and cytokines are micro respect in studying Hb. Here, we are from another macroscopical aspect and directly watching the electrophoresis status of whole blood and RBC.

2 Material and method

The blood samples were from the eighth hospital of Bao Tou, Inner Mongolia, China. All the samples were got from clinical test use in clinical laboratory. The samples were collected from all kinds of common diseases in clinical diagnosis. The whole blood was abstracted in anticoagulant tube with Ethylene Diamine Tetraacetic Acid(EDTA)and stored in 4℃ for no more than 24 hours. Each type of Five electrophoresis was performed one-direction HRT and Double-direction HRT. This study has been cleared by our Institution Ethics Review Board for human studies and that patients have signed an informed consent.

2.1 Preparation of RBC suspension and starch-agarose mixed gel RBC suspension was prepared as our routine method with saline[9]. First, the anti-coagulated blood centrifuged at 1500 rpm for 15 minutes and aspirated out the upper plasma. Then, 200μl of the bottom of RBCs was added to 1 ml saline, gently mixed the solution, centrifuged at 1500 rpm for 10 minutes and aspirated out the supernatant. This washing operation was repeated for five times and then the prepared RBCs was used to perform electrophoresis after 1: 1 dilution with saline.

For one-direction HRT starch-agarose mixed gel, dissolve 0.15g agarose and 1.0g starch with 55ml TEB buffer(pH8.6)in a flask, heat the solution in boiled water for 8 minutes and then cooling the mixed melt to about 50 degrees celsius, immediately lay the gel on a 10cm*17cm glass. After solidification, 10μl RBC suspension and 10μl whole blood were applied on the left and right cathodic side of the gel respectively using 3mm filter paper as described earlier[10, 11].

For double-direction HRT starch-agarose mixed gel, the amount of TEB buffer(pH8.6)was 90ml and other soluble solids added correspondingly, heat the solution in boiled water for 8 minutes and then cooling the mixed melt to about 50 degrees celsius, in the same condition, lay the gel on a 17cm*17cm glass. After solidification, 10μl RBC suspension and 10μl whole blood were applied in the middle and under cathodic side of the gel respectively using 3mm filter paper.

2.2 colouration Stained the starch-agarose mixed gels with Ponceau Red solution for a whole night after electrophoresis immediately.In the next day, put the gels in Benzidine solution Note, Benzidine is highly carcinogenic, it could also be substituted by leucobase of malachite green which in non-carcinogenic, but its staining specificity is lower than Benzidine for a whole night to make sure the gels were stained sufficiently by two dyestuff mentioned above. In the third day, rinsed the gels under the flow slowly and then fixed the gels upon an alcohol lamp in a few seconds and then dried the gels in room temperature for another day[12].at last, when the gels dried completely, took a photograph for them and stored pictures in the following form as the Fig. 2A, Fig. 2B, et al.

2.3 Fixed-point re-released There were always some red sediments which stayed at the original site, these sediments did not draw our attention at fist. Then we assumed that they might be some insoluble components of the RBC. For those sediments could be stimulated out through the power on and power off. Another Hbs could be observed in the gels especially after colouration of Ponceau Red and Benzidine for the whole night.

For one-direction HRT, the electrophoresis was carried from cathode to positive pole, ran at 6V/cm for 2 hours and 30 minutes, then paused for 15 minutes and ran for another 15 minutes by turns.

For double-direction HRT, one-direction HRT was performed as described above. Then, change the direction of electric field, which is vertical to the original direction. The second directional electrophoresis was ran for another 1 hour and 30 minutes.Then stained the gels.

2.4 Multi-band re-released In fixed-point re-released electrophotesis, the power on and power off was executed for one time only. In multi-band re-released electrophotesis, the power on and off was repeated for several times. That is to say, more kinds of Hbs could be stimulated out for the original site. The amount of the re-released Hb depends on the binding the Hb to membrane. The more tightly of the Hb binds to RBC membrane, the fewer the amount of Hb could be re-released for the original site. In different sorts of clinical diseases, the binding of Hb to RBC is varied. That means the disease affects the status of Hb and membrane.

For one-direction HRT, in the same electrophoresis condition as fixed-point re-released, the electrophoresis was carried from cathode to positive pole, ran at 6V/cm.But the electrophoresis was run for 15 minutes and paused for 15 minutes, and then ran for another 15 minutes and paused for 15 minutes by turns, repeatedly for 6 times.

For double-direction HRT, one-direction HRT was performed as described above. Then change the direction of electric field, which is vertical to the original direction. The second directional electrophoresis was ran for 1 hour and 30 minutes.Then stained the gels.

2.5 Retreat re-released The retreat re-released electrophoresis was only took place in one-direction HRT and the first direction of double-direction HRT. It did not perform in the second direction electrophoresis which was vertical to the original direction.But not all the types of electrophoresis was suitable for retreat re-released. The retreat re-released is presumed to happen in such situation, the most tightly of the Hb and membrane. As after the fixed-point re-released electrophoresis and multi-band re-released electrophoresis, still, sediments was left in the original point.

The one-direction and the first direction of double-direction HRT electrophoresis was carried from cathode to positive pole, ran at 6V/cm.At the last turn of the electrophoresis mentioned above, the electrophoresis was performed for 30 minutes instead of 15 minutes, and then, changed the direction of electric field Reversely, and then continue performed retreat re-released electrophoresis for another 15 minutes. The second directional electrophoresis was without retreat re-released.

3 Result

3.1 Re-released type Ⅰ of One-direction HRT and Double-direction HRT Such type of electrophoresis Including normal people and several diseases.Such as polycythemia vera, porphyria, favism, chronic myeloid leukemia, hepatolenticular degeneration, hepatitis B, liver cirrhosis, IgA nephropathy, Multiple myeloma, and so on.

3.2 One direction HRT In one-direction HRT(Fig. 2A), from cathode to positive pole, the RBC suspension and whole blood were displayed in the following way, the plasma albumin ran the fastest in whole blood. No plasma albumin could be observed in RBC suspension in electrophoresis.

The same amount of HbA_1 and HbA_2 in RBC suspension and whole blood could be observed. That means the status of HbA_1 and HbA_2 were alike.

With the power off and then on in one-direction HRT, the fixed-point re-released Hb ladder in RBC suspension is more than that in whole blood. That is to Say, the plasma that existing in whole blood affected the binding way of Hb to membrane.

3.3 Double direction HRT In double direction HRT(Fig. 2B), one-direction HRT was

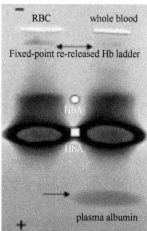

Fig. 2A

Note: The re-released hemoglobin of RBC is more than whole blood
Electrophoresis condition, fixed point re-released
Upper is positive, below is cathode
Left side: RBC   Right side: whole blood
○two sides, HbA_2; □two sides, HbA_1

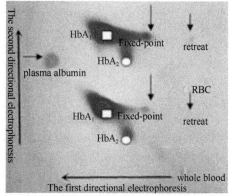

Fig. 2B

Note: The re-released hemoglobin of RBC is more than whole blood
Electrophoresis condition: fixed-point re-released+retreat
The first directional was from right to left,
The second directional was changed it vertical to the first directional
Upper is RBC, below is whole blood
○: HbA_2; □HbA_1

performed first. Then, change the direction of electric field, which is vertical to the original direction. The RBC suspension and whole blood were inserted in gel as follows, the plasma albumin and HbA_1 and HbA_2 and original point were in a line in whole blood.

The same phenomenon could also be observed in RBC suspension, excluding for plasma albumin. The same amount of HbA_1 and HbA_2 in RBC suspension and whole blood could be observed. That means the status of HbA_1 and HbA_2 were alike.

The fixed-point re-released Hb ladder in RBC suspension was more than that in whole blood. That is to say, the plasma that existing in whole blood affected the binding way of Hb to membrane. Retreat re-released could be observed in in Fig. 2B, the amount of retreat Hb in RBC is more than that in whole blood.

3.4 Re-released type II of One-direction HRT and Double-direction HRT Such type of electrophoresis Including thalassemia(light-β-thalassemia, α-thalassemia, hemoglobin H disease, southeast asian type thalassemia), obstructive jaundice, polycythemia vera, Lacunar cerebral infarction, primary thrombocythemia, sicca syndrome, and so on.

See also Table1.

3.5 one direction HRT In one-direction HRT(Fig. 3A), from cathode to positive pole, the RBC suspension and whole blood were displayed in the following way, the plasma albumin ran the fastest in whole blood. No plasma albumin could be observed in RBC suspension in electrophoresis.

The same amount of HbA_1 and HbA_2 in RBC suspension and whole blood could be observed. That means the status of HbA_1 and HbA_2 were alike.

With the power off and then on in one-direction HRT, the fixed-point re-released Hb ladder in RBC suspension is as more as that in whole blood. We already know that, the plasma existing in whole blood affected the binding way of Hb to membrane, as in normal person showed in Fig. 2A. In illness status, the plasma in whole blood may be pathological changed so as to affect the binding way of Hb to membrane.

3.6 two direction HRT In double direction HRT(Fig. 3B), one-direction HRT was per

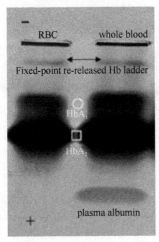

Fig. 3A
Note: The re-released hemoglobin of RBC is equal to whole blood
  Electrophoresis condition: Fixed point re-released
  Upper is positive, below is cathode
  Left side: RBC   Right side: whole blood
  ○two sides; HbA_2; □two sides: HbA_1

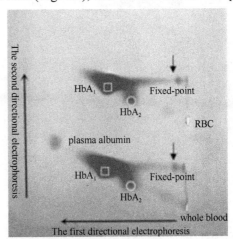

Fig. 3B
Note: The re-released hemoglobin of RBC is equal to whole blood
  Electrophoresis condition: Fixed point re-released
  The first directional was from right to left
  The second directional was changed it vertical to the first direction
  Upper is RBC, below is whole blood
  ○: HbA_2; □: HbA_1

formed first. Then, change the direction of electric field, which is vertical to the original direction.The RBC suspension and whole blood were inserted in gel as follows: the plasma albumin and HbA_1 and HbA_2 and original point were in a line in whole blood.

The same phenomenon could also be observed in RBC suspension, excluding for plasma albumin.The same amount of HbA_1 and HbA_2 in RBC suspension and whole blood could be observed. That means the status of HbA_1 and HbA_2 were alike.

The fixed-point re-released Hb ladder in RBC suspension was as more as that in whole blood. That is because in such kinds of diseases, the pathological change in plasma was the same as described in one-direction.

3.7 Re-released type Ⅲ of One-direction HRT and Double-direction HRT Such type of electrophoresis Included hepatocellular jaundice. See also Table1.In one-direction HRT(Fig. 4A), from cathode to positive pole, the RBC suspension and whole blood were displayed in the following way, the plasma albumin ran the fastest in whole blood. No plasma albumin could be observed in RBC suspension in electrophoresis.The same amount of HbA_1 and HbA_2 in RBC suspension and whole blood could be observed. With the power off and then on for six times, in one-direction HRT, the multi-band re-released Hb ladder in RBC suspension is less than that in whole blood. We already know that, the plasma existing in whole blood affected the binding way of Hb to membrane, as in normal person showed in Fig. 2A.In illness status, the plasma in whole blood may be pathological changed so as to affect the binding way of Hb to membrane. Retreat re-released could be observed in in Fig. 4A, the amount of retreat Hb in whole blood is more than that in RBC suspension.

4 Two Direction HRT

In double direction HRT(Fig. 4B), one-direction HRT was performed first. Then, change the

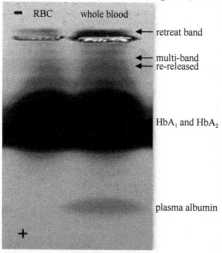

Fig. 4A
Note: The re-released hemoglobin of RBC is less than whole blood
Electrophoresis condition: multi-band re-released+retreat
Upper is positive, below is cathode
Left side: RBC   Right side: whole blood

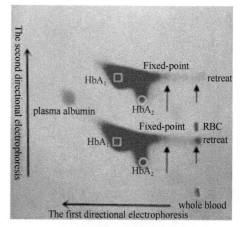

Fig. 4B
Note: The re-released hemoglobin of RBC is less than whole blood
Electrophoresis condition: multi-band re-released+retreat
The first directional was from right to left
The second directional was changed it vertical to the first direction
Upper is RBC, below is whole blood
○: HbA_2; □: HbA_1

direction of electric field, which is vertical to the original direction.The RBC suspension and whole blood were inserted in gel as follows, the plasma albumin and HbA_1 and HbA_2 and original point were in a line in whole blood.

The same phenomenon could also be observed in RBC suspension, excluding for plasma albumin. The same amount of HbA_1 and HbA_2 in RBC suspension and whole blood could be observed. That means the status of HbA_1 and HbA_2 were alike.

The fixed-point re-released Hb ladder in RBC suspension is also less than that in whole blood as in Fig. 4A.That is because in such kinds of diseases, the pathological change in plasma was the same as described in one-direction in Fig. 4A.

4.1　Re-released type IV of One-direction HRT and Double-direction HRT

4.2　one direction HRT　Such type of electrophoresis Including hemolytic jaundice, autoimmune hemolytic anemia, diabetes, acute intermittent porphyria.

In one-direction HRT(Fig. 5A), from cathode to positive pole, the RBC suspension and whole blood were displayed in the following way, the plasma albumin ran the fastest in whole blood. No plasma albumin could be observed in RBC suspension in electrophoresis.

The amount of HbA_1 and HbA_2 in RBC suspension and is more than that in whole blood. With the power off and then on in one-direction HRT, the fixed-point re-released Hb ladder in RBC suspension is obviously, but in whole blood, there is no fixedpoint re-released Hb ladder. That is to Say, the plasma that existing in whole blood affected the binding way of Hb to membrane, no Hb could be re-released when the power off and on.

4.3　one direction HRT　In double direction HRT(Fig. 5B), one-direction HRT was performed first. Then, change the direction of electric field, which is vertical to the original direction. The RBC suspension and whole blood were inserted in gel as follows, the plasma albumin and HbA_1 and HbA_2 and original point were in a line in whole blood. The same phenomenon could also be observed in RBC suspension, excluding for plasma albumin.

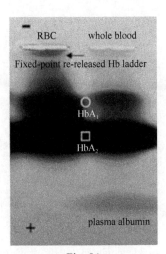

Fig. 5A
Note: The re-released of whole blood is zero
Electrophoresis condition: Fixed point re-released
Upper is positive, below is cathode
Left side: RBC　　Right side: whole blood
○two sides: HbA_2; □two sides: HbA_1

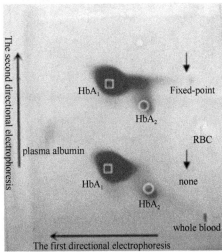

Fig. 5B
Note: The re-released of whole blood is zero
Electrophoresis condition: Fixed point re-released
The first directional was from right to left
The second directional ws changed it vertical to the first direction
Upper is RBC, below is whole blood
○: HbA_2; □: HbA_1

The amount of HbA_1 and HbA_2 in RBC suspension and is more than that in whole blood. the fixed-point re-released Hb ladder in RBC suspension is obviously, but in whole blood, there is no fixed-point re-released Hb ladder.That is to say, the plasma that existing in whole blood affected the binding way of Hb to membrane, no Hb could be re-released in whole blood.

4.4 Re-released type V of One-direction HRT and Double-direction HRT Such type of electrophoresis Included hereditary spherocytosis.See also Table1. None re-released of RBC or whole blood in both one and double-direction HRT. Including hereditary spherocytosis. See also attached list.

4.5 One Direction HRT In one-direction HRT Fig. 6A, from cathode to positive pole, the RBC suspension and whole blood were displayed in the following way, the plasma albumin ran the fastest in whole blood. No plasma albumin could be observed in RBC suspension in electrophoresis.

The same amount of HbA_1 and HbA_2 in RBC suspension and whole blood could be observed. That means the status of HbA_1 and HbA_2 were alike.

With the power off and then no fixed-point re-released Hb could be observed in Fig. 6A.

5 Two Direction HRT

In double direction HRT(Fig. 6B), one-direction HRT was performed first. Then, change the direction of electric field, which is vertical to the original direction.The RBC suspension and whole blood were inserted in gel as follows, the plasma albumin and HbA_1 and HbA_2 and original point were in a line in whole blood. The same phenomenon could also be observed in RBC suspension, excluding for plasma albumin.

The amount of HbA_1 and HbA_2 in RBC suspension and is as more as that in whole blood. No fixed-point re-released Hb could be observed in Fig.6B.

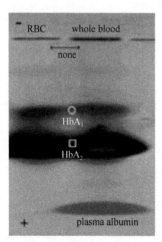 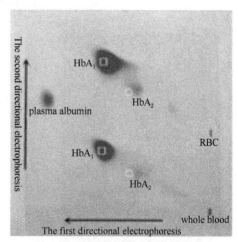

Fig. 6A

Note: No re-released hemoglobin in RBC or whole blood
Electrophoresis condition: Fixed point re-released
Upper is positive, below is cathode
Left side: RBC   Right side: whole blood
○two sides: HbA_2; □two sides: HbA_1

Fig. 6B

Note: No re-released hemoglobin in RBC or whole blood
Electrophoresis condition: Fixed point re-released
The first directional was from right to left
The second directional was changed it vertical to the first direction
Upper is RBC, below is whole blood
○: HbA_2; □: HbA_1

6 Discussion

To investigate the relation between HRT and several kinds of diseases, and establish a method of diagnosis in clinical practice. The HRT was carried in room temperature, after electrophoresis, the colouration was executed follow on.Fixed-point re-released, Multi-band re-released, Retreat re-released were performed to study HRT.We observed re-released electrophoresis of RBC suspension and whole blood. The influence of plasma on RBC is obvious when compared to RBC suspension. This is what we called "plasma-cellula interation(the interaction between plasma and RBCs).The extraordinary examples are of three types of jaundice When obstructive jaundice, the re-released Hb of RBC is equaled to whole blood(see also type Ⅱ).When hepatocellular jaundice, The re-released hemoglobin of RBC is less than whole blood(see also typeⅢ), which is different from obstructive jaundice. The case of hemolytic jaundice is totally opposite, none re-released Hb could be found in whole blood(see also typeⅣ).We suspect that plasma may promote the re-released of whole blood. Bilirubin composed of plasma, which may played an important role in whole blood HRT. Bilirubin including direct bilirubin and indirect bilirubin, they had infection on RBC within whole blood together, and therefore enhanced re-released of whole blood. To our further research, we shall use direct bilirubin and indirect bilirubin separately on the RBC to observe re-released HRT.

Genetically, thalassemia and hereditary spherocytosis are two interesting examples. The re-released Hb of RBC and whole blood are both strengthened in thalassemia(see also type Ⅱ)[15,10].On the contrary, The re-released Hb of RBC and whole blood are disappeared in hereditary spherocytosis(see also type Ⅴ).This visible difference of thalassemia and hereditary spherocytosis could also be deduced from red cell morphology[16].The RBCs of Light beta thalassaemia patients altered to target shape while RBCs of hereditary spherocytosis patients altered to sphere shape. Different RBCs shapes caused different re-released results, this interesting phenomenon reminded us the correlation of RBCs shape and re-released electrophoresis. "Re-released HRT" referring RBC membrane, it is a kind of mechanism that illustrating Hb existential state within RBC. It is more complex than our imagination, the circumstance of blood(oxygen supply, metabolite accumulation)may also affect re-released HRT beyond genetic factors.Oxidative damage may destroy the RBC membrane and Hb that binding to membrane is an maker of injury in RBCs.The genesis of tumor is related to oxidative damage of radicals[19-21], but whether it will affect re-released Hb is unclear.

In normal people and hepatitis B, the re-released Hb of RBC is more than whole blood, the retreat was appeared in double-direction HRT(type Ⅰ).In hepatocellular jaundice, The re-released Hb of RBC is less than whole blood, and the retreat was observed in both one and double-direction(typeⅢ).The mechanism may correlated to the binding of Hb and RBC membrane.

HRT is a new eyesight to reveal the insight status of hemoglobin in whole blood and RBC suspension. It is helpful to assist researchers to understand the hemoglobin binding way with RBC membrane and the correlation between diseases and re-released hemoglobin. The experiment was started in several patients limited by working conditions. Whether these samples stands for the overall examples among patients is still controversial. But our HRT observes a true phenomenon as we had ever done before in laboratory. The re-released electrophoresis of Hb within RBCs is a new field to be worthy explored.Other diseases could also be involved in HRT in the following step.

Acknowledgment

The work is supported by participators from Baotou Medical College and the eighth hospital.

All the patients involved in the research were informed and signed informed consent. The authors were calmed of no conflict of interests.

Reference

[1] Y. Su, L. Gao, Q. Ma, L. Zhou, L. Qin, L. Han and W. Qin, Interactions of hemoglobin in live red blood cells measured by the electrophoresis release test, Electrophoresis 2010, 31, pp, 2913-2920.
[2] Y. Su, J. Shen, L. Gao, Z. Tian, H. Tian, L. Qin, and W. Qin, Molecular Interactions of re-released protein in electrophoresis of human erythrocytes, electrophoresis 2012, 33, pp, 1402-1405.
[3] in. W. B. J. Baotou Medical College(special issue)1990, 7, pp, 1-76.
[4] E.M. Pasini, H.U. Lutz, M. Mann and A.W. Thomas, Red blood cell(RBC)membrane proteomics-Part Comparative proteomics and RBC patho-physiology, Journal of Proteomics 2010, 73, pp, 421-435.
[5] P.I. Margetis, M.H. Antonrlou, I.K. Petropoulos, L.H. Margaritis and I.S. Papassideri, Increased protein carbonylation of red blood cell membrane in diabetic retinopathy, Experimental and Molecular Pathology 2009, 87, pp, 76-82.
[6] I.K. Petropoulos, P.I. Margetis, M.H. Antonelou, J.X. Koliopoulos, S.P. Gartaganis, L.H. Margaritis and I.S. Papassideri, Structural alterations of the erythrocyte membrane protein in diabetic retinopathy, Graefe's Archive for Clinical and Experimental Ophthalmology 245(2007), 1179-1188.
[7] N. Mikirova, H.D. Riordan, J.A. Jackson, K. Wong, J.R. Miranda-Massari and M.J.Gonzalez, Erythrovyte membrane fatty acid composition in cancer patients, Puerto Rico Health Sciences Journal J 23(2004), 107-113.
[8] Wenbin Qin.Electrophoresis release of hemoglobin from living red blood cells.SCIENTIA SINICAVitae.2011 41(8): 597-607.
[9] Yan Su, Guo Shao, Lijun Gao, Lishe Zhou, Liangyi Qin, Wenbin Qin. RBC electrophoresis with discontinuos power supply—a newly established hemoglobin release test. Electrophoresis 2009, 30, 3041-3043.
[10] Qin, W. B., Chinese Journal of Biochemistry Molecular Biology. 1991, 7, 583-584.
[11] Qin, W. B., progress in biochemistry and biophysics. 1991, 18, 286-287.
[12] Lisha Han, Lijun Gao, Jun Guo, Xiaorong Sun, Yan Su, Wenbin Qin.Intrahepatic bile duct carcinoma and hemoglobin release test.Electrophoresis release of Hb from RBC.2015.196-199.
[13] BAO Wu-ren-bi-li-ge, WANG Cui-feng, GAO Li-jun, et al.Particularly abnormal result of hemoglobin release test in blood samples of patients with liver cirrhosis.Journal of Clinical and Experimental Medicine.2011, 12, 10(24): 1915-1918.
[14] Yan Su, Lisha Han, Lijun Gao, Jun Guo, Xiaorong Sun, Jiaxin Li, Wenbin Qin. Abnormal increased re-released Hb from RBCs of an intrahepaticbile duct carcinoma patient was detected by electrophoresis release test. Bio-medical Materials and Engineering 26(2015)S2049-S2054.
[15] Wenbin Qin, Lijun Gao, Yan Su, Guo Shao, Lishe Zhou, Liangyi Qin.Hemoglobin Release Test and β-Thalassem in Trait.Journal of Baotou Medical college .2007, 23(6): 261-263.
[16] MA Hongjie, JIA Guorong, GAO Lijun, et al.Absence of fast moving Hb in RBC of patients with hereditary spherocytosis. Journal of Baotou Medical college.2015, 31(4): 13-15.
[17] D.Chiu, F.Kuypers and B.Lubin, Lipid peroxidation in human red cells, Seminars in Hematology 26(1989), 257-276.
[18] R.Sharma and B.R. Premachandra, Membrane-bound hemoglobin as a marker of oxidative injury in adult and neonatal red blood cells, Biochemical Medicine and Metabolic Biology 46(1991), 33-44.
[19] M.M. Abdel-Daim, M.A. Abd Eldaim and A.G. Hassan, Trigonella foenum-graecum ameliorates acrylamide-induced toxicity in rats: roles of oxidativestress, proinflammatory cytokines, and DNAdamage, Biochemistry and Cell Biology II(2014), 1-7.
[20] R. Cardin, M. Piciocchi, M. Bortolami, A. Kotsafti, L. Barzon, E. Lavezzo, A. Sinigaglia, K.I. Rodriguez-Castro, M.Rugge and F. Farinati, Oxidative damage in the progression of chronic liver disease to hepatocellular carcinoma: An intricate pathway, World Journal of Gastroentrerology 20(2014), 3078-3086.
[21] B. Tekiner-Gulbas, A.D. Westwell and S.Suzen.Oxidative stress in carcinogenesis: New synthetic compounds with dual effects upon free radicals and cancer, Current Medicinal Chemistry 20(2013), 4451-4459.

第十三章 血浆成分对红细胞释放血红蛋白的影响

王彩丽[1※] 刘丽萍[1※] 韩丽红[2※] 苏 燕[2#] 秦文斌[2#] 高丽君[2]
高雅琼[2] 王翠峰[2] 张晓燕[3] 闫 斌[2] 孙 刚[3]

(包头医学院 1.第一附属医院；2.血红蛋白研究室；3.第二附属医院，包头 014010)

摘 要

血红蛋白释放试验(HRT)，包括两大类型，即初释放和再释放，前者来自第一次通电，后者来自再次通电。初释放是将红细胞加在凝胶的原点，一次通电后由红细胞释放出来的血红蛋白 A_2(HbA$_2$)与溶血液者不同，又称"HbA$_2$ 现象"。再释放是再次通电后又有血红蛋白由原点释放出来，它是血红蛋白 A_1(HbA$_1$)。以上研究所用的标本只是由血液中分离出来的红细胞，如果将含有红细胞的全血与红细胞并列加在凝胶的两个原点，重复上述实验，看二者有什么差异？此时我们发现了本文所探讨的内容：全血中血浆成分与红细胞之间的相互作用。在这里，全血中红细胞与游离红细胞的血红蛋白释放情况出现不同：①初释放方面，多数情况是全血中红细胞释放出来的 HbA$_2$，与由游离红细胞中释放出来的 HbA$_2$ 电泳位置不同，来自全血的 HbA$_2$ 稍靠后，位于来自红细胞 HbA$_2$ 的阴极侧。②再释放方面，全血中红细胞释放出来的 HbA$_1$ 与来自红细胞 HbA$_1$ 数量不同，多数情况是来自全血的 HbA$_1$ 相对少于来自红细胞的 HbA$_1$。由此可知，全血中的红细胞受到血浆成分的影响，不仅影响到初释放，也影响到再释放，说明血浆成分与红细胞发生了相互作用。本文通过一系列实验论证这种相互作用的存在，包括正常人和多种疾病。为了深入研究，我们又人为地比较研究各种独立血浆成分对红细胞的影响，并发现多数物质为负影响(释放减弱)而少数物质为正影响(释放增强)，证明全血中多种成分联合起来共同完成了与红细胞的相互作用。

关键词 游离红细胞；全血中红细胞；血浆成分与红细胞之间的相互作用

1 前言

人们早就知道血浆成分能够影响红细胞的存在状态，这就是医院化验室常用的红细胞沉降率测定，简称"血沉"[1-3]。血沉的观察指标是红细胞的沉降速度，有些疾病使血沉加快，有些疾病使血沉减慢，这都是血浆成分与红细胞相互作用的结果。此法的研究手段通常是比较同一标本的血沉值和其他血浆成分化验结果的关系，从而分析哪些成分使血沉加快或减慢。

我们在长期血红蛋白释放实验研究过程中，也注意到全血中的血浆成分能够影响到其中红细胞的血红蛋白释放，它们之间存在相互作用，我们的观察指标是红细胞释放血红蛋白的情况。由于红细胞释放血红蛋白有两种类型(初释放和再释放)，所以血浆影响也要涉及这两方面。为了弄清哪些血浆成分参与这个相互作用，我们让各种独立的血浆成分与红细胞混合

※ 并列第一作者
\# 两个通讯作者

后进行释放电泳,观察它们的具体影响。本章要用一系列电泳实验来发现血浆成分与红细胞相互作用的规律和机制。

2 材料与方法

2.1 材料 血液标本来自许多单位,其中包括包头医学院第一附属医院及第二附属医院的检验科、临床科室及健康体检者。各种独立血浆成分(如血浆蛋白、激素、氨基酸、葡萄糖、维生素等)来自包头医学院第一附属医院肾内科病房。纤维蛋白原购自北京索莱宝科技发展有限公司。

2.2 方法 血红蛋白释放试验操作过程详见文献[4-6].本文中具体操作略述如下。

2.2.1 单向电泳

(1) 红细胞的分离:向 1.5ml EP 管中加入全血 50μl,低速离心 3 分钟后去掉血浆,加入生理盐水至1ml处,混匀后再低速离心并去掉盐水。如此重复 5 次,最后,向红细胞内加等体积的生理盐水,混匀备用。

(2) 全血的配制:取上述不含生理盐水的红细胞,加入等体积的本人血浆,混匀备用。

(3) 插入凝胶:取新华 3 号滤纸两条,各 4mm×1mm,其一加入上述备用红细胞 4μl,另一滤纸条加入备用全血 4μl,插入淀粉-琼脂糖混合凝胶。

(4) 单带释放电泳:电势梯度 6V/cm 通电 2.5～3 小时,停电 15 分钟,再通电 30 分钟。

(5) 多带释放电泳:电势梯度 6V/cm 通电 15 分钟,停电 15 分钟,再通电 15 分钟,停电 15 分钟,交替开闭电源,共 4～5 小时。此过程可用 24 小时定时器来完成。

(6) 染色:电泳后取出凝胶,先染丽春红,必要时再染联苯胺或考马斯亮蓝,拍照留图。

2.2.2 双向电泳 第一向同上述单向电泳,第二向电泳时改变方向,倒极并调转 90°,进行常规电泳,电压 5V/cm,电流 2mA/cm,通电 3 小时。染色等同上述单向电泳。

3 结果

3.1 正常人及患者血液标本单向电泳结果

3.1.1 正常人血液标本的单向电泳结果 见图 13-1。

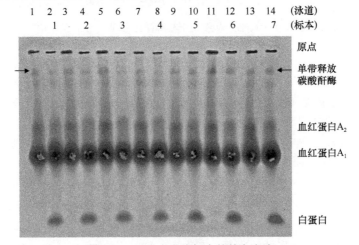

图 13-1 正常人血液标本的单向电泳

注释:正常人 7 名,每人都有红细胞和全血,单数泳道为红细胞,双数泳道为全血

由图 13-1 可以看出，每个标本的单带释放都是全血明显弱于红细胞，而且多数全血中的 HbA_2 电泳位置靠后(阴极侧)，而红细胞 HbA_2 则靠前(阳极侧)。

3.1.2 高血压患者血液标本的单向电泳结果　见图 13-2。

由图 13-2 可以看出，每个标本都是全血中多带释放明显弱于红细胞，同样的，多数全血中的 HbA_2 电泳位置靠后(阴极侧)，而红细胞 HbA_2 则靠前(阳极侧)。

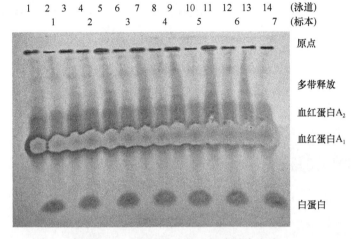

图 13-2　高血压患血液者标本单向电泳

注释：共 7 个标本，1~6 为高血压患者标本，7 为正常人标本。每人都有红细胞和全血，单数泳道为红细胞，双数泳道为全血

3.1.3 糖尿病患者血液标本单向电泳结果　见图 13-3。

由图 13-3 可以看出，血糖明显升高的红细胞多带释放增强，其全血中红细胞则多带明显减弱，甚至消失。血糖不太高的标本，其红细胞基本没有多带或不明显，其全血中红细胞更是看不见多带。全血中 HbA_2 电泳位置情况复杂，有些多带明显标本全血中的 HbA_2 电泳位置靠后，另一些多带明显标本全血中的 HbA_2 电泳位置则不靠后。

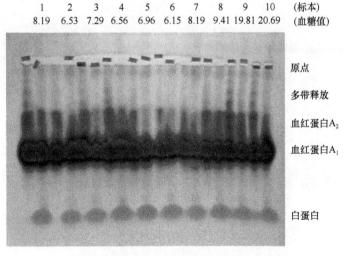

图 13-3　糖尿病患者血液标本单向电泳

注释：共有 10 个标本，都来自糖尿病患者。每个标本都包括红细胞(单数泳道)和全血(双数泳道)

3.1.4 黄疸患者血液标本单向电泳结果　见图 13-4。

由图 13-4 可以看出，黄疸患者的情况比较特殊，不同于上述一系列结果。黄疸标本基本上都是全血的多带释放增强，绝大多数标本红细胞的多带释放都弱于相应的全血标本。阻塞性黄疸的红细胞多带释放增强，其全血的多带释放也增强，有时稍弱于红细胞，有时可达到与红细胞相同的强度。此时，全血中的 HbA_2 的电泳位置多数靠后，有的不明显。

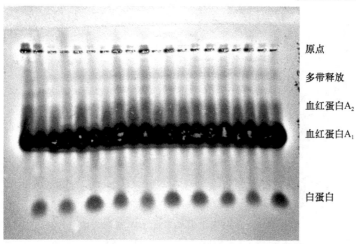

图 13-4　黄疸患者血液标本单向电泳

注释：10 个标本都来自黄疸患者，其中标本 1 明确为阻塞性黄疸，标本 2 明确为溶血性黄疸，标本 6 明确为肝细胞性黄疸。每个标本都包括红细胞(单数泳道)和全血(双数泳道)

3.1.5　皮肤癌和甲状腺癌患者血液标本的单向电泳结果　见图 13-5。

由图 13-5 可以看出，皮肤癌患者的全血中单带释放明显弱于红细胞，甲状腺癌样品结果则相反，它的全血中单带释放明显强于红细胞。皮肤癌全血中血红蛋白 A_2 电泳位置靠后于红细胞的血红蛋白 A_2。由于甲状腺癌患者红细胞中有慢泳异常血红蛋白 D(见〇处)，因此看不清血红蛋白 A_2 的位置区别。

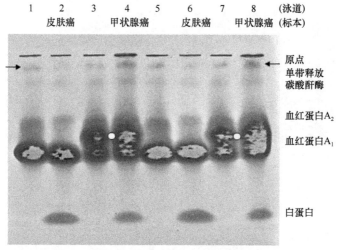

图 13-5　皮肤癌与甲状腺癌患者血液标本单向电泳

注释：两个患者(皮肤癌、甲状腺癌)，8 个泳道(单数泳道为红细胞，双数泳道为全血)。箭头(→←)所指位置为单带释放。甲状腺癌患者红细胞存在慢泳异常血红蛋白 D(见〇处)

3.2 各种血浆成分对红细胞再释放的影响

3.2.1 血浆对红细胞多带释放的影响结果 见图 13-6。

由图 13-6 可以看出,血浆可使红细胞的多带释放减弱。

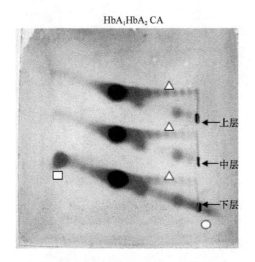

图 13-6 血浆对红细胞多带释放的影响电泳

注释:上层为红细胞,中层为红细胞加等量盐水,下层为红细胞加等量血浆,各层的多带释放见△处各层的原点见←处(□为血浆白蛋白,○为血浆丙种球蛋白)

3.2.2 白蛋白对红细胞多带释放的影响结果 见图 13-7。

由图 13-7A、图 13-7B 可以看出,白蛋白可使红细胞的多带释放减弱。

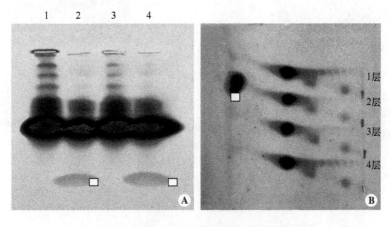

图 13-7 白蛋白对红细胞多带释放的影响电泳

注释:A 图为单向电泳,泳道 1、3 为红细胞,泳道 2、4 为红细胞加白蛋白(见图中□处);B 图为双向电泳,由上向下:第 1、4 层为红细胞,第 2 层为红细胞加白蛋白(见图中□处)

3.2.3 丙种球蛋白对红细胞多带释放的影响结果 见图 13-8。

由图 13-8A、图 13-8B 可以看出,丙种球蛋白可使红细胞的多带释放减弱。

3.2.4 比较血浆与血清对红细胞再释放的影响 见图 13-9。

由图 13-9 可以看出,血浆标本单带释放较强(向下箭头↓所指处),血清标本单带释放较弱(向左箭头←所指处)。

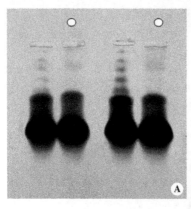 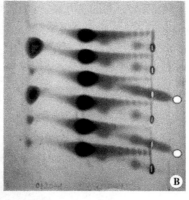

图 13-8 丙种球蛋白对红细胞多带释放的影响电泳

注释：A 为单向电泳，泳道 1、3 为红细胞，泳道 2、4 为红细胞加丙种球蛋白(见图中○处)；B 为双向电泳，由上向下：第 1、6 层为红细胞，第 3、5 层为红细胞加丙种球蛋白(见图中○处)

3.2.5 纤维蛋白原对红细胞再释放的影响 见图 13-10。

由图 13-10 可以看出，四个泳道的单带释放差不多，也就是说纤维蛋白原对再释放没有增强作用，减弱作用也不明显。

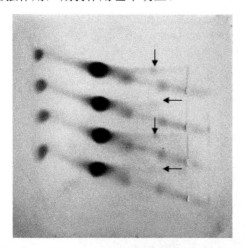 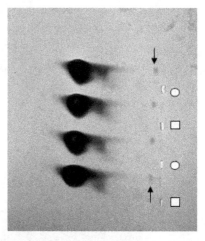

图 13-9 比较血浆与血清对红细胞单带再释放影响电泳

注释：由上向下，第 1、3 层为红细胞加血浆，第 2、4 层为红细胞加血清

图 13-10 纤维蛋白原对红细胞单带再释放的影响电泳

注释：双向电泳，由上向下：第 1、3 层为红细胞(原点见图中○处)，第 2、4 层为红细胞加纤维蛋白原(原点见图中□处)。箭头(↑↓)所指处为单带释放

3.2.6 胰岛素对红细胞多带释放的影响结果 见图 13-11。

由图 13-11 可以看出，胰岛素可使红细胞的多带释放减弱。

3.2.7 氨基酸对红细胞单带释放的影响结果 见图 13-12。

由图 13-12 可以看出，氨基酸可使红细胞的单带释放减弱。全部氨基酸与必需氨基酸无明显差异。

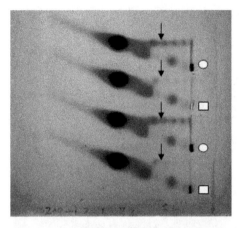 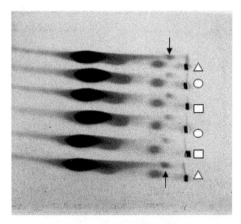

图 13-11 胰岛素对红细胞多带释放的影响电泳

注释：由上向下：第 1、3 层为红细胞(原点在○所指处)，第 2、4 层为红细胞加胰岛素(原点在□所指处)。多带释放见相应箭头(↓)处

图 13-12 氨基酸对红细胞单带释放的影响电泳

注释：由上向下，第 1、6 层为红细胞(原点见△所指处)，第 2、4 层为红细胞加全部氨基酸(原点见○所指处)，第 3、5 层为红细胞加必需氨基酸(原点见□所指处)。上下箭头(↓↑)对应处为单带释放带

3.2.8 葡萄糖对红细胞多带释放的影响结果 见图 13-13。

由图 13-13 可以看出，葡萄糖可使红细胞的多带释放增强，增强程度与葡萄糖浓度成比例。此时全血中红细胞的多带释放明显减弱。

3.2.9 水溶性维生素对红细胞多带释放的影响结果 见图 13-14。

由图 13-14 可以看出，维生素 B_1、维生素 B_6 和维生素 B_{12} 都使红细胞的多带释放增强，但维生素 B_1 特殊，其多带中的第一梯带很强，其后的多带明显减弱(见箭头↓所指处)，维生素 C 可使红细胞的多带释放减弱。

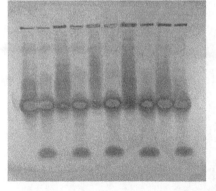 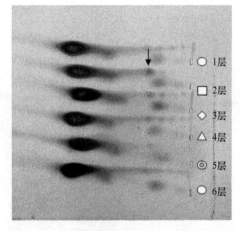

图 13-13 葡萄糖对红细胞多带释放的影响电泳

注释：泳道 1 为红细胞，2 为全血，3 为红细胞加 5% 葡萄糖，4 为全血加 5% 葡萄糖，5 为红细胞加 10% 葡萄糖，6 为全血加 10% 葡萄糖，7 为红细胞加 50% 葡萄糖，8 为全血加 50% 葡萄糖，9 同 1，10 同 2

图 13-14 水溶性维生素对红细胞多带释放的影响电泳

注释：由上向下，第 1 层为红细胞(原点见○所指处)第 2 层为红细胞加维生素 B_1(原点见□所指处)，第 3 层为红细胞加维生素 B_6(原点见◇所指处)，第 4 层为红细胞加维生素 B_{12}(原点见△所指处)，第 5 层为红细胞加维生素 C(原点见◎所指处)，第 6 层同第 1 层

3.2.10 激素对红细胞多带释放的影响结果 见图 13-15。

由图 13-15 可以看出，地塞米松使红细胞的多带释放稍减弱，甲泼尼龙使红细胞的多带释放更弱。

3.2.11 无机物对红细胞多带释放的影响结果　见图 13-16。

由图 13-16 可以看出，氯化钠使红细胞多带释放稍减弱，氯化钾使红细胞多带更弱。氯化钙使红细胞多带相对增强。碳酸氢钠使红细胞多带减弱。

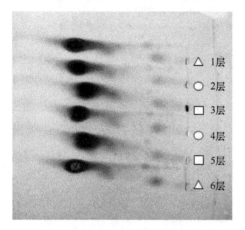 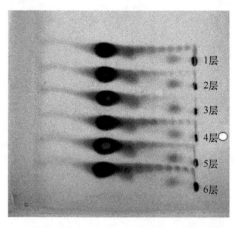

图 13-15　激素对红细胞多带释放的影响电泳
注释：由上向下，第 1、6 层(原点见△泳道处)为红细胞，第 2、4 层为红细胞加地塞米松(原点见○泳道处)，第 3、5 层为红细胞加甲泼尼龙(原点见□泳道处)

图 13-16　无机物对红细胞多带释放的影响电泳
注释：由上向下，第 1、6 层为红细胞，第 2 层为红细胞加氯化钠，第 3 层为红细胞加氯化钾，第 4 层，红细胞加氯化钙(原点见○泳道处)第 5 层，红细胞加碳酸氢钠

3.2.12 血沉值与血红蛋白单带释放值之间关系的比较研究结果　见图 13-17 及表 13-1。

由图 13-17 及表 13-1 可以看出，血沉值与血红蛋白单带释放值之间没有明显相关关系（$R = -0.08997$，$P > 0.05$）。

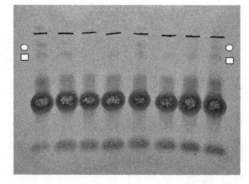

图 13-17　血沉值与单带释放值关系的比较研究电泳
注释：○与○之间的是单带释放，□与□之间的是 CA(碳酸酐酶)

利用 BANDSCAN 软件测单带/CA 比值后与血沉值测相关关系，见表 13-1。

表 13-1　血沉值与单带/CA 比值的关系

编号	血沉值	单带值	CA 值	单带/CA
1	1	330 23	319 14	1.0347
2	9	276 76	327 39	0.8454
3	20	271 07	292 40	0.9271
4	7	275 34	313 45	0.8784
5	30	323 41	327 39	0.9878
6	35	273 06	308 62	0.8848
7	10	289 56	318 29	0.9097
8	68	325 97	342 18	0.9526

4　讨论

血沉和血红蛋白电泳释放都有血浆与红细胞之间的相互作用,但二者之间没有明显的相关性,说明它们的机制不同。我们在长期的实验研究中逐渐认识到血浆成分能够影响红细胞释放血红蛋白,也就是说这里边也存在血浆与红细胞之间的相互作用,简称"浆胞互作"。这种浆胞互作的总趋势是使全血中的红细胞血红蛋白释放减弱,正常人血液标本如此,多种疾病也这样,少数疾病相反(全血释放大于红细胞)。浆胞互作对初释放的影响,多数情况是全血中的 HbA_2 电泳位置靠后(阴极侧),红细胞中 HbA_2 则靠前(阳极侧)。浆胞互作对再释放的影响,总的规律是全血中的红细胞再释放减弱,但也有增强者。

糖尿病患者的结果比较特殊,血糖明显升高的红细胞多带增强,其全血中红细胞则多带明显减弱,甚至消失。血糖不太高的标本,其红细胞基本没有多带或不明显,其全血中红细胞更是看不见多带。HbA_2 的电泳位置也与血糖值有关,血糖明显升高者,与游离红细胞比较全血中血红蛋白 A_2 靠后(阴极侧),血糖不太高的标本则不靠后。黄疸患者的结果更特殊,黄疸标本基本上都是全血的多带释放增强,绝大多数标本红细胞的多带释放都弱于相应的全血标本。

综上所述,血浆成分能够影响红细胞释放血红蛋白,而且有各种情况。这些差异受到血浆中的哪些成分的影响,这是本章要研讨的问题。实验证明,血浆中使再释放增强的成分有葡萄糖、维生素 B_1、维生素 B_6、维生素 B_{12} 和氯化钙等,使再释放减弱的成分有白蛋白、丙种球蛋白、激素、氨基酸、维生素 C、氯化钠、氯化钾、碳酸氢钠等。可以认为,这两大类物质相互对应,调节全血中红细胞的再释放。根据我们的系列研究[4, 5, 7-15],多数疾病的全血都有较明显的再释放,但很少超出其红细胞的释放水平。所以,浆胞互作是人体的自我调节、自我平衡过程,血浆成分控制红细胞的再释放、可能对机体起到保护作用。

参 考 文 献

[1] Sox HC, Liang MH. The erythrocyte sedimentation rate. Ann Intern Med, 1986, 4: 515-523.
[2] Brigden ML. Clinical utility of the erythrocyte sedimentation rate. Am Fam Physician, 1999, 60: 1443-1450.
[3] Vennapuss B, De La Cruz L, Shah H, et al. ESR measured by the streck ESR-auto plus is higher than with the sediplst westergren method. Am J Clin Pathol, 2011,135: 386-390.

[4] 秦文斌. 活体红细胞内血红蛋白的电泳释放. 中国科学生命科学, 2011, 41(8): 597-607.
[5] Su Y, Shen J, Gao LJ, et al. Molecular interactions of re-released proteins in electrophoesis of human erythrocytes. Electrophoresis, 2012, 33: 1042-1045.
[6] Su Y, Gao LJ, Qin WB. Interactions of hemoglobin in lived red blood cells measured by the electrophoesis release test//Kurien BT, Scofield RH. Protein Electrophoresis: Methods and Protocols, Methods in Molecular Biology. New York: Springer Science Busibess Media, 2012.
[7] 秦文斌, 高丽君, 苏燕, 等. 血红蛋白释放试验与轻型 β-地中海贫血. 包头医学院学报, 2007, 23(6): 561-563.
[8] Su Y, Shao Guo, Gao LJ, et al. RBC electrophoresis with discontinuous power supply – a newly established hemoglobin release test. Electrophoresis, 2009, 30: 3041-3043.
[9] 韩丽红, 闫斌, 高雅琼, 等. 普通外科患者血红蛋白释放试验的比较研究. 临床和实验医学杂志, 2009, 8(7): 67-69.
[10] Su Y, Gao LJ, Ma Q, et al. Interactions of hemoglobin in lived red blood cells measured by the electrophoesis release test. Electrophoresis, 2010, 31: 2913-2920.
[11] 张晓燕, 高丽君, 高雅琼, 等.血糖浓度和血红蛋白释放试验的比对研究.国际检验医学杂志, 2010, 31(6): 524-525.
[12] 高雅琼, 王彩丽, 高丽君, 等. 尿毒症患者低渗血红蛋白释放试验的初步结果. 临床和实验医学杂志, 2010, 9(1): 12-13.
[13] 王翠峰, 高丽君, 乌兰苏燕, 等. 血红蛋白释放试验与血液流变学检测结果相关性的研究.中国医药导报, 2010, 7(4): 64-66.
[14] 张咏梅, 高丽君, 苏燕, 等.陈旧血电泳出现快泳红带: 红细胞释放"高铁血红素？".现代预防医学杂志, 2011, 38(15): 3040-3042.
[15] 宝勿仁必力格, 王翠峰, 高丽君, 等.肝硬化患者血液标本的血红蛋白释放试验结果明显异常. 临床和实验医学杂志, 2011, 10(24): 1915-1917.

附 研究的英文翻译

The effect of plasma components on hemoglobin release test in red blood cells(RBC)

Cai-Li Wang, Li-Ping Liu, Li-Hong Han, Yan Su, Wen-Bin Qin, Li-Jun Gao, Ya-Qiong Gao, Cui-Feng Wang, Xiao-Yan Zhang, Bin Yan & Gang Sun

Abstract

Electrophoresis release test(ERT)was established by our lab to observe the re-released hemoglobin(Hb)from red blood cells(RBCs)and whole blood. In this study, ERT was performed to study the effects of different plasma components including plasma, serum, albumin, globulin, fibrinogen, glucose, amino acid, vitamin, insulin, hormone, and inorganic ions on re-released Hb from RBC and whole blood samples during ERT. The results showed that plasma, serum, albumin, globulin, compound amino acid, essential amino acid, vitamin C, insulin, hormone, NaCl, KCl, $CaCl_2$, and $NaHCO_3^-$ decreased re-released RBC Hb; while glucose, vitamin B_1, vitamin B_2, and vitamin $B12$ elevated re-released RBC Hb. The differing effects of various plasma components on re-released Hb of RBC may play a significant role in blood conservation.

Introduction

Red blood cells(RBCs)constitute the most common type of blood cell. In humans, mature RBCs are flexible appearing as oval biconcave disks. RBCs lack cell nucleus and most organelles in order to accommodate maximal space for hemoglobin(Hb), which has important oxygen transporting function. Hb makes up approximately 96% of the RBC dry content(by weight), and 35% of total content including water(Weed et al. 1963). Hb is an assembly of two α-globin family chains(including α and ξ chain)and two β-globin family chains(including β, γ, δ and ε chains). These subunits are bound to each other by salt bridges, hydrogen bonds, and hydrophobic interactions. Three Hb variants exist in normal adult RBC, which are HbA($α_2β_2$, over 95%), HbF($α_2γ_2$, <1%), and HbA2($α_2δ_2$, 1.5%-3.5%))(Ribeil et al. 2013; Su et al. 2010). Plasma, the extracellular matrix of blood cells, is the pale yellow liquid component of blood that makes up approximately 55% of the body's total blood volume. Water(up to 95% by volume), dissolving proteins(6%-8%, such as albumins, globulins, and fibrinogen), glucose, clotting factors, electrolytes(Na^+, Ca^{2+}, Mg^{2+}, HCO_3^-, and, and Cl^-), hormones, and carbon dioxide, constitutes the main components of plasma. Plasma also serves as a protein reservoir and plays a vital role in an intravascular osmotic effect that maintains electrolytes balance and protects the body from infection and other blood disorders.

Electrophoresis release test(ERT), performed by electrophoresing RBCs directly on the starch-agarose mixed gel with intermittent electric current, was established by our lab(Su et al. 2009). According to the number of power outages, ERT is divided into initial release electrophoresis and re-release electrophoresis. The regular starch-agarose mixed gel electrophoresis, which is also termed initial release electrophoresis, only needs to turn on and turn off the power once. During initial release, most of the Hb is released from RBC, and the electrophoretic mobility of HbA_2 was found to differ between RBC and hemolysate groups. This phenomenon was termed "HbA_2 phenomenon"(Su et al. 2010).

In 2007, a sudden power outage was encountered during the electrophoresis of RBC, and another new Hb band was found to be released from the origin after the power was restored(Su et al. 2009). This phenomenon was later named as re-release electrophoresis as opposed to initial release electrophoresis. When the power outages were simulated more than once, multiple Hb bands appeared between HbA and origin, and this phenomenon was termed "multiple-band re-release" or "ladder-band re-release"(Qin 2011). Based on these experiments, isotonic and hypotonic ERT and double-dimensional ERT were subsequently developed. Then, the re-released Hb from some patients' RBC was examined, and the amount varied in different patients(Qin et al. 2007; Han et al. 2009; Su et al. 2015a, 2015b; Gao et al. 2010). Some patients displayed increased Hb re-release, such as in β-thalassemia(Qin et al. 2007), general surgery patients(Han et al. 2009), cirrhosis, and gastroenteric tumor patients(Su et al. 2015a), while hereditary spherocytosis patient(Su et al. 2015b)showed decreased Hb re-release. Therefore, abnormal Hb re-release is not only related to morphologic and metabolic altered RBC, but also to differing plasma components. Specific screening experiments have not yet been done and the exact mechanism underlying this phenomenon remains unclear.

The RBC membrane or cytoskeleton binding Hb was postulated to play a role in this phenomenon. To further study the mechanisms underlying Hb re-release, the effects of blood type(Wei et al. 2011), blood viscosity(Wang et al. 2010), differing membrane destroying methods(Wei et al. 2013), exogenous hydrogen peroxide(Du et al. 2015), glucose(Zhang et al. 2010), and glutaraldehyde(Wei et al. 2014)treatments on the amount of re-released Hb were examined, and re-released Hb was found to be associated with the abnormality of membrane binding Hb. In this study, ERT was performed to determine the influence of different plasma components on re-released Hb.

Materials and methods

Specimens

All experimental protocols of this study were approved by Ethics Committee of Baotou Medical College. The fresh normal anti-coagulated blood was collected from the First Affiliated Hospital of Baotou Medical College. Before collection of blood samples, informed consents were obtained from all subjects. The anti-coagulated blood was stored at 4℃ and used within 24 hr.

Preparation of RBC suspension and hemolysate

The anti-coagulated blood was first centrifuged at 2500 g for 10 min to isolate RBCs from plasma, and then the packed RBCs were washed with saline 4-5 times until the supernatant was clear. RBCs were then used to perform electrophoresis after 1:1 dilution with saline(Qin 2011; Su et al. 2009, 2012). Hemolysate was prepared by continuously adding 200 ml saline and 100 ml carbon tetrachloride(CCl_4)to the RBC. After thorough mixing, the sample was centrifuged at 10,000 g for 10 min and the upper red hemolysate was pipetted out carefully and stored at 4℃ for later use(Su et al. 2010, 2012).

One-direction ERT

Five microliters RBC suspension and whole blood were added to the starch-agarose mixed gel. The electrophoresis was run at 6 V/cm for 15 min with alternating 15 min pause for 2 hr. After electrophoresis, red bands on the gel were initially visually observed and then sequentially stained with Ponceau Red and benzidine(Su et al. 2009, 2012).

Two-direction ERT

First, 5 ml RBC suspension (approximately 1.5×10^9 RBCs) and whole blood were added to the starch-agarose mixed gel for one-direction ERT as described earlier. Then, the direction of electric field was changed, vertical to the original direction. Each directional electrophoresis was conducted for 15 min with alternating 15 min pause for 4 hr. After electrophoresis, the red bands on the gel were visually observed initially and then sequentially stained with Ponceau Red and benzidine.

Results

One-dimensional single-band ERT of normal blood samples

Each blood sample was prepared as previously described, and then Hb single-band re-release of different samples was compared against each other. As indicated in Figure 1, there were five bands in each lane. Each sample displayed a re-released band; however, the band intensity of RBC group(R) was consistently stronger than that of whole blood group(B) in each sample. In addition, the electrophoretic velocity of RBC HbA_2 was consistently faster(near the anode side) than that of whole blood HbA_2, and this phenomenon has already been described as "HbA_2 phenomenon"(Su et al. 2010).

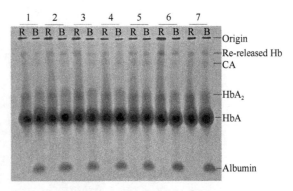

Figure 1. (Color online). One-dimensional single-band ERT of normal blood samples. There are seven normal blood samples(1—7), each of them was divided into RBC group(R) and whole blood group(B)

The effects of plasma on the re-released Hb during multi-band ERT

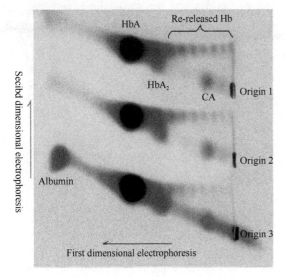

Figure 2. (Color online). The effects of plasma on the re-released Hb during ERT. Origin 1 is RBCs, origin 2 is RBCs saline suspension, origin 3 is RBCs plasma suspension

Two microliters packed RBC, RBC saline suspension(2 ml packed RBC plus 2 ml saline), and RBC plasma suspension(2 ml packed RBC plus 2 ml plasma) were added at the origin.

Two-dimensional(2D)ERT was run, and results showed that re-released Hb was decreased significantly by adding plasma into the RBC(Figure 2).

The influence of albumin and globulin on re-released Hb during ERT

Equal volume of 5% albumin(or 2.5% globulin) were added into the packed RBC for 1 hr, then 2 ml packed RBC and 4 ml 5% albumin(or 2.5% globulin) treated RBC were loaded onto the agarose-starch mixed gel to perform one-dimensional(1D)or 2D ERT. Data demonstrated that both albumin and globulin diminished re-released Hb significantly either in 1D[Figure 3(A, B)]or 2D ERT[Figure 3(C)].

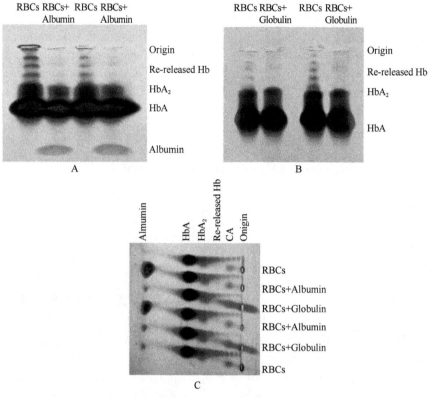

Figure 3. (Color online). The influence of human albumin and globulin on the re-released Hb during ERT. A. 5% albumin on re-released Hb during multi-band one-dimensional ERT; B. 2.5% globulin on the re-released Hb during multi-band one-dimensional ERT; C. 2.5% globulin on the re-released Hb during multi-band two-dimensional ERT

The effects of human plasma and serum on hemoglobin re-release test

Packed RBCs were treated with equal volume of plasma and serum for 1 hr, and then 2D ERT was performed(Figure 4). It was found that re-released Hb from serum treated RBC was decreased

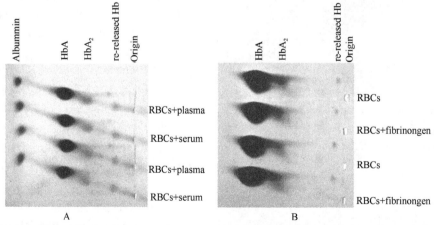

Figure 4. (Color online). Effects of plasma and serum on the re-released Hb during ERT. A. Plasma and serum on re-released Hb during two-directional multiband ERT. B. Fibrinogen on re-released Hb during two-directional one-band ERT

significantly in comparison with plasma treated RBC. As the only difference between plasma and serum is fibrinogen, re-released Hb from RBC was determined after treating RBC with 4 mg/ml fibrinogen solution for 1 hr; however, no significant difference was found between these two groups.

The influence of glucose on the re-released Hb during ERT

In Figure 5, data showed that glucose markedly increased the re-released Hb from RBC, and the rise was proportional to glucose concentration. It is of interest that no marked change of re-released Hb was observed in whole blood.

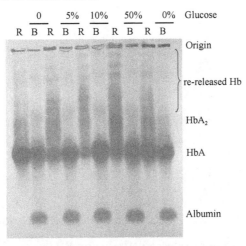

Figure 5. (Color online). Influence of glucose on the re-released Hb of RBC and whole blood during one-directional multi-band ERT. There were four groups: control, 5% glucose, 10% glucose, and 50% glucose. Each group includes RBC and whole blood sample

The effects of amino acids on the re-released Hb during ERT

The results in Figure 6 demonstrated that not only compound amino acid injection(18AA)(12.5 g/250 ml)but also essential amino acids injection significantly lowered re- released Hb of RBCs. No significant difference was found between these amino acid preparations.

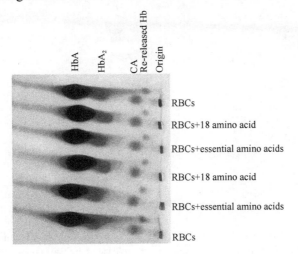

Figure 6. (Color online). Effects of amino acids on re-released Hb during two-directional multi-band ERT

The influence of different water-soluble vitamins on re-released Hb during two-directional multi-band ERT

The results presented in Figure 7 showed that vitamin B_1, vitamin B_6, and vitamin B_{12} slightly elevated re-released Hb of RBC. In contrast to other samples, the first re-released band of vitamin B_1 treated RBC was stronger than other vitamins, and subsequent re-released bands became weaker. However, the re-released Hb from vitamin C treated RBC was markedly decreased.

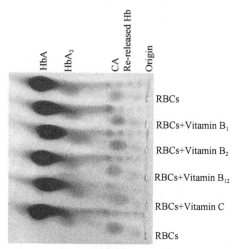

Figure 7. (Color online). Influence of vitamins on re-released Hb during ERT

The effects of insulin on the re-released Hb during ERT

Packed RBC were treated with 40 U/ml insulin for 1 hr, and then ERT was performed to detect the re-released Hb. Data in Figure 8 showed that re-released Hb was reduced significantly by adding insulin into RBC.

The influence of hormone on re-released Hb during ERT

As depicted in Figure 9, RBC treated with 5 mg/ml dexamethasone sodium phosphate or 40 mg/ml

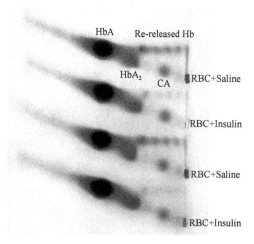

Figure 8. (Color online). Effects of insulin on re-released Hb during two-directional multi-band ERT

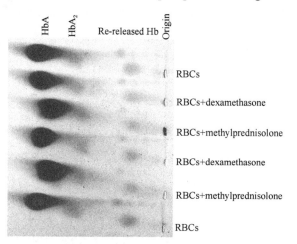

Figure 9. (Color online). Influence of hormones on re-released Hb during two-directional multi-band ERT

methylprednisolone sodium succinate for 1 hr resulted in a marked reduction in Hb re-release. It appeared that methylprednisolone was more effective.

The RBC were treated with 0.9% sodium chloride(NaCl)injection, 1 g/10 ml potassium chloride(KCl)injection, calcium chloride($CaCl_2$)injection, or sodium bicarbonate($NaHCO_3^-$)for 1 hr. NaCl, $CaCl_2$, and $NaHCO_3^-$ produced a slight decrease in re-released Hb, but KCl significantly lowered the re-released Hb(Figure 10).

The results of our study showed that plasma, serum, albumin, globulin, amino acid, vitamin C, insulin, hormones, NaCl, KCl, $CaCl_2$ and $NaHCO_3^-$ significantly lowered re- released Hb; however, glucose, vitamin B_1, vitamin B_2, and vitamin B_{12} increased re-released Hb. These components are involved in regulating the release of RBC and. Thus, re-released Hb of RBC might be used as an indicator of the state of RBC. The general trend of plasma-RBC interaction is that re-released Hb from whole blood rarely exceeds re-release levels of RBC. Therefore, plasma-RBC interaction is a self-regulation and self- balancing process, which may exert a protective effect on the organism.

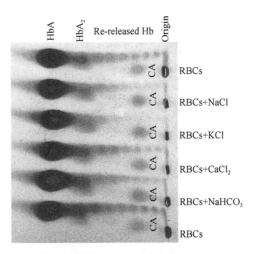

Figure 10. (Color online). Effects of inorganic substances on re-released Hb during two-directional multi-band ERT

Discussion

ERT is a method used to detect pathological and physiological state of RBC. Plasma and RBC are the main components of blood, and interact with each other continuously to maintain the stability of RBC. Plasma, which constitutes 55% of blood fluid, is essentially an aqueous solution containing 92% water, 8% blood plasma proteins, and trace amounts of other materials. Plasma circulates dissolved nutrients, such as glucose, amino acids, vitamins, and fatty acids(dissolved in the blood or bound to plasma proteins), and removes waste products, such as carbon dioxide, urea, and lactic acid. Generally, erythrocyte sedimentation rate(ESR)is used to study the interaction between plasma and RBC, and the change of plasma components might alter the ESR(Sox and Liang 1986; Brigden 1999; Vennapuss et al. 2011). In this study, the interactions between different plasma components and RBC were examined utilizing ERT.

Acknowledgments

This work was supported by grants from Natural Science Foundation of China(81160214), Key Science and Technology Research Project of the Ministry of Education(210039), Natural Science Foundation of Inner Mongolia(2010BS1101), and Doctoral Scientific Research Foundation of Bao- tou Medical College(BSJJ201615). We also especially acknowledge all of the people who donated their blood samples for our research.

Disclosure statement

No potential conflict of interest was reported by the authors.

Funding

Natural Science Foundation of China [grant number 81160214]; Key Science and Technology Research Project of the Ministry of Education [grant number 210039]; Natural Science Foundation of Inner Mongolia [grant number 2010BS1101]; Doctoral Scientific Research Foundation of Baotou Medical College [grant number BSJJ201615].

References

Brigden, M.L. 1999. "Clinical Utility of the Erythrocyte Sedimentation Rate." *American Family Physician* 60: 1443-1450.

Du, X., L. An, C. Wei, X. Li, C. Xie, J. Li, W. Qin, X. Wu, and Y. Su. 2015. "The Protective Effects of Melatonin on the Antioxidative Damage of RBCs." *China Journal of Blood Transfusion* 6: 624-626.

Gao, Y., C. Wang, L. Gao, Y. Su, L. Qin, L. Zhou, and W. Qin. 2010. "A Preliminary Study on Hypotonic Hemoglobin Release Test in Patients with Uremia." *Chinese Journal of Clinical and Experimental Medicine* 9: 12-13.

Han, L., B. Yan, Y. Gao, L. Qin, and W. Qin. 2009. "The Comparative Study of Hemoglobin Release Test in Different General Surgical Patients." *Journal of Clinical and Experimental Medicine* 8: 67-69.

Qin, W. 2011. "Electrophoresis Release of Hemoglobin from Living Red Blood Cells." *Scientia Sonica Vitae* 41: 597-607.

Qin, W., L. Gao, Y. Su, G. Shao, L. Zhou, and L. Qin. 2007. "Hemoglobin Release Test and b-Thalassemia Trait." *Journal of Baotou Medical College* 23: 561-563.

Ribeil, J.A., J.B. Arlet, M. Dussiot, I.C. Moura, G. Courtois, and O. Hermine. 2013. "Ineffective Erythropoiesis in b-Thalassemia." *Scientific World Journal* 2013: 394295.

Sox, H.C., and M.H. Liang. 1986. "The RBC Sedimentation Rate." *Annals of Internal Medicine* 4: 515-523.

Su, Y., G. Shao, L. Gao, L. Zhou, L. Qin, L. Han, and W. Qin. 2009. "RBC Electrophoresis with Dis- continuous Power Supply—a Newly Established Hemoglobin Release Test." *Electrophoresis* 30: 3041-3043.

Su, Y., H. Ma, H. Zhang, L. Gao, G. Jia, W. Qin, and Q. He. 2015b. *Inherited Hemoglobin Disorders: Comparative Study of the Amount of Re-Released Hemoglobin from α-Thalassemia and Hereditary Spherocytosis Erythrocytes*. Croatia: InTech.

Su, Y., J. Shen, L. Gao, H. Tian, Z. Tian, and W. Qin. 2012. "Molecular Interaction of Re-Released Proteins in Electrophoesis of Human RBCs." *Electrophoresis* 33: 1042-1045.

Su, Y., L. Gao, Q. Ma, L. Zhou, L. Qin, L. Han, and W. Qin. 2010. "Interactions of Hemoglobin in Live Red Blood Cells Measured by the Electrophoresis Release Test." *Electrophoresis* 31: 2913-2920.

Su, Y., L. Han, L. Gao, J. Guo, X. Sun, J. Li, and W. Qin. 2015a. "Abnormal Increased Re-Released Hb from RBCs of an Intrahepatic Bile Duct Carcinoma Patient was Detected by Electrophoresis Release Test." *Bio-Medical Materials and Engineering* 26: S2049-S2054.

Vennapuss, B., L. De La Cruz, H. Shah, V. Michalski, and Q.Y. Zhang. 2011. "Erythrocyte Sedimentation Rate(ESR)Measured by the Streck ESR-Auto Plus is Higher than with the Sediplast Westergren Method: A Validation Study." *American Journal of Clinical Pathology* 135: 386-390.

Wang, C., L. Gao, L. Wu, Y. Su, J. Xu, L. Zhou, and W. Qin. 2010. "Study on the Correlation Between Results of Hemoglobin Release Test and Hemorheology." *China Medical Herald* 7: 64-66.

Weed, R.I., C.F. Reed, and G. Berg. 1963. "Is Hemoglobin an Essential Structural Component of Human Erythrocyte Membranes?" *Journal of Clinical Investigation* 42: 581-588.

Wei, C., Y. Su, L. Yang, and X. Li. 2014. "Effect of Difierent Concentrations of Glutaraldehyde on Hemoglobin Release of Red Blood Cells." *Progress in Veterinary Medicine* 2: 16-19.

Wei, C., Y. Su, L. Yang, Q. Ma, X. Li, L. Gu, H. Ding, L. Gao, W. Qin, and Y. Ma. 2013. "The Effects of Different Destroying Red Blood Cell Membrane Methods on HbA_2 Phenomenon." *Journal of Baotou Medical College* 4: 1-3.

Wei, C., Y. Su, L. Yang, X. Li, L. Gu, L. Gao, and W. Qin. 2011. "The Effect of Blood Type on the Hemoglobin Release in Red Blood Cells of Normal Adults." *Journal of Baotou Medical College* 5: 1-3.

Zhang, X., L. Gao, Y. Gao, L. Zhou, Y. Su, L. Qin, and W. Qin. 2010. "Comparative Study on the Blood Glucose Concentration and Hemoglobin Release Test." *International Journal of Labora-tory Medicine* 31: 524-525.

(*Toxicological & Environmental Chemistry*, 99: 3, 448-459.)

第十四章 纤维蛋白原现象的发现和研究

秦文斌[1] 高丽君[1] 郭 俊[2] 王彩丽[3] 马宏杰[4] 周立社[1]
秦良宜[1] 韩丽红[1] 苏 燕[1] 邵 国[1] 王占黎[1] 于 慧[1]

(包头医学院 1. 血红蛋白研究室；2. 第二附属医院 介入科；3. 第一附属医院 肾内科；4. 第一附属医院 血液科，包头 014010)

摘 要

当用人类血浆做淀琼胶对角线电泳时，其中的纤维蛋白原处于对角线上。如果用全血做此双向电泳，其中的纤维蛋白原可以不在对角线上(脱离对角线上)。纤维蛋白原在血浆与全血之间的上述差异，特别是全血中纤维蛋白原脱离对角线，我们把这种现象称为"纤维蛋白原现象"。此现象的机制不明，但因为血浆没有无此现象而全血才有，于是推测与全血里的有形成分有关，特别是全血溶血时此现象消失，由此推测可能与红细胞有关。我们推测，全血中纤维蛋白原与红细胞膜上某种非蛋白成分 X 结合存在(相互作用)，第一向电泳时二者一起离开红细胞，第二向电泳时二者彼此分开，X 看不见(蛋白染色无效)，能看到纤维蛋白原，它脱离了对角线。以上想法需要证实，我们制备并比较研究血浆和全血里的纤维蛋白原，通过质谱分析，看二者的化学组成是否不同、有何不同，来验证我们的上述推测。

关键词 纤维蛋白原现象；全血；血浆；双向电泳；对角线

1 前言

纤维蛋白原是一种由肝脏合成的具有凝血功能的蛋白质，是纤维蛋白的前体，分子量 340 000，半衰期 5~6 日，血浆中参考值 2~4g/L。纤维蛋白原由 α、β、γ 三对不同多肽链所组成，多肽链间以二硫键相连。在凝血酶作用下，α 链和 β 链分别释放出 A 肽和 B 肽，生成纤维蛋白单体。在此过程中，由于释放了酸性多肽，负电性降低，单体易于聚合成纤维蛋白多聚体。但此时单体之间借氢键与疏水键相连，尚可溶于稀酸和尿素溶液中。进一步，在 Ca^{2+} 和活化的XIII 因子作用下，单体之间以共价键相连，变成稳定的不溶性纤维蛋白凝块，完成凝血过程。肝功严重障碍或先天性缺乏，均可使血浆纤维蛋白原浓度下降，严重时可有出血倾向。

综上所述，有关纤维蛋白原的知识主要来源于血浆，全血里纤维蛋白原的情况所知很少。以前没有人用全血直接做电泳，当然无法得知电泳结果如何。1981 年，我们发现"血红蛋白 A_2 现象"，就是来自红细胞或全血电泳[1-6]，那时注意力主要集中于 HbA_2 的电泳行为，后来我们开始注意到纤维蛋白原的情况，于是才发现本章的"纤维蛋白原现象"，详见正文。

2 材料及方法

2.1 材料 血液标本来自包头医学院第一附属医院各临床科室。
2.2 方法 参见文献[7，8]。

3 结果

3.1 单向双排电泳比较全血溶血液和全血中的纤维蛋白原 结果见图14-1。

3.1.1 观察结果 观察全血溶血中液纤维蛋白原与全血中纤维蛋白原的电泳位置关系。全血溶血液的纤维蛋白原比较集中,稍靠前(阳极侧),全血的纤维蛋白原不太集中,稍靠后(阴极侧)。这就是"纤维蛋白原现象"的单向电泳特点。

3.1.2 讨论 全血溶血液里红细胞已经溶血,此时纤维蛋白原电泳位置发生变化,推测全血中纤维蛋白原可能与完整红细胞存在相互作用。红细胞溶血后,这种互作遭到破坏,纤维蛋白原电泳位置才发生变化。红细胞的什么部位参与互作,有待深入研究。

3.2 单向双排电泳比较全血与血浆中的纤维蛋白原 结果见图14-2。

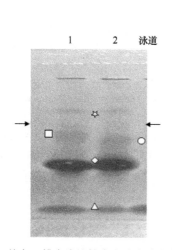

图14-1 单向双排电泳比较全血溶血液和全血中的纤维蛋白原

注释:泳道1为全血溶血液,2为全血;箭头→所指处为全血溶血液的纤维蛋白原(FG),←所指处为全血的纤维蛋白原(FG);□右侧的区带是全血溶血液里的HbA₂,○左侧的区带是全血里的HbA₂;△两侧的区带是白蛋白,◇两侧的区带是HbA₁,☆两侧的区带是CA(碳酸酐酶)

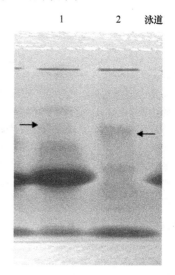

图14-2 单向双排电泳比较全血和血浆中的纤维蛋白原

注释:泳道1为全血,2为血浆;箭头→所指处为血浆中的纤维蛋白原FG,←所指处为血浆中的纤维蛋白原FG;全血中其他成分参见图16-1,血浆中成分,最下为白蛋白,往上依次为$α_1$球蛋白、$α_2$球蛋白、$β_1$球蛋白、$β_2$球蛋白、FG、γ球蛋白。血浆里没有CA碳酸酐酶

3.2.1 观察结果 观察全血中纤维蛋白原与血浆中纤维蛋白原的电泳位置关系。全血的纤维蛋白原不太集中,稍靠后(阴极侧),血浆的纤维蛋白原稍集中,稍靠前(阳极侧)。

3.2.2 讨论 血浆里FG的电泳位置有一点像全血溶血液者,血浆里没有红细胞,不涉及FG与红细胞互作问题。全血溶血液里红细胞已经溶血,与红细胞的互作遭到破坏,所以互作问题也不存在。血浆FG与全血溶血液FG的组成可能相同,但需证实。

3.3 双向三层电泳比较IgA肾病患者全血溶血液、全血和血浆中的纤维蛋白原 见图14-3。

3.3.1 观察结果 箭头↓所指的红色区带为全血溶血液里的FG,它在对角线上;箭头↑所指的红色区带为血浆里的FG,它在对角线上;☆下边的红色区带为全血中的FG,它横过来,脱离了对角线。

3.3.2 讨论 ☆下边全血里的纤维蛋白原脱离对角线,这就是"纤维蛋白原现象"!临床标本中,大多数显示此现象,也有FG不脱离对角线者,详见下文。

3.4 双向三层电泳比较剖宫产产妇全血溶血液、全血和血浆中的纤维蛋白原 见图 14-4。

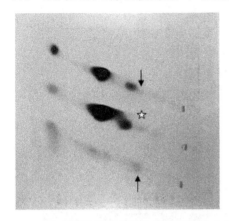

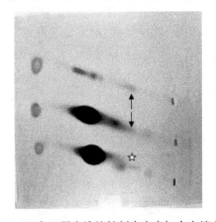

图 14-3 双向三层电泳比较 IgA 肾病患者全血溶血液、全血和血浆中的纤维蛋白原 FG

注释：血液标本来自包头医学院第一附属医院肾内科；双向三层电泳，上层为全血溶血液，中层为全血，下层为血浆

图 14-4 双向三层电泳比较剖宫产产妇全血溶血液、全血和血浆中的纤维蛋白原 FG

注释：血液标本来自包头医学院第一附属医院妇产科；双向三层电泳，上层为血浆，中层为全血溶血液，下层为全血

3.4.1 观察结果　箭头↓所指的红色区带为全血溶血液里的 FG，它在对角线上；箭头↑所指的红色区带为血浆里的 FG，它在对角线上；☆下边的红色区带为全血中的 FG，它横过来，脱离了对角线。

3.4.2 讨论　剖宫产后产妇全血中 FG 也有"纤维蛋白原现象"。

3.5 双向单层电泳检测 1 例骨髓瘤标本中没有"纤维蛋白原现象"的病例 见图 14-5。

3.5.1 观察结果　如图 14-5 所示，患者全血里的纤维蛋白原没有脱离对角线，M-蛋白与 FG 靠近。

3.5.2 讨论　看来此骨髓瘤标本没有"纤维蛋白原现象"，为什么？现在还不明确。推测与 M-蛋白有关，是 M-蛋白与 FG 互作，还是 M-蛋白与红细胞互作，仍有待研究。

3.6 双向双层梯带电泳比较肝癌患者（1 号）红细胞与全血的电泳结果 见图 14-6。

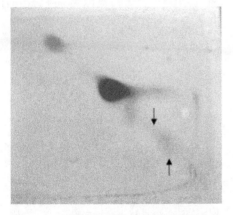

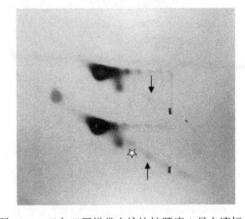

图 14-5 骨髓瘤标本全血的双向单层电泳

注释：血液标本来自包头医学院第一附属医院血液科，箭头↓所指的红色区带为全血里的 FG，箭头↑所指的红色区带为 M-蛋白

图 14-6 双向双层梯带电泳比较肝癌 1 号血液标本中红细胞与全血的电泳结果

注释：血液标本来自包头市肿瘤医院介入科；双向双层电泳上层为红细胞，下层为全血

3.6.1 观察结果　箭头↓所指的红色区带为红细胞里的 CA，它在对角线上；箭头↑所指的红色区带为全血里的 CA，它在对角线上。☆上边的红色区带为全血中的 FG，它横过来，

脱离了对角线。

3.6.2　讨论　肝癌1号的血液标本也有"纤维蛋白原现象"。

3.7　双向双层梯带电泳比较肝癌患者（2号）红细胞与全血的电泳结果　见图14-7。

3.7.1　观察结果　全血中的FG横过来，不在对角线上。

3.7.2　讨论　肝癌2号的血液标本也有"纤维蛋白原现象"。

3.8　双向双层梯带电泳比较慢性粒细胞白血病患者红细胞与全血的电泳结果　见图14-8。

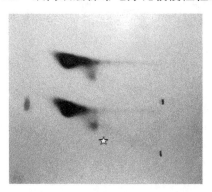 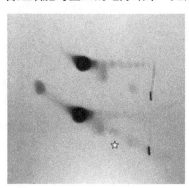

图14-7　双向双层梯带电泳比较肝癌2号血液标本中红细胞与全血的电泳结果
注释：血液标本来包头市肿瘤医院介入科；双向双层电泳上层为红细胞，下层为全血，☆上边的红色区带为全血中的FG

图14-8　双向双层梯带电泳比较红细胞与全血的电泳结果
注释：血液标本来包头医学院第一附属医院血液科；双向双层电泳上层为红细胞，下层为全血，☆上边的红色区带为全血中的FG

3.8.1　观察结果　全血中的FG横过来且变窄，不在对角线上。

3.8.2　讨论　慢性粒细胞白血病的血液标本也有"纤维蛋白原现象"。

3.9　双向双层梯带电泳比较癌症介入患者（4号）红细胞与全血的电泳结果　见图14-9。

3.9.1　观察结果　全血中的FG在对角线上。

3.9.2　讨论　介入4号癌症患者没有"纤维蛋白原现象"。原因目前尚不明确。

3.10　双向双层梯带电泳　比较癌症介入7号患者红细胞与全血的电泳结果见图14-10。

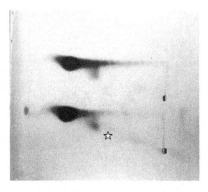 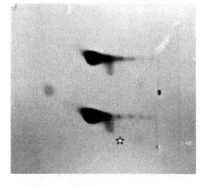

图14-9　双向双层梯带电泳比较红细胞与全血的电泳结果
注释：血液标本来包头市肿瘤医院介入科；双向双层电泳上层为红细胞，下层为全血，☆上边的红色区带为全血中的FG

图14-10　双向双层梯带电泳比较红细胞与全血的电泳结果
注释：血液标本来包头市肿瘤医院介入科术后患者；双向双层电泳上层为红细胞，下层为全血，☆上边的红色区带为全血中的FG

3.10.1　观察结果　全血中的FG变细且横过来，不在对角线上。

3.10.2　讨论　介入7号癌症患者有"纤维蛋白原现象"！手术能否改变FG的存在状态目前尚不明确。

3.11 IgA 肾病患者冻化血液对其中纤维蛋白原的影响 见图 14-11。

3.11.1 结果 全血中的 FG 横过来，不在对角线上；冻化后全血的 FG 呈"V"字形，也不在对角线上。

3.11.2 讨论 此 IgA 肾病患者 FG 脱离对角线，有"纤维蛋白原现象"。有趣的是，冻化溶血后 FG 变成"V"字形，同时脱离对角线，出现奇特的"纤维蛋白原现象"。由此可知，冻化能改变 FG 带型但是原因尚不明确。

3.12 双向双层梯带电泳比较梅花鹿红细胞与全血 EDTA 抗凝的电泳结果 见图 14-12。

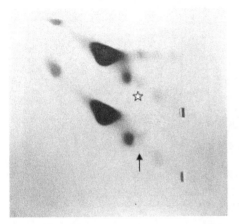 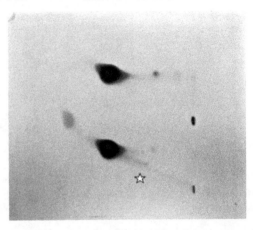

图 14-11 IgA 肾病患者冻化血液对其中纤维蛋白原的影响

注释：血液标本来自包头医学院第一附属医院肾内科；双向双层电泳上层为全血，下层为冻化后全血，☆上边的红色区带为全血中的 FG，箭头↑所指的红色区带为冻化后全血的 FG，呈"V"字形

图 14-12 双向双层梯带电泳比较梅花鹿红细胞与全血的电流结果

注释：梅花鹿血液标本来自上海野生动物园；双向双层电泳上层为红细胞，下层为全血，☆上边的红色区带为全血中的 FG

3.12.1 结果 全血中的 FG 变细且横了过来，不在对角线上。

3.12.2 讨论 梅花鹿 EDTA 抗凝血也有"纤维蛋白原现象"，但梅花鹿肝素抗凝血就没有这种现象，内容详见下文。

3.13 双向双层梯带电泳比较梅花鹿红细胞与全血肝素抗凝的电泳结果 见图 14-13。

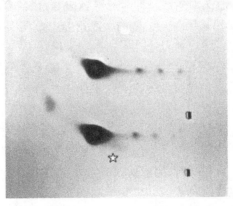

图 14-13 双向双层梯带电泳比较梅花鹿红细胞与全血的电泳结果

注释：梅花鹿血液标本来自上海野生动物园；双向双层电泳上层为红细胞，下层为全血，☆上边的红色区带为全血中的 FG

3.13.1 结果 全血中的 FG 不变细，也没横过来，没脱离对角线。

3.13.2 讨论 梅花鹿肝素抗凝血与 EDTA 抗凝不同，没有"纤维蛋白原现象"。说明抗凝剂能影响纤维蛋白原现象，再对比观察两种抗凝剂的影响，见图 14-14。

肝素是一种大分子的黏多糖硫酸酯，它是如何改变 FG 的存在状态呢？目前推测它能与 FG 相互作用，影响了 FG 与红细胞膜上成分的互作，但此推测仍有待证实。

3.14 并排比较两种抗凝剂对纤维蛋白原的影响 见图 14-14。

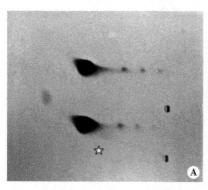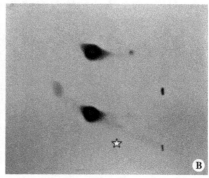

图 14-14 并排比较两种抗凝剂对纤维蛋白原的影响

注释：A 图为梅花鹿用肝素抗凝，B 图为梅花鹿血用 EDTA 抗凝；双向双层电泳上层为红细胞，下层为全血，☆上边的红色区带为全血中的 FG

3.14.1 结果 肝素抗凝血中的 FG 不变细，没横过来，也没脱离对角线；EDTA 抗凝血中的 FG 变细，横了过来，同时脱离对角线。

3.14.2 讨论 梅花鹿肝素抗凝血没有"纤维蛋白原现象"。梅花鹿 EDTA 抗凝血有"纤维蛋白原现象"。说明抗凝剂能影响纤维蛋白原现象。

4 讨论

有关纤维蛋白原的知识主要来源于血浆[9-18]，而全血里纤维蛋白原的情况所知很少。过去没有人用全血直接做电泳，当然无法得知其电泳结果如何。1981 年，我们直接用红细胞做对角线电泳，发现 HbA_2 脱离对角线，即"血红蛋白 A_2 现象"(图 14-15)，后来证明红细胞内血红蛋白 A_2 与 A_1 存在相互作用[1-8]。那时，注意力主要集中于红细胞内 HbA_2 的电泳行为，后来开始注意到全血对角线电泳里纤维蛋白原的情况，才发现此时纤维蛋白原也脱离对角线，推测它也与某种物质存在相互作用，简称为"纤维蛋白原现象"。

此现象的机制不明，但因为血浆没有、全血才有，估计与全血里的有形成分有关，特别是全血溶血时此现象消失，它可能与红细胞有关。因此我们推测，全血中纤维蛋白原与红细胞膜上某种非蛋白成分 X 结合存在(相互作用)，第一向电泳时二者一起离开红细胞，第二向时二者彼此分开，X 观察不到(蛋白染色无效)，但能观察到纤维蛋白原脱离了对角线(图 14-16)。以上想法需要证实，我们制备比较研究血浆和全血里的纤维蛋白原，通过质谱分析，看二者的化学组成是否不同、有何不同，以验证我们的上述假说。

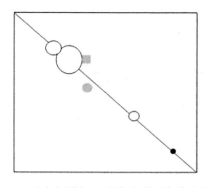

图 14-15 "血红蛋白 A_2 现象"的对角线电泳结果　　图 14-16 "纤维蛋白原现象"的对角线电泳结果

参 考 文 献

[1] 秦文斌, 梁友珍. 血红蛋白 A_2 现象 1 A_2 现象的发现及其初步应用. 生物化学与生物物理学报, 1981, 13(2): 199-205.
[2] 秦文斌. 红细胞外血红蛋白 A 与血红蛋白 A_2 之间的相互作用. 生物化学杂志, 1991, 7(5): 583-587.
[3] 秦文斌. 血红蛋白的 A_2 现象发生机制的研究——"红细胞 HbA_2" 为 HbA_2 与 Hb 的结合产物. 生物化学与生物物理进展, 1991, 18(4): 286-288.
[4] Su Y, Shao G, Gao LJ, et al. RBC electrophoresis with discontinuous power supply – a newly established hemoglobin release test. Electrophoresis, 2009, 30: 3041-3043.
[5] Su Y, Gao LJ, Ma Q, et al. Interactions of hemoglobin in lived red blood cells measured by the electrophoesis release test. Electrophoresis, 2010, 31: 2913-2920.
[6] Su Y, Shen J, Gao LJ, et al. Molecular interactions of re-released proteins in electrophoresis of human erythrocytes. Electrophoresis, 2012, 33: 1402-1405.
[7] 秦文斌. 活体红细胞内血红蛋白的电泳释放. 中国科学生命科学, 2011, 41(8): 597-607.
[8] 秦文斌. 红细胞内血红蛋白的电泳释放——发现和研究. 北京: 科学出版社, 2015.
[9] 苏庆军, 王一男, 陈建国, 等. 纤维蛋白原的临床意义及测定. 实用医技杂志, 2006, 13(23): 4273-4274.
[10] 张丽中, 赵智, 吴滨. 恶性肿瘤患者血浆纤维蛋白原和 D 二聚体含量的测定及其临床意义. 肿瘤研究与临床, 2006, 18(11): 759-760.
[11] 钮心怡, 周盛杰, 陈泽英. 肝癌患者纤维蛋白原和 D 二聚体含量的检测结果分析. 分子诊断与治疗杂志, 2009, 1(2): 108-110.
[12] 张永超, 邱贺. 胃癌患者血浆 D 二聚体和纤维蛋白原的测定及其临床意义. 中国实用医刊, 2008, 35(20): 22-23.
[13] 叶锦俊. 重症肝炎患者血浆 D 二聚体和纤维蛋白原的检测结果分析. 检验医学与临床, 2008, 5(21): 1317-1318.
[14] 刘春红, 王宏, 马雅静. 纤维蛋白原对缺血性脑卒中的影响. 中国动脉硬化杂志, 2004, 12(4): 477-478.
[15] 孟晓彬. 缺血性脑血管疾病患者血浆纤维蛋白原水平与颈动脉粥样硬化的关系. 中国老年学杂志, 2013, 33(14): 3498-3499.
[16] 季芳, 罗美芳, 李蕾. 短暂性缺血性发作、单/多发性腔隙性脑梗死患者血浆纤维蛋白原水平与颈动脉及下肢动脉粥样硬化的相关性研究. 卒中与神经疾病, 2011, 18(2): 86-89.
[17] 刘毅. 急性缺血性脑卒中 TOAST 分型对血浆纤维蛋白原的影响. 实用心脑肺血管杂志, 2012, 20(9): 1445-1447.
[18] 张锡清, 张静, 侯亚萍. 孕晚期血浆纤维蛋白原和 D 二聚体检测的临床意义. 检验医学与临床, 2012, 9(13): 1615-1616.

第三篇　初释放电泳

前　言

1　什么是"初释放电泳"

将完整红细胞加入淀粉-琼脂糖混合凝胶，第一次通电后由红细胞释放出血红蛋白，这一过程称为"初释放"。此时，如果将红细胞裂解液(也称红细胞溶血液或溶血液)与完整的红细胞并排电泳，就会发现二者的血红蛋白 A_2 位置不同，即溶血液的血红蛋白 A_2 靠前(阳极侧)，红细胞的血红蛋白 A_2 靠后(阴极侧)。当时，弄不清怎么回事，暂时命名为"血红蛋白 A_2 现象"。也就是说，初释放电泳的结果就发现了"血红蛋白 A_2 现象"。本篇的核心内容就是论述这种现象的发现和研究。

2　传统的研究方法和观点

传统的血红蛋白研究方法，几乎都是先裂解红细胞制备其溶血液，然后进行各种分析，发现成人溶血液中有血红蛋白 A_1 和 A_2。有人用红细胞进行过"自由电泳"，此时，红细胞移动，血红蛋白并不被释放出来。也有人做过单细胞(红细胞)凝胶电泳，要用显微镜进行观察。此时，红细胞被固定，血红蛋白泳出细胞，也能看到大、小两个红色斑点。但是，在显微镜下，人们不容易将溶血液注射到单个红细胞的旁边，不能平行对比二者血红蛋白电泳行为的差异，所以，无法发现此条件下的"血红蛋白 A_2 现象"

3　建立淀粉-琼脂糖混合凝胶电泳，发现"血红蛋白 A_2 现象"

红细胞中血红蛋白 A_2 与溶血液中血红蛋白 A_2 究竟有何不同？事先我们也不知道。只是在用淀粉-琼脂糖混合凝胶电泳大量筛查异常血红蛋白过程中，经常遇到以下情况：红细胞溶血完全与否，其 HbA_2 的电泳位置似乎有轻微不同。为了弄清这个问题，作者将红细胞与其本身溶血液并排进行电泳，对比观察二者的" HbA_2 "电泳行为有无差异。结果是，由红细胞泳出来的红细胞 HbA_2 与红细胞溶血液泳出来的 HbA_2 在电泳位置上明显不同。这时，我们开始意识到可能遇到了一个有趣的现象，而它能引导我们走进红细胞，发现里边的奥秘。

4　从现象到本质，发现血红蛋白之间的相互作用

"血红蛋白 A_2 现象"来自对实验结果的直观感觉，是观察到的一种现象，而不是事物的

本质。经过一系列实验，我们用双向淀粉-琼脂糖混合凝胶电泳证明，红细胞中的 HbA_2 是与部分 HbA_1 结合存在的，再用淀粉-琼脂糖混合凝胶交叉电泳证明，在红细胞外血红蛋白 A_1 能与血红蛋白 A_2 发生相互作用。

5 从现象到本质，发现硫氧还蛋白过氧化物酶(peroxiredoxin-2，Prx-2)参与血红蛋白相互作用

如果红细胞内的血红蛋白 A_2 是血红蛋白 A_1 与 A_2 的结合产物，那么它的电泳位置应当在血红蛋白 A_1 与 A_2 中间。实际情况并非如此(参见封面图)，说明可能另有其他物质参与相互作用。这次，我们从血红蛋白 A_1 与 A_2 之间的凝胶区带中富集蛋白成分，然后通过 SDS-PAGE 分离蛋白，找到了红细胞与溶血液样本中的差异蛋白，最后通过质谱(Q-TOF-MS)检测发现了 Prx-2，而且证明它与血红蛋白 A_1 和 A_2 共同存在。

6 "血红蛋白 A_2 现象"发现和研究的 36 年(1981—2017 年)

30 多年过去了，血红蛋白 A_2 现象的图像逐渐清晰起来。它告诉我们，在活体红细胞内各种血红蛋白不是孤立存在的，血红蛋白之间有相互作用、血红蛋白与 Prx-2 之间也有相互作用。已知，Prx-2 的功能是抗氧化，它对保护血红蛋白及红细胞膜免受氧化损伤具有非常重要的作用。这样一来，活体红细胞内的奥秘初步揭开，那就是，红细胞中存在一种由多个蛋白质组成的复合体，它既能完成血红蛋白的运氧功能，又能保护血红蛋白和红细胞膜免于氧化损伤。

从进化角度来看，血红蛋白 A_2(含 δ 珠蛋白链的血红蛋白)只出现于人类和类人猿(参见下图)，它是人类进化的标志物。"血红蛋白 A_2 现象"让我们看到了这个标志物的特殊功能。

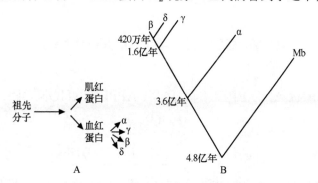

肌红蛋白、血红蛋白分子进化系统树

第十五章　血红蛋白 A_2 现象——Ⅰ. A_2 现象的发现及其初步应用

秦文斌　梁友珍

(包头医学院　血红蛋白研究室，包头　014010)

摘　要

1. 本文报告作者实验室所发现的一种自然现象，并暂时命名为"血红蛋白 A_2 现象"。

2. 所谓"血红蛋白 A_2 现象"，就是溶血液中 HbA_2 与红细胞中"HbA_2"的差异。这里主要指的是电泳行为不同。

3. 将正常成人红细胞与其本身的溶血液并排电泳(薄层淀粉胶电泳)时，二者的 HbA、碳酸酐酶电泳位置相同，只有 HbA_2 不一样。与溶血液中 HbA_2 相比，红细胞中的"HbA_2"电泳速度(向阳极)稍快。

4. 红细胞及溶血液中的 HbF、HbN 包头、HbD 包头及 HbE，都没有上述差异，而且它们对"HbA_2 现象"没有影响。

5. "HbA_2 现象"的机制尚待阐明，但已经发现了它的应用。众所周知，HbE 与 A_2 的电泳行为及层析性质是相同或相近的，鉴别起来比较困难，现在我们可以利用"HbA_2 现象"来区分这两种血红蛋白。如上所述，HbA_2 有"HbA_2 现象"，而 HbE 本身则没有这类现象。

HEMOGLOBIN A_2 PHENOMENON

—Ⅰ. DISCOVERY AND PRELIMINARY APPLICATION OF THIS PHENOMENON

QIN WEN-BIN　LIANG YOU-ZHEN

(Department of Hemoglobin, Baotou Medical College, Baotou 014010)

Abstract

1. A natural phenomenon discovered in our laboratory is here tentatively described as "Hemoglobin A_2 phenomenon".

2. HbA_2 of hemolysate differs from "HbA_2" of RBC in their electropheoretic behavior, this is the main feature of the "HbA_2 phenomenon".

3. When erythrocytes from normal adults were run parallel with their hemolysates in thin layer starch gel electrophoresis, it was found that HbA and carboanhydrase from the above two sources have identical electrophoretic position, only "HbA_2" of erythrocytes differs from HbA_2 of hemolysates, "HbA_2" moves faster to the anode than HbA_2.

4. HbF, HbN-Baotou, HbD-Baotou and HbE, do not show the above described difference between

erythrocytes and hemolysates, and do not show the "HbA_2 phenomenon".

5. The mechanism of "HbA_2 phenomenon" is not yet clarified, but its preliminary application has found. It is known that HbE has identical or similar electrophoretic behavior and chromatograpthic properties to HbA_2, but their differentiation is diffcult. Now, "HbA_2 phenomenon" may be used to distinguish HbE from A_2. As mentioned above, HbA_2 shous the "HbA_2 phenomenon", while HbE does not.

HbA_2 是 Kumkel 与 Wallenius 二人于 1955 年发现的[1]，它的等电点比较高，其含量约占正常成人血红蛋白总量的 2.5%，现在已经知道它的亚单位组成为 $α_2δ_2$，δ 链的一级结构也已清楚[2]。δ 链的氨基酸组成及顺序与 β 链非常相似(146 个残基中只有 10 个不同)[2]，而且类似 HbA_2 成分似乎只出现于高级灵长类[3]，所以人们将 δ 链的起源放在进化程序的末尾[4]。过去曾经有人报道，HbA_2 对氧的亲和力高[5]，最近 Debruin 等证明它的亲和力与 HbA 相同[6]，这种次要血红蛋白成分的存在意义尚不清楚。

我们在自己的实验过程中注意到 HbA_2 的一些特殊行为，为了探讨方便暂称其为"血红蛋白 A_2 现象"，初步估计，由这一现象出发深入研究下去，也许有助于我们弄清 HbA_2 存在的特殊意义。

本文所说的"血红蛋白 A_2 现象"，主要是指以下内容：当用红细胞与其本身的溶血液并排进行电泳时，二者的 HbA 泳动位置相同，二者碳酸酐酶(以下简称为 CA)的电泳行为也一样，只有 HbA_2 特殊，二者的这个成分不在同一位置。与溶血液的 HbA_2 比较，红细胞中没有与其相对应的成分。在许多情况下，是在相当于溶血液 HbA_2 的稍前方(阳极侧)出现一个红色区带，而且它不如 HbA_2 整齐。为了叙述方便，我们暂将这个红色成分称为红细胞 HbA_2 或"HbA_2"。此时还能看到，HbA 与"HbA_2"之间的拖尾现象相当明显。为了观察这个现象的客观性，我们比较研究了正常成人、HbA_2 增多的β-地中海贫血患者、HbF 增多的血液病患者及各种异常血红蛋白杂合体的血样。结果发现，这一现象是普遍存在的。在此项观察中还发现"HbA_2 现象"能帮助鉴别 HbA_2 和 HbE，这样也就逐渐给这个自然现象找到了应用。

1 材料及方法

1.1 血样的来源
(1) 正常成人血样：来源于包头市血站。
(2) β-地中海贫血杂合体血样：来自广西平南县人民医院送检患者。
(3) HbF 增多患者血样：来自我室检查的患者血。
(4) HbDβ包头杂合体血样：来自我室以前发现的此杂合体者[7]。
(5) HbE 杂合体血样：来自广西平南县人民医院送检患者。
(6) HbNβ包头杂合体血样：来自我室以前发现的此杂合体者[8]。

1.2 红细胞的收集
用生理盐水洗红细胞三次，每次离心后吸除一部分上层细胞成分(其中含较多的白细胞)，最后向红细胞中加入等体积生理盐水，混匀备用。

1.3 溶血液制备
取一部分上述红细胞生理盐水液，加入 1/4 体积四氯化碳，振荡约 200 次，离心(3000r/min，20 分钟)，取上层溶血液备用。

1.4 薄层法淀粉胶电泳
方法基本同前[9]。制胶及电极槽中都用 pH 9.0 的 TEB 缓冲液，条件同 Efremov 等的方法[10]。电泳时使用 20cm×10cm 玻板，刀片切口，插滤纸加样，通电时电势梯度 8V/cm，时间 4～6 小时。

1.5 染色
氨基黑 10B 法。

2 结果

2.1 正常成人血样的"HbA_2 现象" 将正常成人的红细胞与其本身的溶血液并排做淀琼胶电泳，染色后结果见图 15-1。由此图可以看出，样品为溶血液时原点(加样处)没有遗留蛋白成分。在原点的阳极侧按泳速快慢为序，可以看到 HbA、HbA_2 及 CA 三个区带。如果不染色只能看到两个红色区带(即 HbA 与 A_2)，HbA_1 表现为快泳扩散成分，这里不专门讨论。样品为红细胞时，也可以看到相应的 HbA 和 CA，但有以下两点和溶血液不同：①没有与溶血液 HbA_2 相对应成分，但在其稍前方(阳极侧)有一蛋白成分，未染色时为红色。这个成分还有一个特点，它往往不易形成规整的区带。本文将此成分暂称为红细胞 HbA_2 或称"HbA_2"，以示与溶血液 HbA_2 的区别。②"HbA_2"与 Hb 之间的拖尾比较明显，而溶血液中 HbA 与 HbA_2 之间此现象就不太显著。

2.2 HbA_2 增多时的"HbA_2 现象" 将β-地中海贫血杂合体患者的红细胞与其本身的溶血液(与正常溶血液比较已证明 HbA_2 增多者)并排电泳，也看到同上"HbA_2 现象"，因其结果与 HbE 者放在一起，故参见后述图 15-5。

2.3 HbF 增多时的"HbA_2 现象" 将怀疑再生障碍性贫血而证明 HbF 增多的患者红细胞与其本身的溶血液并排电泳，再配合上正常对照，得到图 15-2 结果。由此图可以看出，正常人红细胞及其溶血液的结果与第一项中描述者相同。HbF 增多者，其红细胞与溶血液比较：HbA、HbF 及 CA 都在相同位置，仍然是"HbA_2"靠前、HbA_2 靠后。此时，正常人与 HbF 增多患者"HbA_2"的电泳行为相同。

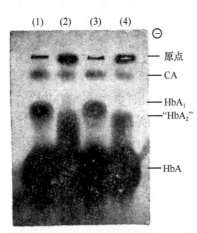

图 15-1 正常成人样品的"HbA_2"现象

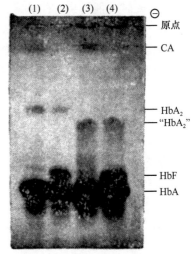

图 15-2 HbF 增多时的"HbA_2 现象"

2.4 快泳异常 Hb 存在时的"HbA_2 现象" 我们用 HbN 包头进行此项实验。将 HbN 包头杂合体的红细胞与其本身的溶血液并排电泳，其结果如图 15-3 所示。由此图可以看出，此时仍然是"HbA_2"快于 HbA_2，而其他成分没有区别。特别应当提到的是 HbN 包头本身，它的红细胞与溶血液结果也相同。

2.5 慢泳异常 Hb 存在时的"HbA_2 现象" 我们用 HbD 包头进行此项实验。将 HbD 包头杂合体的红细胞与其本身的溶血液并排进行电泳分析，得到图 15-4 结果。由此图可以看出，HbA、HbD 及 CA 的电泳位置都一样，仍然是 HbA_2 与"HbA_2"不同。

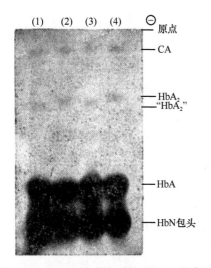

图 15-3 HbN 包头存在时的"HbA$_2$ 现象"

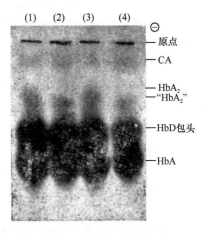

图 15-4 HbD 包头存在时的"HbA$_2$ 现象"

2.6 泳速与 HbA2 相同异常 Hb 存在时的"HbA$_2$ 现象"
此项实验用 HbE 进行。将 HbE 杂合体的红细胞与其本身的溶血液并排电泳，同时配合上 β-地中海贫血杂合体(HbA$_2$ 增多)的相应样品，得到图 15-5 结果。由此图可以看出，β-地中海贫血患者的"HbA$_2$ 现象"与正常人相同，HbA$_2$ 与"HbA$_2$"也不在一个位置。对 HbE 杂合体来说"HbA$_2$ 现象"不好观察。首先是溶血液中 HbA$_2$ 与 HbE 共同存在，其次是红细胞中"HbA$_2$"也没有与 HbE 分开。但是，有一点必须指出，那就是二者的 HbE 电泳位置相同。也就是说，通过溶血液与红细胞对比，可以看出 HbA$_2$ 与 HbE 的差异：HbA$_2$ 在用红细胞或溶血液做电泳分析时表现出不同行为，而 HbE 则没有这种情况。

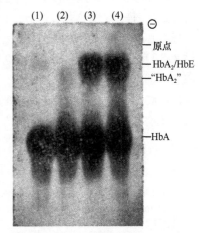

图 15-5 HbE 存在时的"HbA$_2$ 现象"

3 讨论

人类及所有脊椎动物红细胞中都含有血红蛋白，这早已是众所周知的事情。但是，这项知识几乎都是来源下述操作，即用某种手段使红细胞破坏(溶血)；然后用各种方法来分析溶血液中的血红蛋白，从而得知各种性质，其中包括电泳行为等。这种电泳行为是否就是红细胞中血红蛋白的真实反应？不易答复。

过去也有人做过"单个细胞中血红蛋白的电泳分析"[11]，但在该法中也进行了溶血处理。本文所用的方法是将红细胞悬浮(生理盐水)液加到半固体的凝胶支持体中，不经溶血处理，直接进行电泳分析。单个红细胞还有一个不方便之处，它很难与溶血液进行并排电泳，不易看出二者的异同。本章方法可与溶血液平行比较，这样才发现了"HbA$_2$ 现象"。

所谓"HbA$_2$ 现象"，最简单的含义就是红细胞与其本身溶血液并排电泳时 HbA$_2$ 在电泳行为方面的差异。这种差异，最主要表现为 HbA$_2$ 与"HbA$_2$"泳速不同，而且往往是"HbA$_2$"比 HbA$_2$ 快。此外，还能看到用红细胞进行电泳时 HbA 与"HbA$_2$"间的拖尾更为明显。

现在看来,"HbA$_2$ 现象"是客观存在的,它不仅存在于正常成人,也存在于已检查过的各种患者。正常成人红细胞中有 HbA 及 CA 时可以看到这种现象,HbA$_2$ 增多(β-地中海贫血)、HbF 增多(怀疑再生障碍性贫血)时这种现象仍然存在。甚至血中含有异常血红蛋白时,包括快泳(HbN 包头)及慢泳变异物(HbD 包头及 HbE),也都不影响"HbA$_2$ 现象"的出现。与 HbA$_2$ 不同,HbA、HbF、HbN 包头、HbD 包头、HbE 及 CA 都没显示出红细胞结果与溶血液者有何差异。这些结果说明 HbA$_2$ 的含量虽少(正常成人时仅为血红蛋白总量的 2.5%左右),但可能有其特殊作用和性质。从这个线索出发,追踪下去,也许能够给我们带来新的认识。

为什么红细胞的"HbA$_2$"比溶血液的 HbA$_2$ 泳速快?这是一个非常有趣的问题。对于这种现象可以有各种推测,并且需要进一步证实,这里不准备深入讨论。但是,可以认为这一现象表明完整红细胞中的情况可能要比溶血液更为复杂,它也许会帮助我们了解到一些由溶血液所不能了解到的情况,它可能帮助我们解决一些由溶血液所不能解决的问题。例如,HbE 和 HbA$_2$,一般认为,它们的电泳及层析行为都非常相近甚至相同,区别起来很困难。在这种情况下,人们采用 HbA$_2$ 电泳成分的含量(%)来判定其为 HbA$_2$ 或 HbE。例如,有人主张,当此处的 Hb 成分超过 10%时就可以认为是 HbE,10%以下者为 HbA$_2$[12]。但是,最近有人报告一例地中海贫血患者,其 HbA$_2$ 竟达 15%之多[13],可见靠百分浓度来确定 HbE 或 HbA$_2$ 是不够严格的。所以,Lehmann 也曾指出,遇到疑难病例时,鉴别这两种 Hb 的唯一有效方法还是"指纹技术"[12]。现在看来,"HbA$_2$ 现象"给我们提供了一种简便方法,它可以比较容易地区分开 HbE 和 HbA$_2$,正如图 17-5 所示,若是 HbA$_2$,就有"HbA$_2$ 现象";如果是 HbE,就没有这类现象,即红细胞电泳中的 HbE 和 HbA$_2$,随着工作的开展,今后也许会发现它有更多的用途。

本文没有涉及"HbA$_2$ 现象"的机制,这一问题正在研究中。

参 考 文 献

[1] Kunkel H G, Wallenius G. New hemoglobins in normal adult blood. Science, 1995, 122: 288.
[2] Ingram V M, Stretton AOW. Human hemoglobin A$_2$: chemistry, genetics and evolution. Nature, 1961, 190: 1079.
[3] Kunkel H G, Ceppellini R, Muller-eberhard U, et al. Observations on the minor basic hemoglobin component in the blood of normal individuals and patients with thalasees mia. J Clin Invest, 1957, 36(11): 1615.
[4] Bunn, H. F. et al. Human Hemoglobins, 1977, 18, W. B. Saunders Company, Philadelphia.
[5] Meyering, C. A.et al. Studies on the heterogeneity of hemoglobin. Clin. Chim, Acta, 1960, 5: 208.
[6] DeBruin S H, Janssen L H M. Comparison of the oxygen and proton binding behavior of hemoglobin A and A$_2$. Biochim Biophys Acta, 1973, 295: 490.
[7] 秦文斌, 等. 我国北方汉族中相对多见的一种慢泳异常血红蛋白-HbDβ包头. 第三次中国生物化学学术会议论文摘要汇编, 1979: 21.
[8] 秦文斌, 等. 包头地区汉族中又发现一种快泳异常血红蛋白-HbNβ包头. 第三次中国生物化学学术会议论文摘要汇编, 1979: 22.
[9] 秦文斌, 等. 改进淀粉胶薄层电泳法及其对血红蛋白不均一性的分辨能力. 生物化学与生物物理学报, 1965, 5: 278.
[10] Efremov G D, Huisman TH, Smith LL, et al. Hemoglobin Richmond, a human hemoglobin which forms asymmetric hybrids with other hemoglobins. J Bid Chem, 1969, 244: 6105.
[11] Matioli G, Niewisch H. Electrophoresis of hemoglobin in single erythrocytes.Science, 1965, 150: 1824.
[12] Lehmann H, Huntsman BG.Man's Haemoglobin. Blut, 1967, 14(4): 255.
[13] Vella F. Variation in hemoglobin A$_2$.Hemoglobin, 1977, 1: 619.

[原文发表于"生物化学与生物物理学报,1981,13(2):199-205"]

第十六章 血红蛋白 A_2 现象——II. α链异常血红蛋白的 A_2 现象及其意义

闫秀兰 秦文斌

(包头医学院 血红蛋白研究室,包头 014010)

摘 要

1. 在过去研究的基础上,本文报告α链异常血红蛋白所表现的 A_2 现象的特点。

2. 将α链异常血红蛋白杂合体红细胞(悬浮液)与其本身的溶血液并排电泳时,二者的血红蛋白 A 及异常血红蛋白的主要成分($\alpha_2^X\beta_2^A$)的电泳位置相同,而红细胞的"HbA_2"及异常血红蛋白的次要成分($\alpha_2^X\delta_2^A$)则与溶血液者不同。此时,红细胞的"HbA_2"及"$\alpha_2^X\delta_2^A$"靠前(阳极),溶血液的 HbA_2 及 $\alpha_2^X\delta_2^A$ 靠后(阴极)。

3. 本文共研究了 3 种α链异常血红蛋白:HbG-Taichung,HbQueens 及 HbI-包头,它们都有上述现象,快泳与慢泳变异物在这方面没有差异。

4. α链异常血红蛋白主要成分没有 A_2 现象,而其次要成分则有此现象,这一事实给鉴别α链变异物增添了新的知识。

关键词 血红蛋白 A_2;α链异常

HEMOGLOBIN A_2 PHENOMENON
—II. HbA_2 PHENOMENON OF α-CHAIN VARIANT AND ITS SIGNIFICANCE

Yan Xiu-Lan, Qin Wen-Bin

(Laboratory of Hemoglobin, Baotou Medical College, Baotou 014010)

Abstract

1. This paper describes the hemoglobin A_2 phenomenon manifested by α-chain mutants.

2. When erythrocyte suspensions from heterozygotes of α-chain mutants were run in parallel with their hemolysates in thin layer starch gel electrophoresis, it was found that HbA and the major component of abnormal hemoglobin($\alpha_2^X\beta_2^A$)from the above two sources have identical electrophoretic positions, but "HbA_2" and the minor component of abnormal hemoglobin("$\alpha_2^X\delta_2^{A_2}$")from erythrocytes differs from those of hemolysates: "HbA_2" and "$\alpha_2^X\delta_2^A$" move faster towards the anode than HbA_2 and $\alpha_2^X\delta_2^A$.

3. Three α-chain variants: HbG-Taichung, HbQueens and HbI_α-Baotou, have been studied. They all show the above phenomenon, there is no difference between fast and slow moving abnormal hemoglobins as far as this phenomenon is concerned.

4. The major components of α-chain variants did not show the "HbA_2 phenomenon", but the mi-

nor components did, this fact may be taken into consideration when differentiating the α–chain variants from those of the β–chain.

Key words HbA$_2$; α-chain abnormally

血红蛋白 A$_2$ 现象的发现及其在β链异常血红蛋白中的应用，前文已叙述[1]。为了继续寻找这一现象的出现规律，本文报道了α链异常血红蛋白杂合体红细胞与其本身溶血液之间的相应关系。当将这两种样品并排电泳时，我们看到α链异常血红蛋白的主要成分($\alpha_2^X\beta_2^A$)的电泳位置相同，而次要成分($\alpha_2^X\delta_2^{A_2}$)则表现出与血红蛋白 A$_2$ 相同的特殊现象。此项结果进一步丰富了我们关于血红蛋白 A$_2$ 现象的知识。它证明，不仅正常血红蛋白 A$_2$ 有此现象，其α链变异物仍保留着这些特点。$\alpha_2^X\delta_2^{A_2}$ 也有 A$_2$ 现象这一事实，又给α链异常血红蛋白次要成分增加了新的特性。

1 材料与方法

1.1 血样的来源 正常成人血样得自包头市血站。HbG-Taichung 杂合体血样来源于以前发现的此杂合体者[2]，HbQueens 杂合体血样也来源于以前发现的此杂合体者[3]，HbI-包头杂合体血样来源于最近发现的此杂合体者，其异常链的鉴定按 PCMB 法。

1.2 红细胞(悬浮液)及其溶血液的制备 方法同第十五章[1]。

1.3 电泳分析及染色 方法也同第十五章[1]。

2 结果

2.1 G-Taichung 杂合体血样的血红蛋白 A$_2$ 现象 由 HbG-Taichung 杂合体血样分出红细胞及其溶血液，并排进行薄层淀粉胶电泳，结果如图 16-1 所示。此图表明，血红蛋白 G-Taichung 杂合体溶血液中有 4 种血红蛋白成分：HbA，HbG$_\alpha$主要成分($\alpha_2^G\beta_2^A$)、HbA$_2$ 及 HbG$_\alpha$次要成分($\alpha_2^G\delta_2^{A_2}$)。众所周知，这是α链异常血红蛋白的共同特点。由此图还可以看出，与溶血液对应，红细胞(悬浮液)的电泳结果中也出现 4 种血红蛋白成分。其中 HbA 及 $\alpha_2^G\beta_2^A$ 与溶血液者电泳位置相同，而其 HbA$_2$ 及 $\alpha_2^G\delta_2^{A_2}$ 则与溶血液电泳行为互异。此时 HbA$_2$ 结果同一般血红蛋白 A$_2$ 现象。值得注意的是，HbG$_\alpha$的次要成分 $\alpha_2^G\delta_2^{A_2}$ 也显示出"HbA$_2$ 现象"。二者的关系也是红细胞"$\alpha_2^G\delta_2^{A_2}$"靠前(阳极)，溶血液的 $\alpha_2^G\delta_2^{A_2}$ 靠后(阴极)。

2.2 Queens 杂合体血样的血红蛋白 A$_2$ 现象 实验安排同上，只是样品改用 HbQueens 杂合体血样，具体结果如图 16-2 所示。此图表明，实验结果与 HbG-Taichung 类似，即 $\alpha_2^{Qu}\beta_2^A$ 及 HbA 没有 A$_2$ 现象，HbA$_2$ 及 $\alpha_2^{Qu}\delta_2^{A_2}$ 表现出红细胞与溶血液结果不同。

2.3 α-包头杂合体血样的血红蛋白 A$_2$ 现象 前述两种异常血红蛋白都是慢泳α链变异物。为了观察不同泳速α链异常血红蛋白的 A$_2$ 现象，我们又用α-包头杂合体血样进行同上实验，得到图 16-3 结果。此图与图 16-1、图 16-2 不同之处在于，溶血液中异常血红蛋白的主要成分(即 $\alpha_2^I\beta_2^A$)位于 HbA 的前方(阳极)、其次要成分($\alpha_2^I\delta_2^{A_2}$)也在 HbA$_2$ 的前方。将其红细胞结果与溶血液者对比，二者的 $\alpha_2^I\beta_2^A$ 及 HbA 电泳位置相同，不同的仍是 $\alpha_2^I\delta_2^{A_2}$ 及 HbA$_2$。不同之点还有，来自红细胞的"$\alpha_2^I\delta_2^{A_2}$"靠前(阳极)，来自溶血液的 $\alpha_2^I\delta_2^{A_2}$ 靠后(阴极)，对应关系与一般 HbA_2 现象相同。

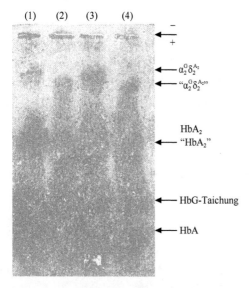 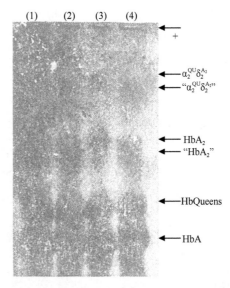

图 16-1　HbG-Taichung 杂合体血样的 HbA_2 现象
注释：(1)，(3)泳道为溶血液；(2)，(4)泳道为红细胞

图 16-2　HbQueens 杂合体血样的 HbA_2 现象
注释：(1)，(3)泳道为溶血液；(2)，(4)泳道为红细胞

3　讨论

本文结果表明，α链变异物与β链者不完全一样。这类异常血红蛋白的主要成分 $\alpha_2^X\beta_2^{A_2}$ (本文中的 $\alpha_2^G\beta_2^{A_2}$、$\alpha_2^{Qu}\beta_2^A$ 及 $\alpha_2^I\beta_2^A$)与β链变异物($\alpha_2^A\beta_2^X$)相同，都没有 HbA_2 现象。但是，α链变异物的次要成分 $\alpha_2^X\delta_2^{A_2}$ (本文中的 $\alpha_2^G\delta_2^{A_2}$、$\alpha_2^{Qu}\delta_2^{A_2}$ 及 $\alpha_2^I\delta_2^{A_2}$)则表现出红细胞与溶血液之间的差异。也就是说，这些血红蛋白成分有 HbA_2 现象。

前文报道，正常成人血红蛋白中 HbA_2 可表现出红细胞与溶血液之间差异(电泳行为)[1]。现在，我们又看到α链异常血红蛋白的次要成分 $\alpha_2^X\delta_2^{A_2}$ 也有这种现象。众所周知，$\alpha_2^G\delta_2^{A_2}$、$\alpha_2^{Qu}\delta_2^{A_2}$ 及 $\alpha_2^I\delta_2^{A_2}$ 都属于 $HbA_2(\alpha_2^A\delta_2^{A_2})$ 的α链变异物。它们都表现出 HbA_2 现象，说明 HbA_2 中α^A 变成α^G、α^{Qu} 或α^I 时并没有影响到产生 HbA_2 现象。

α链异常血红蛋白有一系列特点[4-6]。例如，它有主要与次要两个异常成分；它们的电泳位置与 HbA 及 HbA_2 相对应；其主要异常成分的含量约为 HbA 的 1/3，次要成分约为 HbA_2 的 1/3 等。现在根据我们的实验结果，可以再增加一项新的特点：其

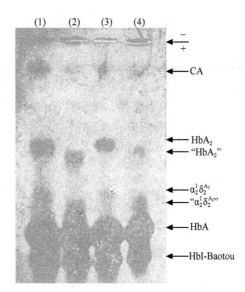

图 16-3　α-包头杂合体血样的 HbA_2 现象
注释：(1)，(3)泳道为溶血液；(2)，(4)泳道为红细胞

次要异常成分有 HbA$_2$ 现象，而其主要异常成分则没有这种情况。

血红蛋白 A$_2$ 现象的产生机制尚不清楚，但根据现有证据可以初步归纳出如下规律：①这种现象可能与δ链有关。因为迄今结果表明，$\alpha_2^A\beta_2^A$、$\alpha_2^A\beta_2^X$ 及 $\alpha_2^X\beta_2^A$ 都没有 HbA$_2$ 现象，而有此情况的则是 $\alpha_2^A\delta_2^{A_2}$ 及 $\alpha_2^X\delta_2^{A_2}$。②此现象可能与 HbA$_2$ 的正电荷较多无关。已知 HbA$_2$ 比 HbA 少大约 4 个负电荷，所以电泳时它移向 HbA 的阴极侧。在完整红细胞中的 HbA$_2$，是否会与其中或膜上负电性较强成分相互作用，从而加快这种"HbA$_2$"泳向阳极的速度(即再现 A$_2$ 现象)？现在看来，这种可能性比较小，初步理由有二：①与 HbA$_2$ 带电情况相同或十分相近的异常血红蛋白——HbE($\alpha_2^A\beta_2^E$)并没有 A$_2$ 现象[1]；②与 HbA$_2$ 带电情况不同的一些变异物——$\alpha_2^G\delta_2^{A_2}$、$\alpha_2^{Qu}\delta_2^{A_2}$ 及 $\alpha_2^I\delta_2^{A_2}$ 都有这种现象。

总之，血红蛋白 A$_2$ 现象的机制尚不清楚，初步认为，它可能与δ链有关、与 HbA$_2$ 的电荷多少可能关系不大，其他规律和影响因素尚在继续研究中。

参 考 文 献

[1] 秦文斌, 梁友珍. 血红蛋白A$_2$现象Ⅰ. 此现象的发现及其初步意义. 生物化学与生物物理学报, 1981, 13: 199.
[2] 秦文斌等. 血红蛋白 G-Taichung. 新医学, 1985, 16: 30.
[3] 睢天林等. 血红蛋白 Queens.包头医学院学报, 1984, 1: 8.
[4] Lehmann H, Carrel RW. Differences between α and β chain mutants of human hemoglobin and between α and β thalassemia. Possible duplication on of the α chain gene. Brit Med J, 1968, 4: 748.
[5] Wasi P. Is the human α chain locus duplicated? Brit J Hematol, 1978, 5: 264.
[6] 秦文斌等. 蒙古族牧民中发现两种 D 型异常血红蛋白. 遗传学报, 1978, 4: 263.

[原文发表于"生物化学与生物物理学报，1989，21(1): 1-5"]

第十七章 血红蛋白 A_2 现象——Ⅲ.δ链异常血红蛋白的 A_2 现象及其意义

秦文斌 雎天林 秦良伟

(包头医学院 血红蛋白研究室, 包头 014010)

关键词 血红蛋白 A_2; A_2 现象; δ 链异常血红蛋白; 红细胞; 溶血

The Hemoglobin A_2 Phenomenon
— Ⅲ. HbA_2 Phenomenon, of δ-Chain Variant and Its Significance

Qin Wen-bin, Ju Tian-lin and Qin Liang-wei

(Laboratory of Hemoglobin, Baotou Medical College, Baotou 014010)

Abstract

1. On the basis of our previous studies, this paper reports the characterisides of the A_2 phenomenon manifested by hemoglobin with δ-chain abnormality.

2. When electrophoresis of erythrocytes(suspension)from heterozygote of δ-chain variant were run parallelly with its hemolysate, it was found that HbA from the above sources, occupy identical electrophoretic position, but "HbA_2" and "HbA_2-X" of erythrocyte differ from HbA_2 and HbA_2-X of the hemolysate. "HbA_2" and "HbA_2-X" move to the anode faster than HbA_2 and HbA_2-X. Their differences are quite clear.

3. δ-Chain variants from 3 different families(HbA_2-Liangcheng, HbA_2-Shaihantala and HbA_2-Xishu)were studied in this paper, the former is of Han nationality and the latter two of Mongol nationality. They all manifest the above described phenomenon, no nationality differences were found.

Key words HbA_2; A_2 phenomenon; δ-Chain variant; Erythrocyte; Hemolysate

1 前言

本文报告 δ 链异常血红蛋白的血红蛋白 A_2 现象。当这种血红蛋白携带者的红细胞(悬浮液)与其本身的溶血液并排电泳时,我们看到,不仅正常的血红蛋白 A_2 有 A_2 现象,δ 链异常血红蛋白也有这种情况。这一结果,使得我们对于血红蛋白 A_2 现象的认识又深入一步。

δ 链异常血红蛋白也有 A_2 现象,这一事实给我们提供一种区别 δ 与 β 链的新的特异方法。根据我室以前研究,β 链异常血红蛋白没有 A_2 现象,现在证明 δ 链变异物有此现象。这样,如果已经知道某种异常血红蛋白不是 α 链变异物,就可能使用本法进一步明确它是何种链异常。最低限度，可以将非α链异常血红蛋白区分为 A_2 现象阳性和阴性两种。

2 材料和方法

2.1 血样的来源 正常成人血液样品来源于包头市血站；血红蛋白 A_2-凉城样品来源于以前发现的汉族家系[1]；血红蛋白 A_2-西苏及血红蛋白 A_2-赛汗塔拉的杂合体样品来源于以前发现的蒙古族家系[2]。

2.2 红细胞(悬浮液)及其溶血液的制备 方法同第十五章[3]。

2.3 电泳分析及染色 方法同第十五章[3]。

3 结果

3.1 血红蛋白 A_2-凉城杂合体样品的 A_2 现象 由血红蛋白 A_2-凉城杂合体血样分出红细胞及其溶血液，并排进行凝胶电泳，结果如图 17-1 所示。由此图的溶血液部分可以看出，血红蛋白 A_2-凉城的电泳位置，是在血红蛋白 A 与 A_2 之间，稍微靠前(阳极)，相当于血红蛋白 G 的位置，只是它的含量非常少。由图中结果来看，血红蛋白 A_2-凉城的含量与血红蛋白 A_2 相近。由此图的红细胞部分可以看出，除了已知血红蛋白 A_2 有 A_2 现象外，血红蛋白 A_2-凉城也有这种现象，即溶血液中的血红蛋白 A_2-凉城靠近后方(阴极)、红细胞中"血红蛋白 A_2-凉城"靠近前方(阳极)，二者不在一个水平线上。与此相反，血红蛋白 A 没有这种情况。

3.2 血红蛋白 A_2-赛汗塔拉杂合体血样的 A_2 现象 用这种异常血红蛋白的杂合体血样，重复上述实验，得到图 17-2 结果。由此图可以看出，血红蛋白 A_2-赛汗塔拉与血红蛋白 A_2-凉城的电泳位置相近似，也是在血红蛋白 A 与 A_2 之间，稍微靠近前方(阳极)，只是本例为蒙古族、前例为汉族，二者来源于不同的民族，没有血缘关系。此时，也是血红蛋白 A_2

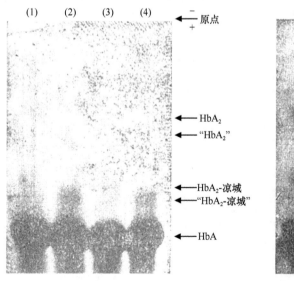

图 17-1 血红蛋白 A_2-凉城杂合体样品的 HbA_2 现象
注释：(1)，(3)泳道为溶血液；(2)，(4)泳道为红细胞

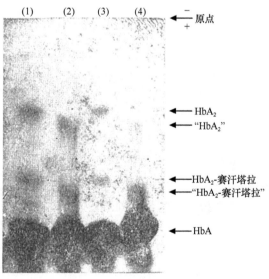

图 17-2 血红蛋白 A_2-赛汗塔拉杂合体样品的 HbA_2 现象
注释：(1)，(3)泳道为溶血液；(2)，(4)泳道为红细胞

及血红蛋白 A_2-赛汗塔拉有 A_2 现象，血红蛋白 A 没有这种现象。此样品中血红蛋白 F 稍明显，位于血红蛋白 A 与血红蛋白 A_2-赛汗塔拉之间(图中未标出)，但是它也没有明显的 A_2 现象。

3.3　血红蛋白 A_2-西苏杂合体血样的 A_2 现象　血红蛋白 A_2-西苏发现于另外一个蒙古族家系，与血红蛋白 A_2-赛汗塔拉没有血缘关系。此样品的实验结果与前二者相似，这一异常血红蛋白也有血红蛋白 A_2 现象(图略)。

4　讨论

我们发现，正常成人血液样品电泳时出现血红蛋白 A_2 现象[3]，并且比较研究了 β 链异常及 α 链异常血红蛋白的相应情况。结果发现，β 链变异物本身并无血红蛋白 A_2 现象[3]，即红细胞及其溶血液之间这类血红蛋白的电泳行为没有差别。α 链变异物与 β 链者略有不同，其主要成分 $\alpha_2^X\beta_2^A$ 没有 A_2 现象，但它的次要成分 $\alpha_2^X\delta_2^{A_2}$ 则有这种现象，并由此提出一种可供鉴别 α 链异常的新指标[4]。

血红蛋白 A_2 的肽链组成为 $\alpha_2^A\delta_2^{A_2}$，α 链异常血红蛋白的次要成分是 $\alpha_2^X\delta_2^{A_2}$，现在二者都有 A_2 现象，自然会想到：δ 链异常血红蛋白 $\alpha_2^A\delta_2^X$ 是否也有这种现象？为了解决这个问题，我们研究了现有的 3 个 δ 链异常血红蛋白：在呼和浩特南部地区汉族中发现血红蛋白 A_2-凉城[1]，以及在锡林郭勒盟蒙古族中发现的血红蛋白 A_2-赛汗塔拉和血红蛋白 A_2-西苏[2]。

本章实验结果表明，这 3 种 δ 链异常血红蛋白都有血红蛋白 A_2 现象。也就是说，和我们预先推测的一样，这类变异物的电泳行为在红细胞与溶血液之间都有着明显的差异。这一发现，使我们对于血红蛋白 A_2 现象的认识又前进一步。

根据我们现在的工作，可以将成人血红蛋白就其 A_2 现象进行分类，具体如图 17-3 所示。

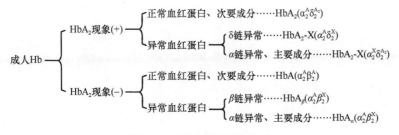

图 17-3　成人血红蛋白分类

由这一分类情况可以看出，A_2 现象阳性的成人血红蛋白，都含有 α 链和 δ 链，都属于血红蛋白 A_2 这一系统。此项发现，一开始就命名为"血红蛋白 A_2 现象"，现在看来，这种提法还是比较合适的。

参 考 文 献

[1] 秦文斌, 睢天林, 崔珊娜, 等. 在内蒙古凉城地区汉族中发现一种 δ 链异常血红蛋白——血红蛋白 A_2 凉城. 第四次中国生物化学学术会议论文摘要汇编, 1981, 37.

[2] 秦文斌等. 在内蒙古西苏旗蒙古族中发现两例 δ 链异常血红蛋白——血红蛋白 A_2-西苏及血红蛋白 A_2-赛汗塔拉. 第四次中国生物化学学术会议论文摘要汇编, 1981: 38.

[3] 秦文斌, 梁友珍. 血红蛋白 A_2 现象 I. 此现象的发现及其初步意义. 生物化学与生物物理学报, 1981, 13: 199.

[4] 闫秀兰, 秦文斌. 血红蛋白 A_2 现象 II. α-链异常血红蛋白的 A_2 现象及其意义. 生物化学与生物物理学报, 1989, 21: 1.

[原文发表于"生物化学与生物物理学报,1991,23(1): 88-89"]

第十八章　红细胞外 HbA$_2$ 与 HbA 间的相互作用

秦文斌

(包头医学院　血红蛋白研究室，包头　014010)

摘　要

1. 为了阐明血红蛋白 A$_2$ 现象的发生机制，研究了红细胞外 HbA$_2$ 与 HbA 有无相互作用。结果表明，这些血红蛋白在离开红细胞后仍能发生相互作用。

2. 我们是用交叉电泳来验证这个问题，主要实验结果如下所示。

(1) 单向交叉电泳结果表明，溶血液 HbA 穿过另一溶血液的 HbA$_2$ 时，出现明显的交叉电泳图像。

(2) 双向交叉电泳实验使上述结果更加明确。

(3) 用提纯的 HbA 做实验，证实了 HbA 与 HbA$_2$ 间的相互作用。

(4) 血浆白蛋白穿过 HbA$_2$、HbA 及 CA 时，没有看到交叉电泳图像。说明 HbA 与 HbA$_2$ 之间的作用是特异的。

3. 初步结论，不仅在红细胞中 HbA$_2$ 是与 HbA 结合存在，就是在离开红细胞的条件下，这两种血红蛋白之间仍能发生相互作用。我们认为，这很可能就是血红蛋白 A$_2$ 现象的发生机制

关键词　交叉电泳；血红蛋白 A$_2$；血红蛋白 A 相互作用；血红蛋白 A$_2$ 现象

Extra-cellular Interaction Between HbA$_2$ and HbA

Qin Wen-bin

(Laboratory of Hemoglobin, Baotou Medical College, Baotou 014010)

Abstract

1. Extra-cellular interaction between HbA$_2$ and HbA was studied in order to clarify the mechanism of hemoglobin A$_2$ phenomenon. The results show that these hemoglobin can still interact on each other after they were removed from erythrocyte.

2. Cross electrophoretic technique was used to test and verify this problem and the following results were obtained:

(1) One-dimensional cross electrophoretic results showed that an obvious "special pattern" was observed during electrophoresis of hemolysate HbA cross that of HbA$_2$.

(2) The above results were confirmed by two-dimensional cross electrophoretic experiment.

(3) The interaction between HbA$_2$ and HbA was finally verified by cross electrophoretic experiment with purified HbA.

(4) No such cross pattern was observed while plasma albumin passed through a mixture of HbA$_2$, HbA and CA. This result showed that the interaction is specific for HbA$_2$ and HbA.

3. Preliminary conclusion: HbA$_2$ can combine with HbA not only in the red blood cell, but they can also interact on each other when they were removed from the erythrocyte. We believe that this is most probably the mechanism of hemoglobin A$_2$ phenomenon.

Key words　　Cross electrophoresis; HbA$_2$; HbA; Hemoglobin A$_2$ phenomenon

1　前言

我们在自己的科学实践过程中发现了血红蛋白 A$_2$ 现象[1]，并且对它的出现规律正在进行系列研究[1-3]。在探讨这种现象的机制时，作者又注意到"红细胞 HbA$_2$"在化学组成上的不均一性，证明在红细胞中 HbA$_2$ 与 HbA 结合、二者之间有相互作用[4]。本文将进一步证明，在离开红细胞的条件下，HbA$_2$ 与 HbA 也能发生相互作用。使我们对血红蛋白 A$_2$ 现象的认识更趋完善。为了在红细胞外证明血红蛋白之间的相互作用，我们采用了交叉电泳技术。先用溶血液做单向和双向交叉电泳，看这种情况下 HbA$_2$ 与 HbA 有无相互作用。然后，用提纯的 HbA 做实验，排除其他可能的干扰。最后，用血浆白蛋白与溶血液做交叉电泳实验，考虑一下上述作用的特异性。这一系列实验都证明，HbA$_2$ 与 HbA 可以在没有红细胞的条件下发生相互作用，而且这种作用是特异的。

2　材料与方法

2.1　血样的来源　　正常成人血样得自包头市血站。
2.2　红细胞(悬浮液)以及其溶血液的制备　　方法同第十五章[1]。
2.3　电泳分析
(1) 淀粉-琼脂混合凝胶薄层电泳(简称淀琼电泳)：方法同第十五章[4]。
(2) 交叉电泳：方法同一般交叉电泳技术[5, 6]，支持体为上述淀琼凝胶。
2.4　染色方法及电泳结果的保存　　同第十五章[4]。

3　结果

3.1　HbA$_2$ 与 HbA 的单向交叉电泳　　淀琼电泳，样品都是溶血液，双排加样，办法如下(图 18-1)：前排(阳极侧)是一条较长的加样线，后排(阴极侧)是几条很短的加样线，互相拉开一定距离，以便让后排(图 18-1 中的 first sample origin)的 HbA 穿过前排(图 18-1 中的 second sample origin)的 HbA$_2$，以便观察二者之间有无相互作用。其他条件同前，结果如图 18-1 所示。

由此图可以看到，在相当于前排溶血液的 HbA$_2$ 附近出现波浪状(或者锯齿状)区带。凡是后排溶血液的 HbA 穿过前排溶血液 HbA$_2$ 的地方(即交叉处)都出现波峰，而没有 HbA 穿过的地方则相对地出现波谷。

3.2　HbA$_2$ 与 HbA 的双向交叉电泳　　实验条件基本同上，只是由单向改成双向。第一次电泳是在凝胶板右下角点状加样(溶血液)，通电可看到 HbA 和 HbA$_2$ 两个红色区带。第二次电泳时，转角 90°。在 HbA 与 HbA$_2$ 的一侧(阳极侧)放一很长的加样线，此线由第一次电泳时的原点(加样处)一直到 HbA 红色区带的前方，仍加溶血液。这样安排的目的，是想让第一次电泳时分出的 HbA 在第二次电泳时穿过第二次溶血液的 HbA$_2$，看它们之间有何情况发生。实验结果如图 18-2 所示。

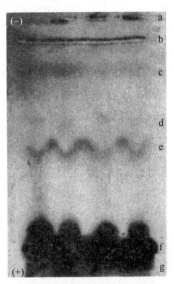

图 18-1　One-dimensional cross electrophores of hemolysate HbA_2 and HbA

Experimental details are given in the text:
Point for attention：The second sample HbA_2 shows sawtooth shape pattern owing the first sample HbA crossed it.
a：first sample origin
b：second sample origin
c：second sample CA
d：first sample HbA_2
e：second sample HbA_2
f：first sample HbA
g：second sample HbA

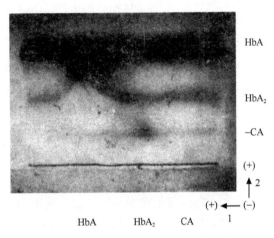

图 18-2　Two-dimensional cross electrophoresis of hemolysate HbA_2 and HbA

Experimental details are given in the text:
Point for attention：A peak-like pattern appeared when HbA of the first dimension crossed with HbA_2 of the second dimension.

由此图可以看到，第一次电泳分出的 HbA 与第二次分出的 HbA_2 接触时出现了交叉电泳所特有的"山峰"样图像。此图中，第一次电泳分出的 HbA 与第二次分出的 CA 也发生了接触。但是，没有看到峰样图像。

3.3　提纯 HbA_2 与 HbA 的单向交叉电泳　方法与条件同"3.1"，只是后排(阴极侧)样品不是溶血液，而是用淀粉板电泳提纯的 HbA。实验结果如图 18-3 所示。此图中，提纯的 HbA 穿过溶血液 HbA_2 时仍然出现了交叉电泳所特有的图像。进一步证实，是 HbA 本身与 HbA_2 发生了作用，和溶血液中其他成分无关。

3.4　白蛋白与溶血液成分的交叉电泳　实验条件同"3.3"，只是用医用血浆白蛋白代替 HbA，结果如图 18-4 所示。由此图可以看到，血清白蛋白穿过了溶血液中的所有主要蛋白质(HbA、HbA_2、CA)，没有出现特殊的交叉电泳图像。

4　讨论

血红蛋白 A_2 现象的发现及它与几种类型异常血红蛋白的关系，前文已述[1-3]。关于这一现象的发生机制，我们报告过红细胞中 HbA_2 与 HbA 相结合的情况[4]。前文中提到，红细胞 HbA_2 中两种 Hb 的联系并非十分密切，一些处理(制备溶血液，电泳过程等)比较容易使它们彼此分离。在红细胞中，会不会是这两种蛋白质胶粒无规律的互相"黏附"、从而可以随便分开？我认为不是这样。我们用各种异常血红蛋白所做的血红蛋白 A_2 现象实验，可以帮助说明问题[1-3]。这些异常血红蛋白，都与正常血红蛋白杂合在同一红细胞中。在用这些红细胞做实验时，并非都能看到血红蛋白 A_2 现象。从数量来看，绝大多数异常血红蛋白(β 链异常血红蛋白和 α 链异常血红蛋白的主要成分)的含量都明显多于 HbA_2，但它们与 HbA 之间没有"黏附"[1, 2]；与此相反，含量少的异常血红蛋白(δ 链异常血红蛋白和 α 链异常血红蛋白次要成分)却有血红蛋白 A_2 现象[2, 3]。另外，含量更少的 CA 又没有这种情况[1]。可见，这不是无规律的"黏附"，是一种特异性的相互作用。因为，到目前为止，所有有"血红蛋白 A_2 现象"的血红蛋白(包括正常和异常)都含有 δ 链：$\alpha_2^A \delta_2^{A_2}$ (HbA_2)、$\alpha_2^A \delta_2^X$ (δ 链异常 Hb)、$\alpha_2^X \delta_2^{A_2}$ (α 链异常 Hb 次要成分)；而不含 δ 链的 Hb($\alpha_2^A \beta_2^A$、$\alpha_2^A \beta_2^X$ 及 $\alpha_2^X \beta_2^A$)都没有血红蛋白 A_2 现象[1-3]。从电荷方面考虑，也是如此。HbE 的净电荷与 HbA_2 相同或者非常相近，但它没有血红蛋白 A_2 现象[1]；而一些净电荷与 HbA_2 不同者，只要含有 δ 链，就能表现出血红蛋白 A_2 现象[3]。这些结果也都说明，红细胞中 HbA_2 与 HbA 的结合并非由"黏附"造成的。还有下边要谈到的白蛋白，它与血红蛋白的分子量相近，都是球形分子、亲水胶体。如果是"黏附"在起作用，电泳中它通过 HbA_2、HbA，甚至通过 CA 时，都应当表现出来特殊图像。但是，实验结果并不支持这种设想。

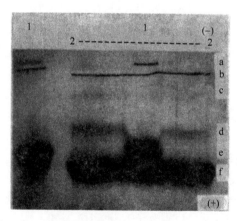

图 18-3　One-dimensional cross electrophores of pure HbA and hemolysate HbA_2

Experimental details are given in the text：
1=pure HbA；2= hemolysate.
A valley-like patten appeared when pure HbA crossed with hemolysate HbA：
a：first origin(pure HbA)
b：second origin(hemolysate)
c：hemolysate CA
d：hemolysate HbA_2
e：pure HbA
f：hemolysate HbA

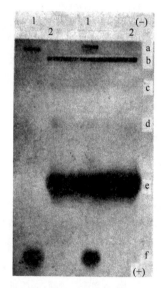

图 18-4　One-dimensional cross electrophores of albumin and hemolysate composition

Experimental details are given in the text.
1= albumin；2= hemolysate.
No：special patten appeared when albumin crossed with hemolysate CA, HbA_2 and HbA.
a：first origin(albumin)
b：second origin(hemolysate)
c：hemolysate CA
d：hemolysate HbA_2
e：hemolysate HbA
f：albumin

以上推理主要来自过去的实验，还没有特异性相互作用的更直接证明。本文的交叉电泳结果，在这方面又提供出新的、重要的可靠依据。图 18-1 结果说明，溶血液中 HbA 穿过 HbA_2 时可能发生二者之间的相互作用。当然，需要排除溶血液中非蛋白成分(泳速快于或稍慢于 HbA、也能穿过 HbA_2 者)的可能影响。图 18-2 的双向交叉电泳结果可以帮助解决这个问题。第一次电泳时 HbA 已经与其他成分分开，第二次电泳时只有 HbA 与 HbA_2 接触的地方出现特殊图像。可见，是 HbA 而不是其前后非蛋白成分在起作用。用提纯的 HbA 所做的实验进一步证实了这种想法，说明在 HbA 与 HbA_2 之间确实存在"相互作用"。这种相互作用是否普遍存在于各种蛋白质之间？前边讨论中 CA 没有血红蛋白 A_2 现象、单向交叉电泳中 HbA 穿过 CA 时也没有出现特殊图像，都已经说明了这个问题。但是，HbA 和 CA 都是红细胞中的蛋白质，是否会有局限性？为了避免这一点，我们又用医用血浆白蛋白进行实验，想用血浆蛋白中白蛋白泳速快的特点，使其穿过红细胞溶血液中的各种蛋白质，观察它的效果。图 18-4 结果表明，白蛋白与 HbA、HbA_2 及 CA 都没有发生"相互作用"。到此，可以得出结论：在我们所做过实验的范围内，HbA 与 HbA_2 的相互作用是特异的！

最后，可以认为，对于血红蛋白 A_2 现象发生机制的认识，又前进了一步。我们不仅证明红细胞中 HbA_2 与 HbA 结合存在，而且又证明脱离红细胞后这两种血红蛋白仍能发生相互作用。这些结果，使我们更加相信，HbA_2 与 HbA 不是孤立地存在于红细胞之中，而是通过特异的相互作用、比较疏松地结合在一起。

参 考 文 献

[1] 秦文斌, 梁友珍. 生物化学与生物物理学报, 1981, 13: 199.
[2] 闫秀兰, 秦文斌. 生物化学与生物物理学报, 1989, 21: 1.
[3] 秦文斌. 生物化学与生物物理学报, 1991, 23: 85.
[4] 秦文斌. 生物化学与生物物理进展, 1991, 18: 282.
[5] Laurell C B. Anal Biochem, 1965, 10: 35.
[6] Nakamura S. Elsevier Amsterdam, 1967.

[原文发表于"生物化学杂志, 1991, 7(5): 583-587"]

第十九章 血红蛋白 A_2 现象发生机制的研究
——"红细胞 HbA_2"为 HbA_2 与 HbA 的结合产物

秦文斌

(包头医学院 血红蛋白研究室，包头 014010)

摘 要

为了弄清血红蛋白 A_2 现象的发生机制，我们对"红细胞 HbA_2"的化学组成进行了分析。"红细胞 HbA_2"的双向电泳结果表明，它含有两种血红蛋白成分：一种相当于 HbA，另一种很可能是溶血液 HbA_2，其单向二次电泳结果也证明，它是由溶血液 HbA_2 和 HbA 所组成。结果初步说明，在红细胞中 HbA_2 可能与 HbA 结合存在，两者可能有相互作用，也许这是产生血红蛋白 A_2 现象的原因。

关键词 血红蛋白 A_2；血红蛋白 A；血红蛋白相互作用；血红蛋白 A_2 现象

MECHANISM OF HEMOGLOBIN A_2 PHENOMENON
—Ⅱ. "ERYTHROCYTE HbA_2" IS A BINDING PRODUCT OF HbA_2 AND HbA

Qin Wen-bin

(Laboratory of Hemoglobin, Baotou Medical College, Baotou 014010)

Abstract

1. Chemical composition of erythrocyte HbA_2 was analyzed in order to clarify the mechanism of hemoglobin A_2 phenomenon. The paper reports the hemoglobin composition of erythrocyte HbA_2.

2. The results of two-dimensional electrophoresis showed that erythrocyte HbA_2 contains two Hb components: one corresponds to HbA and the other most probably is hemolysate HbA_2.

3. The results of one-dimensional re-electrophoresis showed definitly that erythrocyte HbA_2 is composed of hemolysate HbA_2 and HbA.

4. Tentative conclusion: HbA_2 may be combined with HbA in erythrocyte, i.e. there is probably interaction between the two hemoglobins. This is perhaps the possible cause of hemoglobin A_2 phenomenon.

Key words HbA_2; HbA, interaction of Hb, HbA_2 phenomenon

1 前言

我们在科学实践中发现了血红蛋白 A_2 现象[1]，并且对它的出现规律进行了研究[1-3]。注意到"红细胞 HbA_2"在化学组成上的不均一性。根据血红蛋白 A_2 现象的一些特点，我们推测血红蛋白之间有相互作用的可能性。本文用普通淀琼凝胶电泳、双向电泳、单向二次电泳等技术证明"红细胞 HbA_2"中含有 HbA 及溶血液 HbA_2。这说明，在红细胞中 HbA_2 和 HbA 呈结合形式。我们认为，血红蛋白 A_2 现象的产生原因很可能是红细胞内 HbA_2 与 HbA 之间有相互作用，并非孤立存在。

2 材料与方法

2.1 血样 正常成人血样取自包头市血站。

2.2 红细胞(悬浮液)及其溶血液的制备 方法同第十五章[1]。

2.3 电泳分析

(1) 淀粉琼脂混合凝胶薄层电泳(简称淀琼电泳)：取淀粉和琼脂按 4∶1 混合，用 pH 9.0 的 TEB 缓冲液[4]配成 2.5%浓度制胶，趁热将 25ml 胶液倒在 10cm × 10cm 玻璃板上(改变玻璃板尺寸时，胶液量也相应增减)，待凝固。电极槽中放入 pH 9.0 硼酸-NaOH 缓冲液。其余条件同前。

(2) 双向淀琼脂电泳：加样、转向等操作同一般双向电泳，做第二向电泳时缓冲液不变。

(3) 单向二次电泳法：支持体仍为淀粉-琼脂混合凝胶。第一次电泳后，将某一区带及其所在凝胶留下待用，切掉其周围的所有凝胶及其所含区带。然后，在此周围处，再倒上新的凝胶。凝固后再电泳(缓冲液不变)，可以保持原来方向，或者改变方向。

2.4 染色方法 可以不染色，靠血红蛋白本身的红色来观察结果。必要时，采用氨基黑 10B 或者联苯胺法进行染色。

2.5 电泳结果的保存 淀琼电泳后氨基黑 10B 染色，室温晾干，电泳结果就被固定在玻璃板上，可以长期保存。染色后的淀琼凝胶，还可转到 X 线胶片上或者滤纸上晾干，这样保存起来更为方便。

3 结果

3.1 正常成人血样的"HbA_2 现象" 将正常成人红细胞与其溶血液并排电泳，然后染色，结果如图 19-1 所示。

由图 19-1 可以看出：样品为溶血液时，由原点(加样处)向阳极按泳速快慢为序，出现 HbA，HbA_2 及 CA(碳酸酐酶)三个区带。样品为红细胞时也可看到 HbA 和 CA，但有两点与溶血液不同：①没有与溶血液 HbA_2 相对应成分，但在其稍前方(阳极侧)出现一蛋白区带，未染色时为红色。这个区带有些特殊，它不整齐、不像溶血液 HbA_2 那样集中，甚至变形。将此成分称为红细胞 HbA_2 或者"HbA_2"，以区别于溶血液 HbA_2。②HbA 与 HbA_2 之间的拖尾现象比溶血液明显。这种情况给人的印象是：红细胞中的 HbA_2 与 HbA 似乎不易分开；而溶血液中的 HbA_2 与 HbA 则分离良好、界限清楚。暂命名为"血红蛋白 A_2 现象(HbA_2 现象)"，以便于讨论和进一步研究。

3.2 红细胞 HbA_2 的双向电泳 用 20cm × 20cm 玻璃板做淀琼凝胶电泳。样品为红细胞悬浮液，加在玻璃板的右下角(靠两边各 2cm 处)，电泳条件同上。第一次电泳后，可看到两个红色区带(红细胞 HbA_2 和 HbA)；然后转 90°进行第二次电泳；用氨基黑 10B 染色结果如

图 19-2。由图 19-2 可以看到，双向电泳后红细胞 HbA 和 CA 停留在对角线上，而红细胞 HbA$_2$ 不在此线上并分成两个区带：主要成分在对角线的稍下方，次要成分在对角线的上方较远处、相当于 HbA 的位置。由于溶血液 HbA 的泳速比红细胞 HbA$_2$ 慢，推测对角线下边的成分很可能是溶血液 HbA$_2$。说明红细胞 HbA$_2$ 可能含有溶血液 HbA$_2$ 和 HbA，它们在第二次电泳时彼此分离出现两个电泳成分。

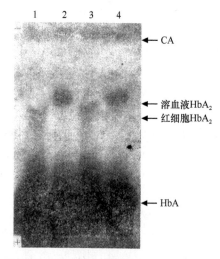

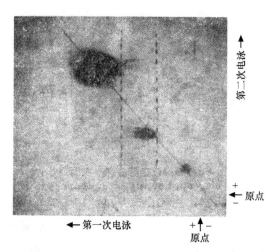

图 19-1　正常成人血样的血红蛋白 A$_2$ 现象
注释：1、3：红细胞；2、4：溶血液

图 19-2　正常成人血样的 HbA$_2$ 现象的双向电泳分析

3.3　红细胞 HbA$_2$ 的单向二次电泳　用淀琼胶电泳做"血红蛋白 A$_2$ 现象"实验。电泳结束后不染色，在玻璃板上留下四块含有红色区带的凝胶(红细胞 HbA$_2$ 和 HbA 两个区带，以及溶血液 HbA$_2$ 和 HbA 两个区带)，切除其余无色淀琼凝胶。含红细胞 HbA$_2$ 的凝胶原位不动，将其他含 Hb 凝胶都推到(在玻璃板上滑动)两旁，并排在一条线上。它们的位置关系是：红细胞 HbA、红细胞 HbA$_2$、溶血液 HbA$_2$、溶血液 HbA。位置调好后，在它们的周围再倒上新的淀琼凝胶，厚度相当，室温放冷，胶液凝固后再通电，条件同前。电泳结束后氨基黑 10B 染色，结果如图 19-3。

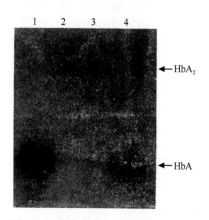

图 19-3　红细胞血红蛋白 A$_2$ 的再电泳
注释：1：溶血液 HbA$_2$；2：溶血液 HbA；
3：红细胞 HbA$_2$；4：红细胞 HbA

由图 19-3 可以看到，再电泳时红细胞 HbA 与溶血液 HbA 的速度仍然相同，和原来血红蛋白 A$_2$ 现象实验时的结果一致。图 19-3 中溶血液 HbA$_2$ 仍然是一个区带，但红细胞 HbA$_2$ 却分成两个区带：一个与溶血液 HbA$_2$ 的电泳位置相同；另一个相当于 HbA。这说明，红细胞 HbA$_2$ 中确实含有溶血液 HbA$_2$ 和 HbA，进一步证实了对双向电泳结果的判断。

4　讨论

血红蛋白 A$_2$ 现象的发现及它与几种类型异常血红蛋白的关系已见前文[1-3]，为了进一步弄清这一现象的发生机制，我们研究了各种可能情况，这里主要报告"红细胞 HbA$_2$"化学组成方面的工作。所谓"血红蛋白 A_2 现象",其关键是"红细胞 HbA_2"与"溶血液 HbA_2"不同。这种差异的主要表现如下所述:在电泳图谱中,"红细胞 HbA_2"比"溶血液 HbA_2"泳速快(向阳极方向)。说明"红细胞 HbA_2"的净电荷中负电性因素较"溶血液 HbA_2"多。假定"溶血液 HbA_2"就是 HbA_2 本身。可以认为,在红细胞中 HbA_2 可能与某种负电性较强的物质 X^- 结合存在。

还有另外一种假设:"红细胞 HbA_2"是 HbA_2 本身在制备溶血液过程中与红细胞中一种带较多正电荷的成分 Y^+ 结合,电泳时表现为"溶血液 HbA_2"与"红细胞 HbA_2"不同。

将本文实验结果与上述两个假设联系起来,可以认为,HbA 就是 X^-,而不是 Y^+。因为 HbA 的负电荷多于 HbA_2,符合对 X^- 的要求。由于 HbA_2 与 HbA 的结合存在,势必影响到结合产物的电荷和分子量。HbA_2-HbA 的负电性大于 HbA_2 的负电性,可以认为红细胞 HbA_2 快于溶血液 HbA_2(阳极方向)。至于红细胞 HbA_2 区带的具体位置比溶血液 HbA_2 靠前一点,可能是 HbA_2-HbA 的分子量大于 HbA_2,凝胶电泳的分子筛作用影响了它的泳速。

红细胞 HbA_2 区带不集中、与 HbA 之间拖尾比较严重。可能是由于红细胞 HbA_2 中两种 Hb(溶血液 HbA_2 与 HbA)结合得不牢固。由红细胞制备成溶血液,血红蛋白 A_2 现象就消失。表明这种结合很不稳定,经过第二次电泳(不管单向或者双向)时这种结合也会遭到破坏,推测在最初的红细胞 HbA_2 中 HbA 的数量可能比较多,随着第一次电泳过程的进行,陆续有少量 HbA 脱离下来,造成了区带不集中和明显的拖尾。第二次电泳时,残余的少量 HbA 与 HbA_2 彻底分开,才出现图 19-2 的情况。

总之 HbA_2 和 HbA 不是孤立地存在于红细胞之中,而是彼此结合存在。不一定全部 HbA 都与 HbA_2 结合,至少有一部分是这样的。这种结合并非十分紧密,比较容易分离,所以过去没有被人发现。

参 考 文 献

[1] 秦文斌, 梁友珍. 生物化学与生物物理学报, 1981, 13: 199.
[2] 闫秀兰, 秦文斌. 生物化学与生物物理学报, 1989, 21: 1.
[3] 秦文斌等. 生物化学与生物物理学报, 1991, 23: 85.
[4] Efremov G D et al. J Biol Chem, 1969, 244: 6105.

[原文发表于"生物化学与生物物理进展,1991,18(4): 286-288"]

第二十章 血红蛋白 A_2 现象与红细胞膜关系的研究
—— Ⅰ.血红蛋白 A_2 现象与溶血的关系

秦文斌 岳秀兰 睢天林

(包头医学院 血红蛋白研究室，包头 014010)

摘 要

1. 在发现"血红蛋白 A_2 现象"的基础上，我们开始对其与红细胞膜的关系进行系列研究，本文先报告此现象与溶血的关系。

2. 本文使用了两种最常用的溶血办法：①用蒸馏水使红细胞溶血；②用皂素使红细胞溶血。

3. 在常规"血红蛋白 A_2 现象"实验方法的基础上，增加一项"红细胞溶血产物"与完整红细胞及无细胞膜溶血液(血红蛋白溶液)的比较，观察这种产物有无 A_2 现象。

4. 实验结果表明，溶血产物中虽未除去红细胞膜残骸，但是它已经没有"血红蛋白 A_2 现象"，或者这种现象已经明显减弱。

5. 本项研究表明，"血红蛋白 A_2 现象"的出现与红细胞膜的完整性有关。

关键词 血红蛋白 A_2；"血红蛋白 A_2 现象"；红细胞膜；溶血

STUDIES ON THE RELATION BETWEEN "HBA$_2$ PHENOMENON" AND ERYTHROCYTE MEMBRANE
— Ⅰ. THE RELATION BETWEEN HBA$_2$ PHENOMENON AND HEMOLYSIS OF ERYTHROCYTE

Qin Wen-bin, Yue Xiu-Lan and Ju Tian-Lin

(Laboratory of Hemoglobin, Baotou Medical College, Baotou 014010)

Abstract

1. On the basis of the discovery of HbA$_2$ phenomenon We start a series of studies on its relation to erythrocyte membrane. The paper reports the relation between A$_2$ phenomenon and erythrocyte membrane.

2. Two common hemolytic methods were used: (1)erythrocyte was hemolyzed by distilled water; (2)Saponin was used to hemolyze erythrocyte.

3. The hemolytic product was analyzed side by side with intact erythrocyte and hemoglobin solution in the routine method of HbA$_2$ phenomenon.

4. The experimental results show that there is no HbA$_2$ phenomenon or this phenomenon is obvi-

ously weakened when above hemolytic products were examined.

5. It suggests that the occurring of HbA$_2$ phenomenon is associated with the integrity of erythrocyte membrane.

Key words Hemolysis; Erythrocyte membrane; HbA$_2$; HbA$_2$ phenomenon

1 前言

"血红蛋白 A$_2$ 现象"来源于血红蛋白溶液(也称溶血液),其与红细胞生理盐水悬浮液之间的电泳对比,详情参见前文[1, 2]。此时,一种样品是完整的红细胞,另一种标本是溶血后已经除掉红细胞膜的血红蛋白溶液。二者的界限是清楚的,对比之下,"血红蛋白 A$_2$ 现象"非常明显。现在的问题是,如果红细胞已经溶血,但其膜的残骸并未除去,这会怎么样?本文的目的就是要观察这种条件下有无"血红蛋白 A$_2$ 现象"。

实验结果表明,红细胞溶血后"血红蛋白 A$_2$ 现象"消失或者基本消失。无论用蒸馏水溶血或者用皂素溶血,所得结果大致相同。这说明,"血红蛋白 A$_2$ 现象"需要有完整的红细胞膜作为基础。如果红细胞膜的完整性遭到破坏,"血红蛋白 A$_2$ 现象"就要受到明显影响。这一实验结果,给"血红蛋白 A$_2$ 现象"机制的研究提出了新的内容,本章就此问题进行了研究和讨论。

2 材料和方法

2.1 血液样品的来源 包头市血站。

2.2 红细胞悬浮液的制备 方法同第十五章[1, 2]。

2.3 血红蛋白溶液(即溶血液)的制备 方法同第十五章[1, 2]。

2.4 红细胞溶血方法 本文采用蒸馏水溶血和皂素溶血两种办法,具体如下。①蒸馏水溶血:用生理盐水洗红细胞三次,向离心压实的红细胞中加入相等体积的蒸馏水,旋涡振荡半分钟后备用。②皂素溶血:先制备红细胞生理盐水悬浮液[1, 2]。再按 0.1%(W/V)浓度加入皂素,同上旋涡振荡后备用。

2.5 电泳分析 采用淀粉-琼脂混合凝胶电泳方法[1]。

2.6 染色及结果的保存 参见文献[1]。

3 结果

3.1 蒸馏水溶血时的"血红蛋白 A$_2$ 现象" 将正常成人红细胞(生理盐水悬浮液)、其本身的蒸馏水溶血产物(包括红细胞膜残骸及血红蛋白)、其本身的血红蛋白溶液(过去常称"溶血液",本文中为了与"溶血产物"相区别,改称为血红蛋白溶液,因为此时红细胞膜残骸已经除去),并排进行凝胶电泳,染色后结果如图 20-1。

由此图可以看出,红细胞及其血红蛋白溶液的电泳结果与一般"血红蛋白 A$_2$ 现象"相同,红细胞溶血产物的电泳结果与完整红细胞者不同,没有明显的"血红蛋白 A$_2$ 现象";与血红蛋白溶液比较,有时似乎有一些差异、有时根本没有区别。

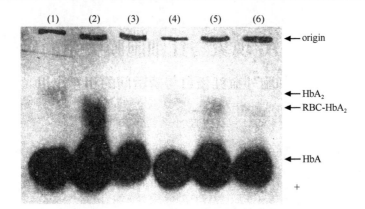

图 20-1　HbA$_2$ phenomenon of dH$_2$O—hemolytic product of erytrocyte.
From left to right：(1)，(4)= hemolysate；(2)，(5)= hemolytic product；(3)，(6)erythrocyte

3.2　皂素溶血时的"血红蛋白 A$_2$ 现象"　实验条件同上，只是用皂素溶血产物代替上述蒸馏水溶血产物，实验结果如图 20-2。此图结果表明，皂素溶血产物的电泳图像与蒸馏水溶血产物者相似，故不再描述。

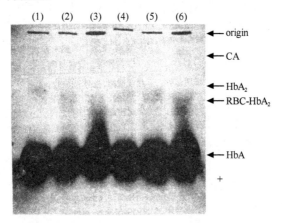

图 20-2　HbA$_2$ phenomenon of saponin —hemolytic product of erythrocyte.
From left to right；(1)，(4)=hemolysate；(2)，(5)=hemolytic product；(3)，(6)= erythrocyte.

4　讨论

"血红蛋白 A$_2$ 现象"发现以来，我们首先进行了结构规律方面的系列研究。实验结果表明，血红蛋白中δ珠蛋白链的存在可能是出现这种现象的决定因素[2-4]，也可称为这种现象对δ珠蛋白链的依赖性。还对这种现象的发生机制进行了研究。这方面的实验结果表明，来自红细胞的"HbA$_2$"中含有 HbA$_2$ 和 HbA，在红细胞外的某种特定条件下还能看到 HbA$_2$ 与 HbA 发生相互作用。因此，我们认为，在红细胞中 HbA$_2$ 与部分 HbA 之间可能通过相互作用而结合存在。用红细胞生理盐水悬浮液进行电泳时，这种结合产物与一般 HbA$_2$ 不同，表现为"血红蛋白 A$_2$ 现象"[5,6]。这就是目前我们对于"血红蛋白 A$_2$ 现象"发生机制的初步认识。

现在的问题是，红细胞中有 HbA$_2$ 和 HbA，血红蛋白溶液中也有这两种血红蛋白，为什么前者有相互作用而后者没有？大家知道，这两种样品的根本差别是有无红细胞膜，看来红细胞膜在产生"血红蛋白 A$_2$ 现象"方面也很重要。不过，红细胞膜还有各种不同存在状态，如果用溶血的办法破坏红细胞膜，但是不除掉(仍留在原来溶液中)，实验结果会这么样？本章就是根据这一想法设计的，由此观察溶血产物中的红细胞膜残骸对"血红蛋白 A_2 现象"有无影响。实验结果表明，红细胞膜一旦遭到破坏，即使其残骸尚在，"血红蛋白 A_2 现象"也明显减弱、甚至消失。这说明，红细胞膜的完整性很重要，破碎的红细胞膜在这方面已经不起作用。

关于血红蛋白与红细胞膜的联系(相关)问题，文献上有过记载，但是还未能得出明确结论。1975 年，Fischer 等人曾经提出，血红蛋白可能与正常及镰状红细胞的细胞膜相结合[7]。1977 年，Shaklai 等人也认为血红蛋白能与红细胞膜发生相互作用[8]。本文结果说明，"血红蛋白 A_2 现象"支持"血红蛋白可能是一种红细胞膜相关蛋白"的概念。不仅如此，它还告诉我们，与红细胞膜发生联系的血红蛋白，不是单独的 HbA 或者 HbA_2，很可能是二者的结合产物("HbA_2-HbA")。现在看来，红细胞膜与"HbA_2-HbA"的联系，要比 HbA_2 与 HbA 之间的联系更加疏远，电泳时更容易断裂并彼此分开。但是必须指出，尽管红细胞膜与"HbA_2–HbA"的相互作用比较微弱，但是它在维系 HbA_2 与 HbA 之间的联系上仍是重要的。如果红细胞膜遭到破坏、它与血红蛋白的相互作用消失，HbA_2 与 HbA 的结合也遭到破坏，"血红蛋白 A_2 现象"随之消失。当然，究竟是红细胞膜上的什么成分与血红蛋白发生联系，如何联系，这是我们应当进一步深入研究的问题。此项工作正在进行中。

参 考 文 献

[1] 秦文斌. 血红蛋白 A_2 现象——A_2 现象的发现及其基本概念. 包头医学院学报, 1990, 7: 6.

[2] 秦文斌, 梁友珍. 血红蛋白 A_2 现象Ⅰ. 此现象的发现及其初步意义. 生物化学与生物物理学报, 1981, 13: 199.

[3] 闫秀兰, 秦文斌. 血红蛋白 A_2 现象Ⅱ. α链异常血红蛋白的 A_2 现象及其意义. 生物化学与生物物理学报, 1989, 21: 1.

[4] 秦文斌等. 血红蛋白 A_2 现象Ⅲ. δ链异常血红蛋白的 A_2 现象及其意义. 生物化学与生物物理学报, 1990, 待发表(已修稿).

[5] 秦文斌. 血红蛋白 A_2 现象机制的研究Ⅰ. "红细胞 HbA_2" 为 HbA_2 与 HbA 的结合产物. 生物化学与生物物理进展, 1991, 待发表(已修稿).

[6] 秦文斌. 血红蛋白 A_2 现象机制的研究Ⅱ. 红细胞外 HbA_2 与 HbA 的相互作用. 生物化学杂志, 1991, 待发表(已修稿).

[7] Fischer, S., Nagel RL, Bookchin RM, et al. The binding of hemoglobin to membranes of normal and sickle erythrocytes. Biochim. biophys. Acta, 1975, 375: 422.

[8] Shaklai N, Yguerabide J, Ranney H M, et al. Interaction of hemoglobin with red blood cell membranes as shown by a fluorescent chromophore. Biochemistry, 1977, 16: 5585.

[原文发表于"包头医学院学报, 1990, 7(3): 47-51"]

第二十一章 血红蛋白 A₂ 现象与红细胞膜关系的研究
——Ⅱ．红细胞外血红蛋白与磷脂间的相互作用

秦文斌

(包头医学院 血红蛋白研究室，包头 014010)

摘 要

1. 在证明血红蛋白 A_2 现象需要红细胞膜完整性的基础上，本文探讨膜上成分与血红蛋白的关系问题。

2. 红细胞膜上成分很多，这里主要研究磷脂与血红蛋白的关系。

3. 本文用脑磷脂做实验，观察其与血红蛋白之间有无相互作用。

4. 通过单向和双向交叉电泳，以及 HbA 或者 HbA_2 与脑磷脂混合实验，都证明脑磷脂可与血红蛋白发生反应而且 HbA 及 HbA_2 都能与脑磷脂相互作用，产生的是既含血红蛋白、又有脂类性质的复合物。这一事实，从细胞外实验支持红细胞膜中磷脂与血红蛋白之间的相互作用，并且进一步提示，在红细胞中不仅 HbA 可能与磷脂结合，HbA_2 也可能与磷脂相互作用。

5. 由此，我们推测，在完整红细胞中，HbA_2 与 HbA(部分)结合、二者还与红细胞膜上的磷脂相互作用。在固态支持体中电泳时"HbA_2-HbA"与膜上磷脂脱离，表现出"血红蛋白 A_2 现象"。

关键词 血红蛋白 A_2 现象；红细胞膜；磷脂；血红蛋白；相互作用

STUDIES ON THE RELATION BETWEEN "HBA₂ PHENOMENON" AND ERYTHROCYTE MEMBRANE
—Ⅱ. INTERACTION OF HEMOGLOBIN WITH PHOSPHOLIPID OUTSIDE OF ERYTHROCYTE

Qin Wen-bin

(Laboratory of Hemoglobin, Baotou Medical College, Baotou 014010)

Abstract

1. After discovering the dependence of HbA_2 phenomenon upon integrity of erythrocyte membrane, the author starts to study the relation between components of erythrocyte membrane and hemoglobin.

2. There are many components on the erythrocyte membrane, the paper reports the relation between phospholipid and hemoglobin. Cephalin was used to study this problem.

3. The results of one—or two—dimensional elctrophoresis and those of experiments mixing hemolysate directly with cephalin, all demonstrate that cephalin can interacts with hemoglobin.

4. All three components of cephalin(phosphatidyl ethanolamine, phosphatidyl serine and phosphatidyl inositol)interact with hemoglobin and two components of hemolysate(HbA and HbA_2)interact with above phospholipids.

5. It suggests that in intact erythrocyte part of HbA combined with HbA_2 to form "HbA_2-HbA" complex and the latter interacts further with phospholipids of erythrocyte membrane. When running electrophoresis on gel support "HbA_2-HbA" complex separated from phospholipid and go out of erythrocyte appearing as "HbA_2 phenomenon".

Key words　HbA_2 phenomenon; RBC membrane; Hb; Phospholipids; Interaction

1　前言

血红蛋白 A_2 现象来源与红细胞与溶血液之间的差异[1, 2]，进一步实验证明这种现象要求红细胞膜的完整性[3]。红细胞膜的完整性涉及膜上成分及其存在状态，哪些成分与血红蛋白 A_2 现象有关？这是本章的研究目的。

红细胞膜上成分很多，这里报告磷脂与血红蛋白的关系。我们用脑磷脂做实验，结果表明，磷脂可以与血红蛋白发生相互作用。文中讨论了此项结果的意义。

2　材料和方法

2.1　血液样品的来源　正常成人血液，来自医院的献血员。
2.2　溶血液的制备　采用 Goldberg 法[4]，此时有机溶剂为氯仿。
2.3　磷脂　用脑磷脂做实验，有以下两种产品，效果相同。
(1) 上海食品公司制药厂生产，生物化学试剂，批号 630427。
(2) 上海工农兵制药厂生产，生物化学试剂，批号 670422。
2.4　电泳　本项研究是用纸上电泳方法进行的，文献同溶血液的制备[4]。
(1) 滤纸：所用滤纸为新华 1 号或者 WhatmanNo.1。
(2) 缓冲液：滤纸用 TEB 缓冲液(pH 9.0，0.03mol/L)浸湿，电极槽中为硼酸-NaOH 缓冲液(pH 9.5)。
(3) 电泳条件：常规方法是电势梯度 6V/cm，通电 4 小时左右。单向交叉电泳时，条件基本如此；双向交叉电泳时，第二向的电势梯度要 7V/cm，通电 5 小时左右。
2.5　染色　采用以下几种常用染色方法。
(1) 蛋白质：使用溴酚蓝(BPB)染色法。
(2) 血红蛋白：联苯胺染色法。
(3) 脂类：苏丹黑 B 染色法。

3　结果

3.1　血红蛋白与脑磷脂的相互作用　将正常成人红细胞溶血液与等体积蒸馏水混合，作为对照组。同前溶血液，加入等体积脑磷脂水溶液(2%，W/V)，混匀后作为实验组。将上述两组样品同时放入 37℃保温箱中，时间为 30 分钟。到时取出，并排进行纸上电泳，得到图 21-1 结果(此时为 BPB 染色)。由此图可以看出，在溶血液与脑磷脂混合后出现新的电泳

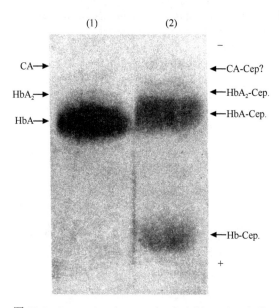

图 21-1 Interactions between hemoglobin and cephalin
Note: (1)=hemolysate; (2)=hemolysate+cephalin

成分。与正常成人红细胞溶血液结果对比，有一个快泳成分出现在 HbA 的前方(阳极侧)，其他成分(HbA，HbA_2，CA)大致相当，但都向阴极后退。图中 Cep 为 Cephalin(脑磷脂)的缩写。在未染 BPB 之前。肉眼观察时，可以看到某些区带显红色，即图中的 HbA，HbA_2，HbA-Cep，HbA_2-Cep 及 Hb-Cep。说明它们是血红蛋白，或者含有血红蛋白。

3.2 血红蛋白–脑磷脂相互作用产物的各种染色特性 上述电泳结果用不同染色方法处理，可归纳为表 21-1。

由表 21-1 可以看出，只有相当于 CA 的成分不含血红蛋白和脂类，其他成分都同时含有血红蛋白和脂类。这说明不仅快泳成分 Hb-Cep 为血红蛋白与脑磷脂相互作用产物，HbA-Cep 和 HbA_2-Cep 也都是二者的结合物。HbA、HbA_2 与 Cep 相互作用后电泳位置后退，但是 CA 似乎未与 Cep 相互作用，它的后退原因不明。

表 21-1 血红蛋白–脑磷脂(Hb-Cep.)反应产物的染色结果

Hb-Cep. 产物	电泳位置	肉眼观察	BPB 染色	联苯胺染色	苏丹黑 B 染色
HbA-Cep.	稍后于 HbA	红色	(+)	(+)	(+)
HbA_2-Cep.	稍后于 HbA_2	红色	(+)	(+)	(+)
Hb-Cep.	最快成分	红色	(+)	(+)	(+)
CA-Cep?	稍后于 NHP	无色	(+)	(−)	(−)

3.3 溶血液与脑磷脂的单向交叉电泳 按图 21-2 设计(A—B 为正常成人溶血液、X—Y 为脑磷脂溶液)，进行单向交叉电泳，结果见图 21-2。

由图 21-2(BPB 染色)可以看出，在两种溶液重叠的部分，电泳后出现新的区带。其与溶血液的未重叠部分比较，它们的关系大致同图 21-1。脑磷脂未与血红蛋白重叠部分为 BPB 染色阴性。与图 21-1 相比，此时，阳极侧出现两个 Hb-Cep.肉眼红色、脂染阳性，HbA-Cep 与 HbA_2-Cep。没有分开，也是肉眼红色、脂染阳性，但其位置仍落后于溶血液 HbA、HbA_2。此时 CA(染 BPB 前无色)明显后退。脑磷脂本身的三个区带，只是在脂类染色时才显阳性。这些结果表明，溶血液与脑磷脂溶液事先不混合，在电泳过程中相遇(或者所谓"交叉")时血红蛋白与脑磷脂也可以发生相互作用。

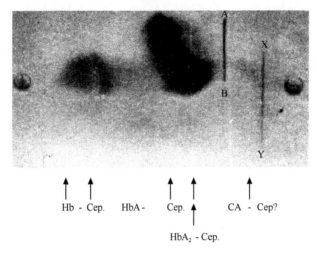

图 21-2　One—dimensional cross electrophoresis of hemolysate and cephalin.
Note：A — B= hemolysate；X — Y = hemolysate+cephalin

3.4　溶血液与脑磷脂的双向交叉电泳　按图 21-3 设计，先在 A—B 处加正常成人溶血液进行第一次电泳；然后在 X—Y 处加脑磷脂溶液，按图中箭头方向进行第二次电泳，结果如图 21-3 所示。

图 21-3 结果表明，第二次电泳后，在溶血液 HbA 及 HbA_2 的前方(图中为上方)都出现了新的电泳成分(肉眼红色、BPB 蓝色)，而且也是两个。这说明不仅 HbA 与脑磷脂有相互作用，同样，HbA_2 与脑磷脂也有作用。它们的作用产物也是类似的。但是，必须指出，此时 HbA 与 HbA_2 没有完全分开，它们与脑磷脂的相互作用产物也彼此界限不清。

3.5　HbA 及 HbA_2 与脑磷脂单独相互作用
用淀粉板电泳由溶血液中分离出来 HbA 及 HbA_2，各自单独与脑磷脂混合，然后进行纸上电泳，结果如图 21-4 所示。

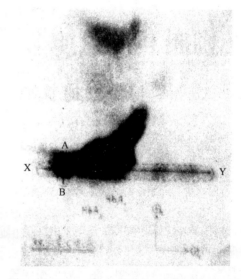

图 21-3　Two—dimensional cross electrophoresis of hemolysate and cephalin.
A—B=hemolysate；X—Y=cephalin. Note：Some new bands(Hb-Cep.)appeared above HbA and HbA_2.

图 21-4 结果进一步明确表明，脑磷脂与 HbA 及 HbA_2 都可以单独发生作用，并且得到类似的结果；脑磷脂与 HbA 作用产生 $HbA-C_1$、$HbA-C_2$ 和 $HbA-C_3$；脑磷脂与 HbA_2 作用产生相应的 HbA_2-C_1、HbA_2-C_2 和 HbA_2-C_3。

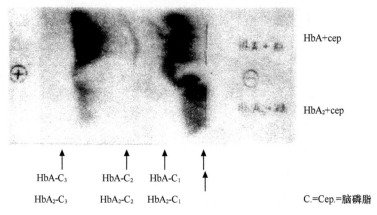

图 21-4　Cephalin interacts with HbA and HbA$_2$.
Note: They all have Hb-C$_1$, Hb-C$_2$ and Hb-C$_3$

4　讨论

本文的一系列实验结果，已经证明磷脂能够与血红蛋白发生相互作用，作用产物中既有血红蛋白、又有脂类成分。现在的问题是脑磷脂表现出来明显的不均一性。脑磷脂，最初是指由脑组织提取出来的一种不溶于醇类的磷脂组分。有人认为，脑磷脂就是磷脂酰乙醇胺；也有人主张，它是一种混合物。后来证明，脑磷脂至少含有三种磷脂成分：磷脂酰乙醇胺(简称 PE)、磷脂酰丝氨酸(PS)及磷脂酰肌醇(PI)[5]。本文，利用纸上电泳分析方法，证明脑磷脂确实含有三种成分(电泳中出现三个区带，都是脂类染色阳性)。根据这些成分在脑磷脂中的含量关系(PE > PS > PI)及它们所带电荷情况(以负电荷多少为序：PS > PI > PE)，可以认为脑磷脂电泳结果中的三个区带的成分如下(参见单向交叉电泳示意图中脑磷脂部分)所示。

本文实验证明，脑磷脂能与血红蛋白相互作用，而且可以在两种状态下进行：①脑磷脂与血红蛋白直接混合；②二者在电场中相遇(交叉电泳时)。不仅如此，深入分析还表明，脑磷脂中的三个成分与溶血液中的两种血红蛋白也都能彼此相互作用。因为它们作用产物都同时具有血红蛋白和脂类特性，故可得到表 21-2。

表 21-2　脑磷脂与血红蛋白的作用产物

Hb/C[※]	PS	PI	PE
HbA	HbA-PS	HbA-PI	HbA-PE
HbA$_2$	HbA$_2$-PS	HbA$_2$-PI	HbA$_2$-PE

注释：※　C=脑磷脂；HbA-PE=HbA-C$_1$；HbA$_2$-PE=HbA$_2$-C$_1$；HbA-PI= HbA-C$_2$；HbA$_2$-PI= HbA$_2$-C$_2$；HbA-PS=HbA-C$_3$；HbA$_2$-PS=HbA$_2$-C$_3$

值得注意的是，虽然这三种磷脂都能与血红蛋白发生相互作用，但在细节上并不完全相同。PS、PI 与血红蛋白的作用好像没有特异性。这里边没有显示出来 HbA 与 HbA$_2$ 的电荷差异；HbA-PS 和 PHbA$_2$-PS 的电泳位置相近；HbA-PI 和 HbA$_2$-PI 也是泳速相同。与此相反，PE 与血红蛋白的作用似乎有特异性。首先是 HbA-PE 和 HbA$_2$-PE 的电泳位置不同，表现出来 HbA 与 HbA$_2$ 的电荷差异。由此可见，脑磷脂与血红蛋白的相互作用应该分为两种类型：①PS、PI 型和②PE 型。前者似乎是一种物理学作用，后者可能属于化学反应范畴。

有关细胞膜的知识告诉我们，红细胞膜分为内外两层(或者内外两个区)：外层磷脂主要是鞘髓磷脂和磷脂酰胆碱；内层主要是磷脂酰丝氨酸(PS)和磷脂酰乙醇胺(PE)。在磷脂与血红蛋白相互作用方面，Szundi 等[6]使用单分子脂层法研究了血红蛋白与磷脂的关系。他们的结论如下：与血红蛋白发生相互作用的是内层磷脂，外层磷脂不起作用。本文所用的磷脂基本上都是内层磷脂，所得结论也与 Szundi 等的结论类似。不过，在红细胞膜内层两种磷脂与血红蛋白相互作用的相对重要性方面，Szundi 等强调了 PS，本文实验结果表明，在相互作用数量方面，用溶血液所做实验中 PE 明显多于 PS；两种血红蛋白单独与磷脂作用时，HbA 的结果是 PS 与 PE 大致相等(也许 PS 略微多于 PE?)；HbA$_2$ 的结果是 PE 明显大于 PS，与溶血液的整体结果一致，由于 Szundi 等没有区分 HbA 和 HbA$_2$，故无法与其比较。

血红蛋白 A$_2$ 现象的机制研究结果表明[7, 8]，红细胞中有一部分 HbA 是以"HbA-HbA$_2$"方式结合存在。通过溶血实验，还证明 A$_2$ 现象与红细胞膜有关[3]，本文证明，这两种血红蛋白都能与磷脂发生相互作用。结合 Szundi 等的实验，我们认为，红细胞中"HbA-HbA$_2$"可能是与红细胞膜内层的磷脂发生相互作用。不过，磷脂与血红蛋白之间的作用，可能比 HbA 与 HbA$_2$ 之间的作用更弱一些。因此，前者在电场中可以分开，使"HbA-HbA$_2$"脱离红细胞表现出来"血红蛋白 A$_2$ 现象"。

以上，是我们根据一系列实验得出的初步结论，准备接受今后实践的考验。

必须指出，CA 似乎没有与 Cep.发生相互作用，但是电泳位置后退，原因不明，有待进一步研究。

参 考 文 献

[1] 秦文斌. 血红蛋白 A$_2$ 现象-A$_2$ 现象的发现及其基本概念. 包头医学院学报, 1990, 7(3): 6.
[2] 秦文斌, 梁友珍. 血红蛋白 A$_2$ 现象Ⅰ. 此现象的发现及其意义. 生物化学与生物物理学报, 1981, 13: 199.
[3] 秦文斌. 血红蛋白 A$_2$ 现象与红细胞膜关系的研究Ⅰ. 血红蛋白 A$_2$ 现象与溶血的关系. 包头医学院学报, 1990, 7(3): 46.
[4] Goldberg C A J. A discontinuous buffer system for paper electrophoresis of human hemoglobins. Clin Chem, 1959, 5: 446.
[5] Ansell G B, Hawthorne J N. Phospholipids—Chemistry, Metabolism and Function. New York: Elsevier, 1964, 14.
[6] Szundi I, Szelényi JG, Breuer JH, et al. Interactions of haemoglobin with erythrocyte membrane phospholipids in monomolecular lipid layers. Biochim biophys Acta, 1980, 595: 41.
[7] 秦文斌. 血红蛋白 A$_2$ 现象发生机制的研究Ⅰ. "红细胞 HbA$_2$"为 HbA$_2$ 与 HbA 的结合产物. 生物化学与生物物理进展, 1991, 待发表(已修稿).
[8] 秦文斌. 血红蛋白 A$_2$ 现象发生机制的研究Ⅱ. 红细胞外 HbA$_2$ 与 HbA 间的相互作用. 生物化学杂志, 1991, 待发表(已修稿).

[原文本文发表于"包头医学院学报，1990，7(3): 52-54"]

第二十二章　血红蛋白 A_2 现象异常症(秦氏病)

秦文斌

(包头医学院　血红蛋白研究室，包头　014010)

摘　要

1. 在发现血红蛋白 A_2 现象、寻找结构规律、研究发生机制及探讨红细胞膜上成分与血红蛋白的关系的基础上，本文报告一例血红蛋白 A_2 现象异常症。

2. 患者患有血液病，怀疑有地中海贫血或者有异常血红蛋白，但诊断不明。

3. 实验室检查结果表明，未见异常血红蛋白，查不出有明确的 α 地中海贫血或者 β 地中海贫血。但是，血红蛋白 A_2 现象明显异常。

4. 用患者的红细胞进行凝胶电泳时，"红细胞 HbA_2" 完全消失，也没有 "溶血液 HbA_2"、HbA 明显减少、原点处留下大量沉淀(而且呈油脂扩散状)。

5. 根据上述结果，作者推测，此时红细胞中 "HbA_2-HbA 复合物"（"红细胞 HbA_2"）是与膜上异常脂溶性成分(不是正常的磷脂成分)牢固结合，电泳时二者不能脱离，表现为原点沉淀、"血红蛋白 A_2 现象异常"。

关键词　血红蛋白 A_2 现象异常症；"秦氏病"；红细胞膜成分异常

HEMOGLOBIN A_2 PHENOMENON ABNORMALITY(CHIN'S DISEASE)

Qin Wen-bin

(Laboratory of Hemoglobin, Baotou Medical College, Baotou 014010)

Abstract

The paper reports one case of HbA_2 phenomenon abnormality. There is a patient suffering from hematological disease suspected as thalassemia or abnormal hemoglobin syndrome. Laboratory results show that no abnormal hemoglobin and no definite α-or β-thalassemia were found, but HbA_2 phenomenon shows obvious abnormality: all "RBC HbA_2" disappeared from electrophoretic pattern of erythrocyte. HbA decreased distinctly and a great quantity of precipitate stayed at the origin appearing as a spread oil. The author suggest from these results that the "HbA_2-HbA" complex in patient's RBC may bound firmly with abnormal fat soluble component on the membrane, thus, in the course of electrophoresis "HbA_2-HbA" complex could'nt separate from the lipid and the abnormality of HbA_2 phenomenon occurred.

Key words　HbA_2 phenomenon abnormality; "Chin's disease"; Abnormality of RBC membrane composition

1 前言

"血红蛋白 A_2 现象"是作者在长期实验过程中发现的。由于它能在指定的实验条件下多次重复,故可以肯定这种现象的客观存在。人们似乎有这样一种习惯,一旦发现某种现象或者规律,马上就想知道它有什么用途、解决什么问题。这种想法很自然,从长远来看用途是会有的,只是时间长短问题。

血红蛋白 A_2 现象刚刚被发现时,我们就曾经注意到它能帮助鉴别 HbE 和 HbA_2[1]。后来,还发现它可以帮助区分异常珠蛋白链[2, 3]。在研究血红蛋白 A_2 现象发生机制的同时,作者继续寻找它与疾病的联系。本文报告一例血红蛋白 A_2 现象异常症,既初步找到了这种现象的直接应用,又促进了对其机制的进一步了解。

2 材料和方法

2.1 血液样品的来源

(1) 正常成人血样:来自包头市血站。
(2) 血红蛋白 A_2 现象异常症血样:来自某省某县医院患者。

2.2 红细胞悬浮液的制备 用生理盐水洗红细胞三次,每次离心后吸除少量上层细胞成分(其中含较多白细胞)。最后,向压积红细胞中加入等体积生理盐水,混合均匀,备用。

2.3 溶血液的制备 方法同一般常规,只是不用蒸馏水使红细胞溶血。取一部分上述红细胞悬浮液,直接加入其 1/4 体积的四氯化碳。振荡约 200 次后离心,取上层溶血液备用。

2.4 电泳分析 薄层淀粉-琼脂混合凝胶电泳(简称淀琼电泳):取淀粉和琼脂,按 4:1 混合,用上述 pH 9.0 TEB 缓冲液配成 2.5%浓度制备凝胶,操作基本同薄层淀粉凝胶电泳[4]。趁热将 25ml 胶液倒在 10cm×10cm 玻璃板上,等待凝固。电极槽中为 pH 9.0 硼酸-NaOH 缓冲液。其余条件同参考文献[4]。

2.5 染色方法 不染色,可以直接观察血红蛋白的红色。必要时,采用氨基黑 10B 或者联苯胺法。

2.6 染色结果的保存 淀琼电泳后用氨基黑 10B 染色,室温放置,结果被固定在玻璃板上,可以长期保存。染色后的淀琼凝胶,还可转到 X 线胶片或者滤纸上,然后晾干。这样保存起来更为方便。

3 结果

3.1 患者的一些情况 某省某县医院寄来的血液标本,共 15 份,目的是要作者帮助解决血红蛋白病的诊断问题。由于他们是住院患者,所以都有不同程度的临床表现。轻者有轻度贫血,有人有红细胞形态异常,甚至有肝脾大等。实验室结果表明,这些患者的红细胞中没有异常血红蛋白,HbA_2 和 HbF 都不增多。只是,其中有一个患者出现"血红蛋白 A_2 现象"异常。

3.2 血红蛋白 A_2 现象异常症的实验结果 将此患者的红细胞(悬浮液)与其本身的溶血液并排电泳,染色后结果如图 22-1。

由此图可以看出,其红细胞的电泳结果比较特殊。此时 HbA 含量明显减少,没有通常出现的"红细胞 HbA$_2$",也没有溶血液 HbA$_2$;原点(加样处)出现大量沉淀,而且呈油脂扩散状。

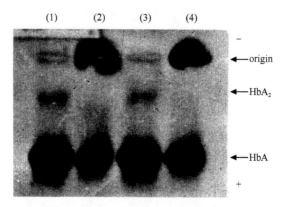

图 22-1　HbA$_2$ phenomenon abnormality(Chin's disease)

(1),(3)=hemolysate;(2),(4)=erythrocyte. Note: For erythrocyte there is no RBC-HbA$_2$ and a great quantity of precipitate appeared at the origin

4　讨论

患者血液标本的实验室结果表明,他没有异常血红蛋白综合征,也没有查出明显的地中海贫血。但是,他的血红蛋白 A$_2$ 现象显著异常。过去,作者检查过属于 α 地中海贫血的 HbH 病,能够看到红细胞 HbA$_2$;也研究过 HbA$_2$ 增多型 β 地中海贫血,其血红蛋白 A$_2$ 现象与正常人相同[1]。现在看来,这个病例可能比较特殊,它的临床诊断不明,作者只好从实验室角度暂时将其命名为"血红蛋白 A$_2$ 现象异常症"(秦氏病),以便今后接受批评、讨论和引用。

发现血红蛋白 A$_2$ 现象以来,作者一直在寻找它的异常情况。因为,一旦发现异常,不仅能够解决医学诊断方面的需要,而且可能有助于弄清这种现象的发生机制。血红蛋白 A$_2$ 现象异常可能有各种情况,例如,红细胞 HbA$_2$ 的电泳速度异常、含量明显增多、显著减少、完全消失……本文所报告的病历属于完全消失类型,并且在原点(加样处)留有大量沉淀。

这个病历的发现,使作者对血红蛋白 A$_2$ 现象机制的认识又深化一步。我们在研究正常"血红蛋白 A$_2$ 现象"的发生机制时,曾经证明红细胞中 HbA$_2$ 与部分 HbA 结合存在(二者之间有相互作用)[5]。在本文中,"血红蛋白 A$_2$ 现象"异常,红细胞 HbA$_2$ 消失,HbA 明显减少,原点出现大量沉淀。患者的红细胞溶血液中 HbA$_2$ 含量很少(属于正常情况),如果只是它留在原点,沉淀物不应太多。另一方面,患者的红细胞溶血液中 HbA 含量很多(也属于正常情况),而红细胞电泳结果中,HbA 明显减少,原点处出现大量沉淀。由此推测,原点沉淀物很可能同时含有 HbA$_2$ 和 HbA。也就是说,此时二者可能还是结合存在,但是在电场作用下没有能够离开红细胞,和它一起留在原点。

为什么此时"HbA$_2$-HbA"没有冲出红细胞膜的包围?我们曾经提出"血红蛋白 A$_2$ 现象"可能与红细胞膜有关[6],现在看来可能是红细胞膜出现异常。至少有两个可能:①红细胞膜上的通路变窄,电泳时 HbA 可以通过,"HbA$_2$-HbA"不能通过。②红细胞膜上的异常成分与 HbA$_2$-HbA 结合牢固,电泳时后者未能脱离下来。红细胞膜的异常情况不明。但是,由于原点成分呈现"油脂扩散状",估计是红细胞膜上脂溶性成分增多或者出现异常的脂溶性成分。我们曾经提出正常红细胞中"HbA$_2$-HbA"与水溶性比较强的磷脂(PL)结合存在的假说[7]，现在可以认为，是磷脂变成脂溶性比较强的成分(X)并与"HbA$_2$-HbA"比较牢固地结合在一起，无法与红细胞膜脱离。

参 考 文 献

[1] 秦文斌, 梁友珍. 血红蛋白 A$_2$ 现象 Ⅰ. 此现象的发现及其初步意义. 生物化学与生物物理学报, 1981, 13: 199.
[2] 闫秀兰, 秦文斌. 血红蛋白 A$_2$ 现象 Ⅱ. α链异常血红蛋白的 A$_2$ 现象及其意义. 生物化学与生物物理学报, 1989, 21: 1.
[3] 秦文斌等. 血红蛋白 A$_2$ 现象Ⅲ. δ链异常血红蛋白的 A$_2$ 现象及其意义. 生物化学与生物物理学报, 1990, 待发表(已修稿).
[4] 秦文斌等. 改进淀粉胶薄层电泳法及其对血红蛋白的分辨能力. 生物化学与生物物理学报, 1965, 5: 278.
[5] 秦文斌. 血红蛋白 A$_2$ 现象发生机制的研究. Ⅰ. "红细胞 HbA$_2$" 为 HbA$_2$ 与 HbA 的结合产物. 生物化学与生物物理进展, 1991, 待发表(已修稿).
[6] 秦文斌. 血红蛋白 A$_2$ 现象与红细胞膜关系的研究. Ⅰ. 血红蛋白 A$_2$ 现象与溶血的关系. 包头医学院学报, 1990, 7(3): 46.
[7] 秦文斌. 血红蛋白 A$_2$ 现象与红细胞膜关系的研究. Ⅱ. 红细胞外血红蛋白与磷脂间的相互作用. 包头医学院学报, 1990, 7(3): 51.

[原文发表于"包头医学院学报，1990，7(3)：59-63"]

第二十三章 初释放研究的国外引用情况

国外科技著作 *Protein Electrophoresis*（图23-1），用了一章(第32章)专门介绍我们的发现。

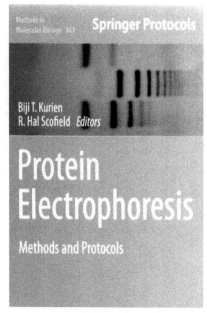

图23-1 *Propein Electrophoresis(Methods and Protocols)*封面及部分目录

Interactions of Hemoglobin in Live Red Blood Cells Measured by the Electrophoresis Release Test

Yan Su, Lijun Gao, and Wenbin Qin

Abstract

Electrophoresis release test(ERT)is the starch-agarose mixed gel electrophoresis of live red blood cells(RBCs). Mixed gel electrophoresis used to be one of classic methods to isolate proteins, and in our laboratory, this technique is usually performed to isolate hemoglobins. Recently, combined with sodium dodecyl sulfate polyacrylamide gel electrophoresis(SDS-PAGE)and(liquid chromatog raphy coupled with tandem mass spectrometry LC/MS/MS), ERT has been used to study the interactions between hemoglobin and other proteins in live RBCs.

Key words red blood cell; Hemoglobin; Interaction; Electrophoresis; Release test

1 Introduction

It is believed that biological processes are carried out through precise protein-protein interactions [1, 2]. A better understanding of these interactions is crucial for elucidating the structural/functional relations of proteins, investigating their roles in the development of associated disease and determining potential drug targets for clinical applications [3, 4]. Now, increasing number of approaches, such as yeast two-hybrid systems, tandem affinity purification(TAP), protein chip, co-immunoprecipitation and glutathione-S-transferase(GST)pull-down methods, have been developed to detect protein-protein interactions [5]. However, functional protein-protein interactions are dynamic processes and many are maintained by non-covalent bonds. Therefore, the detection of interactions in live cells remains difficult. Red blood cell(RBC)is the simplest human cell, and the loss of internal organelles becomes the great limitation to study its protein- protein interaction by the usual detection approaches such as TAP and GST pull-down in the RBC [4]. The electrophoresis release test(ERT)is a newly developed experiment to study the interactions between hemoglobin(Hb)and other proteins in live RBCs [4]. During ERT, live RBCs can be added directly onto the gel and the electric current perforates the membrane instantaneously [6]. The protein complexes in live RBCs can be thus released directly to the electric field and the different electrophoresis behaviors of RBC proteins can be directly compared with hemolysate, in which the protein-protein interactions would have been damaged during preparation, especially interactions mediated by membrane proteins [4]. Thus, in vivo interactions in the RBC can be identified by ERT.

2 Materials

Prepare all solutions using ultrapure water(prepared by purifying deionized water to attain a sensitivity of 18 MΩ cm at 25°C)and analytical grade reagents. Prepare and store all reagents at room temperature(unless indicated otherwise). Diligently follow all waste disposal regulations when disposing waste materials.

2.1 Apparatus

1. Horizontal electrophoresis cell: DYY-9B, Beijing Liuyi instrument factory, Beijing, China.
2. Electrophoresis power supply: DYY-2C, Beijing Liuyi instrument factory, Beijing, China.
3. Glass plate: 17 cm×10 cm or 17 cm×17cm; about 2~3 mm thick(*see* **Note 1**).
4. High-speed centrifuge: TGL-16G, Shanghai Anting Instrument Factory, Shanghai, China.
5. Vortex mixer.

6. Sprit levels.

7. Whatman® 3MM filter paper.

2.2 Fresh Anti-coagulated Blood Venous blood samples were anti-coagulated with EDTA(*see* **Note 2**), stored at 4 °C, and generally analyzed within 24 h.

2.3 RBCs and Hemolysate Preparation Components

1. Physiological saline: Dissolve 0.9 g NaCl in 100 ml water. Store at 4°C.

2. Carbon tetrachloride(CCl_4).

2.4 Starch-agarose Mixed Gel Electrophoresis Components

1. Gel preparation buffer($1 \times$ TEB pH8.6): Weigh 5.1 g Tris, 0.5 g EDTA, 0.4 g boric acid and transfer them to a cylinder, add water to a volume of 900 ml. Mix and adjust pH to 8.6 with 1 N HCl, and then make up the volume to 1 L with water.

2. Electrophoresis buffer(Boric acid buffer, pH 9.3): Weigh 18.55 g boric acid, 4.0 g NaOH and transfer them to a cylinder, add water to a volume of 900 ml. Mix and adjust pH to 9.3 with 1 N NaOH, and then make up the volume to 1 L with water.

3. Potato starch($C_6H_{10}O_5$): Molecular weight is 162(*see* **Note 3**).

4. Agarose: Biowest(*see* Note 4).

2.5 Staining solution Components

1. Ponceau red staining solution: Dissolve 1.0 g Ponceau(*see* Note 5), 50 ml glacial acetic acid and 20 ml glycerol(*see* **Note 6**)in 1 L water.

2. Benzidine staining solution: 0.6 g benzidine(*see* Note 7), 25 ml glacial acetic acid, 10 ml glycerol, add water to 500 ml, then heat the solution in 75℃ water bath for 1 h until the benzidine is dissolved completely. Sodium nitroprusside and 30% H_2O_2 should be added to this solution before use(*see* **Note 8**).

3. Rinsing solution: 50 ml glacial acetic acid, 20 ml glycerin, adds water to 1 L.

3 Methods

3.1 Preparation of RBCs suspension

1. Centrifuge the anti-coagulated blood at 700 g for 5 minutes and aspirate the upper plasma(*see* **Note 9**).

2. Transfer 200 μl of the lower RBCs to 1.5 ml Eppendorf tube(*see* **Note 10**), and then add 1 mL physiological saline to the Eppendorf tube, mixed gently.

3. Centrifuge at 700 g for 4 minutes and aspirate the supernatant.

4. Repeat this washing operation 4～5 times until the supernatant becomes colorless and transparent.

5. Finally, aspirate the supernatant and 1∶1 dilute the RBCs with saline to prepare RBCs suspension.

3.2 Preparation of hemolysate

1. Add 200 μl CCl_4 to the well prepared RBCs suspension.

2. Mix turbulent on the vortex mixer for at least 5 minutes.

3. Centrifuge at 10 000 G for 10 minutes.

4. Transfer the upper red hemolysate carefully to a new 1.5 ml Eppendorf tube(see **Note 11**), and stored it at 4°C for later use.

3.3 Preparation of the starch-agarose mixed gel

1. Clean a 17 cm×10 cm glass, and then put it onto the horizontal desk top(*see* **Note 12**).

2. Weigh 1.0 g potato starch, 0.15 g agroase and transfer them to a 250 mL triangle flask, then add 50 mL 1×TEB(*see* **Note 13**), mix together.

3. Heat the triangle flask on the electric stove repeatedly(*see* **Note 14**)until the starch and agarose resolved completely(*see* **Note 15**). Cool the gel solution at room temperature to about 60 ℃(*see* **Note 16**).

4. Pour the gel solution onto the horizontal glass plate(*see* **Note 17**), then solidify at room temperature for at least 20 minutes(*see* **Note 18**).

3.4 Sample loading

1. Cut the 3MM filter paper to 0.1 cm×1 cm small pieces(*see* **Note 19**).

2. Add 5 μl RBC suspensions to the 0.1 cm×1 cm small piece of 3MM filter paper(*see* **Note 20**).

3. Clamp the filter papers with tweezers, and insert(*see* **Note 21**)them one by one into the gel which is 1.5 cm to 2 cm away from one end of the gel(*see* **Note 22**), the intervals between every two samples are about 0.5 cm to 1 cm.(Fig. 1)

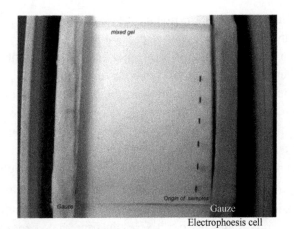

Fig. 1 The methods of sample loading and electrophoresis. The origin of samples are 1.5 cm to 2 cm away from one end of the gel, and the intervals between every two samples are about 0.5 cm to 1 cm. On each side, 4~6 layers of gauze are put on the gel. The origin should be placed on the negative side

3.5 RBCs release electrophoresis

1. Add 500 ml electrophoresis buffer(boric acid buffer, pH 9.3)to each side of the electrophoresis cell.

2. Immerse proper size of the gauze into the electrophoresis buffer.

3. Put the gel into the horizontal electrophoresis cell, the origin of sample should be placed on the negative side.

4. Put 4~6 layers of gauze over each end of the gel(*see* **Note 23**).

5. Electrophoresis is performed in borate buffer at constant voltage(6 V/cm)for 2~3 hours.(*see* **Note 24**).

6. After the red hemoglobin band move to the other side, turn off the power.

3.6 Staining and rinsing

1. First, observe the red bands on the gel with our eyes directly(*see* **Note 25**), and take pictures or scan the gel to keep the results.

2. Put the gel into Ponceau red staining solution for at least 4 h.

3. Rinse the gel with rinsing solution repeatedly until the background is clean.

4. Dry the gel in the air or in the oven.

5. Put the gel into Benzidine staining solution for about 20 mins, until the Hb bands are stained blue black.

6. Take out the gel and wash it with water, and then dry it in the air.

7. Take pictures or scan the gel to keep the results.

3.7 Results analysis

All of protein can be stained red by Ponceau red, but only hemoglobin can be stained black by Benzidine specifically. Compare the electrophoresis behavior of Hbs in RBC and hemolysate group(see **Note 26**). If there are some differences, proteins can be recovered and enriched from the bands of starch-agarose mixed gel(see **Note 27**), and further isolated by SDS-PAGE. The important band can be cut down and detected by LC/MS/MS to disclose the protein components. This will help us to find the important protein which has interaction with Hbs in live RBCs.

4 Notes

1. The size of the glass is determined both by the size of the electrophoresis tank and the requirements of the experiment.

2. Venous blood can be anti-coagulated by a variety of anti-coagulants, such as EDTA, heparin, sodium citrate, but hemolytic blood should not be used here.

3. Potato starch should be placed for many years before it is used. It can also be replaced by lotus root starch and corn starch sold in the market.

4. In the past, the mixed gel was made up of starch and agar, but now agar has been substituted by agarose which can increase the electrophoretic resolution.

5. The dissolution progress of Ponceau is slow, so the $10\times$ stock solution should be prepared and stored at room temperature. Before use, it should be 1∶10 diluted with water.

6. Proper concentration of glycerol can avoid the gel crack after it is dry, but if too much, the gel will reduce the gel's ability to dry. Also, if the results needn't be kept for a long time, glycerol can be omitted in the staining or rinsing solutions.

7. Benzidine is highly carcinogenic, it could also be substituted by leucobase of malachite green which is non-carcinogenic, but its staining specificity is lower than Benzidine.

8. 1 g sodium nitroprusside and 1.4 ml 30% H_2O_2 should be added to 500 ml well-prepared benzidine staining solution before use.

9. Normal ratio of blood cells to plasma in the whole blood is about 1∶1.

10. When RBCs are transferred, the pipette tip must be placed under the liquid surface at least 2~3 minutes to avoid the contamination of white blood cells and platelets.

11. After centrifugation, three layers of the solution can be visible: the lower layer is CCl_4, the middle layer is membrane fragments, and the upper layer is hemolysate. Pipe out the upper supernatant carefully, and avoid touching the middle layer.

12. Level must be placed on the horizontal desktop. If the glass is not horizontal, the gel is easy to leak from the edge of glass when gel solution is poured on the glass.

13. The concentrations of potato starch and agroase are 2% and 0.3% respectively. Different size of glasses need different volume of $1\times$ TEB, the average volume is about 3.2 ml/cm^2.

14. When the triangle flask is heating on the electric stove, the flask must be shaken slowly o avoid the starch and agarose attach on the bottom of flask. In addition, when the gel solution be-

gins boiling, overflow must be avoided at any time. When the solution is boiling, the flask should be removed from the electric stove immediately and be cooled at room temperature for 1~2 minutes, and then continue to get heated, and so forth, until the gel is completely dissolved.

15. When the solution bubbles become larger during heating, the gel should have been well dissolved.

16. 60℃ is the proper temperature to pour the gel. If the temperature is too high, the gel solution is easy to leak from the edge of glass; if too low, the gel solution will become solid during pouring, which will cause incomplete and uneven surface of gel.

17. Pour the gel solution at the center of glass slowly and constantly. The gel solution will stop flowing at the edge of glass because of the tension of edge.

18. If the room temperature is high, it will take a longer time for the gel to get solid; if the gel is not used immediately, please put it in the 4℃ refrigerator.

19. The length of filter paper is determined by the volume and number of samples, generally the size of a 17 cm×10 cm small plate can load up to 8 samples(5 μl/sample).

20. If the size of filter paper is 0.1 cm×2 cm, it can load 10~15 μl samples.

21. 3MM filter paper has certain degree of hardness, and can be inserted into the gel directly. Also, the comb of proper size and thickness can be used to groove on the gel firstly, and then insert the filter paper into them.

22. Keep each sample located at the same level as far as possible.

23. When putting the gauze on the gel, the action must be gentle to avoid torn the gel. The gauze should closely contact with the gel surface, and the edge of gauze should be neat and be parallel with the sample origin.

24. We call this kind of electrophoresis primary release electrophoresis of RBCs, there are also point release electrophoresis, ladder release electrophoresis and bidirectional ladder release electrophoresis.

(1) Point release electrophoresis of RBCs: Electrophoresis is performed in borate buffer at(6 V/cm)for 1 h, then pause for 5 minutes and run for another 1hour, at this time, a Hb band released from the origin will appear between HbA_2 and carbonic anhydrase(CA)(Fig.2). Change the time of the first and second electrophoresis can change the location of the released Hb band.

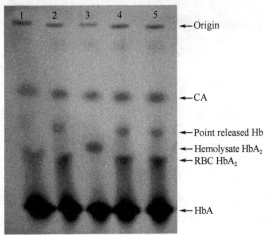

Fig. 2 Point release electrophoresis of RBCs. Sample 1, 2, 4 and 5 are RBCs of different patients; sample 3 is hemolysate. All of the RBC samples have different degree of released Hbs which locate between HbA_2 and CA, but hemolysate sample dosen't have released Hb band and its electrophoretic speed of HbA_2 is slower than that of RBCs obviously.

(2) Ladder release electrophoresis of RBCs: Electrophoresis is performed in borate buffer at(6 V/cm)for 1 hour, then pause for 5 minutes and run for 15 minutes by turns. It took about 3~4 hours for the entire electrophoresis(Fig.3).

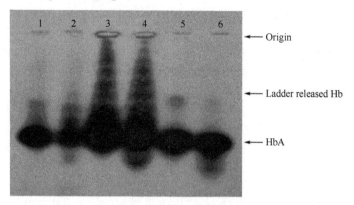

Fig. 3 Ladder release electrophoresis of RBCs. Sample 1~6 are RBCs come from different patients; the ladder released Hbs of sample 3 and 4 are more than the other samples significantly.

(3) Bidirectional ladder release electrophoresis of RBC: two vertical-direction electrophoresis runs were performed at 6 V/cm. The first directional electrophoresis is run for 15 minutes with a break of 15 minutes alternately. After the first-direction eletrophoresis, the direction of the electric field is altered vertically and continued to electrophoresis for about 1 hour(Fig.4).

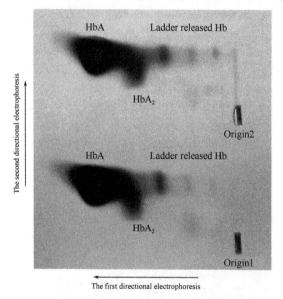

Fig. 4 Bidirectional ladder release electrophoresis of RBC. Sample 1 and 2 are RBCs come from different patients and the contents of ladder released Hb is different with each other.

25. Hbs is red, so we can see the main Hb bands directly without staining.

26. RBCs come from different patients; different treatments and different physiological and pathological conditions can be analyzed by this method.

27. When recovering and enriching the protein from the starch-agarose mixed gel, the gel should not be stained by any dye after starch-agarose mixed gel electrophoresis. The red band of HbA, HbA_2 and the gel located between HbA and HbA_2(HbA-HbA_2)of RBC and hemolysate can

be cut out separately and frozen at –80°C for at least 30 minutes. Before use, the gels should be taken out and thawed at room temperature. After centrifuging at 10 000 g for 10 minutes, the supernatants can be piped out into new Eppendorf tubes and dried in a vacuum freeze drier for about 6~8 hours.

Acknowledgements

This work was supported by grants from the Key Science and Technology Research Project of the Ministry of Education(210039), the Major Projects of Higher Education Scientific Research in the Inner Mongolia Autonomous Region(NJ09157), Natural Science Foundation of Inner Mongolia(2010BS1101)and Inner Mongolia Health Department Fund. We also especially acknowledge all of the people who donated their blood samples for our research.

References

[1] Kim, K. K. and Kim, H. B.(2009)Protein interaction network related to Helicobacter pylori infection response. *World J. Gastroenterol* 15, 4518-4528.

[2] Bu, D., Zhao, Y., Cai, L., Xue, H., Zhu, X., Lu, H., Zhang, J., Sun, S., Ling, L., Zhang, N., Li, G., and Chen, R.(2003)Topological structure analysis of the protein-protein interaction network in budding yeast. *Nucleic Acids Res* 31, 2443-2450.

[3] Hase, T., Tanaka, H., Suzuki, Y., Nakagawa S., and Kitano H.(2009)Structure of protein interaction networks and their implications on drug design. *PLoS. Comput. Biol* 5, e1000550.

[4] Su Y., Gao L., Ma Q., Zhou L., Qin L., Han L., and Qin W.(2010)Interactions of hemoglobin in live red blood cells measured by the electrophoresis release test. *Electrophoresis* 31, 2913-2920.

[5] Drewes, G., and Bouwmeester, T.(2003)Global approaches to protein-protein interactions. *Curr. Opin. Cell Biol* 15, 199-205.

[6] Su Y., Shao G., Gao L., Zhou L., Qin L., and Qin W.(2009)RBC electrophoresis with discontinuous power supply - a newly established hemoglobin release test. *Electrophoresis* 30, 3041-3043.

第四篇　再释放电泳

前　言

1　什么是"血红蛋白(电泳)再释放现象"

　　红细胞裂解液(也称红细胞溶血液或溶血液)与未裂解的完整红细胞并排进行淀粉琼脂糖混合凝胶电泳时，第一次通电后，能看到有血红蛋白由原点泳出来[这是"血红蛋白(电泳)初释放现象"]。如果再次通电，或者人为地停电—再通电，又会有新的血红蛋白由红细胞的原点释放出来，而溶血液的原点却没有新的血红蛋白泳出来。这种停电—再通电时又由红细胞释放出来新血红蛋白的现象，我们称之为"血红蛋白(电泳)再释放现象"，以下简称"血红蛋白再释放现象"，下同。

2　"血红蛋白再释放现象"的单向电泳结果

　　2.1　"单带再释放"　如上所述，在第一次通电的基础上，人为地停电—再通电，又会有新的血红蛋白由红细胞的原点释放出来，如果停电—再通电只有一次，再释放出来的新血红蛋白只有一个区带，称之为"单带再释放"。

　　再释放出来的单带，可因再通电时间的长短而电泳位置不同，调节再通电时间可使血红蛋白区带到达预定地点(位置)，这样又有"定时再释放"和"定点再释放"之称。常规"单带再释放"中最常用的再通电时间为 15 分钟(电势梯度 6V/cm)，此时再释放的血红蛋白区带离开原点不远，位于原点与碳酸酐酶 CA 之间，便于测定再释放血红蛋白与碳酸酐酶的比值(Hb/CA)，来判定再释放血红蛋白的多少。

　　2.2　"多带再释放"　前边说的是"单带再释放"，停电—再通电一次，再释放出来一个新的血红蛋白区带。如果反复多次进行停电—再通电，会看到再释放出来多个新的血红蛋白区带，我们称之为"多带再释放"。正常人红细胞也有"多带再释放"，但强度不大，各种疾病的"多带再释放"情况不同，最早发现"多带再释放"明显增强的是遗传性疾病——轻型 β 地中海贫血。

　　2.3　"等低渗全程再释放"　上述"单带再释放"和"多带再释放"都是在等渗条件下进行的，也就是说，红细胞处于等渗状态，红细胞存在于生理盐水之中，或者直接用全血做实验。"等低渗全程再释放"实验是在等渗的基础上加入低渗因素，即向红细胞或全血中加入不同数量的蒸馏水，造成不同程度的低渗状态，然后进行电泳并完成停电—再通电过程，观察血红蛋白的再释放情况。每次实验是一个标本(红细胞或全血)分成 10 种等级，除了有等渗标本外，还有轻度低渗标本(9 份标本加 1 份蒸馏水)、8 份标本加 2 份蒸馏水、7 份标本加

3 份蒸馏水……重度低渗标本(1 份标本加 9 份蒸馏水)，然后做再释放电泳.此时，得到的信息量较大，不同人之间容易看到差别。最明显的例子是阻塞性黄疸，此时，等低渗 10 个等级都显示再释放增强。必须指出，上述"等低渗全程再释放"是在室温下进行的，此外还有保温"等低渗全程再释放"，通常是在 37℃过夜(12 小时)，可以提供更多信息。

3 "血红蛋白再释放现象"的双向电泳结果

上述各项都是来自单向电泳，双向电泳时再释放出来的新血红蛋白处于什么位置，它与对角线的关系如何，也是"血红蛋白再释放现象"的重要组成部分。实验结果表明，再释放出来的新血红蛋白离开对角线较远，只出现于对角线的右上方，左下方没有再释放成分。这一点，它与"血红蛋白初释放现象"的双向对角线电泳结果不同，初释放出来的血红蛋白同时出现在对角线的上下两侧，相互对应。这也说明，再释放与初释放的血红蛋白在红细胞内存在状态不一样，在临床方面初释放不如再释放应用面广，可能与此有关。

前面提到"等低渗全程再释放单向电泳"的信息量大，如果在此基础上做双向电泳，其信息量更大，我们称之为"再释放指纹图"，试图从红细胞蛋白质角度进行"个体识别"。此时，每个人有一组共四个"再释放指纹图"：红细胞全程室温、全血全程室温、红细胞全程 37℃和全血全程 37℃，推测两个人一组四个图都相同的可能性很小。这方面的研究刚刚开始，初步看到一例轻型 α 地中海贫血标本的"再释放指纹图"明显异常。

4 "血红蛋白再释放现象"的应用

与"血红蛋白初释放现象"("血红蛋白 A_2 现象")比较，"血红蛋白再释放现象"的应用范围较广。首先，在突然停电而来电后继续电泳时，我们发现轻型 β-地中海贫血患者全血中红细胞又(再)释放出来新血红蛋白，而且释放量明显多于正常人。后来对普通外科住院患者血液标本进行再释放研究，结果发现，低渗优于等渗，37℃低渗优于室温低渗。10 个住院患者中有 4 例出现再释放增强，他们的原始诊断不尽相同，这说明再释放实验能够揭示与红细胞有关的更深刻问题。我们证明再释放与血液流变学结果有一定相关性，而血液流变学结果与脑梗死关系密切，故推测脑梗死血液标本再释放增强。糖尿病患者血液标本的研究结果很有趣，它在等渗条件下就能发现问题，此时游离红细胞的再释放强度与血糖浓度相关，餐后血糖高于空腹血糖，其再释放结果也如此，与游离红细胞相比，全血没有这种效果。尿毒症患者与正常人血液标本的比较研究结果表明，尿毒症患者标本的血红蛋白再释放量多于正常人。肝硬化患者与正常人血液标本的比较研究结果是，患者全血标本及红细胞标本的等低渗全程再释放量都多于正常人。

5 "血红蛋白再释放现象"的机制研究

过去关于"血红蛋白 A_2 现象"的机制研究中，我们发现，红细胞内除了血红蛋白 A_2 与 A_1 相互作用外，还有 Prx-2(过氧化物还原酶-2)参与。现在，"血红蛋白再释放现象"的机制如何？再释放出来的新血红蛋白是哪种血红蛋白(血红蛋白 A_1、A_2、A_3、F)，是否还有其他蛋白质参与作用？为此我们进行下述一系列研究。

首先，从单带再释放的双向电泳结果来看，再释放出来的新血红蛋白位于对角线的右上方，与对角线左上方的血红蛋白 A_1 处于同一水平，推测它很可能是血红蛋白 A_1 或与血红蛋白 A_1 相近的物质。为了深入研究再释放机制，我们从淀琼胶抠取红细胞的再释放血红蛋白及溶血液的相应对照，然后进行 SDS-PAGE 分析。结果发现，红细胞出现三条蛋白带：16kDa、28.9kDa、29.3kDa，溶血液只出现一条蛋白带：29.3kDa，最后进行 MALDI-TOF MS 分析。结果表明，16kDa 为 β-珠蛋白，28.9kDa 为碳酸酐酶 $1(CA_1)$，29.3kDa 为碳酸酐酶 $2(CA_2)$，由于红细胞与溶血液都有 CA_2，我们的结论是：单带再释放血红蛋白主要由 HbA_1 与 CA_1 组成。众所周知，红细胞内碳酸酐酶参与 CO_2 的运输、血红蛋白参与 O_2 的运输，而且二者之间存在协同作用，现在，这两种蛋白质同时出现于再释放血红蛋白，从而从电泳释放角度证明红细胞内二者之间的结构性相互作用。碳酸酐酶有 CA_1、CA_2 之分，二者在功能上差异何在？目前还不明确，本文电泳再释放实验结果告诉我们，与再释放血红蛋白相互作用的是 CA_1，不是 CA_2。

单带电泳再释放实验证明，血红蛋白 A_1 和碳酸酐酶 1 共同存在于红细胞内，互相配合，联合起来完成运送氧气和二氧化碳的任务。

第二十四章　血红蛋白释放试验与轻型 β-地中海贫血

秦文斌　高丽君　苏燕　邵国　周立社　秦良谊

(包头医学院　血红蛋白研究室，包头　014010)

摘　要

目的： 建立血红蛋白释放试验，明确方法，探讨其意义。

方法： 用轻型 β-地中海贫血患者的红细胞做淀粉-琼脂糖混合凝胶电泳；第一次电泳后，再进行第二次或更多次电泳，观察红细胞中血红蛋白的释放情况。

结果： 在第二轮电泳过程中，由原点又释放出血红蛋白，轻型 β-地中海贫血患者释放出来的血红蛋白明显多于其他人。

结论： 红细胞中血红蛋白可能具有不同的存在状态，大部分血红蛋白在第一次电泳时就释放出来，少量还残留在红细胞中，它在第二轮电泳才被释放出来。轻型 β-地中海贫血患者标本帮助我们揭开了其中奥秘，这一发现可能具有一定的理论和实践意义。

关键词　轻型 β-地中海贫血；红细胞；血红蛋白；残留；释放

Hemoglobin Release Test and β-Thalassemia in Trait

QIN Wenbin, GAO Lijun, SU Yan, SHAO Guo, ZHOU Lishe, QIN Liangyi

(Laboratory of Hemoglobin, Baotou Medical College, Inner Mongolia Baotou 014010, China)

Abstract

Objective: To establish a hemoglobin release test(HRT)and to approach its significance.

Methods: The RBC from β-thalassemia trait was analyzed by starch- agarose mixed gel electrophoresis. After first run of electrophoresis the second and more times of electrophoresis were performed and the second release of hemoglobin from RBC was observed.

Results: During the second run of electrophoresis some Hb were released from the origin and those with β-thalassemia trait give more Hb than other persons.

Conclusion: Hb may have different existing state in RBC and thus it can be mostly released from RBC with first time of electrophoresis, with some residue in RBC released in the second time of electrophoresis. The blood sample of β-thalassemia trait disclosed its mystery. The discovery of this phenomenon let us learn that the existing state of Hb in RBC is more complex than we expected and we believe it may have some theoretical and practical significance.

Key words　β-thalassemia trait; RBC; Hb; Residue; Release

1 前言

红细胞(RBC)中含有大量血红蛋白(Hb)，这是众所周知的。成人 RBC 中的 Hb 主要是

HbA、HbA$_2$，也是早有记载。但是，这些 Hb 在 RBC 内是如何存在的，则未见报道。1981年，我们用淀粉-琼脂糖混合凝胶电泳检测 RBC 时，发现有 Hb 释出，而且此时的 HbA$_2$ 与溶血液者不同，故称之为"HbA$_2$ 现象"，并推测在 RBC 中 HbA$_2$ 是与 HbA 结合存在的[1-4]。当时曾看到，电泳后原点还残留一些 Hb，对它的存在和意义并不清楚。最近，在用轻型β-地中海贫血患者 RBC 进行电泳分析时，偶然发现由原点释放出大量 Hb，而且与其他人不同，从而引起注意。本项研究旨在论证"残留-再释放"这一现象的存在，提出对它的检测方法，探讨其理论和实践意义。

2 材料与方法

2.1 血液学资料 轻型β-地中海贫血患者为汉族，祖籍四川，现居住在包头市。全血计数及各种 RBC 指数的测定结果来自包头医学院第一附属医院检验科：WBC $4.9×10^9$/L，RBC $5.24×10^{12}$/L，HGB 115g/L，HCT 0.365/L，MCV 69.7fl.，MCH 21.9pg，MCHC 315g/L，PLT $212×10^9$/L，LYM% 0.454，MXD% 0.093，NEUT% 0.453，LYM $2.2×10^9$/L，MXD $0.5×10^9$/L，NEUY $2.2×10^9$/L，RDW 43.0 fl，PDW 14.3fl。MPV 10.0 fl。P-LCR 0.276。常规 Hb 凝胶电泳 HbA 28%，未见异常 Hb，HbF 未见增多。

2.2 单向 Hb 释放试验 使用我研究室的常规淀粉琼脂糖混合凝胶电泳[1-2]，将患者的 RBC 借助 3mm 滤纸加在凝胶的阴极侧。第一次电泳 2 小时(5V/cm)。停电 30 分钟后再电泳，通电 30 分钟、停电 30 分钟，交替进行，共 6 小时。然后用联苯胺染色，拍照留图。

2.3 双向 Hb 释放试验 方法基本同上，只是第二次电泳时改变方向，调转 90°，还是停电—通电交替进行，各 30 分钟、共 6 小时。染色、拍照同上。

3 结果

3.1 轻型β-地中海贫血患者 RBC 标本，在单向第一次电泳后原点处残留较多 Hb，它们在第二轮电泳时逐步释放出来，表现为由上向下明显的梯带或竹节带，见图 24-1。

3.2 轻型β-地中海贫血患者 RBC 标本，在双向第一次电泳后原点处残留较多 Hb，它们在第二轮电泳时逐步释放出来，表现为由上向下明显的梯带或竹节带，见图 24-2。

3.3 临床应用的初步结果 在实验室结果明确的基础上，对一些临床标本进行初步筛查。血液病患者中，轻型β-地中海贫血又有同一家族的两名患者，都查到"残留和释放"增多。有 1 例急性淋巴细胞白血病患者，也出现类似结果。传染科患者中，有的肝硬化患者查到"残留和释放"增多，但并非都如此，原发性肝癌也如此，有的增多，有的不增多。普通外科患者中，个别胆囊炎、胆石症患者也出现类似结果，多数此类患者"残留和释放"并不增多。举例见图 24-3。

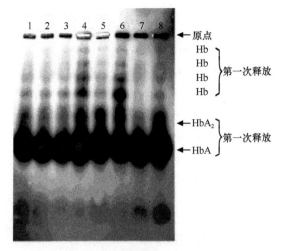

图 24-1 单向 Hb 释放试验结果

注释：淀粉-琼脂糖混合凝胶电泳，联苯胺染色。4、6 号标本为轻型β-地中海贫血患者 RBC。其余均为健康成人对照

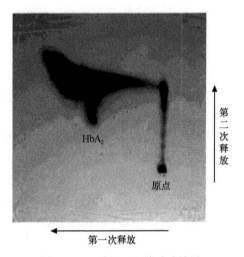

图 24-2 双向 Hb 释放试验结果

注释：原点在右下角，患者 RBC 加在此处。第一轮电泳，按箭头方向由右向左进行。结束后调转 90°进行第二轮电泳，泳动方向由下向上。此时是通电 30 分钟、停电 30 分钟，交替进行

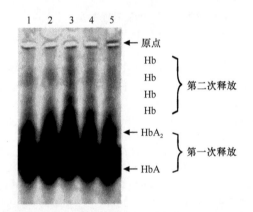

图 24-3 急性淋巴细胞白血病患者单向第一次 Hb 电泳结果

注释：淀粉-琼脂糖混合凝胶电泳，联苯胺染色。3 号标本为急性淋巴细胞白血病患者 RBC。1、2、4、5 为其他患者 RBC

由图 24-3 可以看出，急性淋巴细胞白血病患者 RBC 标本在单向第一次电泳后原点处也残留较多 Hb，与图 24-1 中轻型β-地中海贫血患者 RBC 标本类似。

4 讨论

细胞电泳或 RBC 电泳有多种[1-3]，它们都是直接分析 RBC 的电泳行为，未涉及 RBC 内的 Hb，无法了解 RBC 内 Hb 的存在状态。我们曾用淀粉-琼脂糖混合凝胶电泳证明 HbA_2 在 RBC 中可能并非游离存在[4-7]。HbA_2 是与 HbA 结合存在，也许还有另外一种膜上成分 X，三者结合($HbA-HbA_2-X$)，形成 RBC 电泳时释放出来的"$RBC-HbA_2$"。X 可能是磷脂类物质，有待进一步研究。

以上是过去发现的"HbA_2 现象"，它是 RBC 在第一次电泳时出现的。此时原点还有红色沉淀，没有泳出来，说明还有 Hb 残留。当时认为，这可能是一些不溶解成分，停留在原点处不动。但是，如果停电—再通电，又会由原点释放出一些 Hb，这在当时并未引起特别

注意。26年后,对1例轻型β-地中海贫血患者RBC标本进行电泳分析时,又遇到上述停电——再通电情况,发现他的血液标本释放出来的Hb明显多于其他人。再故意制造多次停电——再通电后,每次都有释放,形成梯带或竹节带,而且总是轻型β-地贫患者标本的释放多于其他人。此时,我们开始意识到有了新发现,好像看到RBC中Hb的更多存在状态:有些Hb与RBC膜结合不太牢固,第一次电泳就能释放出来;另外一些Hb则不然,与RBC膜结合比较牢固,需要第二次电泳(再第二次电击)才能释放出来。残留和释放的机制不明,但是轻型β-地贫患者标本的释放情况特殊,这才使它暴露出来,这是非常有趣的。

值得注意的是,此现象发现于轻型β-地中海贫血,但不仅限于此病。初步的临床筛查,看到一些与地中海贫血无关的疾病也可出现"残留和释放"增多。这里有急性淋巴细胞白血病、原发性肝癌等,但所有这类疾病并未出现相同结果。是否与疾病的进展情况或严重程度有关有待进一步研究。"残留和释放"是RBC中Hb存在状态的一种机制,涉及RBC膜。看来,这种机制比想象的要复杂,除了遗传因素外,血液的环境情况(如氧气供应,各种代谢产物的堆积等)也可能影响到它。是否还有基因多态性等因素都是今后要深入研究的内容。

参 考 文 献

[1] 秦文斌. 血红蛋白 A_2 现象. 包头医学院学报, 1990, 7(3): 1-76.
[2] 秦文斌, 梁友珍. 血红蛋白 A_2 现象 I. 此现象的发现及其初步应用. 生物化学与生物物理学报, 1981, 13(2): 199.
[3] 秦文斌. 红细胞外血红蛋白A与血红蛋白A_2之间的相互作用. 生物化学杂志, 1991, 7(5): 583.
[4] 秦文斌. 血红蛋白的A_2现象发生机制的研究 "红细胞HbA_2" 为HbA_2与HbA的结合产物. 生物化学与生物物理进展, 1991, 18(4): 286.
[5] Baskurt OK, Tugra l E, Meiselman H I. Particle electrophoresis as a tool to understand the aggregation behavior of red blood cells. Electrophoesis, 2002, 23(13): 2103-2109.
[6] Wilk A, Rośkowicz K, Korohoda W. A new method for the preperative and analytical electrophoresis of cells. Cellular and Molecular Biology Letters, 2006, 11(4): 578-593.
[7] Snyder RS, Rhodes PH, Herren BJ, et al. Analysis of freezone electrophoresis of fixed erythrocytes performed in microgracity. Electrophoresis, 2005, 6(1): 3-9.

[原文发表于"包头医学院学报,2007,23(6): 261-263"]

第二十五章 不连续通电的红细胞电泳
——一种新的血红蛋白释放试验

RBC electrophoresis with discontinuous power supply —a newly established hemoglobin release test

Yan Su[1], Guo Shao[1], Lijun Gao[1], Lishe Zhou[1], Liangyi Qin[2], Wenbin Qin[1]

(1. Laboratory of Hemoglobin, Baotou Medical College, Baotou, P. R. China; 2. Clinical laboratory of Nanhui Central Hospital, Shanghai, P. R. China)

Abstract

In this paper, we aimed to introduce a newly established red blood cells(RBCs)electrophoresis method hemoglobin release test(HRT)and tried to determine its significance. Human blood samples from β-thalassemia patients and healthy controls were analyzed with HRT, which was carried out on starch–agarose mixed gel. First, the whole blood samples were electrophoresed for 2 h, then paused for 15 min and ran for 15 min by turns. This "pause-run-pause" experiment was performed for several turns and the total electrophoresis time lasted for about 6 h. The results showed that some other hemoglobin(Hb)components were released from the origin of each sample during the HRT, and the samples from β-thalassemia patients released more Hb than the healthy controls. This finding demonstrates that Hb may exist differently associated in RBCs, and it may have an important theoretical and clinical significance in Hb and RBC research.

Key words β-thalassemia / Hemoglobin / Hemoglobin release test / Red blood cell Abbreviations: **Hb** or **HGB**, hemoglobin; **HCT**, hematocrit; **HRT**, hemoglobin release test; MCH, mean corpuscular hemoglobin; **MCHC**, mean corpuscular hemoglobin concentration; **MCV**, mean corpuscular volume; **RBC**, red blood cell

It is well known that there are a large amount of hemoglobins(Hbs)in human red blood cells(RBCs)and the major components of adult Hb are HbA(96%)and HbA_2(2%–3%). The structure and function of Hb has been well studied, but the actual state of Hb in RBC is not very clear till date. In the past, the binding of Hb with erythrocyte membrane had been studied using Hb quenching of the fluorescence intensity of 12-(9-anthroyl)stearic acid embedded in the lipid membrane [1–2]. The inner side of the RBC membrane is composed of phosphatidylserine lipids, but these are normally masked by membrane proteins. In cases where abnormal Hbs are increased or when membrane lipids are abnormally exposed, Hbs might interact irreversibly with the lipid layer and distort the membrane [3–6]. Apart from the membrane components, Sitdikova [7] demonstrated that the Hb could also interact with serum proteins of mice with Ehrlich carcinoma. Ni-

---

Correspondence: Dr. Wenbin Qin, Laboratory of Hemoglobin, Baotou Medical College, Baotou, Inner Mongolia 014060, P. R. China
E-mail: qinwenbinbt@sohu.com
Fax: 86-472-5152442

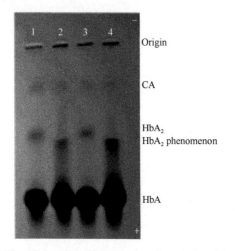

Fig. 25-1 HbA$_2$ phenomenon. Sample 1 and 3 are hemolysate, sample 2 and 4 are RBC, all of these samples come from nomal adult. HbA$_2$ released from RBC moves faster than that from hemolysate.

koli[8] was the first to demonstrate that cholesterol could bind to Hb in the normal human erythrocytes. Our laboratory observed that the released HbA$_2$ from RBC moved faster than that from the hemolysate when RBC electrophoresis was performed on starch–agarose mixed gel. We named this differential electrophoresis behavior of HbA$_2$ as "HbA$_2$ phenomenon" in 1981(Fig. 25-1)[9]. From this result, we primarily speculated that the HbA$_2$ might exist by binding with HbA or other membrane components in RBC [10], but the exact mechanism is not yet well clarified. In this research, we aimed to develop a hemoglobin release test(HRT), by describing its method of discovery, demonstrating the existence of a "residue–release" phenomenon, and observing its theoretical and clinical significances.

This study had been approved by our local ethics committee. Blood samples from healthy adult people and two β-thalassemia patients(mother and her daughter)were collected randomly from the first affiliated hospital of Baotou Medical College. Before the blood samples were collected, all the people who took part in this experiment were asked to sign the consent information. Blood samples were anti-coagulated with heparin, stored at 41℃, and generally analyzed within 24 h. Routine hematological examinations including RBC count, hemoglobin(HGB), hematocrit(HCT), mean corpuscular volume(MCV), mean corpuscular hemoglobin(MCH)and mean corpuscular hemoglobin concentration(MCHC)were performed with SYSMEX XE-2100 Hematology Analyzer(made in Japan). HbA$_2$ quantification was performed by routine elution method after electrophoresis of hemolysate on cellulose acetate membrane in Tris-EDTA borate buffer(pH 8.6), the absorbance of Hb eluant was detected with spectrophotometer at 540 nm. Dissolve 0.2 g agarose and 0.8 g starch in 50 ml TEB buffer(pH8.6), heat the solution until the agarose melt and then lay the gel on a 10 cm×20 cm glass while hot. After solidification, about 10 mL blood samples were applied on the cathodic side of the gel using 3MM filter paper as described earlier[11–12]. The first electrophoresis was performed in borate buffer(0.3 mol/L boracic acid, 0.06mol/L NaOH, pH9.0)at 5 V/cm for 2 h, then paused for 15 min and ran for 15 min by turns. It took about 6 h for the entire electrophoresis. After electrophoresis, we first observed the red bands on the gel directly, and then sequentially stained the gel with Ponceau Red and Benzidine(Note: Benzidine is highly carcinogenic, it could also be substituted by leucobase of malachite green which is non-carcinogenic, but its staining specificity is lower than Benzidine). The sample addition, first electrophoresis and staining methods of double-direction HRT were similar to that of the single-direction HRT. The only difference was that after eletrophoresis the direction of the electric field was altered vertically and electrophoresis was run for 15 minutes with a break of 15 minutes alternately. The routine hematological examination and HbA$_2$ quantification results are shown in Table 25-1. Sample 1, 2, 3, 5, and 7 were the blood from healthy control, sample 4 and 6 were the blood from mother and daughter β-thalassemia patients, whose hematological examination results of MCV, MCH, and MCHC were all lower than the normal standard, but the quantification of HbA$_2$ was higher than the normal standard. With these blood samples, we performed the single-direction and double-direction HRT. After electrophoresis, we first observed the red band on the gel without staining, and then the gel was stained with Ponceau Red, which stained all the proteins red, followed by staining with Benzidine that specifically stained the Hb blue. In these experiments, there were many red sediments that stayed at the origin after the first electrophoresis, and additional Hbs were released from these sediments during each cycle of the "run-pause-run" electrophoresis, and were located between HbA and the origin on the electrophoresis gel, resembling the ladder(Fig. 25-2A). Unlike the healthy blood samples, β-thalassemia blood samples released more Hb components during each HRT-step. To further demonstrate this phenomenon, we performed the double-direction HRT with β-thalassemia and healthy blood samples. As shown in Fig. 25-2B, besides the single- direction release, there were also many other released Hb bands upon the original position after the second round of electrophoresis. Again, more Hb was released from β-thalassemia patients when compared with their healthy counterparts.

Table 25-1 Routine hematological examination results of the subjects

Sample	1	2	3	4	5	6	7
RBC($\times 10^{12}$/L)	4.61	4.90	5.23	4.24	5.31	5.05	4.98
HGB(g/L)	140	149	150	115	148	118	147
HCT(L/L)	0.462	0.433	0.425	0.365	0.428	0.383	0.432
MCV(fL)	86.2	86.1	86.3	69.7 ↓	86.6	70.6 ↓	85.5
MCH(Pg)	30.4	30.4	28.7	21.9 ↓	27.9	21.1 ↓	29.5
MCHC(g/L)	352	353	332	315 ↓	322	312 ↓	345
HbA$_2$(%)	2.3	2.4	2.1	8.9 ↑	2.1	9.1 ↑	2.5

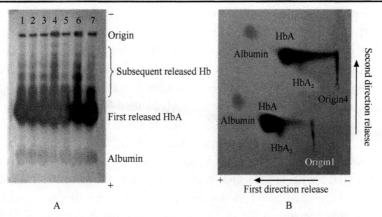

Fig. 25-2 HRT of blood from healthy people and β-thalassemia patients.(A)Single-direction HRT of the blood from healthy people and β-thalassemia patients.(B)Double direction HRT of the blood from healthy people(sample 1)and β-thalassemia patient(sample 4). Electrophoresis was performed on mixed starch–agarose gel, samples 4 and 6 were blood from daughter and mother β-thalassemia patients, and the other samples were from normal healthy adults.

There are many kinds of RBC electrophoresis methods and all of them can be used to analyze the behavior of RBC [13]; however, none of them can be used to study the actual state of Hb in RBC. In our previous study, we observed the "HbA$_2$ phenomenon, " which made us speculate that HbA$_2$, HbA, and perhaps some other membrane components(X)might bind with each other and form HbA$_2$-HbA-X complex in RBC [10]. Furthermore, we also found that there were always some red sediments, which stayed at the original site, but it did not draw our attention, as we presumed that the red sediments might be some insoluble components of the RBC. Recently, when we carried out the whole-blood electrophoresis of β-thalassemia patients, the power was suddenly cut off for about 15 min and then switched on again. A strange phenomenon appeared, when some other Hb components were released again from the origin of each sample, and the samples of β-thalassemia patients released more Hb than the other samples(Fig. 25-2). Hence, we made the "power off and then on" deliberately during the electrophoresis, and later named it as HRT. Each time, we found that Hb was released and its release from the β-thalassemia patients was always more than the healthy adults. This result made us realize that we encountered a new problem, and we believe that the mechanism of Hb "residue-release" perhaps dealt with the Hb existence state in RBC.

Discontinuous power supply during electrophoresis is the most important innovation of HRT technique. As to the mechanism of HRT, we primarily speculate that the electric pulse from turning-on and -off the power supply could create plasmatorrhexis of RBCs. Also, some Hbs in the RBC might bind with the membrane loosely and would be released in the first cycle of electrophoresis, whereas the other Hbs might bind with the RBC membrane tightly and would be released during

The subsequent electrophoresis. β-thalassemia is caused by the decreased or defective synthesis of β-globin chain, whereas the synthesis of α-globin chain is normal. The excess or free α-globin chains are structurally intact, but can damage the erythrocyte membrane not only by sedimentation on membrane, but also by oxidative damage. Furthermore, both the membrane skeleton of erythrocyte and the metabolism of RBC are abnormal in β-thalassemia patients. Erythrocyte osmotic fragility tests confirmed that the RBCs of β-thalassemia are more rigid than the normal ones, however, the deformability of RBCs is reduced and they can easily be destroyed in the blood circulation. In our experiment, the lower osmotic fragility of β-thalassemia may delay the first release of Hb, and the subsequent outflow of Hb might increase. In addition, the increase of released Hb of β-thalassemia patients may be caused by the abnormal membrane structure and the different binding state of the abnormal Hbs. Hbs from β-thalassemia patients are usually instable and can combine with the membrane proteins, such as band 3 protein, in the membrane or form heinz body in cytoplasm. During each HRT, these Hbs could be continuously released from these Hb complexes. The released Hb might be HbA, because in double-direction HRT(Fig. 25-2B), we did not see the released ladders below the HbA_2 band. Finally, it must be pointed out that the "residue-release" phenomenon of β-thalassemia patients observed with pure RBC is not as obvious as with whole blood, and this increased "residue release" phenomenon can also appear in some other diseases, so the exact mechanism of HRT needs further research. This work was supported by grants from the Wenbin Qin Foundation for Scientific Research. We especially acknowledge all of the people who donate their blood samples for our research. The authors have declared no conflict of interest.

References

[1] Shaklai, N., Ranney, H. R., *Isr. J. Med. Sci*, 1978, *14*, 1152-1156.
[2] Yamaguchi, T., Kuranoshita, K., Harano, T., Kimoto, E., *J. Biochem*, 1993, *113*, 513-518.
[3] Ideguchi, H., *Rinsho Byori*, 1999, *47*, 232-237.
[4] Rees, D. C., Clegg, J. B., Weatherall, D. J., *Blood*, 1998, *92*, 2141-2146.
[5] Shaklai, N., Sharma, V. S., Ranney, H. M., *Proc. Natl. Acad. Sci. USA*, 1981, *78*, 65-68.

[6] Datta, P., Chakrabarty, S., Chakrabarty, A., Chakrabarti, A., *Biochim. Biophys. Acta*, 2008, *1778*, 1-9.
[7] Sitdikova, S. M., Amandzholov, B. S., Serebryakova, M. V., Zhdanovich, M. Y., Kiselevskii, M. V., Donenko, F. V. *Bull. Exp. Biol. Med*, 2006, *141*, 624-627.
[8] Nikolić, M., Stanić, D., Antonijević, N., Niketić, V., *Clin. Biochem*, 2004, *37*, 22-26.
[9] Qin, W. B., Liang, Y. Z., *Acta Biochem. Biophys. Sin*, 1981, *13*, 199-201.
[10] Qin, W. B., J. Baotou Medical College(Special issue), 1990, 7, 1-76.
[11] Qin, W. B., *Chin. Biochem. J*, 1991, *7*, 583-584.
[12] Qin, W. B., Prog. Inorg. Biochem. Biophys, 1991, 18, 286-287.
[13] Baskurt, O. K., Tugral, E., Neu, B., Meiselman, H. J., *Electrophoresis*, 2002, *23*, 2103-2109.

(*Electrophoresis* 2009，30，3041-3043 Short Communication)

第二十六章 普通外科患者血红蛋白释放试验的比较研究

韩丽红[1]　闫　斌[3]　高雅琼[2]　高丽君[2]　秦艳晶[2]　秦文斌[2]

(包头医学院 1. 生化教研室；2. 血红蛋白研究室；3. 第一附属医院普外科，包头　014010)

摘　要

目的：研究一些普外患者血红蛋白释放试验(HRT)结果的差异，探讨其临床意义。

方法：普通外科住院患者，取静脉血做 HRT，采用 3 种 HRT 定点释放技术，即等渗室温 HRT、等渗 37℃ HRT、低渗 37℃ HRT。各项处理结束后，进行淀粉-琼脂糖混合凝胶薄层电泳，第一次电泳后，停电一段时间后继续第二次电泳，丽春红、联苯胺染色后观察血红蛋白的释放情况。

结果：①等渗室温 HRT，8 例患者中有 1 例出现阳性结果；②等渗 37℃ HRT，10 例患者中有 3 例出现阳性结果；③低渗 37℃ HRT，同上 10 例患者中有 5 例出现明显阳性结果。

结论：各种类型 HRT 的结果不同，灵敏度有差异，临床意义也有区别，可以提供多种疾病信息。HRT 的基本原理是观察不同状态下红细胞膜释放血红蛋白的能力，详细机制有待于进一步研究。

关键词　普通外科；血红蛋白释放试验 HRT；等渗室温 HRT；等渗 37℃ HRT；低渗 37℃ HRT

The Comparative Study On Hemoglobin Release Test Between Different General Surgical Patients

Han Lihong[1], Yan Bin[3], Gao Yaqiong[2], Gao Lijun[2], Qin Yanjing[2], Qin Wenbin[2]

(1. Department of Biochemistry, Baotou Medical College; 2. Institute of Genetic Diagnosis, Baotou Medical College; 3. General Surgery, The First Affiliated Hospital of Baotou Medical College, Baotou 014010)

Abstract

Objective: To study the hemoglobin release test on different general surgical patients, and to approach it's significance. **Methods:** Blood were drawn from general surgery patients and the hemoglobin release test(HRT)was carry out. There are three kinds of technique: (1)isotonic HRT at room temperature; (2)isotonic HRT at 37℃; (3)hypotonic HRT at 37℃. Then the blood samples were analyzed by starch–agarose mixed gel electrophoresis. After first run of electrophoresis, second run of electrophoresis was performed after had shut down the power for some time, stain with ponceau and benzidine, then the results of release of hemoglobin from RBC was observed. **Results:** (1)isotonic HRT at room temperature: there is 1 positive outcome in 8 patients; (2)isotonic pressure at 37℃ HRT: there are 3 positive outcomes in 10 patients; (3)hypotonic at 37℃ HRT: there are 5 positive outcomes in 10 patients ibid. **Conclusion:** Various types of HRT have different results, different sensitivity, the clinical significance. It may provide a wide range of disease information. The basic principles of HRT is observation of different state of red cell membrane to release the ability of hemoglobin. The detailed mechanism is to be further studied.

Key words General surgery; Hemoglobin release test(HRT); isotonic HRT at room temperature; isotonic HRT at 37℃ HRT; hypotonic HRT at 37℃

1 前言

1981年，本研究室利用淀粉-琼脂糖混合凝胶电泳技术发现"血红蛋白 A_2 现象"，证明红细胞内血红蛋白 A_2 是以血红蛋白 A-血红蛋白 A_2(HbA-HbA$_2$)或 HbA-HbA$_2$-X 形式存在，X为红细胞膜上成分[1, 2]。2007年初，我研究室在研究1例轻型β-地中海贫血病例红细胞电泳过程中偶然发现，该患者红细胞第一次电泳后，比正常人红细胞电泳原点残留有更多的红色物质(加样量相等)；在停电一段时间后再次电泳，该患者标本从原点电泳出来红色物质(后经证实为血红蛋白)比正常成人的多出很多；连续多次停电再通电后，轻型β-地中海贫血患者红细胞释放多条血红蛋白区带(称梯状带或竹节带)[3]。在这一现象的启示之下，我们开始研究其他临床标本的血红蛋白释放情况。本文研究的是一些普通外科标本。

2 材料与方法

2.1 血液样品来源 患者全血标本来自包头医学院第一附属医院普通外科病房，时间是2007年5月至2008年7月。第一批患者8例，是在本项研究早期，临床诊断记录不全，只知其中第5例为等渗 HRT 阳性。第二批患者10例，临床诊断记录全面。

2.2 方法

2.2.1 等渗室温血红蛋白释放试验(HRT) 使用本研究室的常规淀粉琼脂糖混合凝胶电泳[1-3]。将患者的血液借助3mm滤纸加在凝胶的阴极侧。第一次电泳2小时(5V/cm)。停电30分钟后再电泳30分钟，共3小时。然后用丽春红、联苯胺染色，拍照留图。

2.2.2 等渗37℃HRT 方法基本同上，只是血液标本先在37℃保温12小时。

2.2.3 低渗37℃HRT 方法基本同等渗37℃HRT，只是在上电泳前血液标本先用蒸馏水稀释(血∶水 = 8∶12)造成低渗。

2.3 结果判定 首先观察有无定点释放，有定点释放者判为阳性结果，否则为阴性。出现阳性结果后要进行数量比较。等渗 HRT 时，将定点释放带与碳酸酐酶进行比较，区带强度大于碳酸酐酶时为强阳性，否则为弱阳性。低渗 HRT 时，将定点释放带与"血红蛋白 A_2"进行比较，区带强度大于"血红蛋白 A_2"时为强阳性，否则为弱阳性。

3 结果

3.1 等渗室温 HRT 结果 第一批8例普通外科患者，其血液标本做等渗室温 HRT，同时上一板凝胶电泳，结果如图26-1。由此图可以看出，这批患者中，上述第5个标本显示等渗室温 HRT 阳性结果。

3.2 等渗 37℃HRT 结果 第二批10名普通外科患者的血液标本，做等渗37℃HRT，同时上一板凝胶电泳，结果如图26-2。由此图可以看出，这批患者中，第1、6、8标本显示等渗37℃HRT 阳性结果。

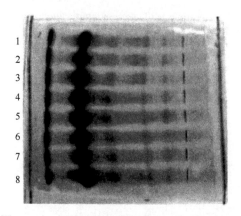
图 26-1　8 例普通外科患者血液标本的等渗室温 HRT 结果
注释：由上向下第 5 个标本，其定点释放区带量明显大于本人的碳酸酐酶

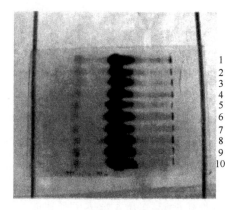
图 26-2　10 例普通外科患者血液标本的等渗 37℃ HRT 结果
注释：由上向下第 1、6、8 标本，其定点释放区带量明显大于本人的碳酸酐酶

3.3 低渗 37℃ HRT 结果　同上 10 例普通外科患者的血液标本，做低渗 37℃ HRT，同时上一板凝胶电泳，结果如图 26-3。由此图可以看出，这批患者中，上述第 1、4、6、8、9 标本显示低渗 37℃ HRT 阳性结果。与等渗 37℃ HRT 相比，此时血红蛋白的释放量明显增多，而且第 4、9 标本出现阳性。

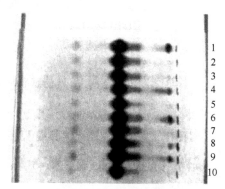
图 26-3　10 例普通外科患者血液标本的低渗 37℃ HRT 结果
注释：由上向下第 1、4、6、8、9 标本，出现定点释放区带，其数量明显大于本人的碳酸酐酶(此时碳酸酐酶量少，区带不明显)。
与"HbA_2"量比较，大于"HbA_2"者为 1、4、6、8，小于或等于"HbA_2"为 9

4　讨论

血红蛋白(Hb)是体内重要的载氧色素蛋白，是由 4 个亚基组成的 4 聚体。

早在 20 世纪 90 年代，包头医学院血红蛋白研究室就发现了"血红蛋白 A_2 现象"，并推测血红蛋白 A_2 在红细胞内不是游离态存在，是以 HbA-HbA_2 或者 HbA-HbA_2-X(X 是膜上成分，可能是磷脂之类的脂类)形式存在[1, 2]。根据文献记载[4-7]，膜上有两种类型的血红蛋白结合位点：高亲和性的带 3(band 3)蛋白，以及低亲和性的血型糖蛋白和某些磷脂的极性头部。上述 X 是否就是某些磷脂的极性头部，需要进一步明确。

这里所说的"血红蛋白释放试验"，是专门研究红细胞在凝胶电泳过程中释放血红蛋白的情况。在常规电泳(不断电)时，红细胞释放出来的血红蛋白 A_2，与溶血液者不同，称为"血

红蛋白 A_2 现象"，实际上，这是血红蛋白的第一次释放。进行一段时间的普通电泳后，人为断电一段时间后再通电电泳，又有一部分血红蛋白从原点红细胞中释放出来，这是血红蛋白的第二次释放，称之为定时释放或定点释放。如果，电泳过程中多次开闭电源，就会陆续释放出多个血红蛋白区带(呈梯状或竹节状)，这是血红蛋白的多次释放，称之为梯带释放。本试验是在第一次电泳断电一段时间后继续电泳，则原点残留的物质又会泳出一部分。这部分物质也是血红蛋白[7]。这说明红细胞中除了以前研究、推断出的 HbA-HbA$_2$-X 结构外，还有另外一部分血红蛋白与细胞膜的关系更密切，与细胞膜结合得更紧密而使其在第一次电泳时并没有从红细胞中释放出来。这是一个很有趣的课题，有待进一步深入研究。

本文的目的是比较研究一些普通外科患者的血红蛋白释放试验结果，分析其与临床病情的关系，探讨 HRT 的可能机制。本文所涉及的患者，其 HRT 结果不尽相同，现对资料相对较全的第二批患者的 HRT 结果进行讨论和分析。第二批患者中 HRT 增强最明显的是 1 号胆石症患者，其次是 6 号肠梗阻患者，其 HRT 增强程度与 1 号相近。再次就是 4 号胆总管结石患者和脾破裂患者，最后是化脓性阑尾炎患者，再往下排：3 号有少许、10 号更少，5 号、7 号接近消失，2 号完全没有释放。1~4 号患者都是胆石症，HRT 结果明显不同，说明 HRT 阳性结果并非来自胆石症，或者并非来自单纯的胆石症，可能另有原因。1 号患者的化验结果初步分析，HRT 增强可能与血脂增高、脂类代谢异常有关。而 2、3 号胆石症患者则无此情况。4 号胆总管结石患者，未见血脂增高，但胆红素代谢及肝功能等出现异常，由此看来，4 号患者的 HRT 增强，不是来自脂类代谢异常，而是可能与胆红素代谢异常等肝功异常有关。6 号患者患有肠梗阻，他的几乎所有的化验结果都降低，给人一种营养不良的印象，这又是一种类型的 HRT 增强。值得注意的是，这个患者有过脑梗死、冠心病、高血压、肺气肿的病史。8 号患者是脾破裂患者，他的化验结果与 6 号患者有类似之处。脾破裂与肠梗阻的 HRT 类型为什么相同或相近，有待进一步研究。最后是 9 号化脓性阑尾炎患者，他的异常化验结果主要是慢性炎症所致。是否所有化脓性阑尾炎都出现此类 HRT 异常，还不是很清楚。

HRT 是我们研究室近期提出的一种实验方法，用来观察各种条件下红细胞释放血红蛋白的情况。释放的动力是通电(电泳)，这是共同因素，不同释放结果则取决于红细胞及其与血浆环境之间的相互作用。根据本次普外患者的化验结果，可将 HRT 初步分成以下几种类型：①脂类代谢异常型；②肝功异常型；③营养不良型；④慢性炎症型。由于标本量有限、化验项目不全，这些分类是比较粗略的。

各种水平 HRT 的灵敏度不同：低渗高于等渗，37℃高于室温。我们用等渗室温 HRT 发现地中海贫血出现"梯带"，注意到淋巴细胞白血病有时出现轻度的 HRT 异常。低渗 37℃ HRT 发现一系列情况，有一些脑梗死患者、尿毒症患者出现异常，有的是脑梗死后遗症患者，仍显示较明显的异常 HRT。灵敏度不同，也可能揭示不同层次的问题，等渗室温 HRT 时，多数临床标本不出现阳性，一旦释放增强，可能病情更严重。总之，HRT 研究工作刚刚开始，初步结果显示它可能显示某些疾病的严重程度。在机制研究方面要做的工作更多，也许在红细胞膜与血红蛋白之间、红细胞膜与血浆环境之间会发现更多的相互作用。

参 考 文 献

[1] 秦文斌. "血红蛋白 A_2 现象"专辑. 包头医学院学报, 1990, 7(3): 1-76.

[2] 秦文斌. 红细胞外 HbA$_2$ 与 HbA 间的相互作用. 生物化学杂志, 1991, 7(5): 583-587.
[3] Shaklai N, Yguerabide Y, Ranney HM. Interaction of hemoglobin with red blood cell membranes as shown by a fluorescent chromophore. Biochemistry, 1977, 16(25): 5585-5592.
[4] Shaklai N, Sharma VS, Ranney HM. Interaction of sickle cell hemoglobin with erythrocyte membranes. Biochemistry, 1981, 78(1): 65-68.
[5] Premachandra BR. Interaction of hemoglobin and its component α and β chains with band 3 protein. Biochemistry, 1986, 25: 3455-3462.
[6] Landolt-Marticorena C, Charuk JHM, Reithmeier RAF. Two glycoprotein populations of band 3 dimers are present in human erythrocytes. Molecular Membrane Biology, 1998, 15(3): 153-158.
[7] 秦文斌, 高丽君, 苏燕, 等. 血红蛋白释放试验与轻型 β-地中海贫血. 包头医学院学报, 2007, 23(6): 561-563.

[原文发表于"临床和实验医学杂志, 2009, 8(7): 67-69"]

第二十七章 血糖浓度和血红蛋白释放试验的比较研究

张晓燕　高丽君　高雅琼　周立社　苏　燕　秦良谊　秦文斌

(包头医学院，包头　014010)

摘　要

目的：探讨血红蛋白释放试验(HRT)与糖尿病的关系，观察血糖浓度对 HRT 的影响。

方法：取各种血糖浓度的血液标本，每例标本同时做全血和红细胞的 HRT(应用梯带释放法)，观察血糖浓度与 HRT 结果的关系；进一步进行体外试验，以研究葡萄糖对 HRT 的影响。

结果：全血 HRT 结果与血糖浓度无关；梯带释放与血糖浓度有一定的相关性；体外试验葡萄糖浓度对红细胞的梯带释放影响明显；葡萄糖能使梯带增强，且有按 5%、10%、50%递增的趋势。

结论：葡萄糖能影响红细胞的 HRT，改变红细胞的通透性。

关键词　血糖；血红蛋白；红细胞；对比研究

Comparative Study on the Blood Glucose Concentration and Hemoglobin Release Test

ZHANG Xiaoyan, GAO Lijun, GAO Yaqiong, ZHOU Lishe, SU Yan, QIN Liangyi, QIN Wenbin

(Baotou Medical College, Inner Mongolia Baotou 014010, China)

Abstract

Objective: To explore the relationship between hemoglobin release test (HRT) and diabetes, and to know the effect of blood glucose on HRT. **Methods**: 10 blood samples was take to do whole blood HRT and erythrocyte HRT (ladder release), simultaneously, analysis the relationship between the concentration of blood glucose and results of HRT. In vitro, the effect of blood glucose on HRT was studied. **Results**: There did not exist relationship between whole blood HRT and blood glucose concentration. But ladder release was associated with blood glucose concentration. In vitro, glucose concentration influenced the ladder release of erythrocyte HRT. Glucose enhanced the ladder release with the trend of 5%、10%、50%. **Conclusion**: Glucose could influence the erythrocyte HRT with changing the permeability of erythrocyte.

Key words　blood glucose; hemoglubins; erythrocytes; comparative study

1　前言

2006 年，本研究室在用轻型 β-地中海贫血患者血液标本进行淀粉-琼脂糖凝胶电泳分析时，发现由原点释放出多量血红蛋白(hemoglobin，Hb)，从而开始建立"血红蛋白释放试验(hemoglobin release test，HRT)"[1-4]。当时的试验中，主要是比较各个标本之间的全血梯带释放，发现轻型β-地中海贫血患者这种释放明显增强，此时，地中海贫血的红细胞梯带释放也增强，但与对照组的比较不如全血明显。在比较研究其他疾病的全血释放时，注意到与地中海贫血类似的情况很少，多数都是全血释放少，红细胞释放较多，规律待进一步研究。糖尿病情况也类似。本研究室近期同时比较血糖和两种释放(全血释放及红细胞释放)的关系，观察到血糖浓度与红细胞释放有一定相关性，需要进一步研究，报告如下。

2 材料与方法

2.1 标本来源 10例血液标本，来自门诊及肾内科住院患者，多数血糖升高，也有正常和稍低者，血糖浓度分别为 8.19mmol/L、6.53mmol/L、7.29mmol/L、6.56mmol/L、6.96mmol/L、6.15mmol/L、8.19mmol/L、9.41mmol/L、19.81mmol/L、20.69mmol/L。

2.2 全血血红蛋白释放试验 取全血5μl，加在5mm×1mm滤纸条上，插入淀粉-琼脂糖混合凝胶内，进行常规梯带试验。电势梯度6V/cm，通电15分钟，断电15分钟，反复进行，共5小时，丽春红-联苯胺复合染色，摄影留图，观察红细胞中血红蛋白释放情况。

2.3 红细胞血红蛋白释放试验 全血低速离心去上清液，用生理盐水洗涤红细胞5次。最后，在红细胞上留等体积盐水，混匀，重复上述全血血红蛋白释放试验。

2.4 体外葡萄糖试验 取血糖正常的全血标本，分别加入等体积的3种葡萄糖溶液(5%葡萄糖、10%葡萄糖、50%葡萄糖)，混匀后37℃保温2小时，健康对照组加等体积盐水，同样处理。2小时后取出，重复上述红细胞血红蛋白释放试验。

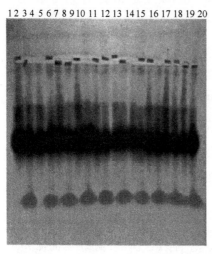

图27-1 部分标本的双释放电泳结果

注释：泳道由左向右1～20，1，2为1个标本，单数为红细胞释放结果，双数为全血释放结果，梯带阳性的泳道是1(血糖浓度8.19mmol/L)、5(血糖浓度6.96mmol/L)、7(血糖浓度8.19mmol/L)、15(血糖浓度9.14mmol/L)、17(血糖浓度19.8mmol/L)、19(血糖浓度20.69mmol/L)

3 结果

3.1 每例标本均同时做全血和红细胞梯带释放试验，观察其与血糖浓度关系，结果见图27-1。由图27-1可见，全血释放几乎都没有梯带，红细胞释放有梯带；而且血糖浓度与红细胞的梯带释放的相关性比较明显。

3.2 3种浓度葡萄糖溶液对红细胞释放的影响，见图27-2。由此图可以看出，全血基本无梯带，红细胞梯带随葡萄糖浓度增加而增加。

4 讨论

目前认为，HRT[1, 2]是以Hb为标志物，通过其电泳释放情况，来了解红细胞膜通透性及其与血浆环境之间的相互关系。人们对糖尿病患者的红细胞膜情况早有了解[3-5]，王桂侠等人[6]证明，老年性糖尿病患者红细胞膜 Na^+-K^+-ATP酶异常；宋皆金[7]提出，糖尿病的发生及发展与红细胞膜 Na^+-K^+-ATP酶及 Ca^{2+}-Mg^{2+}-ATP酶的活性有密切关系。最近，蒋明等[8]运用蛋白质组学技术研究 2 型糖尿病患者红细胞膜蛋白异常,结果显示,精氨酸酶 1(Arginase-1)和脂筏蛋白-1(Flotillin-1)在糖尿病患者红细胞膜上的表达都增加。他们推测,Flotillin-1 通过与 Arginase-1 的相互作用,将后者结合到红细胞膜上,从而调节其活性。本组研究表明这 2 种物质影响了红细胞膜的通透性,从而造成红细胞梯带释放增强,从另一角度探讨了血糖浓度与红细胞蛋白释放的关系。

高血糖标本的游离红细胞梯带释放增强,而包括这种红细胞在内的全血为什么梯带不增强?与此对应,地中海贫血时,红细胞释放增强,其全血释放也增强。说明糖尿病患者血浆中有些物质,能够抑制红细胞梯带释放,即所谓的"浆胞互作"(血浆成分与红细胞之间的相互作用)。但这种抑制作用也是有限的。图 27-1 显示,血糖最高、红细胞梯带最强的全血标本,也能看到轻微的梯带,其他全血则没有。这说明,血糖高到一定程度,血浆成分的抑制作用开始失效。"浆胞互作"中,血浆里哪些成分起作用,如何起作用,还需要深入研究。

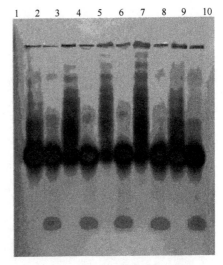

图 27-2　体外葡萄糖试验结果

注释:泳道自左向右,1(红细胞、生理盐水),2(全血、生理盐水),3(红细胞、5%葡萄糖),4(全血、5%葡萄糖),5(红细胞、10%葡萄糖),6(全血、10%葡萄糖),7(红细胞、50%葡萄糖),8(全血、50%葡萄糖),9(红细胞、生理盐水),10(全血、生理盐水)

参 考 文 献

[1] 秦文斌,高丽君,苏燕,等. 血红蛋白释放试验与轻型 β-地中海贫血. 包头医学院学报, 2007, 23 (6): 561-563.

[2] Su Y, Shao G, Gao LJ, et al. RBC electrophoresis with discontinous power supply—a newly established hemoglobin release test. Electrophoresis, 2009, 30: 3041-3043.

[3] 吕达嵘,鲁科峰. 2 型糖尿病患者红细胞免疫功能与血脂过氧化的关系. 放射免疫学杂志, 2009, 22 (2): 162-164.

[4] 韦春玲,姜秋芬,张松筠,等. 伴不同并发症 2 型糖尿病患者红细胞膜流动性的改变. 基础医学与临床, 2004, 24 (4): 473-474.

[5] 兰帆,张素华,邱鸿鑫,等. 2 型糖尿病患者红细胞膜钠-钙逆转换活性改变. 中国糖尿病杂志, 2000, 8 (3): 131-134.

[6] 王桂侠,张兵华,郝守才,等. 老年糖尿病患者红细胞膜 Na^+-K^+-ATP 酶活性及临床意义. 中国老年学杂志, 1999, 19 (2): 55-57.

[7] 宋皆金. 2 型糖尿病肾病患者血清脂联素水平与红细胞膜 ATP 酶的关系. 放射免疫学杂志, 2008, 21 (6): 468-471.

[8] 蒋明,胡小键. 2 型糖尿病人红细胞膜蛋白异常的蛋白质组学研究. 生物化学与生物物理进展, 2007, 34 (suppl): 96-101.

[原文发表于"国际检验医学杂志,2010,31(6):524-525"]

第二十八章　尿毒症患者低渗血红蛋白释放试验的初步结果

高雅琼　王彩丽　高丽君　苏　燕　秦良谊　周立社　秦文斌

(包头医学院，包头　014010)

摘　要

目的：建立低渗血红蛋白定时释放试验，比较研究慢性肾衰竭(CRF)患者红细胞中血红蛋白的存在状态，探讨其机制和临床意义。

方法：用低渗血红蛋白释放试验(HRT)五管法比较研究 CRF 患者与正常人的血红蛋白释放情况。再用低渗 HRT 一管法比较研究 5 例 CRF 与 4 例非 CRF 患者的血红蛋白释放结果。

结果：与正常对照比较，CRF 患者标本出现以下特点——白蛋白前移、血红蛋白 A_2 稍前移、定时释放增多、原点残留和泄漏明显。5 例 CRF 的血红蛋白释放强于 4 例非 CRF 患者。

结论：与正常人或非 CRF 患者比较，CRF 患者血液成分的电泳释放行为变化较大。白蛋白前移可能是它与血中毒素结合有关、血红蛋白释放方面的变化也可能来自毒素对红细胞的影响。

关键词　尿毒症；血红蛋白释放试验；低渗；红细胞；血浆

A preliminary study on hypotonic hemoglobin release test in patients with uremia

GAO Yaqiong, WANG Caili, GAO Lijun, et al.

(Baotou Medical College, Inner Mongolia Baotou 014010)

Abstract

Objective: To establish hypotonic hemoglobin release test (HRT) for comparative study of existing status of hemoglobin in RBC from normal controls and patients with uremia, and to explore its mechanism and clinical significance. **Methods:** Hypotonic HRT with five tubes method was used for comparative study on results of hemoglobin release between normal persons and uremic patients. One tube method was used for comparative study in hemoglobin release results between non-uremic patients and uremic patients. **Results:** In comparison with normal controls, samples of uremic patients showed following features: antelocation of albumin, slight antelocation of HbA_2, increased timed-release, obvious original remains and leakage. The HRT results of five patients with chronic renal failure (CRF) were stronger than those of patients with non-CRF. **Conclusion:** The behavior of electrophoretic release of blood components from uremia is different from that of normal controls or non-uremic patients. Antelocation of albumin may be due to its binding with uremic toxin. Change in Hb release perhaps may be due to the effect of toxin on RBC membrane. But its mechanism in detail needs further investigation.

Key words　Uremia; Hemoglobin release test/HRT; Hypotension; RBC; Albumin

1 前言

2006年，秦文斌等[1, 2]在用轻型β-地中海贫血患者血液标本进行淀粉-琼脂糖混合凝胶电泳分析时，发现由原点释放出多量血红蛋白(Hb)，从而开始建立"血红蛋白释放试验(HRT)"。当时的实验中血液标本未经任何处理，后来发现将血液用蒸馏水适当稀释(低渗处理)可以明显改进实验效果。我们将此项技术称之为"低渗HRT"，与此对应，上次的实验应当称为"等渗HRT"。本研究观察"低渗HRT"在尿毒症研究中的应用，比较尿毒症患者与正常对照之间红细胞膜或红细胞内血红蛋白存在状态的差异，同时探讨其机制和意义。

2 材料及方法

2.1 标本来源 尿毒症患者，女性，50岁，2008年6月包头医学院第一附属医院肾内科住院患者。健康对照为我室的研究生。

2.2 临床资料 患者的化验结果：尿素(UREA)32.3mmol/L↑(正常值3.2～7.1mmol/L)，肌酐(CREA)259.0μmol/L(正常值62～106μmol/L)，尿酸(UA)520.9μmol/L↑(正常值208～506μmol/L)，二氧化碳结合力(CO_2CP)17.6mmol/L↓(正常值22～30mmol/L)，钾(K)8.30mmol/L↑(正常值3.6～5.0mmol/L)，钠(Na)133.0mmol/L↓(正常值137～145mmol/L)，氯(Cl)98.0mmol/L(正常值98～107mmol/L)。诊断为尿毒症，肾功能衰竭，已经透析。健康对照者：上述指标化验无异常。

2.3 低渗Hb释放试验

2.3.1 单向电泳：取5支大EP管，用蒸馏水稀释血液造成低渗，蒸馏水与血液的比例为水：全血(μl)=①18：2、②16：4、③14：6、④12：8、⑤10：10。混匀，各取3μl分别加入5个滤纸条(3mm×1mm)中，然后做淀粉-琼脂糖混合凝胶电泳(电压5V/cm, 电流2mA/cm)。第一次电泳2.5小时，然后停电0.5小时，再通电0.5小时(此为"半小时定时释放")。电泳后先染丽春红，再染联苯胺，拍照留图，观察红细胞中血红蛋白的释放情况。

2.3.2 叠层双向对角线电泳：选单向电泳时效果最明显的稀释比例(水：全血=14：6)做双向电泳。第一向电泳基本同上，只是第二轮电泳时改变方向，调转90°，进行常规电泳：电压5 V/cm，电流2mA/cm，通电3小时左右。染色等同单向电泳。

3 结果

3.1 单向电泳结果 由图28-1可以看出，尿毒症患者的原点残留明显大于对照标本。定时释放(区带在原点的稍前方)也是患者明显增多。"血红蛋白A_2"前移，主要出现在尿毒症患者的第6泳道，其次是第8泳道。血红蛋白泄漏与定时释放结果一致。对照标本，只在第3泳道，有一点定时释放。还有白蛋白，与对照标本比较，尿毒症患者的血浆白蛋白都突向前方。

3.2 双向叠层对角线电泳结果 由图28-2可以看出，对照标本和尿毒症患者标本，对角线上的电泳成分差别不太大，尿毒症患者的原点残留大于对照标本，但不如单向电泳时

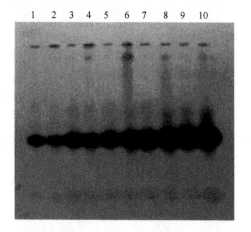

图 28-1　尿毒症患者单向低渗释放电泳结果
注释：低渗 HRT 单向淀粉-琼脂糖混合凝胶电泳，丽春红-联苯胺染色。由左向右，1、3、5、7、9 泳道来自对照标本；2、4、6、8、10 来自尿毒症患者标本

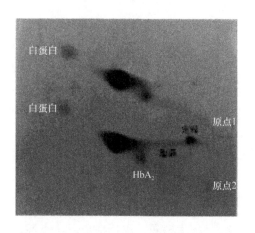

图 28-2　尿毒症患者双向低渗释放电泳结果
注释：低渗 HRT 双向叠层对角线淀粉-琼脂糖混合凝胶电泳，丽春红-联苯胺染色。两个原点：上边的是对照标本，下边的是尿毒症患者标本。第一向，电泳成分由右向左泳动。第二向时，第一向展开的电泳成分全部由下向上泳动，得到以上图像

明显。定时释放，患者更明显。"血红蛋白 A_2"前移，不如单向式明显。血红蛋白泄漏非常清楚。

4　讨论

　　细胞电泳或红细胞(RBC)电泳有多种[3-5]，它们都是直接分析 RBC 的电泳行为，未涉及 RBC 内的 Hb，因此无法了解 RBC 内 Hb 的存在状态。我们曾用淀粉-琼脂糖混合凝胶电泳证明[6-9]，HbA_2 在 RBC 中可能并非游离存在。HbA_2 是与 HbA 结合存在，也许还有另外一种膜上成分 X，三者结合($HbA-HbA_2-X$)，形成 RBC 电泳时释放出来的"$RBC-HbA_2$"。X 可能是磷脂类物质，有待进一步研究。

　　以上是过去发现的"HbA_2 现象"，它是 RBC 在第一次电泳时出现的。此时原点还有红色沉淀，没有泳出来，说明还有 Hb 残留。当时认为，这可能是一些不溶解成分，停留在原点处不动。但是，如果停电—再通电，又会由原点释放出一些 Hb，当时并未引起特别注意。27 年后，对 1 例轻型β-地中海贫血患者 RBC 标本进行电泳分析时，又遇到上述停电—再通电情况，发现他的血液标本释放出来的 Hb 明显多于其他人。再故意制造多次停电—再通电后，每次都有释放，形成梯带或竹节带，而且总是轻型β-地贫患者的释放多于其他人。此时，我们开始意识到有了新发现，好像看到 RBC 中 Hb 的更多存在状态：有些 Hb 与 RBC 膜结合不太牢固，第一次电泳就能释放出来；另外一些 Hb 则不然，与 RBC 膜结合比较牢固，需要第二次电泳(再第二次电击)才能释放出来。残留和释放的机制不明，但是轻型β-地贫患者标本的释放情况特殊，这才使它暴露出来，这是非常有趣的。

　　值得注意的是，此现象发现于轻型β-地中海贫血，但不仅限于此病。初步的临床筛查，看到一些与地中海贫血无关的疾病也可出现"残留和释放"增多。这里有急性淋巴细胞白血病、原发性肝癌等，但所有这类疾病并未出现相同结果。是否与疾病的进展情况或严重程度有关有待进一步研究。"残留和释放"是 RBC 中 Hb 存在状态的一种机制，涉及 RBC 膜。看来，这种机制比想象的要复杂，除了遗传因素外，血液的环境情况(如氧气供应，各种代

谢产物的堆积等)也可能影响到它。是否还有基因多态性等因素都是今后要深入研究的内容。

参 考 文 献

[1] 秦文斌, 高丽君, 苏燕, 等. 血红蛋白释放试验与轻型β-地中海贫血. 包头医学院学报, 2007, 23 (6): 561-563.

[2] Su Y, Shao G, Gao L J, et al. RBC electrophoresis with discontinuous power supply-a newly established hemoglobin release test[J]. Electrophoresis, 2009, 30 (17): 3041-3043.

[3] Baskurt OK, Tugral E, Meiselman H I. Particle electrophoresis as a tool to understand the aggregation behavior of red blood cells[J]. Electrophoresis, 2002, 23 (13): 2103-2109.

[4] Wilk A, Roskowicz K, Korohoda W. A new method for the preperative and analytical electrophoresis of cells. Cell Mol Biol Lett, 2006, 11 (4): 579-593.

[5] Snyder RS, Rhodes PH, Herren BJ, et al. Analysis of free zone electrophoresis of fixed erythrocytes performed in microgravity. Electrophoresis, 2005, 6 (1): 3-9.

[6] 秦文斌, 梁友珍. 血红蛋白 A_2 现象 I. 此现象的发现及其初步应用. 生物化学与生物物理学报, 1981, 13 (2): 199-205.

[7] 秦文斌. 红细胞外血红蛋白 A 与血红蛋白 A_2 之间的相互作用. 生物化学杂志, 1991, 7 (5): 458-464.

[8] 秦文斌. 血红蛋白的 A_2 现象发生机制的研究 "红细胞 HbA_2" 为 HbA_2 与 HbA 的结合产物. 生物化学与生物物理进展, 1991, 18 (4): 282-288.

[9] 韩丽红, 闫斌, 高雅琼, 等. 普通外科患者血红蛋白释放试验的比较研究. 临床和实验医学杂志, 2009, 8 (7): 67-69.

[原文发表于"临床和实验医学杂志, 2010, 9(1): 12-13"]

第二十九章 血红蛋白释放试验与血液流变学检测结果相关性的研究

王翠峰 高丽君 乌兰 苏燕 徐军 周立社 秦文斌

(包头医学院，包头 014010)

摘 要

目的：研究血红蛋白释放试验(HRT)与血液流变学检测结果之间的相关性，探讨有关机制和临床意义。

方法：对 48 例门诊随机标本配对做低渗 37℃ HRT。将血液流变学测试报告中的 3 项全血黏度(低切、中切、高切)值与 HRT 测定值进行比较，分析两者的相关性。

结果：HRT 与血液流变学结果呈一定的正相关，其与各种黏度值的相关系数分别为低切 $r=0.642(P<0.001)$，中切 $r=0.636(P<0.001)$，高切 $r=0.641(P<0.001)$。

结论：HRT 的反应机制中，至少有一部分与血液黏度增加有关。

关键词 血红蛋白释放试验 HRT；血液流变学；全血黏度；低切；高切

Study on the correlation between results of hemoglobin release test and hemorheology

WANG Cuifeng, GAO Lijun, WU Lan, SU Yan, XU Jun, ZHOU Lishe, QIN Wenbin

(Baotou Medical College, Inner Mongolia Baotou 014010, China)

Abstract

Objective: To study the correlation between results of hemoglobin release test and hemorheology, and approach the mechanism of HRT and its clinical significance. **Methods:** 48 blood samples randomly obtained from OPD were mated with 37℃ HRT. The result of hemorheology was compared with that of HRT and their correlation was analyzed. **Results:** The correlation coefficients between HRT and hemorheology were $r=0.642$ of low shear ($P<0.001$), $r=0.636$ of middle shear ($P<0.001$) and $r=0.641$ of high shear ($P<0.001$) respectively. **Conclusion:** It indicates that blood viscosity may play a part in the reaction mechanism of HRT.

Key words Hemoglobin release test; Hemorheology; Blood viscosity; Low shear; High shear

1 前言

2007 年，本研究组在研究 1 例轻型β-地中海贫血病例的红细胞电泳过程中偶然发现，在停电一段时间后再次电泳，该患者标本从原点释放出较多的血红蛋白，从而开始创建血红蛋白释放试验(HRT)[1]。利用这种技术筛查一些普通外科住院患者的血液，其中有 4 例胆结石标本的 HRT 结果有差异，另 1 例标本是脑梗死后遗症，其 HRT 异常最明显[2]。在大量筛查过程中，笔者也注意到脑梗死及其后遗症患者 HRT 增强者较多。由于血液流变学检测在脑梗死辅助诊断中有一定的参考意义[3-6]，故决定研究HRT与血液流变学结果的关系，探讨前者的可能作用机制。

2 材料与方法

2.1 血液样品来源 患者全血标本均来自包头医学院第一附属医院检验科2008年12月至2009年2月共48例门诊样本。所用仪器为普利生N6C全自动血液流变学分析仪。本文使用的数据来自血液流变学检测报告单。

2.2 低渗37℃ HRT 每个标本取全血8μl，加双蒸水12μl，混匀后37℃水浴保温2 h。使用我科研究室的常规淀粉琼脂糖混合凝胶电泳[1]，将保温后的标本借助3mm滤纸加在凝胶的阴极侧。第一次电泳2h(5V/cm)。停电30分钟后再电泳30分钟，共3小时。然后用丽春红、联苯胺染色，拍照留图。

2.3 结果判定 首先观察第2次释放(定点释放)时有无血红蛋白泳出，有定点释放血红蛋白者判为阳性结果，无则为阴性(定为0级)。出现阳性结果者进行数量比较。定点释放的血红蛋白数量可参考碳酸酐酶、血红蛋白A_1、A_2区带，由少到多定为1、2、3、4级。

2.4 统计学分析 使用SPSS11.0软件进行分析。用Spearman相关分析计算出HRT与全血黏度的高切、中切、低切的相关性，$P<0.05$表示有显著性差异。

3 结果

3.1 HRT结果 一次电泳检测12个定点释放结果见图29-1。

3.2 相关分析结果 HRT与各种全血黏度值的相关系数分别如下所示：HRT与低切 $r=0.642(P<0.001)$、HRT与中切 $r=0.636(P<0.001)$、HRT与高切 $r=0.641(P<0.001)$。

4 讨论

HRT创建不久，由于它能在诸多临床标本中发现差异、查出问题，所以引起极大关注。地中海贫血是一种遗传性疾病，本法能查出患者及其家族成员[1]，这在理论和实践上都有重要意义。接着又将HRT深化和细分为室温等渗HRT(上述HRT属于此类)、室温低渗HRT、等渗37℃ HRT、低渗37℃ HRT，灵敏度逐级提升，效果更加明显。笔者利用低渗37℃ HRT筛查一些普通外科住院患者的血液，其中有4个胆结石标本，其HRT结果有明显差异，另1例脑梗死后遗症患者，其HRT最明显[2]。从遗传性疾病到普通外科患者，都能进入本项技术的检测范围，它的作用机制如何，这是迫切需要解决的问题。HRT机制研究涉及面很广，本文从它与

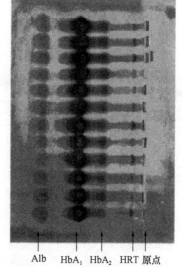

图29-1 一次电泳检测12个定点释放结果

注释：由上向下共12个标本泳道，每个泳道中的区带(由左向右)为清蛋白(丽春红染色)、血红蛋白A_1(黑色，联苯胺染色，下同)、A_2和定点释放血红蛋白区带(箭头所指)。一些标本定点释放的血红蛋白较多，如2、3标本泳道，一些标本释放很少，如7标本泳道

血液流变学关系角度来探讨 HRT 的作用机制。

　　本实验证明，HRT 与血液流变学结果有一定的相关性。几种全血黏度值中，低切反应细胞的聚集性、高切反应细胞的变形性，均与 HRT 有关，说明红细胞的变形能力降低，聚集性增强，黏度加大，造成第一次释放后原点残留增多，第二次释放时才离开红细胞，表现为再释放血红蛋白增多。已知与红细胞变形性有关的疾病很多，地中海贫血也是其中之一，血红蛋白 H 病、重型 β-地中海贫血等红细胞的变形能力明显降低，而轻型 β 地中海贫血红细胞的变形能力接近正常[7]。这在总体上与本研究结果相符，但在细节上略有不同，本研究组的研究能够非常清楚地查出轻型 β-地中海贫血[1]，可见，HRT 的反应能力更强，其作用机制可能还包括红细胞变形性和聚集性以外的因素，HRT 的具体作用机制尚待深入研究。

参 考 文 献

[1] 秦文斌, 高丽君, 苏燕, 等. 血红蛋白释放试验与轻型 β-地中海贫血. 包头医学院学报, 2007, 23 (6) : 561.
[2] 韩丽红, 闫斌, 高雅琼, 等. 普通外科患者血红蛋白释放试验的比较研究. 临床和实验医学杂志, 2009, 8 (6) : 67-68.
[3] 秦任甲. 临床血液流变学. 北京: 北京大学医学出版社, 2006.
[4] 董曼莉, 逄淑华, 翟关中, 等. 脑梗塞病人血液流变学的变化. 中国血液流变学杂志, 2004, 14 (2) : 221.
[5] 高中芳, 席向红. 老年脑梗塞患者血液流变学指标检测分析. 中国血液流变学杂志, 2005, 15 (1) : 73.
[6] 丁翊, 赵红红. 脑梗塞患者血液流变学和血脂指标测定. 微循环杂志, 2004, 14 (3) : 88, 90.
[7] 金永娟. 红细胞可变形性与溶血性贫血. 中华血液学杂志, 1987, 8 (2) : 85.

[原文发表于"医学检验, 2010, 7(4): 64-65"]

第三十章 肝硬化患者血红蛋白释放试验明显异常

宝勿仁必力格　王翠峰　高丽君　高雅琼　苏　燕　秦文斌

(包头医学院，包头　014010)

摘　要

目的： 利用血红蛋白释放试验对肝硬化患者血液标本进行比较研究。

方法： 对化验室38例检测凝血酶原的全血标本进行筛查，发现1例反常标本，回查大生化全项，发现它是肝功多项异常的肝硬化、腹水标本。在此基础上，从上述38个标本中取一例肝功正常的标本，对比之下做等低渗全程试验和双向双层对角线电泳。

结果： 与对照标本相比，肝硬化标本的等低渗全程试验和双向双层对角线电泳明显不同。

结论： 肝硬化患者血液标本的血红蛋白释放试验结果明显异常。

关键词　血红蛋白释放试验；肝硬化；等低渗全程试验；双向双层对角线电泳

Particularly abnormal result of hemoglobin release test in blood samples of patients with liver cirrhosis

BAO Wu renbi li ge, WANG Cuifeng, GAO Lijun, et al.

(Baotou Medical College, Inner Mongolia Baotou 014010)

Abstract

Objective: To make comparative study on blood samples with hemoglobin release test (HRT) in patients with liver cirrhosis. **Methods:** In screening thirty-eight clinical samples for prothrombin, one abnormal sample was picked out. Re-examination of results of its biochemical tests, multi-indexes of liver function were abnormal. The patient was diagnosed as liver cirrhosis accompanied with ascites. Based on these information, one sample with almost normal liver function tests was picked out. These two samples were compared with isotonic-hypotonic full distance test and bidirection-bilayer diagonal electrophoresis. **Results:** The result of sample with liver cirrhosis was quite different from that of control in isotonic-hypotonic full distance test and bidirection-bilayer diagonal electrophoresis. **Conclusion:** The result of blood sample for HRT from patient with liver cirrhosis is obviously abnormal.

Key words　Hemoglobin release test; Liver cirrhosis; Isotonic-hypotonic full distance test; Bidirection-bilayer diagonal electrophoresis

1　前言

肝硬化与红细胞的关系已有许多研究。肝硬化患者红细胞体积分布宽度(RDW)改变可以反映肝功能损害程度[1]。肝硬化患者的红细胞免疫功能的测定表明肝硬化患者红细胞C3b受体花环率低于正常组,而免疫复合物花环率高于正常组,红细胞免疫黏附抑制活性明显高于正常组[2]。肝硬化患者的红细胞变形性及能量代谢变化的研究显示:肝硬化患者红细胞 2,3-二磷酸甘油酸(2,3-DPG)含量增加,而 ATP 含量正常;当患者并发肝性脑病时 ATP 和 2,3-DPG 含量均明显下降,而在肝性脑病缓解后又恢复到并发前的肝性脑病水平[3]。肝硬化患者的红细胞膜结构与功能的病理性改变与正常对照组比较,肝硬化患者红细胞分布宽度(RDW)、刚性指数和电泳指数、红细胞渗透脆性显著增高($P<0.01$),红细胞变形指数和聚集指数、C3b 受体花环率明显下降($P<0.05$)[4]。肝硬化患者存在缺钾缺镁,且随病情加重而加重,缺钾缺镁可能为病情加重的原因之一,活性降低可导致细胞内低钾低镁和钠钙蓄积,缺镁为酶活性在失代偿期进一步降低的原因之一[5]。可见,肝硬化患者的红细胞出现多种变化。本文利用我室建立的"血红蛋白释放试验(HRT)"[6-8],来研究肝硬化患者红细胞内血红蛋白的电泳释放。

2 材料及方法

2.1 材料为了用全血进行 HRT 筛查,由 2011 年 7 月 11 日包头医学院第一附属医院检验科取来检测凝血酶原的全血标本 38 份。筛查出 HRT 明显异常为肝硬化标本,再配以 HRT 正常标本(对照组),标本的化验记录见表 30-1。

表 30-1 肝硬化与对照标本的化验记录

检查项目	肝硬化患者	对照组	参考正常值
直接胆红素(μmol/L)	48.1	5.3	0～7
间接胆红素(μmol/L)	69.3	9.4	2～19
谷丙转氨酶(U/L)	140.0	21.0	0～42
白蛋白(g/L)	27.2	43.9	35～55
球蛋白(g/L)	47.9	26.6	20～30
白蛋白/球蛋白	0.57	1.64	1.5～2.5
胆碱酯酶(U/L)	1448.0	1591.0	5000～14 000
γ-谷氨酰转肽酶(U/L)	90.0	29.0	11～61
谷草转氨酶(U/L)	154.0	20.0	0～37
钠(mmol/L)	134.0	141.0	135～145
胆固醇(mmol/L)	1.45	3.53	3～6

2.2 方法

(1) 单向筛查性 HRT:使用我研究室的常规淀粉-琼脂糖混合凝胶电泳,将患者的全血借助 3mm 滤纸加在凝胶的阴极侧。第一次电泳 2h(电压 5V/cm,电流 2mA/cm),停电 5 分钟后再电泳 35 分钟,第二次阴极与阳极对调电泳 15 分钟(此为 15 分钟的后退),电泳后用丽春红、考马斯亮蓝染色,拍照留图。

(2) 等低渗全程试验

1) 全血等低渗全程试验:用蒸馏水稀释肝硬化腹水患者的全血造成低渗血。取 10 支 0.5ml 的离心管,按表 30-2 进行处理。

对照组全血处理方法与患者一致。此时第 1 管为等渗,其余都是低渗(程度不同)。将这 20 支离心管轻轻混匀,用微量移液器各取 6μl 分别加入 20 个滤纸条(4mm×1mm),插入淀粉-琼脂糖混合凝胶,进行电泳(电压 5V/cm,电流 2mA/cm)。第一次电泳 2 小时,停电 5 分钟,再通电 35 分钟(此为 35 分钟的定时释放),第二次阴极与阳极对调电泳 15 分钟。电泳后用丽春红、考马斯亮蓝染色,拍照留图。

表 30-2 肝硬化患者血液稀释表

管号	1	2	3	4	5	6	7	8	9	10
全血(μl)	20	18	16	14	12	10	8	6	4	2
蒸馏水(μl)	0	2	4	6	8	10	12	14	16	18

2) 红细胞等低渗全程试验:在 1.5ml 离心管中加入全血 200μl,低速离心 3min 后去掉血浆,加入生理盐水至 1ml 处,混合均匀,再低速离心 3 分钟后去掉生理盐水。此为洗一次红细胞,共洗 2 次。最后,留下与红细胞等体积的生理盐水,混匀。然后取 10 支 1.5ml 的离心管,用蒸馏水稀释肝硬化腹水患者的红细胞造成低渗血。

3) 低渗双向双层对角线电泳:在单向电泳结果图中选择电泳效果最明显的第 8 管(水:全血稀释比例为 14:6)做双向电泳。第一向电泳同上,即"35 分钟定时释放"和"15 分钟后退",第二向电泳改变方向,倒极并调转 90°进行常规电泳 1 小时,共通电 4 小时左右。电泳后用丽春红、考马斯亮蓝染色,拍照留图。

3 结果

3.1 单向筛查性 HRT 结果见图 30-1。由图 30-1 可以看出,原点残留物在第 2 号标本明显大于其他标本。第 2 泳道的标本的定释带、后退带都增强,拖泄明显,其他标本定释带、后退带较弱或没有。

3.2 等低渗全程试验结果

(1) 全血等低渗全程试验结果见图 30-2。由图 30-2 可以看出,肝硬化标本的第 8 泳道定释带、后退带都增强,拖泄明显。

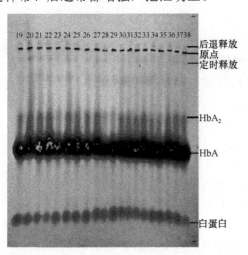

图 30-1 19~38 凝血酶原标本的单向筛查 HRT 结果
注释:此筛查图为序号 19~38 的全血标本,1~18 标本的定时带和后退带弱或不明显所以图未收录。第 2 泳道是肝硬化腹水患者的标本

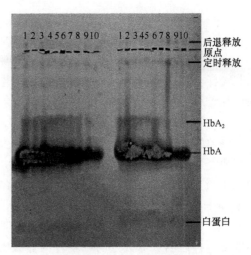

图 30-2 对照与肝硬化全血标本的等低渗全程 HRT 结果
注释:左边 10 个泳道是对照全血标本,右边 10 个泳道是肝硬化全血标本

(2) 红细胞等低渗全程试验结果见图 30-3。由图 30-3 可以看出,与对照标本明显不同,

肝硬化标本大多数泳道定释带都增强、后退带增强不明显，大多数泳道拖泄明显。

3.3 双向双层对角线电泳结果 由图30-4可以看出，肝硬化全血标本在对角线上丙种球蛋白增多，对角线外，前进与后退成分都是HbA。与对照相比，肝硬化患者的后退释放和泄漏明显增强。

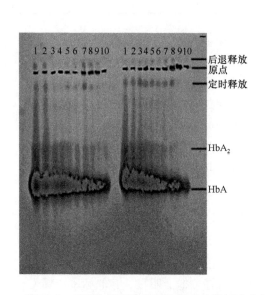

图30-3 对照与肝硬化红细胞标本的等低渗全程HRT结果

注释：左边10个泳道是对照红细胞，右边10个泳道是肝硬化红细胞标本

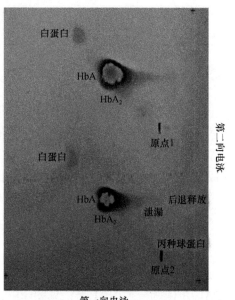

图30-4 对照与肝硬化红细胞标本的双向对角线电泳结果

注释：上层为对照全血标本(原点1)，下层是肝硬化全血标本(原点2)

4 讨论

血红蛋白释放试验(HRT)的完整概念应当是"红细胞内血红蛋白的电泳释放"[8]。将红细胞加样于淀粉-琼脂糖混合凝胶，一次通电就有血红蛋白释放出来，称之为"初释放"，如果电泳过程中加入一次停电—再通电，又会有血红蛋白释放出来，称之为"再释放"。再释放可以一次，也可多次。一次再释放，放出一个血红蛋白区带，称之为"单带再释放"。"单带再释放"可以认为定时、定点，也称"定点再释放"，简称为"定释"。多次再释放，放出多个血红蛋白区带，称之为"多带再释放"，由于像梯子蹬，又称为"梯带再释放"。初释放研究中我们发现了"血红蛋白A_2(HbA$_2$)现象"，证明红细胞内血红蛋白之间的相互作用。筛查性HRT中初释放实验的作用不太大，只有溶血性贫血时才能看到HbA$_2$现象消失。再释放实验的情况不同，它与临床联系较广[9-12]。过去的单向再释放，都是"前进型"的，即停电—再通电后电极方向不变，再释放出来的血红蛋白继续泳向正极。本次实验，除了有"前进型"再释放外，开始引入"后退型"再释放，即停电—再通电后改变电极方向，阴极与阳极对调，此时再释放出来的血红蛋白并不泳向正极，而是退向负极。本文肝硬化患者标本就是前进和后退再释放都增强。在此基础上再做等低渗全程，就能看到它与对照标本的区别。

双向对角线电泳能清楚看到后退成分的存在，并且证明它主要是血红蛋白。也就是说，肝硬化患者全血红细胞中血红蛋白存在状态明显异常，前进再释放放出很多血红蛋白后，还有较多血红蛋白残留，后退时又释放出来。

肝硬化有代偿期与失代偿期之分，失代偿期患者肝功能检测明显异常而且可能有腹水等，本文病例可能属于此类型。如果是这样，肝硬化失代偿期患者除肝功能检测明显异常之外，还涉及红细胞内血红蛋白电泳释放异常。HRT 的前期研究，已经显示出临床应用前景(如糖尿病、尿毒症等)，现在又可检测肝硬化，今后还有许多内容需要进一步研究。

参 考 文 献

[1] 吴诗品, 黄自存. 慢性肝病红细胞体积分布宽度变化及其临床意义. 临床肝胆病杂志, 1997, 13 (1) : 36-37.
[2] 何浩明, 冯岚, 田小平, 等. 肝硬化患者的红细胞免疫功能. 蚌埠医学院学报, 2000, 25 (1) : 67-68.
[3] 彭友清, 吕敏和, 吕红, 等. 肝硬化患者红细胞变形性及能量代谢的变化.临床肝胆病杂志, 2002, 18 (5) : 277-278.
[4] 吕霞飞, 郭希超, 王佩佩. 肝硬化患者红细胞膜结构与功能的病理性改变. 浙江中西医结合杂志, 2004, 14 (7) : 421-423.
[5] 王方剑, 刘安立, 曹洁, 等. 肝硬化红细胞钠钾 ATP 酶、钙镁 ATP 酶及钠钾钙镁改变. 世界华人消化杂志, 2006, 14 (7) : 722-726.
[6] Su Y, Shao G, Gao L, et al. RBC electrophoresis with discontinuous power supply -a newly established hemoglobin release test. Electrophoresis, 2009, 30 (17) : 3041-3043.
[7] Su Y, Gao L, Ma Q, et al. Interactions of hemoglobin in lived red blood cells measured by the electrophoresis release test. Electrophoresis, 2010, 31 (17) : 2913-2920.
[8] 秦文斌. 活体红细胞内血红蛋白的电泳释放. 中国科学, 2011, 41 (8) : 597-607.
[9] 韩丽红, 闫斌, 高雅琼, 等. 普通外科患者血红蛋白释放试验的比较研究. 临床和实验医学杂志, 2009, 8 (7) : 67-69.
[10] 张晓燕, 高丽君, 高雅琼, 等. 血糖浓度和血红蛋白释放试验的比对研究. 国际检验医学杂志, 2010, 31 (6) : 524-525.
[11] 高雅琼, 王彩丽, 高丽君, 等. 尿毒症患者低渗血红蛋白释放试验的初步结果. 临床和实验医学杂志, 2010, 9 (1) : 12-13.
[12] 王翠峰, 高丽君, 乌兰, 等. 血红蛋白释放试验与血液流变学检测结果相关性的研究. 中国医药导报, 2010, 7 (4) : 64-66.

[原文发表于"临床和实验医学杂志，2011，10(24)：1915-1917"]

第三十一章 人红细胞电泳中再释放蛋白质的分子互作

Molecular interactions of re-released proteins in electrophoresis of human erythrocytes

Yan Su[1], Jing Shen[2], Lijun Gao[1], Huifang Tian[2], Zhihua Tian[2], Wenbin Qin[1]

1. Department of Biochemistry and Molecular Biology, Baotou Medical College, Baotou, China; 2. Key laboratory of Carcinogenesis and Translational Research (Ministry of Education), Central laboratory, Peking University Cancer Hospital & Institute, Beijing, China

Abstract

Recently, we found that hemoglobin (Hb) could be re-released from live erythrocytes during electrophoresis release test (ERT). The re-released Hb displays single-band and multiple-band re-release types, but its exact mechanism is not well understood. In this article, the protein components of the single-band re-released Hb were examined. First, the re-released band of erythrocytes and the corresponding band of hemolysate, which was used as control, were cut out from starch-agarose mixed gel. Next, proteins were recovered from the starch-agarose mixed gel by freeze-thaw method. After condensing in a vacuum freeze drier, the samples were loaded onto a 5%~12% SDS-PAGE. After electrophoresis, three protein bands (16, 28.9, and 29.3 kDa) emerged from the erythrocytes re-released Hb single-band (R-R), but only one band (29.3 kDa) emerged from the corresponding hemolysate control band (H-R). Finally, these bands were analyzed by MALDI–TOF MS. The results showed that these proteins were beta-globin (16 kDa), carbonic anhydrase 1 (CA_1, 28.9 kDa), and carbonic anhydrase 2 (CA_2, 29.3 kDa). Because CA_2 exists in both erythrocytes re-released band and hemolysate control band, we conclude that the single-band re-released Hb is mainly composed of HbA and CA_1. Studying the possible interaction between HbA and CA_1 will help us further understand the in vivo function of Hb.

Key words Carbonic anhydrase; Erythrocyte; Hemoglobin; Interaction

Electrophoresis release test (ERT), which is performed by electrophoresing live erythrocytes directly on the starch-agarose mixed gel with intermittent electric current, was established by our lab [1,2]. In the past, most cell electrophoreses were erythrocyte electrophoreses because the color of erythrocytes could be easily observed under a microscope. There are many kinds of methods used in cell electrophoresis, such as thin-layer electrophoresis [3], capillary electrophoresis [4-10], isoelectric focusing electrophoresis [11], and micro-gel electrophoresis [12, 13]. A majority of stud-

---

Received October 13, 2011 Revised February 13, 2012 Accepted February 17, 2012
Correspondence: Dr. Wenbin Qin，31# Jianshe Road，Donghe District, Laboratory of Hemoglobin, Baotou Medical College, Baotou. Inner Mongolia, 014060, China
E-mail: qinwenbinbt@sohu.com
Fax: +86-472-7167857
DOI 10.1002/elps.201100644

ies about cell electrophoresis emphasized on surface charge and electrophoretic mobility of erythrocytes [4-7], while some focused on electrophoretic behavior of hemoglobin (Hb) [8-11], and only a handful of them were concerned with the Hb from undestroyed erythrocytes [12, 13]. Matinli and Niewisch [12] added a single erythrocyte that contained normal or abnormal Hb into the micro-gel, and during electrophoresis, the erythrocyte was fixed in the gel, and Hb moving out of the erythrocyte could be observed by a microscope. Anyaibe et al. [13] further designed a device that permitted several erythrocytes to be electrophoresed side-by-side at the same time, thus their electrophoretic behavior could be compared with each other.

Our erythrocytes electrophoresis experiment began in 1981, when the "HbA$_2$ phenomenon" was discovered [14]. At that time, electrophoresis was performed on the starch-agarose mixed gel with continuous power supply, and Hb within the erythrocytes would be released only once. Now, this phenomenon is also named "initial release" by us. During the initial release, the difference in mobility of HbA$_2$ was found between erythrocytes group and hemolysate group. In 2007, a sudden power outage was encountered during the electrophoresis of erythrocytes, however, the experiment was not abandoned and electrophoresis was continued after the power was restored. To our surprise, another new Hb band was found to be released from the origin [1], which was named "single-band re-release" as opposed to the "initial release". When the power outages were simulated more than once, multiple Hb bands appeared between HbA and origin, and this phenomenon was named "multiple-band re-release" or "ladder-band re-release". Using this method, re-released Hb from many patients' erythrocytes had been observed, and its amount varied in different patients [1, 15, 16]. Therefore, detecting the different amount of re-released Hb might have important clinical significance. In this article, the protein components of the single-band re-released Hb were detected to help us better understand the mechanism of ERT.

This study was approved by our local ethics committee. Fresh blood samples were collected from healthy adult donors who had been asked to sign the consent information before blood collection. Venous blood samples were anti coagulated with heparin, stored at 4 ℃, and generally analyzed within 24 h.

Packed erythrocytes and hemolysate were prepared according to our standard protocol [1,2], and then the packed erythrocytes were used to perform electrophoresis after dilution with saline at the ratio of 1 ∶ 1.

Two percent starch-agarose mixed gel (starch: agarose = 4: 1) was prepared with Tris-EDTA-Boric acid buffer (pH 8.6) as described previously [1, 2]. After adding 8 μl of erythrocytes and hemolysate at the indicated position, the electrophoresis was performed at 5 V/cm for a total of 120 min. This applied voltage was turned off for 5 min during electrophoresis to simulate power interruption at different times. The interruption point varied from 60 min, 90 min, and 105 min into the electrophoresis that is the resumed electrophoresis time after interruption was 60 min, 30 min, and 15min, respectively. After electrophoresis, the gel was photographed.

The red re-released Hb bands of erythrocytes sample (RR) located at the anode side of origin and the corresponding bands of hemolysate sample (H-R) were cut out and frozen at –80 ℃ for at least 30min. Before use, the gel was taken out and thawed at room temperature. After centrifuging at 12 000 rpm for 10 min, the supernatant was transferred into a new Eppendorf tube and dried in a vacuum freeze drier (Multi-drier, FROZEN IN TIME Ltd, England) for about 6～8 h. The freeze-dried sample was dissolved with ultra-pure water (1/10 volume of the original freeze-thaw supernatant volume). After boiling the sample with loading buffer for 5 min, 5%～12%

SDS-PAGE was performed as before [2].

The targeted polyacrylamide gel bands were excised into 1~2 mm pieces, and then destained and digested with trypsin as described previously [2]. The peptides generated from tryptic digestion were spotted onto a sample plate and cocrystallized with α-cyano-4-hydroxycinnamic acid (CHCA) (5mg/ml in 50% CH_3CN/0.1% TFA) and then air dried for MALDI-TOFMS (BrukerDaltonics, Bremen, Germany) analysis. The obtained peptide mass fingerprints were acquired in a positive reflector mode and analyzed using the FlexAnalysis v3.0 software and Biotools v3.2 software (Bruker Daltonics). Peptide calibration standard (Bruker Daltonics) was used for external calibration for each spectrum.

Peptide mass fingerprinting (PMF) data were analyzed by searching through Swiss-Prot and NCBInr database with the protein search engine MASCOT (v2.3, Matrix Sciences, London, UK), with a tolerance of approximately ±100ppm, one missed cleavage site, and peptide modifications by acrylamide adducts with cysteine and methionine oxidation. Proteins identified by PMF were further evaluated by comparing the calculated and observed molecular mass and pI, as well as the number of peptides matched and percent sequence coverage.

In our experiments, the red re-released single-band was observed in the erythrocyte sample during single-band rerelease electrophoresis (Fig. 31-1A-b), but not observed during initial release electrophoresis (Fig. 31-1A-a). Thus, the re-released Hb was speculated to come from the erythrocyte residue left in the origin and bind with the erythrocyte membrane. The distance between origin and re-released Hb band was found to depend on the resumed electrophoresis time (Fig. 31-1B): the longer the resumed time, the further the distance. To identify the protein components of the re-released Hb, the re-released bands from R-R and the corresponding control bands from H-R were cut out (Fig. 31-1A-b) and the recovered and enriched proteins from these bands were separated by SDS-PAGE. Finally, 16 kDa, 28.9 kDa, and 29.3 kDa bands were observed in R-R, but only one 29.3 kDa band in H-R (Fig. 31-1C-a).

To further determine the identity of these proteins, four bands (H_1, H_2, R_1, and R_2) indicated in Fig. 31-1 C-b were cut out from the polyacrylamide gel for MS analysis. Due to the difficulty in physically separating the 28.9 kDa and 29.3 kDa bands completely, both bands were included in the R_1 band. MALDI-TOF MS detection results showed that H_1 band was mainly composed of 29.3 kDa human carbonic anhydrase 2 (CA_2) (Table 31-1), R_1 band was made up of 29.3 kDa CA_2 and 28.9 kDa CA_1 (Table 31-1), R_2 band was mainly composed of 16 kDa human beta-globin (Table 31-1), and there was no detectable protein in H_2 band. Because CA_2 existed in both R_1 and H_1 band, we speculated that CA_2 might come from the initial release of erythrocytes, but was overlapped by the re-released Hb band. Therefore, we concluded that the singleband re-released Hb was mainly composed of beta-globin of HbA and CA_1.

CA, which catalyzes the reversible interconversion of carbon dioxide (CO_2) and water to bicarbonate (HCO_3^-) and protons, is the second most abundant protein in erythrocyte, and plays a crucial role in CO_2 transportation with the HCO_3^-/Cl^- exchange membrane protein, band 3 [17]. There are two cytoplasmic CA isozymes (CA_1 and CA_2) in human erythrocyte. CA_2 is composed of 260 amino acids and is a high turnover isozyme found virtually in every tissue, while CA_1 is composed of 261 amino acids and is a low turnover isozyme found mostly in the erythrocyte and intestine [18]. It has been proven that both Hb [19] and CA [20] have interactions with band 3. The interaction between band 3 and CA would increase efficiency by delivering the processed HCO_3^- directly to band 3 for outward transport, and the interaction between band 3 and Hb may make the transportation of O_2 easier. Our results showed for the first time that CA_1 might also have interaction with HbA, which was speculated to make the band 3-CA-HbA complex stronger to adapt to the transportation of O_2, CO_2, and proton. However, the reason why HbA interacts with CA_1, but not CA_2, and the exact molecular mechanism of the interaction still need further research. As to the method, ERT should be a simple and special method to find new in vivo protein–protein interaction. In the future, further information will be provided by using this method, and this will open up new ways to explore the mystery of erythrocytes.

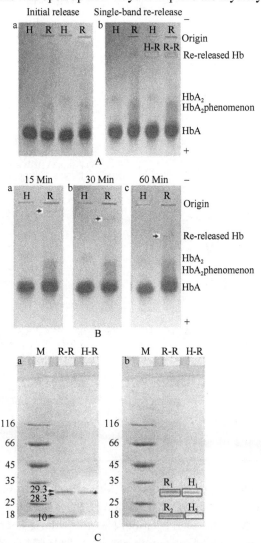

Fig. 31-1　Electrophoresis release test and the protein component isolation from single-band re-released Hb by 5%～12% SDS-PAGE. (A) Electrophoresis release test on starch-agarose mixed gel— (a) is the initial release electrophoresis; (b) is the single-band rerelease electrophoresis. H represents hemolysate sample, which is used as a control; R represents erythrocyte sample. R-R represents the re-released Hb band from erythrocytes; H-R represents the corresponding band from hemolysate. (B) The different location of re-released Hb band from erythrocytes at different resumed electrophoresis time after interruption— (a) the resumed electrophoresis time is 15 min; (b) the resumed electrophoresis time is 30 min; (c) the resumed electrophoresis time is 60 min. (C) The protein separation result of the single-band re-released Hb using 5%～12% SDS-PAGE— (a) the SDS-PAGE result of the R-R and H-R samples; (b) is the same as (a), only labels the name of the bands for MS detection. M represents protein marker; R_1 and R_2 represent the ～29 kDa and ～16 kDa bands of R-R sample respectively; H_1 and H_2 represent those corresponding bands of H-R sample, respectively.

Table 31-1 Proteins identified from three bands by MALDI-TOF MS

ID no.	NCBI accession no.	Name	Mass	pI	Score	Series matched	Coverage
H$_1$	GI: 4557395	Carbonic anhydrase2, *Homo saplliens*	29 285	6.87	95	KYDPSLKPLSVSYDQATSLR.I RLNNGHAFNVEFDDSQDKAVLKG KGGPLDGTYR.L KYAAELNLVNWNTK.Y KAVOQPDGLAVLGIFLK.V KGKSADFTNFDPRG KSADFTNFDPRG	35%
R$_1$	GI: 4557395	Carbonic anhydrase2, *Homo saplliens*	29 285	6.87	114	KHNGPEHWHKDFPIAKG KYDPSLKPLSVSYOQATSLRI RILNNGHAFNVEFD DSODKA KGGPLDGTYRL RUOFHFHWGSLDGOGSEHTVDKK KKYAAELHLVHWNTKY KAVQQPDGLAVLGIFLKV KSADFTNFDPRG RKLNFNGEGE- PEELMVDNWRPAOPLK.N KLNFNG EGEPEELMVDNWRPAOPLK.N	56%
	GI: 4502517	Carbonic anhydrase1, *Homo saplliens*	28 909	6.59	85	KTSETKHDTSLKPISVSYNPATAKE KHDTSLKPISVSYNPATAKE KEIINVGHSFHVNFEONDNRS KGPFSDSYRL RLFQFHFHWGSTNEHGSEHTV DGVKY KYSAELHVAHWNSAKY KESISVSSEQLAQFRS RSLLSNVEGONAVPMQHNNRPRP- TOPLKG	49%
R$_2$	GI: 4504349	HBB-HUMAN	16 102	6.75	158	KSAVTALWGKV KVNVOEVGGEALGRL RLLWYPWTORF RFFESFGOLSTPOAVMGNPKV KKVLGAFSOGLAMLDNLKG KVLGAFSOGLAHLDNK.G KGTFATLSELHCDKL KLHVDPENFRL RLLGNVLVCVLAHHFGKE KEFTPPVOAAYOKV KVVAGVANALAHKYH	89%

This work was supported by grants from the Major Projects of Higher Education Scientific Research in the Inner Mongolia Autonomous Region (NJ09157), the Key Science and Technology Research Project of the Ministry of Education, Natural Science Foundation of Inner Mongolia (2010BS1101), and Natural Science Foundation of China (81160214). We acknowledge all of the people who donated their blood samples for our research. The authors have declared no conflict of interest.

References

[1] Su, Y., Shao, G., Gao, L., Zhou, L., Qin L, Qin W., *Electrophoresis* 2009, *30*, 3041-3043.
[2] Su, Y., Gao, L., Ma, Q., Zhou, L., Qin, L., Han, L., Qin, W., *Electrophoresis* 2010, *31*, 2913-2920.
[3] Akagi, T., Ichiki, T., *Anal. Bioanal Chem.* 2008, *391*, 2433-2441.
[4] Wilk, A., Ro´skowicz, K., Korohoda, W., *Cell Mol. Biol. Lett.* 2006, *11*, 579-593.
[5] Korohoda, W., Wilk, A., *Cell Mol. Biol. Lett.* 2008, *13*, 312-326.

[6] Wilk, A., Urban′ska, K., Woolley, D. E., Korohoda, W., *Cell Mol. Biol. Lett.* 2008, *13*, 366-374.
[7] Kabanov, D. S., Ivanov, A. Yu, Melzer, M., Prokhorenko, I. R., *Biochemistry (Moscow) Supplemental Series A: Membrane and Cell Biology* 2008, 2, 128-132.
[8] Lillard, S. J., Yeung, E. S., Lautamo, R. M., Mao, D.T., *J. Chromatogr A.* 1995, *718*, 397-404.
[9] Lu, W. H., Deng, W. H., Liu, S. T., Chen, T. B., Rao, P. F., *Anal. Biochem.* 2003, *314*, 194-198.
[10] Zhang, H., Jin, W., *Electrophoresis* 2004, *25*, 480-486.
[11] Chen, H. W., Lii, C. K., *Methods Mol. Biol.* 2002, *186*, 139-146.
[12] Matinli, G., Niewisch, H., *Science* 1965, *150*, 1824-1828.
[13] Anyaibe, S. I., Headings, V. E., *Am. J. Hematol.* 1977, *2*, 307-315.
[14] Qin, W. B., Liang, Y. Z., *Chin. J. Biochem. Biophys.* 1981, *13*, 199-201.
[15] Zhang, X., Gao, L., Gao, Y., Zhou, L., Su, Y., Qin, L., Qin, W., *Int. J. Lab. Med.* 2010, *31*, 524-525.
[16] Han, L., Yan, B., Gao, Y., Gao, L., Qin, L., Qin, W., *J. Clin. Exp. Med.* 2009, *8*, 67-68.
[17] Tufts, B. L., Esbaugh, A., Lund, S. G., *Comp. Biochem. Physiol. A. Mol. Integr. Physiol.* 2003, *136*, 259-269.
[18] Esbaugh, A. J., Tufts, B. L., *J. Exp. Biol.* 2006, *209*, 1169-1178.
[19] Chu, H., Breite, A., Ciraolo, P., Franco, R. S., Low, P. S., *Blood.* 2008, *111*, 932-938.
[20] Kifor, G., Toon, M. R., Janoshazi, A., Solomon, A. K., *J. Membr. Biol.* 1993, *134*, 169-179.

(*Electrophoresis*. 2012，33，1402-1405. Short Communication)

第三十二章 剖宫产前后血红蛋白释放试验的连续观察

葛 华　高丽君　罗 劲　李丕宇　苏 燕　秦文斌

(包头医学院，内蒙古 包头 014010)

摘　要

目的： 观察剖宫产手术前后红细胞释放血红蛋白的情况。
方法： 1 例剖宫产孕妇，分别于剖宫产术前、术后 1、5、13 天采血，对这 4 份血液标本进行双向多层电泳，对其中术后 5 天的血液标本进行等低渗全程电泳，再对术后 1 天和 5 天的标本的血浆进行单向电泳试验。
结果： 剖宫产术后 5 天标本血红蛋白释放的变化较大。即在血红蛋白 A 前方出现快泳联苯胺阳性成分，血浆中也有此联苯胺阳性成分，在白蛋白前出现丽春红阳性物质，涉及急性期反应蛋白。
结论： 剖宫产手术前后血红蛋白释放结果有差异。

关键词　剖宫；产血红蛋白释放试验；连续观察

Study on the significance of continuous observation on hemoglobin release test before and after cesarean delivery

GE Hua, GAO Lijun, LUO Jin, et al.

(Baotou Medical College, Baotou, Inner Mongolia 014010, China)

Abstract

Objective: To observe the condition of hemoglobin release test before and after cesarean section. **Methods:** In a pregnant woman performed cesarean section, blood samples were taken before operation and 1, 5 and 13 days after operation for two—way multilayer electrophoresis, and the blood sample at 5th day after operation was additionally examined with low permeability electrophoresis, then plasma specimens of this patient taken at 1st and 5th day after operation were examined with one—way electrophoresis. **Results:** There was marked change in hemoglobin release test in blood sample taken at 5th day after operation, the fast aniline positive element appeared in the front of hemoglobin A, and the benzidine positive element was also presented in plasma, beautiful spring red positive material appeared before albumin, and it involved in the acute phase response of protein. **Conclusion:** There is difference in results of hemoglobin release test before and after delivery with cesarean section.

Key words　Cesarean delivery;　Hemoglobin release test;　Continuous observation

1 前言

剖宫产是妇产科常见手术之一，但查阅文献未见剖宫产手术前后血液成分变化的报道，更没有与血红蛋白释放试验有关的研究。本文报告我们的研究结果。

2 资料与方法

2.1 资料血液标本来自 2012 年 3 月至 2012 年 5 月包头医学院第一附属医院妇产科同一产妇。不同时间采血 4 次。编号如下：0 号为术前，1 号为术后 1 天，2 号为术后 5 天，3 号为术后 13 天标本。

2.2 方法实验方法基本同文献[1]，具体如下。

(1) 4 份标本的双向多层电泳采集血液标本后均为当天做实验。取上述四份标本，做双向多层电泳。最上层(第一层)为红细胞溶血液，下一层(第二层)是红细胞，再下层(第三层)为全血，最下层(第四层)是血浆(图 32-1)。电泳后染丽春红，再染联苯胺。血浆蛋白成分显红色，血红蛋白成分显蓝黑色。

(2) 术后 5 天标本的等低渗全程电泳取 3 号标本，操作同文献[1]，先做室温等低渗全程电泳，再做 37℃保温等低渗全程电泳，血红蛋白的释放条件为定时(15 分钟)释放，染色条件同上。

(3) 术后 1 天与术后 5 天血浆成分的单向电泳取 1 号和 2 号的血浆标本，做普通单向电泳，电势梯度 6V/cm，电泳时间 3 小时，染色条件同上。

3 结果

3.1 4 次标本的双向多层电泳结果 由图 32-1 的 A～D 可以看出，四次标本的溶血液，血红蛋白释放没有差异。红细胞部分，产后 13 天者梯带增强。全血部分，产后 5 天者在血红蛋白 A 前方出现联苯胺阳性成分，见图 32-1 和表 32-1。

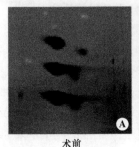 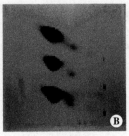 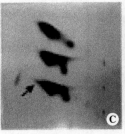 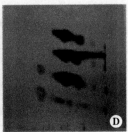

　　　　 术前　　　　　　　　 术后1天　　　　　　　　术后5天　　　　　　　　 术后13天

图 32-1 双向多层血红蛋白释放

注释：A、B、C、D 由上向下，第一层为红细胞溶血液，第二层为红细胞，第三层为全血，第四层为血浆，C 术后 5 天中的箭头所指为快泳联苯胺阳性成分

3.2 术后 5 天标本等低渗全程电泳释放结果如图 32-2A 显示，红细胞的每个泳道都有定时释放，先弱后强，最明显的是倒数第 3 泳道。全血各泳道的定时释放普遍减弱，相对最明显的是倒数第 2 泳道，与此同时，在血红蛋白 A 前方出现联苯胺染色阳性成分(参见箭头处)。图 32-2A 与图 32-2B 全血结果基本相同，只是在图 32-2B 的白蛋白部分出现高铁血

红素白蛋白(MHA methemalbumin)。由图 32-2A、2B 可以看出,红细胞的定释比全血明显一些,37℃保温时出现 MHA。两图中都能看到血红蛋白 A 前有联苯胺阳性成分。

3.3 术后 1 天与 5 天血浆的比较电泳 产后 5 天的血浆标本中出现两个异常成分：①在血红蛋白 A_1 之前(阳极侧)出现联苯胺染色阳性物质(箭头 1)；②在白蛋白之前(阳极侧)出现丽春红染色阳性物质(箭头 2),见图 32-3。

表 32-1 剖宫产前后四次标本电泳结果的综合比较

实验内容	术前	术后 1 天	术后 5 天	术后 13 天
红细胞梯带释放	+−	−	+−	+
全血梯带释放	+−	−+	−+	+
血浆：HbA_1 前联苯胺阳性成分	−	−	+	−
白蛋白前丽春红阳性成分	−	−	+	−

注释：表中的"+"为阳性,"−"为阴性,"−+"为微弱阳性,"+−"为弱阳性。

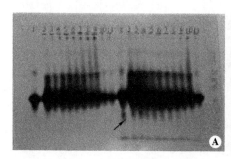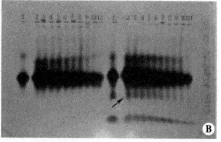

图 32-2 等低渗全程血红蛋白电泳释放

注释：图 A 为等低渗全程电泳,室温结果；图 B 为等低渗全程电泳,37℃结果。每个电泳图的左侧 11 个泳道为红细胞的全程结果,右侧 11 个泳道为全血的全程结果。每 11 个泳道的第 1 泳道为对照,分别为红细胞溶血液和全血溶血液。图 A、B 中箭头指示均为快泳联苯胺阳性成分

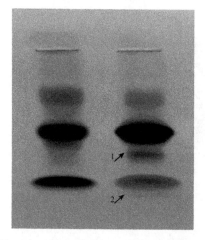

图 32-3 术后 1 天与 5 天血浆电泳的比较
注释：左侧为术后 1 天标本、右侧为术后 5 天标本；产后 5 天标本在血红蛋白 A 之前出现快泳联苯胺染色阳性物质(箭头 1)；产后 5 天标本在白蛋白之前出现丽春红染色阳性物质(箭头 2)

4 讨论

血红蛋白释放试验是我们自己创建的研究手段[1],1981 年开始发现血红蛋白 A_2 现象[2],属于红细胞内血红蛋白的初释放范畴。2007 年又发现红细胞内血红蛋白的再释放[3, 4]。后来又应用于其他疾病。例如,普通外科住院患者与血红蛋白释放试验[5],糖尿病[6],尿毒症[7],血红蛋白释放试验与血液流变学检测结果相关性的研究[8],陈旧血电泳出现快泳红带；红细胞释放"高铁血红素白蛋白"[9],血小板成分电泳释放的初步研究[10],肝硬化与血红蛋白释放试验[11]。以上几乎都是其他各种临床疾病,没有涉及剖宫产。剖宫产是妇产科里的一种常见手术,也是血红蛋白释放试验没有涉足的领域,本研究旨在剖宫产前后血红蛋白释放试验的情况。观察结果表明,产后 5 天变化较大,产后 13 天也有变化。产后 13 天红细胞变化明显,说明此

时红细胞内血红蛋白的存在状态异常,电泳时表现为产后 13 天,深入机制有待继续研究。产后 5 天变化较大,主要表现为 HbA$_1$ 前出现联苯胺染色阳性成分和白蛋白前出现丽春红染色阳性成分。关于出现白蛋白前的蛋白质,首先想到前白蛋白(prealbumin,PA)。PA 于肝脏合成,属于"急性期反应蛋白质"中的一种,HbA$_1$ 前出现的联苯胺染色阳性成分是何种物质?根据电泳位置,它应当是血红蛋白与结合珠蛋白(Hp)的复合物。Hp 也是一种急性期反应蛋白,它与 PA 同时出现是可以理解的。剖宫产第 5 天出现 PA 和血红蛋白与结合珠蛋白(Hp)的复合物,可以认为它是对剖宫产手术的一种急性期反应。至于为什么手术后第 5 天出现,这是我们的发现,也是需要进一步研究的内容。如上所述,血红蛋白释放试验已经比较广泛地应用于非手术疾病,本研究是它在手术疾病中的首次尝试。仅此一例,抛砖引玉,希望能有更多发现。

参 考 文 献

[1] 秦文斌. 活体红细胞内血红蛋白的电泳释放. 中国科学: 生命科学, 2011, 41 (8): 597-607.
[2] Su Y, Gao L, Ma Q, et al. Interactions of hemoglobin in live red blood cells measured by the electrophoresis release test. Electrophoresis, 2010, 31 (17): 2913-2920.
[3] 秦文斌, 高丽君, 苏燕, 等. 血红蛋白释放试验与轻型β-地中海贫血. 包头医学院学报, 2007, 23 (6): 561-563.
[4] Su Y, Shao G, Gao L, et al. RBC electrophoresis with discontinuous power supply-a newly established hemoglobin release test. Electrophoresis, 2009, 30 (17): 3041-3043.
[5] 韩丽红, 闫斌, 高雅琼, 等. 普通外科患者血红蛋白释放试验的比较研究. 临床和实验医学杂志, 2009, 8 (7): 67-69.
[6] 张晓燕, 高丽君, 高雅琼, 等. 血糖浓度和血红蛋白释放试验的比对研究. 国际检验医学杂志, 2010, 31 (6): 524-525.
[7] 高雅琼, 王彩丽, 高丽君, 等. 尿毒症患者低渗血红蛋白释放试验的初步结果. 临床和实验医学杂志, 2010, 9 (1): 12-13.
[8] 王翠峰, 高丽君, 乌兰, 等. 血红蛋白释放试验与血液流变学检测结果相关性的研究. 中国医药导报, 2010, 7 (4): 64-65.
[9] 张咏梅, 高丽君, 苏燕, 等. 陈旧血电泳出现快泳红带: 红细胞释放"高铁血红素白蛋白". 现代预防医学, 2011, 38 (15): 3040-3042.
[10] 乔姝, 沈木生, 韩丽红, 等. 血小板成分电泳释放的初步研究. 现代预防医学, 2011, 38 (4): 684-685, 688.
[11] 宝勿仁必力格, 王翠峰, 高丽君, 等. 肝硬化患者血红蛋白释放试验明显异常. 临床和实验医学杂志, 2011, 10 (24): 1915-1917.

[本文发表于"临床和实验医学杂志, 2013, 12(5): 362-363, 365"]

第三十三章　肝内胆管癌与血红蛋白释放试验

韩丽莎 [1#]　高丽君 [2#]　郭　俊 [3#]　孙晓荣 [1]　苏　燕 [2]　秦文斌 [2※]

(1. 包头医学院　病例生理教研室；2. 包头医学院　血红蛋白研究室；3. 包头市　第七医院　介入科, 包头 014010)

摘　要

目的：研究肝内胆管癌患者红细胞及全血的血红蛋白释放情况。

方法：单向双释放试验、双向双释放试验、等低渗全程双释放试验。结果：七例癌症血液标本中肝内胆管癌的单向双释放显著增强，明显区别于其他标本。双向双释放结果也是显著增强，梯带中除大量血红蛋白 A_1 外，还能看到血红蛋白 A_2。等低渗全程双释放结果是，开始时全面增强，随放置时间延长全血结果减弱，甚至出现部分梯带缺失情况。

结论：与其他标本不同，肝内胆管癌患者双释放试验结果显著增强，血红蛋白 A_2 的相应结果也增强，全程双释放试验中出现一些奇怪现象，需要进一步弄清。

关键词　肝内胆管癌；血红蛋白释放试验；梯带

1　前言

肝内胆管癌发病率和病死率有逐年增高的趋势，目前缺乏早期诊断的方法，对其发病机制了解较少。一般认为，胆管癌的发生与其他肿瘤一样，也是一个渐变的过程，都要经历暴露于高危因素，细胞增殖调控异常，逃避凋亡与免疫监视，最后发展成癌的过程[1-6]。

血红蛋白释放试验是我们近期建立的一种检测手段，开始时发现它能辅助轻型β-地中海贫血的诊断[7-8]，后来注意到一些普通外科疾病也出现释放异常[9]。鉴于癌症是人类健康的最大威胁，此时血红蛋白释放试验如何，成为本项研究的动力。

2　材料与方法

2.1　标本来源　血液标本来自包头市肿瘤医院介入科，EDTA 抗凝。

2.2　单向双释放试验　用患者的全血和由此全血分离出来的红细胞，并排做淀粉-琼脂糖混合凝胶电泳，电势梯度 6V/cm，通电 15 分钟、停电 15 分钟，交替进行，共 3 小时。丽春红-联苯胺复染，血红蛋白呈蓝黑色，其他蛋白显红色。

2.3　双向双释放试验　第一向同上，到时转第二向，电势梯度 6V/cm，通电 1.5 小时，中间不停电。染色同上。

2.4　等低渗全程双释放试验　总体同上述单向双释放试验，只是红细胞或全血都要分

\# 并列第一作者

※ 通讯作者：秦文斌，电子邮箱：qinwenbinbt@sohu.com

成10管并做如下处理(表33-1)。

表33-1 10管红细胞或全面标本的处理方法

管号	1	2	3	4	5	6	7	8	9	10
标本(μl)	10	9	8	7	6	5	4	3	2	1
蒸馏水(μl)	0	1	2	3	4	5	6	7	8	9

混合后,按单向双释放试验进行操作。此时,第1管为等渗,其余为低渗,程度不同。

3 结果

3.1 一批癌症患者标本的筛查结果　见图33-1。

由图34-1可以看出,肝内胆管癌患者红细胞及全血的梯带释放明显增强。

3.2 肝内胆管癌患者红细胞及全血的双向双释放电泳结果　见图33-2。

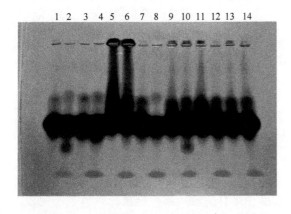

图33-1　肝内胆管癌患者红细胞及全血电泳
注释:泳道(由左向右)5、6(箭头)为肝内胆管癌患者红细胞及全血

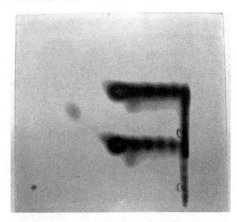

图33-2　肝内胆管癌患者红细胞及全血的双向双释放电泳结果
注释:上层为患者的红细胞,下层为患者的全血

由图33-2可以看出,肝内胆管癌患者红细胞及全血中HbA_1的梯带释放明显增强,而且HbA_2的梯带释放也增强。二者相比,红细胞的梯带释放更强一些。

3.3 肝内胆管癌患者红细胞及全血的等低渗全程双释放电泳　室温结果见图33-3。

由图33-3可以看出,室温条件下,红细胞梯带释放全程增强,全血全程梯带释放明显弱于红细胞。此时,全血的白蛋白处未见MHA(高铁血红素白蛋白)。

3.4 肝内胆管癌患者红细胞及全血的等低渗全程双释放电泳(37℃结果)　见图33-4及图33-5。

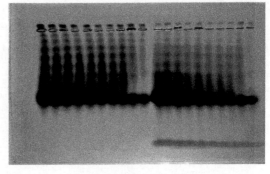

图33-3　肝内胆管癌患者红细胞及全血的等低渗全程双释放电泳(室温)
注释:全程双释放(室温)左侧为红细胞,右侧为全血

由图 33-4 可以看出，37℃时红细胞梯带释放全程增强，全血全程都没有梯带，此时全血的白蛋白处出现 MHA(高铁血红素白蛋白)。

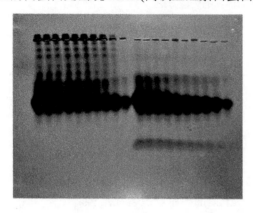 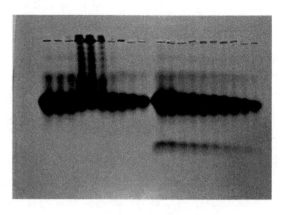

图 33-4　肝内胆管癌患者红细胞及全血的等低渗全程双释放电泳(37℃)
注释：全程双释放(37℃)左侧为红细胞，右侧为全血

图 33-5　两天后全程双释放电泳(37℃)
注释：两天后再做图 33-4 实验，全程双释放(37℃)。左侧为红细胞，右侧为全血

由图 33-5 可以看出，37℃时红细胞出现部分梯带释放缺失，全血全程都没有梯带，此时全血的白蛋白处出现 MHA(高铁血红素白蛋白)加重。

4　讨论

双释放中红细胞与全血梯带都增强病例首先见于轻型β-地中海贫血[10, 11]，在其他疾病中这种情况非常少见。遗传性球形红细胞增多症的双释放结果与此相反，其红细胞与全血梯带都明显减弱、甚至消失。糖尿病时，全血梯带释放不太明显，红细胞梯带释放较强，而且与血糖浓度成正比[12]。正常人的双释放结果是红细胞梯带释放较明显，全血梯带释放不明显。必须指出，肝内胆管癌的双释放，在梯带强度上与地贫有类似之处，但也有明显不同。首先，在双向双释放时，此病出现血红蛋白 A_2 的梯带，其他疾病从未见过。这说明，此病的红细胞再释放机制特殊，通常只有血红蛋白 A_1 的梯带、没有血红蛋白 A_2 的梯带。现在有了血红蛋白 A_2 的梯带，推测红细胞内血红蛋白的再释放可能有两个机制，血红蛋白 A_1 的再释放机制和血红蛋白 A_2 的再释放机制，地贫时启动 A_1 机制，肝内胆管癌时启动 A_2 机制。情况是否如此，有待进一步研究。另外，等低渗全程实验中出现部分梯带缺失，也是其他疾病从未见过。这说明，在这些反应管里出现了一种物质，它能使红细胞溶血，失去释放基础。这种物质是什么，目前尚不明确。我们推测，试管内血液中有这种物质，开始数量较少，作用不明显，室温放置几天后数量增多或者效应累加，才出现这种现象。但是，无论是增多还是累加，效应应该是均匀的，不会"跳管"，这是无法解释的。但是由于种种原因，患者未作介入而离开医院，失去联系，未能再取血核对。之后一系列的深入研究，只好留给未来。

参 考 文 献

[1] 黄志强. 肝内胆管结石与肝胆管癌. 中华外科杂志, 1981, 19 (7) : 403-404.
[2] 田成武, 朱华文, 于永山, 等. 肝胆管结石并发胆管癌. 中国普通外科杂志, 2001, 110 (1) : 21-23.
[3] 何德云. 肝胆管结石合并肝胆管细胞癌的诊断. 实用肿瘤学杂志, 2002, 16 (3) : 208-209.

[4] 梁力建, 邓伟. 肝胆管结石和肝胆管癌. 临床外科杂志, 2005, 13 (7) : 409-410.
[5] Nakanuma Y, Harada K, Ishikawa A, et al. Anatomic and molecular pathology of intrahepatic cholangiocarcinoma. J Hepatobiliary Pancreat Surg, 2003, 10 (4) : 265-281.
[6] Ahrendt SA, Rashid A, Chow JT, et al. p53 overexpression and K-rans gene mutations primary sclerosing cholangitis-associated biliary tract. J Hepatobiliary Pancreat Surg, 2000, 7 (4) : 426-431.
[7] 秦文斌, 高丽君, 苏燕, 等. 血红蛋白释放试验与轻型β-地中海贫血. 包头医学院学报, 2007, 23 (6) : 561-563.
[8] Su Y, Shao G, Gao L J, et al. RBC electrophoresis with discontinuous power supply-a newly established hemoglobin release test. Electrophoresis 2009, 30: 3041-3043
[9] 韩丽红, 闫斌, 高雅琼, 等. 普通外科患者血红蛋白释放试验的比较研究. 临床和实验医学杂志, 2009, 8 (7) : 67-69.
[10] 秦文斌, 高丽君, 苏燕, 等. 血红蛋白释放试验与轻型 β-地中海贫血. 包头医学院学报, 2007, 23 (6) : 561-563.
[11] Su Y, Shao G, Gao L J, et al. RBC electrophoresis with discontinuous power supply - a newly established hemoglobin release test. Electrophoresis, 2009, 30: 3041-3043.
[12] 张晓燕, 高丽君, 高雅琼, 等. 血糖浓度和血红蛋白释放试验的比对研究. 国际检验医学杂志, 2010, 31 (6) : 524-525.

附 依据研究发表的论文

Abnormal increased re-released Hb fromRBCs of an intrahepatic bile duct carcinoma patient was detected by electrophoresis release test

Yan Su[a], Lisha Han[a], Lijun Gao[a], Jun Guo[b], Xiaorong Sun[a], Jiaxin Li[a] and Wenbin Qin[a*]

[a]Laboratory of Hemoglobin, Baotou Medical College, Baotou, 014060, China

[b]Department of Invasive Technology, the Seventh Hospital of Baotou, Baotou, 014010, China

Abstract. In this paper, the hemoglobin (Hb) re-released from red blood cells (RBCs) and whole blood of 7 carcinoma patients were studied by using electrophoresis release test (ERT), which was established by our lab. Among the 7 carcinoma patients, the re-released Hb was distinctively increased from an intrahepatic bile duct carcinoma patient during one-dimension isotonic ERT. Different from the others, the result of double-dimension Hb re-release of this intrahepatic bile duct carcinoma patient showed that not only HbA but also HbA_2 could be re-released from both RBCs and whole blood. The result of isotonic & hypotonic ERT which was performed at room temperature showed that more Hb could be re-released from both RBCs and whole blood of the intrahepatic bile duct carcinoma patient than that of the normal control. After keeping the samples at 37℃ for 1 hour, the re-released Hb from RBCs could still be found more than that of the normal control, but was disappeared completely from the whole blood sample. To our surprise, when the isotonic & hypotonic ERT was repeated 2 days later at 37℃, the re-released Hb from RBCs of the intrahepatic bile duct carcinoma patient was increased only in tube 4-6, and disappeared in the other tube. Further mechanism research work cannot be continued because of the patient's leave, but ERT is speculated to be a useful and effective technology to observe the physiological or pathological change of RBCs, blood or body in the future.

Key words Hemoglobin, red blood cell, electrophoresis release test, intrahepatic bile duct carcinoma

1 Introduction

Electrophoresis release test (ERT) has been established by our laboratory[1,2]. During ERT, live red blood cells (RBCs) are added directly onto the starch-agarose mixed gel and the electric current perforates the membrane instantaneously. Discontinuous power supply during electrophoresis is the most important innovation of the ERT technique. As to the mechanism of ERT, we primarily speculate that the electric pulse from turning-on and -off the power supply could create plasmatorrhexis of RBCs. Hb, existing free in the cytoplasma, would be released during the first cycle of electrophoresis; while the other Hb, binding with the RBC membrane, would be released during the subsequent cycle of electrophoresis[3]. The amount of re-released Hb may be associated with the status of both Hb and RBC membrane, which can be affected by many diseases, such as hematonosis, tumor, diabetes and so on[4-8]. Generally, little re-released Hb can be observed during ERT. Our previous experiments proved that the re-released Hbs were increased distinctively from RBCs of β-thalassemia, diabetic and some general surgical patients[3]. As we known, oxidative damage is the important factor to destroy the RBC membrane[9] and membrane-bound Hb is an important marker of oxidative injury in RBCs[10]. The genesis of tumor is closely related to the oxidative damage of radicals[11-13], but whether it will affect the amount of re-released Hb has not been studied. In this study, the amount of re-released Hb from some upper gastrointestinal cancer patients will be observed.

2 Materials and methods

2.1 Specimens

Our research was approved by the Ethics Committee of Baotou Medical College. Blood samples were collected from the intervention department of the seventh hospital of Baotou. Before the blood samples were collected, all the people who took part in this experiment were asked to sign the consent information. Blood samples were anti-coagulated with heparin and stored at 4℃. Hematology examinations including blood routine, liver function and renal function were performed by the clinical laboratory of the seventh hospital of Baotou.

2.2 Preparation of the RBC suspension and starch–agarose mixed gel

The anti-coagulated blood was firstly centrifuged at 3000 rpm for 10 minutes to isolate RBCs from the plasma, and then wash the RBCs with saline 4 to 5 times until the supernatant was clear. RBCs were then used to perform electrophoresis after 1∶1 dilution with saline. Hemolysate was prepared by continuously adding 200 μl saline and 100 μl CCl_4 to the RBCs. After turbulent mixing, the sample was centrifuged at 12 000 rpm for 10 minutes and the upper red hemolysate was pipetted out carefully and stored at 4℃ for later use[1-3]. The 2% starch–agarose mixed gel (4∶1) was prepared with TEB buffer (pH 8.6) as described previously[1].

2.3 One-direction ERT

5 μl of RBC suspension and whole blood were added on the starch–agarose mixed gel. The electrophoresis was ran at 6 V/cm for 15 minutes, then paused for 15 minutes and ran for another 15 minutes by turns, and the total electrophoresis time was about 2 hours. After electrophoresis, the red bands on the gel were firstly observed with eyes and then sequentially stained with Ponceau Red and Benzidine[1].

2.4 Double-direction ERT

First, 5 μl of RBC suspension (about 1.5×10^9 RBCs) and whole blood were added on the starch–agarose mixed gel and one-direction ERT was performed as described above. Then change the direction of electric field, which is vertical to the original direction. Each directional electrophoresis was ran for 15 min, and then paused for 15 min by turns, and the total electrophoresis time was about 4 hours.

2.5 One-direction *ERT*

The electrophoresis method was the same as one-direction ERT, but the RBC suspension and whole blood needed to be diluted with H_2O in the proportion from 10∶0 to 1∶9 (named as tube 1 to 10 respectively), and then kept them at room temperature or 37℃ for 1 hour.

3 Results

3.1 One-direction ERT of blood samples from 7 upper gastrointestinal carcinoma

In this experiment, the re-released Hb from 7 upper gastrointestinal carcinoma patients were detected by one-direction ERT. As shown in Figure 1, the main electrophoretic bands are albumin (exits in whole blood, but not in RBCs), HbA and HbA_2. After the first cycle of electrophoresis, there was some red sediment stayed at the origin. Then during each cycle of the "run–pause–run" electrophoresis, there was other Hb re-released from the sediments. Among these 7 patients, patient 1 and 7 were cardia carcinomapatient 2, 3, 4, 5 and 6 were hepatocellular carcinoma. The re-released Hb ladder of patient 3, 5, 6 and 7 were increased, but the increase of patient 3 (intrahepatic bile duct carcinoma) was especially distinctive.

3.2 Double-direction ERT and isotonic & hypotonic ERT were performed with the blood of patient 3

Double-direction HRT result (Fig. 2A) showed that not only HbA but also HbA_2 could be re-released from the origin of the intrahepatic bile duct carcinoma patient. The result of isotonic & hypotonic ERT at room temperature (Fig. 2B) showed that more Hbs could be re-released from both whole blood and RBCs samples of intrahepatic bile duct carcinoma patient than the normal control. After keeping the samples at 37℃ for 1 hour (Fig. 2C), the re-released Hbs from RBCs could still been found more than that of normal control, but the re-released Hb ladder from whole blood sample was disappeared completely. In addition, the red albumin bands stained by Ponceau Red were found to be stained blue slightly by Benzidine. The most interesting result appeared 2 days later, when the isotonic & hypotonic ERT was repeated at 37℃, the released Hb ladder from tube 4-6 of RBC samples were found to be increased, but those from the other tubes were disappeared completely. Also, in the whole blood sample, the red albumin bands stained by Ponceau Red were stained blue by Benzidine distinctively.

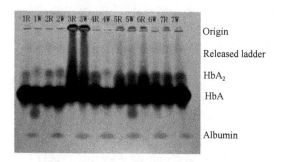

Fig. 1 One-direction ERT of 7 upper gastrointestinal carcinoma blood samples. There were 7 samples; each sample was divided into whole blood group (W) and RBC group (R) correspondingly. Patient 1 and 7 were cardia carcinoma, patient 2, 3, 4, 5 and 6 were hepatocellular carcinoma.

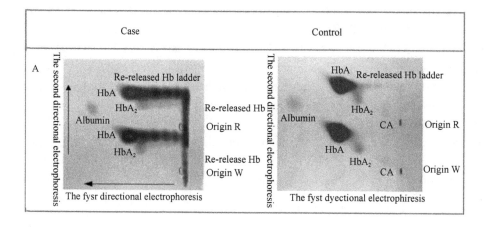

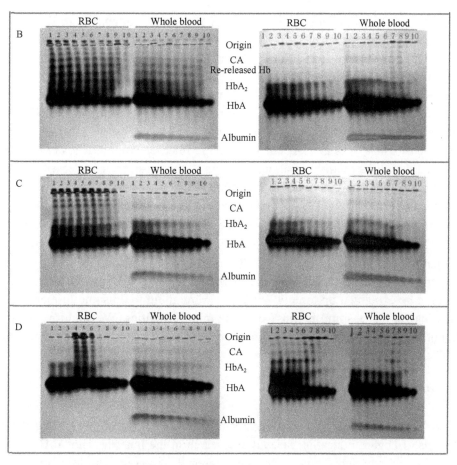

Fig. 2 Different kind of ERTs of the intrahepatic bile duct carcinoma patient. A. Double-direction ERT; B. Isotonic & hypotonic ERT at room temperature; C. Isotonic & hypotonic ERT at 37℃; D. Isotonic & hypotonic ERT at 37℃ two days later.

4 Discussion

ERT is a new method established by our laboratory to study the membrane binding Hb of RBC. During experiments, this technology had been continuously optimized and some new ERT methods were developed, such as double-direction ERT and isotonic & hypotonic ERT. Double-direction ERT could help us to observe the re-released HbA_2, which usually cannot be observed easily during one-direction ERT. Isotonic & hypotonic ERT could help us observe the resistance of RBC membrane to the change of osmotic pressure. In this experiment, the red Hb bands can be observed directly without any staining. In order to observe the trace hemoglobin band better and distinguish hemoglobin band with the other protein, the gel was sequentially stained with Ponceau Red and Benzidine after ERT. Ponceau Red can stain all the protein red (including Hb), but Benzidine can specifically stain Hb blue. So where the blue band exists, there must have hemoglobin. In our results, the blue bands are Hb, and the two main red bands are albumin (fast moving) and carbonic anhydrase (slow moving) respectively. Hemolysis of RBCs leads to Hb leakage, so the albumin bands of whole blood samples can be double stained by Ponceau Red and Benzidine in Fig. 2D.

The increased Hb re-release was firstly observed in ß-thalassemia patient[1], and then it was observed in diabetes patients and some general surgical patients[3]. To the contrary, the re-released Hb could also decrease distinctively or disappear from RBCs of hereditary spherocytosis patient. We have proved that the membrane integrity and oxidative damage could affect the amount of re-released Hb. Lose of RBC membrane leads to the decrease of re-released Hb, but the oxidative damage can increase the amount of re-released Hb. In this study, not only HbA but also HbA_2 could be re-released from the intrahepatic bile duct carcinoma patient distinctively, and this phenomenon had not been observed in any other patients as yet. In the past, HbA_2 was also speculated to be re-released from RBCs, but it was difficult to be observed due to its relative small amount. As to this case, the re-released HbA_2 was increased distinctively. During isotonic & hypotonic ERT, the re-released Hb ladder of the intrahepatic bile duct carcinoma was also increased distinctively than normal control not only at room temperature but also at 37℃. However, the most interesting phenomenon was that some re-released Hb bands were disappeared from the isotonic & hypotonic ERT (Figure 2D). We speculate that the membrane structure of RBCs might be destroyed after keeping the blood at room temperature for 2 days, but why the increased Hbs only disappeared in tube 1-3 and 7-10 could not be explained up to now. When we were going to do some further research, the patient had leaved the hospital because of economic difficulties. We could not continue our research, but this case report makes our mind to clarify the mechanism and clinical application of ERT in the future.

Acknowledgments

This work was supported by grants from Natural Science Foundation of China (81160214), Major Projects of Higher Education Scientific Research in the Inner Mongolia Autonomous Region (NJ09157), Key Science and Technology Research Project of the Ministry of Education (210039), Natural Science Foundation of Inner Mongolia (2010BS1101), Scientific Research Projects of Inner Mongolia Education Department (NJZY14265) and Doctoral scientific research foundation of Baotou Medical College. We also especially acknowledge all of the people who donated their blood samples for our research.

References

[1] Y. Su, G. Shao, L. Gao, L. Zhou, L. Qin and W. Qin, RBC electrophoresis with discontinuous power supply-a newly established hemoglobin release test, Electrophoresis 30 (2009), 3041-3043.

[2] Y. Su, L. Gao, Q. Ma, L. Zhou, L. Qin, L. Han and W. Qin, Interactions of hemoglobin in live red blood cells measured by the electrophoresis release test, Electrophoresis 31 (2010), 2913-2920.

[3] Y. Su, J. Shen, L. Gao, Z. Tian, H. Tian, L. Qin and W. Qin, Molecular interactions of re-released proteins in electrophoresis of human erythrocytes, Electrophoresis 33 (2012), 1402–1405.

[4] E.M. Pasini, H.U. Lutz, M. Mann and A.W. Thomas, Red blood cell (RBC) membrane proteomics--Part II: Comparative proteomics and RBC patho-physiology, Journal of Proteomics 73 (2010), 421-435.

[5] P.I. Margetis, M.H. Antonelou, I.K. Petropoulos, L.H. Margaritis and I.S. Papassideri, Increased protein carbonylation of red blood cell membrane in diabetic retinopathy, Experimental and Molecular Pathology 87 (2009), 76-82.

[6] I.K. Petropoulos, P.I. Margetis, M.H. Antonelou, J.X. Koliopoulos, S.P. Gartaganis, L.H. Margaritis and I.S. Papassideri, Structural alterations of the erythrocyte membrane proteins in diabetic retinopathy, Graefe's Archive for Clinical and Experimental Ophthalmology 245 (2007), 1179-1188.

[7] A. Hernández-Hernández, M.C. Rodríguez, A. López-Revuelta, J.I. Sánchez-Gallego, V. Shnyrov, M.

Llanillo and J. Sánchez-Yagüe, Alterations in erythrocyte membrane protein composition in advanced non-small cell lung cancer, Blood Cells, Moleculars & Diseases 36 (2006), 355-363.

[8] N. Mikirova, H.D. Riordan, J.A. Jackson, K. Wong, J.R. Miranda-Massari and M.J. Gonzalez, Erythrocyte membrane fatty acid composition in cancer patients, Puerto Rico Health Sciences Journal J 23 (2004), 107-113.

[9] D. Chiu, F. Kuypers and B. Lubin, Lipid peroxidation in human red cells, Seminars in Hematology 26 (1989), 257-276.

[10] R. Sharma and B.R. Premachandra, Membrane-bound hemoglobin as a marker of oxidative injury in adult and neonatal red blood cells, Biochemical Medicine and Metabolic Biology 46 (1991), 33-44.

[11] M.M. Abdel-Daim, M.A. Abd Eldaim and A.G. Hassan, Trigonella foenum-graecum ameliorates acrylamide-induced toxicity in rats: roles of oxidativestress, proinflammatory cytokines, and DNA damage, Biochemistry and Cell Biology 1 (2014), 1-7.

[12] R. Cardin, M. Piciocchi, M. Bortolami, A. Kotsafti, L. Barzon, E. Lavezzo, A. Sinigaglia, K.I. Rodriguez-Castro, M. Rugge and F. Farinati, Oxidative damage in the progression of chronic liver disease to hepatocellular carcinoma: An intricate pathway, World Journal of Gastroenterology 20 (2014), 3078-3086.

[13] B. Tekiner-Gulbas, A.D. Westwell and S. Suzen, Oxidative stress in carcinogenesis: New synthetic com pounds with dual effects upon free radicals and cancer, Current Medicinal Chemistry 20 (2013), 4451-4459.

(*Bio-Medical Materials and Engineering*. 2015, 26: S2049–S2054)

第三十四章 血红蛋白释放试验鉴别黄疸类型的初步研究

王翠峰[1] 高丽君[2] 郭 俊[4] 秦文斌[2※] 韩丽莎[3※] 孙晓荣[3]
高雅琼[2] 宝勿仁必力格[2] 苏 燕[2]

(包头医学院 1. 第一附属医院 检验科；2. 血红蛋白研究室；3. 病理生理教研，包头 014010 室；4 包头市肿瘤医院 介入科，包头 014010)

摘 要

背景与目的：利用我研究室创建的血红蛋白释放试验(参见中国科学：生命科学 2011 第 8 期)对黄疸患者血液标本进行比较研究。

方法：血液标本来自胆红素指标异常者。用血红蛋白释放试验比较研究三种类型黄疸(阻塞性黄疸、溶血性黄疸、肝细胞性黄疸)。

结果：阻塞性黄疸使释放增强；溶血性黄疸使释放减弱；肝细胞性黄疸结果类似于溶血性黄疸，但二者之间有区别。

结论：三种类型黄疸患者血液标本在血红蛋白释放试验结果方面有差异。

关键词 血红蛋白释放试验；黄疸；阻塞性黄疸；溶血性黄疸；肝细胞性黄疸

1 前言

黄疸是指由于血中胆红素浓度升高而引起的皮肤、黏膜、巩膜和其他组织的黄染现象，是多种疾病的临床表现。黄疸主要分为阻塞性黄疸、溶血性黄疸、肝细胞性黄疸。要判断黄疸的类型，最好的方法就是进行肝功能检查，因为检查中的胆红素代谢指标是判断黄疸类型最好的手段。胆红素代谢有三个指标，分别是总胆红素、直接胆红素和间接胆红素。这三个指标升高的情况不同可以反映出不同的黄疸类型。临床上，根据上述胆红素指标，再结合其他检查，鉴别三种类型黄疸并不难[1]；本文的目的是从血红蛋白释放试验角度[2, 3]来观察它们的释放特点和鉴别意义。

2 材料及方法

2.1 材料

(1) 胆红素结果异常全血标本 10 份，来自包头医学院第一附属医院检验科。10 份黄疸标本的化验记录见表 34-1。

※通讯作者：秦文斌，电子邮箱：qwb5991309@tom.com； 韩丽莎，电子邮箱：lishahan@sina.com

表 34-1　10 份黄疸标本的化验记录

指标 标本号	总胆红素 正常参考值 2～20	直接胆红素 正常参考值 0～6	间接胆红素 正常参考值 2～19	诊断
1	139.0	107.0	32.0	阻塞性黄疸
2	83.2	22.9	60.3	溶血性黄疸
3	20.7	12.9	7.8	
4	34.4	9.4	25.0	
5	32.1	16.4	15.7	
6	49.4	35.0	24.4	肝细胞性黄疸*
7	32.1	7.8	24.3	
8	30.4	6.4	24.0	
9	36.6	9.6	27.0	
10	24.9	5.7	19.2	

注释：*谷丙转氨酶升高 177，γ-谷氨酰转肽酶增加 596，谷草转氨酶升高 149。

(2) 肝内胆管癌(重症阻塞性黄疸)患者血液标本来自第七医院介入科。它的胆红素化验记录如下：总胆红素 237.4，直接胆红素 91.9，间接胆红素 145.5。

2.2 方法

(1) 全血和由全血分离红细胞：取全血，低速离心(3000 转/分钟)，去上清液，用生理盐水洗涤红细胞 5 次。最后，在红细胞上留等体积盐水，混匀备用。

(2) 血红蛋白释放试验：取红细胞或全血 5μl，加在 5mm×1mm 滤纸条上。插入淀粉-琼脂糖混合凝胶内，进行梯带再释放试验。电势梯度 6V/cm，通电 15 分钟，断电 15 分钟，反复进行，共 4 小时。最后正负极对调，再通电 15 分钟。丽春红-联苯胺复合染色，或丽春红-考马氏亮蓝染色，照相留图。

3　结果

3.1　10 个胆红素异常标本的筛查结果　见图 34-1。

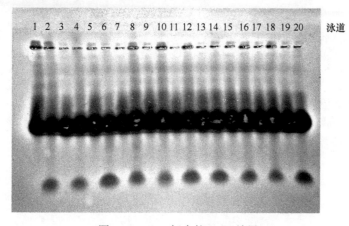

图 34-1　1-10 标本的 HRT 结果

注释：泳道由左向右：单数为红细胞，双数为全血。泳道 1、2 为阻塞性黄疸的红细胞和全血；泳道 3、4 为溶血性黄疸的红细胞和全血；泳道 11、12 为肝细胞性黄疸的红细胞和全血

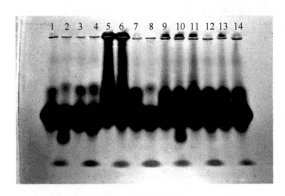

图 34-2　肝内胆管癌标本的 HRT 结果

注释：泳道由左向右，单数为红细胞，双数为全血。泳道 5、6 为肝内胆管癌(重度阻塞性黄疸)的红细胞和全血。

由图 34-1 可以看出，1 号标本为阻塞性黄疸(见泳道 1、2)；红细胞及全血 HRT 都增强，血红蛋白后退也明显。2 号标本为溶血性黄疸(见泳道 3、4)；红细胞及全血 HRT 都减弱，血红蛋白后退也是很弱或无。6 号标本为肝细胞性黄疸(见泳道 11、12)。

3.2　肝内胆管癌的 HRT 筛查结果

见图 34-2。

由图 34-2 可以看出，肝内胆管癌标本红细胞及全血 HRT 都显著增强，血红蛋白后退也非常明显。

4　讨论

从血红蛋白释放试验 HRT 角度来看，三种类型黄疸的结果互不相同，特别是阻塞性黄疸，阻塞越严重差异越明显，肝内胆管癌的结果非常突出。由此可见，HRT 在鉴别黄疸，尤其是阻塞性黄疸与非阻塞性黄疸方面，具有一定的参考价值。正常人的两种释放，是红细胞的再释放稍强于其全血的再释放；糖尿病患者血糖升高时，红细胞的再释放明显强于其全血的再释放[4]。黄疸时结果相反，全血的再释放强于其红细胞的再释放，阻塞性黄疸时更为明显，这说明，血浆成分促进了全血中红细胞的再释放。血浆成分很多，此时首先考虑的是胆红素，很可能是它在起作用。胆红素又分间接胆红素(又称游离胆红素)和直接胆红素(又称结合胆红素，为游离胆红素与葡萄糖醛酸结合产物)，是它们作用于全血中的红细胞，增强其再释放。从机制方面推测，阻塞性黄疸时结合胆红素对全血中红细胞的作用更强，进一步研究时需要专门用结合胆红素单独作用于红细胞，观察其实验效果。当然，还可以专门用游离胆红素单独作用于红细胞，这也是需要我们深入研究的内容。

参 考 文 献

[1] 黄宗干, 陈运贞. 临床症状鉴别诊断学. 4 版. 上海: 上海科学技术出版社, 2001.
[2] Su Y, Shao G, Gao L J, et al. RBC electrophoresis with discontinuous power supply-a newly established hemoglobin release test. Electrophoresis, 2009, 30: 3041-3043.
[3] 秦文斌. 活体红细胞内血红蛋白的电泳释放. 中国科学生命科学, 2011, 41 (8) : 597-607.
[4] 张晓燕, 高丽君, 高雅琼, 等. 血糖浓度和血红蛋白释放试验的比对研究. 国际检验医学杂志, 2010, 31 (6) : 524-525.

第三十五章 中药穿心莲和当归对小鼠红细胞血红蛋白再释放的影响

周成江[1] 高丽君[2] 王宗霞[1] 夏一铭[1] 秦文斌[2#]

(包头医学院 1. 基础学院；2. 血红蛋白研究室，包头 014010)

摘 要

目的：探讨穿心莲和当归对小鼠红细胞血红蛋白再释放的影响。
方法：将实验用小鼠雌雄各半按体重均匀分为 7 组(6 个用药组和 1 个对照组)，按实验设计分别对穿心莲和当归高、中、低浓度用药组小鼠用穿心莲和当归水煎液进行灌胃，对照组用生理盐水灌胃，每天一次，连续灌胃 15 天，然后分别采血分离红细胞观察其血红蛋白再释放的情况，再用 BandScan 扫描定量，进行统计分析。
结果：中高浓度的穿心莲和高浓度的当归都能使小鼠红细胞血红蛋白的再释放增强。
结论：穿心莲和当归水煎液均能影响实验小鼠的血红蛋白再释放，但影响程度随浓度不同而有差异。

关键词 小鼠红细胞；血红蛋白再释放；穿心莲；当归

The effect of Andrographis paniculata and Angelica sinensis on the hemoglobin release of mice blood cell

ZHOU Cheng-jiang[1], GAO Li-jun[2], WANG Zong-xia[1], XIA Yi-ming[1], QIN Wen-bin[2#]

(1. Basic Medical School, Baotou Medical College; 2. Laboratory of Hemoglobin, Baotou Medical College, Baotou 014060, China)

Abstract

Objective: To study the effect of water decoction of Andrographis paniculata and Angelica sinensis on the hemoglobin re-release of mice blood cell. **Methods:** 70 mice were randomly divided into six groups designated as the low and median and high concentration group of both Andrographis paniculata and Angelica sinensis and one control group. The six groups mice were subjected to continuous gastric perfusion with water decoction of Andrographis paniculata and Angelica sinensis respectively for 15 days. Then the control group were managed with saline. Then taking the blood of the mice and separating the blood cell to observe the hemoglobin release by BandScan scanning and statistics analysis. **Results:** Andrographis paniculata and Angelica sinensis both can increase the hemoglobin re-release, but the effect is indistinct for the low concentration of Angelica sinensis. **Conclusion:** The different concentration water decoction of Andrographis paniculata and Angelica sinensis all can affect hemoglobin re-release of mice blood cell, then the level of effect is different with the different concentration of water decoction of Andrographis paniculata and Angelica sinensis.

Key words mice blood cell; hemoglobin re-release; Andrographis paniculata; Angelica sinensis

---

通讯作者：秦文斌，电子邮箱：qwb5991309@tom.com

1 前言

中医中药在我国医药卫生事业中占据非常重要的位置，中药更在中西医结合治疗中起到非常重要的作用。研究发现中药对红细胞的主要作用表现为能提高红细胞 C3b 受体花环率，降低红细胞免疫复合物花环率，增强正常小鼠红细胞免疫功能[1]；改善膜磷脂、膜蛋白及抗氧化损伤[2]；能够减缓运动造成的红细胞参数下降，使血清中促红细胞生成素增多[3]。但到目前为止，关于中药对红细胞释放血红蛋白的影响还未见有报道，因此，在本研究中，我们检测了中药穿心莲和当归对小鼠红细胞血红蛋白再释放的影响。

2 材料及方法

2.1 材料 动物为昆明种小白鼠，成熟雌雄兼用，体重 20～25g，由内蒙古大学实验动物中心提供。实验用药为穿心莲和当归水煎液。

2.2 方法

(1) 实验用药的制备：分别将穿心莲和当归加适量水煮沸 1 小时，过滤，共煮两次，合并滤液，浓缩至 1mg，相当于 1g 生药；之后再用蒸馏水稀释，配成高(0.08g/ml)、中(0.04g/ml)、低(0.02g/ml)三个浓度。

(2) 动物实验：将 70 只小鼠雌雄各半随机分为穿心莲和当归高(0.08g/ml)、中(0.04g/ml)、低(0.02g/ml)浓度用药组和对照组，每组 10 只，分别连续 15 天灌胃，每次 1ml，每日一次。之后处死小鼠，取 1ml 全血保留于抗凝管中，4℃冰箱保存备用。

(3) 血红蛋白试验：参照文献[4-6]，操作如下，在 1.5ml 的 EP 管中加入全血 50μl，低速离心 3 分钟后去掉血浆，加入生理盐水至 1ml，混合均匀，再低速离心 3min 后去掉生理盐水，此为洗一次的红细胞，共洗 5 次。最后，留下与红细胞相同体积的生理盐水，混匀。取此红细胞 5μl，加在 5mm×1mm 滤纸条上，插入淀粉-琼脂糖混合凝胶内，进行我室常规定时释放试验，电势梯度 6V/cm，第一次电泳 2.5 小时，然后停电 15 分钟，再通电 15 分钟。电泳后丽春红染色，拍照留图，BandScan 扫描定量。

3 结果

3.1 穿心莲对小鼠红细胞血红蛋白再释放的影响 低浓度穿心莲造成的定释带很弱，中浓度者稍强，高浓度者更强一些。扫描定量结果显示，中高浓度的穿心莲能够影响小鼠红细胞血红蛋白的再释放，影响程度与穿心莲的浓度成正比，各剂量组间的差异具有统计学意义。详见图 35-1、表 35-1。

表 35-1 不同浓度穿心莲对小鼠血红蛋白再释放的影响($\bar{x} \pm s$)

组别	例数	灰度值
对照组	5	46 105.26±88.34
中浓度组	5	49 294.66±894.51*
高浓度组	5	50 830.00±1691.252

注释：*与对照组和低浓度组相比，均 $P<0.05$；与中浓度组相比，$P<0.05$。

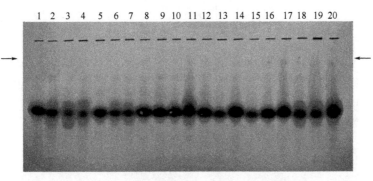

图 35-1 穿心莲对小鼠红细胞血红蛋白再释放的影响
注释：1～5 为对照组，6～10 为低浓度组，11～15 为中浓度组，16～20 为高浓度组

3.2 低中浓度当归对小鼠红细胞血红蛋白再释放的影响 低中浓度当归造成的定释带都较弱，差别不明显。扫描定量结果显示，对照组及两种浓度之间的差异无统计学意义。详见图 35-2、表 35-2。

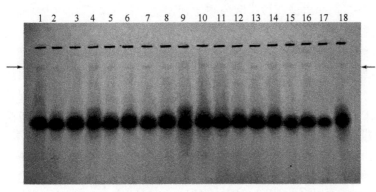

图 35-2 低、中浓度当归对小鼠红细胞血红蛋白再释放的影响
注释：1～6 为对照组，7～12 为低浓度组，13～18 为中浓度组

表 35-2 低中浓度当归对小鼠血红蛋白再释放的影响($\bar{x} \pm s$)

组别	例数	灰度值
对照组	6	52 334.59±1486.25
低浓度组	6	51 198.62±1166.11
中浓度组	6	52 538.80±2191.58

3.3 中高浓度当归对小鼠红细胞血红蛋白再释放的影响 中浓度与高浓度当归造成的定释带强度不同，前者较弱，后者增强。扫描定量结果显示，两种浓度之间的差异有统计学意义。详见表 35-3、图 35-3。

表 35-3 当归中高浓度对小鼠血红蛋白再释放的影响($\bar{x} \pm s$)

组别	例数	灰度值
中浓度组	7	52 005.28±1091.53
高浓度组	9	53 517.55±1504.75*

注释：*与中浓度组相比，$P<0.05$。

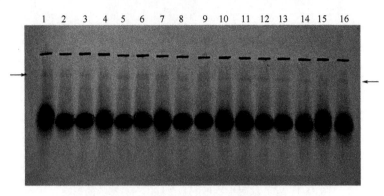

图35-3 中高浓度当归对小鼠红细胞再释放的影响
注释：1~7 为当归中浓度，8~16 为当归高浓度

4 讨论

红细胞内血红蛋白的电泳释放是我室自主研发的一种实验手段和理论体系，是从电泳释放角度来研究红细胞内血红蛋白的存在状态及其生理和病理意义[1-16]。扩大到红细胞以外，还可以研究任何细胞内各种蛋白质的存在状态及其生理和病理意义[15]。但是，迄今的研究几乎都是人类红细胞，对动物红细胞研究很少。此外，大多数研究都是各种疾病时血红蛋白的电泳释放，很少涉及药物治疗时的情况，中药治疗更是一个空白。本文研究了两种中药穿心莲、当归对小鼠红细胞再释放血红蛋白的影响。结果显示三种浓度穿心莲对小鼠红细胞血红蛋白的再释放的影响程度不同，随浓度增加而增强。三种浓度当归的情况稍有不同，低浓度与中浓度当归之间差别不大，中浓度与高浓度有差异，浓度越高再释放越强。然而关于中药造成的血红蛋白再释放增强的意义目前还不清楚。目前已知，人类红细胞的再释放往往来自某些疾病，例如糖尿病[11]尿毒症[12]、肝硬化[16]、胆石症[10]、地中海贫血[8-9]等。因此，我们推测高浓度中药对小鼠红细胞血红蛋白再释放的影响可能是一种损伤性毒性作用。提示中药用量要适度，过量时也会影响健康。以后应进一步研究中药对人类红细胞的影响，观察有效治疗量时中药的效果，以便进一步探讨中药对红细胞血红蛋白再释放影响的作用机制。

参 考 文 献

[1] 李桂海, 刘绛琴. 红细胞免疫及中药对红细胞免疫的影响. 山东中医学院报, 1991, 15 (5)：61-62.

[2] 王艳伟, 林娜, 乔利. 中药对红细胞结构与功能影响的研究进展. 中国实验方剂学杂志, 2006, (11)：64-67.

[3] 张敏, 刘洪珍. 复方中药对大鼠红细胞参数及促生长因子影响的研究. 吉林体育学院学报, 2012, 28 (5)：77-80.

[4] 秦文斌. 活体红细胞内血红蛋白的电泳释放. 中国科学生命科学, 2011, 41 (8)：597-607.

[5] Su Y, Gao L J, Tian HF, et al. Molecular interaction of re-released proteins in electrophoesis of human erythrocytes. Electrophoresis, 2012, 33: 1042-1045.

[6] Su Y, Gao LJ, Qin WB. Interactions of hemoglobin in lived red blood cells measured by the electrophoesis release test [C]// Biji T. Protein Electrophoresis: Methods and Protocols, Methods in Molecular Biology. Springer Science and Busibess Media, 2012: 393-402.

[7] Su Y, Gao J, Ma Q, et al. Interactions of hemoglobin in lived red blood cells measured by the electrophoesis release test. Electrophoresis, 2010, 31: 2913-2920.

[8] Su Y, Shao G, Gao LJ, et al. RBC electrophoresis with discontinuous power supply-a newly established he-

moglobin release test. Electrophoresis, 2009, 30: 3041-3043.

[9] 秦文斌, 高丽君, 苏燕, 等. 血红蛋白释放试验与轻型 β-地中海贫血. 包头医学院学报, 2007, 23 (6): 561-563.

[10] 韩丽红, 闫斌, 高雅琼, 等. 普通外科患者血红蛋白释放试验的比较研究. 临床和实验医学杂志, 2009, 8 (7): 67-69.

[11] 张晓燕, 高丽君, 高雅琼, 等. 血糖浓度和血红蛋白释放试验的比对研究. 国际检验医学杂志, 2010, 31 (6): 524-525.

[12] 高雅琼, 王彩丽, 高丽君, 等. 尿毒症患者低渗血红蛋白释放试验的初步结果. 临床和实验医学杂志, 2010, 9 (1): 12-13.

[13] 王翠峰, 高丽君, 乌兰, 等. 血红蛋白释放试验与血液流变学检测结果相关性的研究. 中国医药导报, 2010, 7 (4): 64-66.

[14] 张咏梅, 高丽君, 苏燕, 等. 陈旧血电泳出现快泳红带: 红细胞释放"高铁血红素?现代预防医学杂志, 2011, 38 (15): 3040-3042.

[15] 乔姝, 沈木生, 韩丽红, 等. 血小板成分电泳释放的初步研究. 现代预防学, 2011, 38 (4): 685-688.

[16] 宝勿仁必力格, 王翠峰, 高丽君, 等. 肝硬化患者血液标本的血红白释放试验结果明显异常. 临床和实验医学杂志, 2011, 10 (24): 1915-1917.

[原文发表于"包头医学院学报, 2013, 29(6): 7-9"]

第三十六章 梅花鹿镰状红细胞内血红蛋白的电泳释放

徐春忠[2] 秦良谊[1] 高丽君[1] 韩丽红[1] 高雅琼[1] 苏 燕[1] 秦文斌[1※]

(1. 包头医学院 血红蛋白研究室，包头 014010；2. 上海野生动物园，上海 201399)

摘 要

背景和目的：我们曾发现地中海贫血患者的血红蛋白释放试验(HRT)异常，此时想到另一种人类血红蛋白病——血红蛋白 S 病(镰状细胞贫血)，它的 HRT 的结果如何？但在国内尚未找到这种标本，在这种情况下我们想到梅花鹿的红细胞也是镰状，故决定取它的红细胞进行 HRT，观察梅花鹿血镰状红细胞的血红蛋白释放情况。

方法：用显微镜观察梅花鹿血中的镰状红细胞，用双向对角线电泳比较研究梅花鹿红细胞与人体红细胞，用低渗全程 HRT 研究鹿与人红细胞之间的差异。建立以偏重亚硫酸钠为介质的淀粉-琼脂糖混合凝胶电泳，比较人血与鹿血在无氧条件下的电泳行为。

结果：显微镜下看到梅花鹿血出现非常漂亮的镰状红细胞。双向对角线电泳和低渗全程 HRT 中，梅花鹿血与人血红细胞的差别都不太明显。偏重亚硫酸钠电泳结果特殊，鹿血红细胞与人血红细胞差别明显，前者的多带再释放明显强于后者。

结论：无氧条件下鹿血红细胞的电泳再释放明显强于人血红细胞，这就是鹿血镰化红细胞的特殊之处。

关键词 梅花鹿；镰状红细胞；无氧；镰状细胞贫血；血红蛋白电泳释放；偏重亚硫酸钠电泳

1 前言

人们早就发现梅花鹿(以下简称为鹿)血含有镰化红细胞。1936 年 Earl C. O'Roke 首次发现有些鹿中有镰状细胞贫血[1]，1960 年 E. Undritz 等在 Nature 上发表文章，明确鹿血红细胞有镰化现象[2]，1964 年 Kitchen H 等在 Science 上发表关于血红蛋白多态性与鹿血红细胞镰化的关系[3]的文章。1967 年 C. F. Whitten 提出鹿血含有镰化红细胞对机体无害[4]，1974 年 W. J. Taylor 和 C. F. Simpson 报道鹿血镰化红细胞的超微结构[5]，Amma E.L.等研究鹿红细胞的镰化机制[6]，1983 年 D. Seiffe. 进行了鹿血镰化红细胞的血液流变学研究[7]。多年来，我们研究室一直进行一种红细胞内血红蛋白的电泳释放研究[8-12]，曾发现地中海贫血时血红蛋白释放异常[13]，准备用这种方法再研究人类的镰状细胞贫血(血红蛋白 S 病)，但在国内拿不到这种标本，此时想到鹿的红细胞也是镰状，故取它的红细胞进行研究，观察鹿血镰状红细胞的血红蛋白释放情况。

※通讯作者：秦文斌，电子信箱：qwb5991309@tom.com

2 材料与方法

2.1 材料 梅花鹿血来自上海动物园。

2.2 方法

(1) 显微镜观察：载物片上加一滴盐水，用 tip 头蘸少许鹿血与盐水混匀，盖上盖玻片，30 分钟后，40×10 倍镜下观察结果，并照相留图。

(2) 双向双层对角线电泳：参见文献[12]。

(3) 室温等低渗全程电泳：参见文献[12]。

(4) 37℃等低渗全程电泳：参见文献[12]。

(5) 淀粉-琼脂糖凝胶偏重亚硫酸钠电泳：我室常规淀粉-琼脂糖凝胶电泳 TEB 缓冲液中加入 0.1%偏重亚硫酸钠，其他操作都不变。

3 结果

3.1 梅花鹿血显微镜结果 见图 36-1。

由图 36-1 可以看出，鹿血涂片中出现许多镰状细胞。

3.2 鹿血与人血的双向双层对角线电泳结果 见图 36-2。

由图 36-2A 和图 36-2B 可以看出，鹿血和人血红细胞都有一些多带释放，他们的全血多带释放不明显。

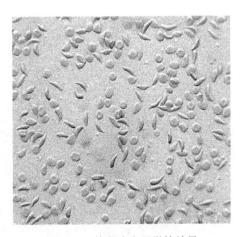

图 36-1 梅花鹿血显微镜结果

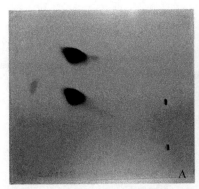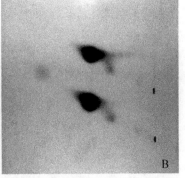

图 36-2 鹿血与人血双向双层对角线电泳

注释：双向双层对角线电泳，上层为红细胞，下层为全血；图 A 为鹿血，图 B 为正常人血

3.3 室温等低渗全程电泳结果 见图 36-3。

由图 36-3 可以看出，人血红细胞都有一些多带释放，鹿血红细胞的多带释放更弱。他们的全血多带释放也如此。

3.4 37℃等低渗全程结果 见图 36-4。

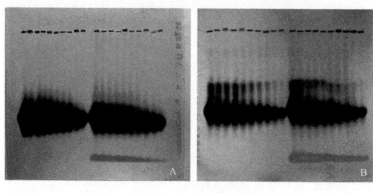

图 36-3　鹿血和人血室温等低渗全程电泳
注释：全程室温，左侧 10 个泳道为红细胞，右侧 10 个泳道为全血；图 A 为鹿血，图 B 为正常人血

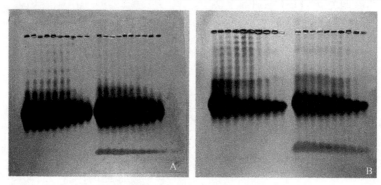

图 36-4　鹿血和人血 37℃等低渗全程电泳
注释：全程 37℃，左侧 10 个泳道为红细胞，右侧 10 个泳道为全血；图 A 为鹿血，图 B 为正常人血

由图 36-4 可以看出，人血红细胞有较强的多带释放，鹿血红细胞的多带释放相对弱一些。它们的全血多带释放都不明显。

3.5　偏重亚硫酸钠电泳结果与常规电泳结果比较　见图 36-5。

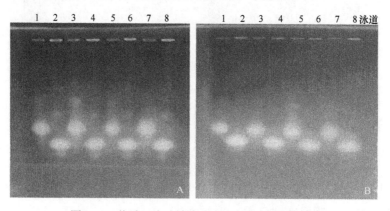

图 36-5　偏重亚硫酸钠淀琼电泳与常规淀琼电泳
注释：图 A 为偏重亚硫酸钠淀琼电泳，图 B 常规淀琼电泳；图 A 与图 B 为相同标本(泳道由左向右，1、3、5、7 为人血红细胞，2、4、6、8 为鹿血红细胞)，比较两种电泳的效果

由图 36-5A 可以看出，在无氧条件下人红细胞多带释放不明显，鹿红细胞多带释放明显。图 36-5B 表明，在有氧条件下人红细胞与鹿红细胞多带释放都不明显。

4 讨论

我们发现轻型 β-地中海贫血时红细胞和全血的电泳释放都增强[13],随即想到另一种血红蛋白病——镰状细胞贫血(血红蛋白 S 病)是否也有异常情况。但是,在国内目前找不到这种疾病的患者,而我们知道梅花鹿红细胞也呈镰刀状,在没有临床标本的情况下先研究梅花鹿的镰状红细胞,既给将来的临床研究打基础,也从动物学角度提供梅花鹿的血红蛋白释放试验资料。现在从双向双层对角线电泳看来,鹿血的电泳释放与地中海贫血不同,它的全血释放比红细胞弱,红细胞本身的释放也不太强,与正常人血差不多。等低渗全程释放方面鹿血与人血略有差别,但鹿血仍无明显独特之处。鉴于镰状细胞贫血患者血红蛋白在还原状态下更容易镰化,我们将镰状细胞贫血快速诊断试验[14]中使用偏重亚硫酸钠的办法引入电泳分析,观察它对鹿血红细胞中血红蛋白的电泳释放情况。结果发现,此条件下鹿血红细胞与人血红细胞的差别明显,而同一电泳槽中制胶不含偏重亚硫酸钠的电泳则未见差异。鹿血红细胞在还原的环境中(偏重亚硫酸钠存在下)多带释放强度明显增强,说明此条件下鹿血镰状红细胞的特殊性才显现出来。现在看来,鹿血的电泳释放情况已经明确,我们推测人类镰状细胞贫血患者的电泳释放结果也可能是这样,实际情况如何,有待拿到临床标本时再来验证。

参 考 文 献

[1] Earl C. O'roke sickle cell anemia in deer. Proceedings of the Society for Experimental Biology and Medicine, 1936, 34: 738-739.
[2] Undritz E, Betke K, Lehmann H. Sickling phenomenon in deer. Nature, 1960, 187: 333-334.
[3] Kitchen H, Putnam F, Taylor W. Hemoglobin polymorphism: its relation to sickling of erythrocytes in white-tailed deer. Science. 1964, 144: 1237-1239.
[4] Whitten C F. Innocuous nature of the sickling (pseudosickling) phenomenon in deer. British Journal of Haematology, 13 (5) : 650.
[5] Taylor W J, Simpson. C F. Ultrastructure of sickled deer erythrocytes. II. The Matchstick Cell Blood, 1974, 43: 907-914.
[6] Amma EL, Sproul GD, Wong S, et al. Mechanism of sickling in deer erythrocytes. Ann N Y Acad Sci, 1974, 241 (0) : 605-613.
[7] Seiffe D. Haemorheological studies of the sickle cell phenomenon in european red deer (Cervus elaphus). Blut, 1983, 47: 85-72.
[8] 秦文斌, 梁友珍. 血红蛋白 A_2 现象 1 A_2 现象的发现及其初步应用. 生物化学与生物物理学报, 1981, 13 (2) : 199-205.
[9] 秦文斌. 红细胞外血红蛋白 A 与血红蛋白 A_2 之间的相互作用. 生物化学杂志, 1991, 7 (5) : 583-587.
[10] 秦文斌. "红细胞 HbA_2" 为 HbA_2 与 HbA 的结合产物. 生物化学与生物物理进展, 1991, 18 (4) : 286-288.
[11] Su Y, Gao LJ, Ma Q, et al. Interactions of hemoglobin in lived red blood cells measured by the electrophoesis release test. Electrophoresis, 2010, 31: 2913-2920.
[12] 秦文斌. 活体红细胞内血红蛋白的电泳释放. 中国科学: 生命科学, 2011, 41 (8) : 597-607.
[13] Su Y, Shao G, Gao LJ, et al. RBC electrophoresis with discontinuous power supply-a newly established hemoglobin release test. Electrophoresis, 2009, 30: 3041-3043.
[14] Itano HA, Pauling L. A rapid diagnostic test for sickle cell anemia. Blood, 1949, 4: 66-68.

第三十七章 遗传性球形红细胞增多症患者红细胞内快泳血红蛋白缺失?

马红杰[2#] 高丽君[1#] 贾国荣[2#] 苏 燕[1#] 卢 艳[2] 李 喆[2]
李 静[2] 贺其图[2※] 秦文斌[1※]

(包头医学院 1. 血红蛋白研究室；2. 第一附属医院 血液科，包头 014010)

摘 要

目的：利用我室创建的血红蛋白释放试验，研究遗传性球形红细胞增多症患者红细胞释放血红蛋白的情况。

方法：利用我室的一系列血红蛋白释放试验：单向释放电泳、双向释放电泳、全血多组分电泳和等低渗全程释放电泳(室温，37℃)，比较研究患者红细胞与正常人红细胞的差异。

结果：患者红细胞经单向电泳出来的血红蛋白快泳部分缺失或明显减少。双向电泳结果，患者除快泳血红蛋白减少或缺失外，还看到红细胞及全血 HbA_2 基本上没有脱离对角线，而且全血出现 MHA。全血多组分实验中看到，患者红细胞及全血都没有再释放。等低渗全程电泳显示，患者红细胞及全血的再释放都明显减弱或消失。

结论：从电泳释放角度证明患者的红细胞明显异常，发现患者红细胞内快泳血红蛋白明显减少或缺失。

关键词 遗传性球形红细胞增多症；红细胞；血红蛋白释放试验/HRT；快泳血红蛋白

1 前言

遗传性球形红细胞增多症(hereditary spherocytosis，HS)是一种红细胞膜先天性缺陷所致的溶血性贫血[1-10]。本病的红细胞膜缺陷主要涉及五种红细胞骨架蛋白：α-血影蛋白，β-血影蛋白，锚定蛋白，带 3 蛋白及蛋白 4.2。它们的单独或联合缺陷，造成骨架蛋白的不稳定。异常的细胞骨架蛋白不能为红细胞脂质膜提供足够的支持，从而造成细胞表面积减少，使红细胞不能维持正常的双凹盘形状而变为球形，出现明显的溶血性贫血。

红细胞膜缺陷影响到红细胞形态并引起一系列病态,此时红细胞内血红蛋白情况如何？未见这方面文献。我们从红细胞电泳释放血红蛋白角度来研究本病，看它与正常人红细胞释放有何不同。

2 材料及方法

2.1 标本来源 遗传性球形红细胞增多症患者血液标本来自包头医学院第一附属医院

#列第一作者
※通讯作者：贺其图，电子邮箱：Heqitu@163.com；秦文斌，电子邮箱：qwb5991309@tom.com

血液科，健康对照标本来自本研究室人员。

2.2 临床资料 遗传性球型红细胞增多症(hereditary spherocytosis，HS)患者，女，45岁，有溶血性贫血、溶血性黄疸(间接胆红素25.6μmol/L)，15岁发病，现在脾稍大。外周血涂片中胞体小、染色深、中央苍白区消失的球形细胞增多(24.5%)，见图37-1。Coombs实验阴性，渗透性脆性试验提示渗透性脆性增加，细胞在0.75%的盐水中就开始溶血，在0.39%时已完全溶血。

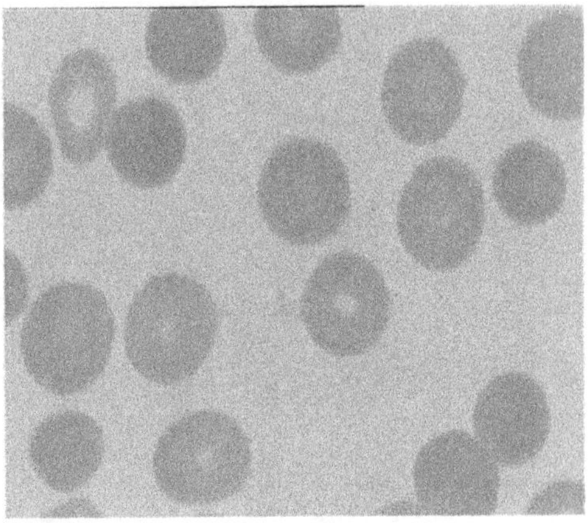

2.3 单向电泳 血红蛋白释放试验方法参见文献[11-15]，本文具体操作如下：取红细胞(含等体积生理盐水)4μl，加在 5mm×1mm 滤纸条上，插入淀粉-琼脂糖混合凝胶，电势梯度6V/cm，通电5小时。电泳过程中直接观察红细胞释放出来血红蛋白的颜色，电泳结束后先染丽春红，再染联苯胺，观察细节。

图 37-1 HS 患者红细胞的显微镜照相图

2.4 双向电泳 做双向双层电泳，上层为上述红细胞，下层为全血。第一向的条件是：电势梯度6V/cm，通电15分钟，停电15分钟，反复进行，共4小时。第二向：调转90°，电势梯度6V/cm，通电3小时。染色情况同前。

2.5 全血多组分血红蛋白释放电泳 8个泳道如下：①全血溶血液=全血用四氯化碳处理后的上清液；②全血基质=全血用四氯化碳处理后的沉淀物；③全血；④红细胞；⑤红细胞溶血液=红细胞用四氯化碳处理后的上清液；⑥红细胞基质=红细胞用四氯化碳处理后的沉淀物；⑦血浆；⑧=1(全血溶血液)。电泳条件：电压6V/cm，第一次电泳2.5小时，然后停电15分钟，再通电15分钟，反复进行，共4小时。染色情况同前。

2.6 等低渗全程血红蛋白释放电泳

(1) 红细胞部分：取10支EP管，按以下比例加入红细胞与蒸馏水。①等渗(红细胞20，水0)；②低渗(红细胞18，水2)；③(红细胞16，水4)；④(红细胞14，水6)；⑤(红细胞12，水8)；⑥(红细胞10，水10)；⑦(红细胞8，水12)；⑧(红细胞6，水14)；⑨(红细胞4，水16)；⑩(红细胞2，水18)混匀后淀琼电泳。将这10个标本通过滤纸条(每条滤纸加液体5μl)加在10个泳道。电泳条件同上，观察各泳道血红蛋白释放情况。

(2) 全血部分：条件完全同上，只是标本为全血。

(3) 两种温度：上电泳前标本处理有两种，室温和37℃保温过夜。

(4) 电泳时，同一标本的红细胞与全血并列放在一个胶板上，以便比较。

3 结果

3.1 患者与正常人红细胞的单向电泳结果 见图37-2。

由图37-2A可以看出，患者红细胞释放出来的血红蛋白，颜色鲜红(参见图中○处)，其

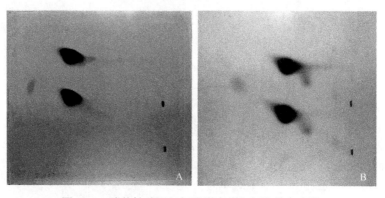

图 37-2 遗传性球形红细胞增多症红细胞单向电泳

注释：图 A 为电泳中染色前，图 B 为电泳后及染色后。泳道 1 为正常人溶血液，2 为正常人红细胞，3 为患者红细胞，4 为正常人红细胞。5=2，6=3，7=4，8=1

他标本都是暗红，而且患者血红蛋白前缘平齐，似乎没有快泳血红蛋白。由图 37-2B 看出，患者红细胞释放出来的血红蛋白中没有快泳成分(参见箭头↑处)。

3.2 患者与正常人双向电泳结果 见图 37-3。

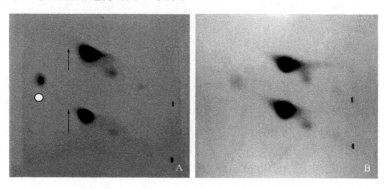

图 37-3 患者和正常人红细胞及全血双向电泳

注释：图 A 为 HS 的双向电泳，图 B 正常人的双向电泳；上层是红细胞，下层为全血

由图 37-3 可以看出，球型红细胞增多症与正常人的双向电泳结果差异明显。正常人的红细胞有一些多带释放，球形红细胞增多症没有。从各成分与对角线关系来看，正常人的血红蛋白 A_2 都离开对角线，而球形红细胞增多症患者的血红蛋白 A_2 则几乎都在对角线上。球形红细胞增多症患者的白蛋白含 MHA(参见图中○处)，而正常人则没有 MHA。这些都说明球形红细胞增多症是一种溶血性疾病。还有，患者红细胞和全血血红蛋白都看不到快泳成分 HbA_1，或者 HbA_1 不明显(参见图中↑处)。

3.3 全血多组分血红蛋白释放电泳结果 见图 37-4。

由此图可以看出，球形红细胞增多症与正常人的电泳结果差异明显。具体如下：泳道 1，在球形红细胞增多症的白蛋白中有轻度的 MHA(高铁血红素白蛋白)，即在红色白蛋白区带里有黑色成分(联苯胺染色阳性)；正常人的白蛋白为红色，没有 MHA。泳道 3、4，正常人有明显的再释放(见□处)，球型红细胞增多症则没有(见○处)。泳道 7，球型红细胞增多症的血浆中 α_1、α_2、β_1、β_2 球蛋白区带清晰可见，对照血浆中 α_1、α_2 球蛋白区带不明显。正常人泳道 1、3、4、5、8 都有明显快泳血红蛋白(见→处)，而患者则几乎看不到 HbA_1。

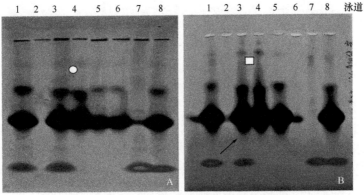

图 37-4　HS 患者和将抗人全血多组分电泳

注释：图 A 为 HS 患者全血多组分电泳，图 B 为健康人全血多组分电泳；由左向右，泳道 1 为全血溶血液，2 为全血基质，3 为全血，4 为红细胞，5 为红细胞溶血液，6 为红细胞基质，7 为血浆，8=1(全血溶血液)

3.4　患者与正常人等低渗全程电泳室温结果　见图 37-5。

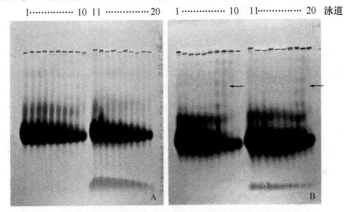

图 37-5　HS 患者和正常人红细胞及全血电泳

注释：图 A 为 HS 患者室温全程电泳图 B 为正常人室温全程电泳；泳道 1…10 为红细胞，泳道 11…20 为全血

由图 37-5 可以看出，球形红细胞增多症与正常人的室温全程电泳结果存在明显差异。正常人的红细胞各泳道都有不同程度的多带释放(如箭头←指示处)，球形红细胞增多症则看不见明显的再释放。

3.5　患者与正常人等低渗全程电泳 37℃结果　见图 37-6。

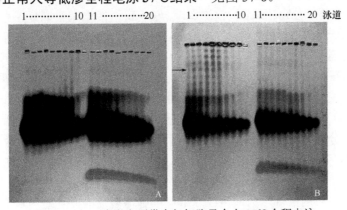

图 37-6　HS 患者和正常人红细胞及全血 37℃全程电泳

注释：图 A 为 HS 患者 37℃全程电泳，图 B 为正常人 37℃全程电泳；泳道 1～10 为红细胞，泳道 11～20 为全血

由图 37-6 可以看出，球形红细胞增多症与正常人的 37℃ 全程电泳结果差异明显。特别是在红细胞方面，正常人的红细胞有明显的多带释放(见箭头→指示处)，球形红细胞增多症的红细胞再释放很弱。

4 讨论

血红蛋白释放试验(hemoglobin release test，HRT)是将活体红细胞放在凝胶电场中，通电后观察其释放出来的血红蛋白的情况[12-16]。本文中的单向电泳结果表明，正常人红细胞存在明显的快泳血红蛋白，而 HS 患者红细胞电泳释放出来的血红蛋白中缺乏这种快泳成分。双向电泳结果表明患者游离红细胞看不见快泳血红蛋白，其全血中红细胞也一样。由于全血中 HbA_2 没有明显脱离对角线，又从释放角度说明患者存在溶血现象。白蛋白中有 MHA，也是血管内溶血的佐证。全血多组分电泳实验中，除了看不清快泳血红蛋白外，全血和红细胞都没有再释放，这一点也较特殊，不同于正常人和许多其他疾病。等低渗全程电泳结果中，各种低渗管几乎都没有再释放，类似于渗透脆性升高结果。

HS 患者血液标本的总的特点是两个缺失，既没有再释放，也看不到快泳血红蛋白。关于再释放，我们过去的研究证明正常人和多种疾病患者的红细胞都有再释放[11, 15]，可有强度不同。全血标本的再释放，多数情况下稍弱于红细胞，但也有增强者。例如，轻型β地中海贫血，患者母女两人的全血多带再释放都明显强于其他人[13, 17]。在对普通外科住院患者全血标本进行研究时发现，等渗条件下的再释放强度不如低渗，低渗时室温不如 37℃[18]，而本文 HS 患者的血液则在上述各种条件下都是没有再释放或是再释放极其微弱。糖尿病患者的全血再释放很弱，但它的红细胞再释放在等渗时很强，而且与血糖浓度相关[19]，而本文的 HS 患者血液标本，则是红细胞与全血都没有再释放。尿毒症患者的情况与普外标本类似，也是低渗条件下全血再释放增强[20]。我们曾对血液流变学检测的全血标本进行低渗再释放比较研究，发现二者之间存在一定相关关系[21]，已知血液流变学异常常见于脑血管疾病。肝硬化患者血液标本的再释放也增强，它的特点是全血和红细胞的再释放都增强[22]。

关于快泳血红蛋白，它有两种命名法，需要加以说明。快泳血红蛋白可以称为 HbA_3，也可以命名为 HbA_1。众所周知，成人血红蛋白主要有三种：一个主要成分，两个次要成分(一个快泳，一个慢泳)。慢泳次要成分已被固定命名为血红蛋白 A_2，快泳次要成分则有两种：血红蛋白 A_3 或 A_1。主要成分也有两种：血红蛋白 A_0 或 A_1。这样一来，就出现两套命名系统，一个是 HbA_0(主要成分)，HbA_1(快泳次要成分)，HbA_2(慢泳次要成分)；另一个是 HbA_1(主要成分)，HbA_2(慢泳次要成分)，HbA_3(快泳次要成分)。多年来，我们一直在使用快泳成分称 HbA_3 的这个命名系统[11-22]，其中，糖尿病时快泳血红蛋白增多，我们曾经使用 HbA_3 这个名称[16]。但是，在糖尿病的诊断和治疗中糖化血红蛋白是一个非常重要的化验指标，现在已经成为诊断糖尿病的金标准[23, 24]。在这里使用的是 HbA_0(主要成分)，HbA_1(快泳次要成分)，HbA_2(慢泳次要成分)命名系统。此时快泳血红蛋白是 HbA_1，人们利用层析技术可将快泳血红蛋白 HbA_1 再细分为 A_{1a}、A_{1b}、A_{1c} 和 A_{1d}，其中主要成分 A_{1c} 就是糖化血红蛋白。由于快泳血红蛋白在命名上存在上述复杂情况，所以本文标题中的快泳血红蛋白，既没有 HbA_1 也没有 HbA_3。

必须指出，本文只是报告一例遗传性球形红细胞增多症的血红蛋白释放情况，还不知道其他同种病例是否也有类似结果。已知此病的病因是红细胞膜蛋白缺陷[1-11]，它主要涉及五种红细胞骨架蛋白：α-血影蛋白、β-血影蛋白、锚定蛋白、带 3 蛋白及蛋白 4.2。本文患者的具体缺陷不明，不同骨架蛋白的缺陷很可能导致不同结果，所以这里只是问题的开始，抛砖引玉，等待今后广泛而深入研究。

参 考 文 献

[1] Delauny J. The molecular basis of hereditary red cell membrane disorders. Blood rev, 2007, 21 (1) : 1-20.
[2] Perotta S, Gallagher PG, Mohandas N. Hereditary spherocytosis. The Lancet, 2008, 327 (9647) : 1411-1426.
[3] Stoya G, Gruhn B, Vogelsang H, et al. Flow cytometry as a diagnostic tool for hereditary spherocytosis. Acta Hematol, 2006, 116 (3) : 186-191.
[4] Bolton-Maggs P H B, Stevens R F, Dodd N J, et al. Guidelines for the diagnosis and management of hereditary spherocytosis. Br J haematol, 2004, 123 (4) : 455-474.
[5] Gallagher PG, Forget BG. Hematologically important mutations: band 3 and protein 4. 2 variants in hereditary spherocytosis. Blood cell mol dis, 1997, 23 (3) : 427-431.
[6] Mohandas N, Gallagher PC. Red cell membrane: past, present and future. Blood, 2008, 112 (10) : 3939-3948.
[7] An X, Mohandas N. Disorders of red cell membrane. Br J Haematol, 2008, 142 (3) : 367-375.
[8] Huq S, Pietroni M AC, Rahman H, et al. Hereditary spherocytosis. J Health Popul Nutr, 2010, 28 (1) : 107-109.
[9] Hassan H, Babadoko AA, Isa AH, et al. Hereditary spherocytosis in a 27-year -old woman: case report. Ann Afr Med, 2009, 8 (1) : 61-63.
[10] Eber S, Lux S E. Hereditary spherocytosis-defects in proteins that connect the membrane skeleton to the Lipid Bilayer. Semin Hematol, 2004, 41: 118-141.
[11] 秦文斌. 活体红细胞内血红蛋白的电泳释放. 中国科学生命科学, 2011, 41 (8) : 597-607.
[12] Su Y, Gao LJ, Qin WB. Interactions of hemoglobin in lived red blood cells measured by the electrophoesis release test. Kurien B T, Scofield R H. Protein Electrophoresis: Methods and Protocols. New York: Springer Science+Busibess Media, 2012.
[13] Su Y, Shao G, Gao LJ, et al. RBC electrophoresis with discontinuous power supply—a newly established hemoglobin release test. Electrophoresis, 2009, 30: 3041-3043.
[14] Su Y, Gao LJ, Ma Q, et al. Interactions of hemoglobin in lived red blood cells measured by the electrophoesis release test. Electrophoresis, 2010, 31: 2913-2920.
[15] Su Y, Shen J, Gao LJ, et al. Molecular interaction of re-released proteins in electrophoesis of human erythrocytes. Electrophoresis, 2012, 33: 1042-1045.
[16] Qin WB, Wang FQ. Diabets mellitus and hemoglobin A_3. Chinese Medical Journal, 1979, 92: 639-642.
[17] 秦文斌, 高丽君, 苏燕, 等. 血红蛋白释放试验与轻型 β-地中海贫血. 包头医学院学报, 2007, 23 (6) : 561-563.
[18] 韩丽红, 闫斌, 高雅琼, 等. 普通外科患者血红蛋白放试验的比较研究. 临床和实验医学杂志, 2009, 8 (7) : 67-69.
[19] 张晓燕, 高丽君, 高雅琼, 等. 血糖浓度和血红蛋白释放试验的比对研究. 国际检验医学杂志, 2010, 31 (6) : 524-525.
[20] 高雅琼, 王彩丽, 高丽君, 等. 尿毒症患者低渗血红蛋白释放试验的初步结果. 临床和实验医学杂志, 2010, 9 (1) : 12-13.
[21] 王翠峰, 高丽君, 乌兰苏燕, 等. 血红蛋白释放试验与血液流变学检测结果相关性的研究. 中国医药导报, 2010, 7 (4) : 64-66.
[22] 宝勿仁必力格, 王翠峰, 高丽君, 等. 肝硬化患者血液标本的血红蛋白释放试验结果明显异常. 临床和实验医学杂志, 2011, 10 (24) : 1915-1917.
[23] 吴捷, 邹大进. 2010 年美国糖尿病学会指南推行用糖化血红蛋白筛查和诊断糖尿病的背景. 中华糖尿病杂志, 2010, 2 (3) : 226-228.
[24] American Diabetes Association. Executive summary: standards of medical care in diabetes-2010. Diabetes Care, 2010, 33 suppl 1: S4-10.

第三十八章 慢阻肺患者红细胞内血红蛋白的电泳再释放——一例报告

孙丽蓉[1] 高丽君[2] 秦文斌[2]※ 王海龙[1] 张敏[1]

(包头医学院 1. 第二附属医院 呼吸科；2. 血红蛋白研究室，包头 014010)

摘 要

目的：观察慢阻肺患者红细胞内血红蛋白的电泳再释放特点。
方法：利用我室建立的红细胞及全血等低渗全程再释放技术(简称全程)，比较研究慢阻肺两期(稳定期和急性加重期)的差异。
结果：两期之间全血红细胞的再释放差异明显。
结论：说明慢阻肺两期之间全血红细胞内部发生深刻变化。

关键词 慢阻肺；稳定期；急性加重期；血红蛋白释放试验

1 前言

慢阻肺，全称为慢性阻塞性肺病(chronic obstructive pulmonary disease，COPD)，是一种逐渐削弱患者呼吸功能的破坏性慢性肺部疾病。据世界卫生组织最新统计，目前全球已有 2.1 亿慢阻肺患者。在中国，40 岁以上人群总患病率高达 8.2%，共有 4300 万人患有慢阻肺。从病情发展来看，慢阻肺可分为"稳定期"(或缓解期)和"急性加重期"(或急性发作期)。由于人体肺脏具有较强的代偿能力，疾病进展隐秘，患者在日常稳定期时疾病特征不明显，却在急性加重期时症状骤然出现或原有症状急剧恶化。每一次慢阻肺急性加重发作后，患者的肺功能大多会进一步下降，即病情不可逆性恶化，从而加快疾病进程，增加死亡的风险。慢阻肺的临床和基础研究很多[1-3]，但未见血红蛋白释放方面的报道，本文则专题研讨慢阻肺稳定期和急性加重期的血红蛋白全程变化。

2 材料和方法

2.1 材料 患者血液标本来自包头医学院第二附属医院呼吸科，慢阻肺患者，男，72 岁，同一人提供稳定期和急性加重期两份标本，属配对资料，有利于对比分析。
2.2 方法 利用我室创建的等低渗全程再释放技术(简称全程)[4-6]，比较分析患者稳定期和急性加重期两份标本的定点再释放(简称定释)，观察二者之间的差异。

※通讯作者：秦文斌，电子邮箱：qwb5991309@tom.com

3 结果

3.1 两期标本室温等低渗全程定释结果 见图38-1。

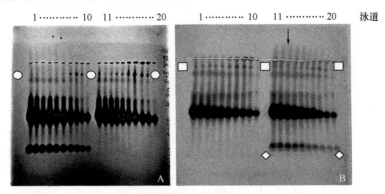

图38-1 慢阻肺患者全血及红细胞电泳

注释：慢阻肺稳定期(图A)与急性加重期(图B)的室温全程实验结果；图A泳道1~10为全血的全程，泳道11~20为游离红细胞的全程。图B泳道1~10为红细胞的全程，泳道11~20为全血的全程。图中主要观察看两个○之间或两个□之间的区带，还有白蛋白前方两个◇之间的前白蛋白，以及原点处的后退成分——丙种球蛋白

由图38-1可以看出，稳定期的红细胞全程(图38-1A○11~20○)，各泳道都有再释放区带，急性加重期红细胞全程定释也都有区带(图38-1B□1~10□)，而且更明显。稳定期全血全程定释区带分布不均，泳道7、8、9较明显(图38-1A○1~10○)，急性加重期全血全程定释带都增强，7、8、9更明显(图38-1B□11~20□)。此外，还能看到急性加重期全血全程中白蛋白前方出现前白蛋白(见图38-1B中◇11~20◇)。再有，急性加重期全血的原点后退区带明显(见图38-1B 11~20↓处)，而稳定期则不明显，这些成分应当属于丙种球蛋白范畴。总之，两期之间在全血部分差别较大，说明血浆成分也起到重要作用。

3.2 两期标本37℃等低渗全程定释结果 见图38-2。

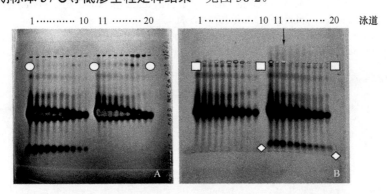

图38-2 慢阻肺患者稳定期和急性加重期37℃全程电泳

注释：慢阻肺稳定期(图A)与急性加重期(图B)的37℃全程实验结果；图A泳道1~10为全血的全程，泳道11~20为游离红细胞的全程。图B泳道1~10为红细胞的全程，泳道11~20为全血的全程。图中主要观察看两个○之间或两个□之间的区带，还有白蛋白前方两个◇之间的前白蛋白，以及原点处的后退成分——丙种球蛋白

由图38-2可以看出，稳定期的红细胞37℃全程(图38-2A○11~20○)，各泳道都有一些再释放区带，前少后多，泳道6、7、8相对多一些。急性加重期红细胞37℃全程定释也都有明显区带(图38-2B□1-10□)，而且是前多后少，泳道1~8都较强。稳定期全血37℃全程定释区带不明显，只有泳道8稍强(图38-2A○1-10○)。急性加重期全血37℃全程定释带都

增强，均匀分布(图 38-2B□11-20□)。此外，与室温全程相同，还能看到急性加重期全血全程中白蛋白前方出现前白蛋白(图 38-2B 中◇11～20◇)，以及急性加重期全血的原点后退区带明显(图 38-2B 11～20↓处)。37℃全血全程中两期的差别更明显。

4 讨论

比较稳定期与急性加重期的实验结果，可以看出二者之间存在明显差异。首先是室温全程：游离红细胞方面，急性加重期红细胞的再释放稍强于稳定期的红细胞。在全血方面，急性加重期全血全程定释带都增强，能看到前白蛋白，还有丙种球蛋白区带明显，而稳定期则不明显。再看 37℃全程：游离红细胞方面，稳定期的红细胞的再释放带是前少后多，急性加重期红细胞则是前多后少，而且后者的再释放带要强一些。在全血方面，稳定期再释放带不明显，急性加重期全血再释放带都增强，与室温全程相同，还能看到急性加重期全血有前白蛋白及丙种球蛋白区带增多。总之，慢阻肺的这两个时期在血红蛋白释放试验方面存在明显差异，给人留下深刻印象。但是，本文只是一例报告，还需大量研究来验证，机制问题也有待深入研究。

参 考 文 献

[1] Gold Executive Committee. Global strategy for the diagnosis, management, and prevention of chronic obstractive pulmonary disease (Revised 2011) , http//www. goldcopd. com.
[2] 中华医学会呼吸病学分会慢性阻塞性肺疾病学组. 慢性阻塞性肺疾病诊疗指南 (2013 年修订版). 结核和呼吸杂志, 2013, 36 (4) : 255-264.
[3] 慢性阻塞性肺疾病急性加重抗菌治疗论坛专家组. 慢性阻塞性肺疾病诊疗指南中急性加重抗菌治疗的地位. 结核和呼吸杂志, 2013, 36 (9) : 712-714.
[4] 秦文斌. 活体红细胞内血红蛋白的电泳释放. 中国科学生命科学, 2011, 41 (8) : 597- 607.
[5] Su Y, Shao G, Gao LJ, et al. RBC electrophoresis with discontinuous power supply-a newly established hemoglobin release test. Electrophoresis, 2009, 30: 3041-3043.
[6] Su Y, Gao LJ, Ma Q, et al. Interactions of hemoglobin in live red blood cells measured by the electrophoresis release test. Electrophoresis, 2010, 31: 2913-2920.

第三十九章 球形红细胞与靶形红细胞电泳释放血红蛋白明显不同

马宏杰　高丽君　贾国荣　苏　燕　秦文斌　贺其图　李　哲
卢　艳　李　静

(包头医学院，包头　014010)

摘　要

红细胞内血红蛋白的电泳释放与红细胞的形态有什么关系是本文的研究内容。现在看来，球形红细胞与靶形红细胞的电泳释放明显不同，在红细胞释放血红蛋白方面，球形红细胞几乎没有多带释放，而靶形红细胞则多带释放非常明显。全血中的红细胞释放也是如此。在基质(相当于红细胞膜残骸)方面二者也不相同，红细胞基质是球形红细胞有较多血红蛋白、靶形红细胞则只有一些非血红蛋白成分，全血中红细胞基质则是球形红细胞看不清、靶形红细胞则出现较多血红蛋白。以上是等渗条件下的实验结果，各种低渗条件下的实验结果也显示这两种红细胞明显不同，此时在全血中红细胞释放血红蛋白方面，球形红细胞所有泳道没有多带释放，靶形红细胞绝大多数泳道多带释放非常明显。双向对角线电泳方面，球形红细胞增多症患者全血中的血红蛋白 A_2 基本处于对角线上。靶形红细胞增多症患者全血中的血红蛋白 A_2 则出现于对角线的下方。上下对比，球形红细胞的血红蛋白 A_2 靠后，靶形红细胞的血红蛋白 A_2 靠前。球形红细胞增多症患者红细胞做普通电泳时，球形红细胞释放出来的血红蛋白 A_1 呈鲜红颜色，而且其前方没有血红蛋白 A_3。

关键词　球形红细；靶形红细胞；血红蛋白释放；基质；血红蛋白 A_3

1　前言

关于球形红细胞增多症和靶形红细胞增多症已有很多研究[1-10]，球形红细胞增多症几乎都是遗传性的[1-6]，靶形红细胞增多症则可与遗传有关，也可来自后天因素，与遗传有关的是地中海贫血[7]，后天因素是某些贫血和肝脏疾病[8-10]。过去的研究无人涉及红细胞内血红蛋白的电泳释放，本文是从这个角度进行研究。本文中的两个标本分别是遗传性球形红细胞增多症和轻型β-地中海贫血[11, 12]，我们发现它们在血红蛋白电泳释放方面出现非常明显的差异。可以这样说，本文遗传性球形红细胞增多症的血红蛋白电泳释放几乎为零(全血中红细胞没有再释放，游离红细胞也没有)，而地中海贫血的再释放接近百分百(全血中红细胞的再释放很强，达到与游离红细胞者相同的强度)。详情参见正文。

2　材料与方法

2.1　标本来源　遗传性球形红细胞增多症血液标本来自包头医学院第一附属医院血液

科，患者，女，45 岁，溶血性贫血、黄疸、脾大，血涂片中球形红细胞大于 20%，渗透脆性增高。靶形红细胞增多症血液标本来自包头医学院第一附属医院检验科，属于轻型β-地中海贫血标本，详情见文献[13，14]。

2.2 显微镜照相

2.3 血红蛋白释放试验方法学的总体情况 参见文献[11，14]，本文实验步骤具体如下所示。

2.4 全血多组分电泳 通常为 8 个泳道。第 1 泳道为全血溶血液：取全血，加等量四氯化碳，震荡后离心，上清液就是全血溶血液(也就是全血里的红细胞已经溶血，故称全血溶血液)。第 2 泳道为全血基质：在制备上述全血溶血液时，在上清液之下、四氯化碳之上有一膜样成分，它就是全血基质(实际上，它就是全血内红细胞膜)。第 3 泳道为全血：将全血直接加在原点，电泳结果代表全血中所有成分对电泳释放的反应。第 4 泳道为红细胞：先由全血分离出红细胞，将此红细胞直接加在原点，电泳结果代表红细胞中所有成分对电泳释放的反应。第 5 泳道为红细胞溶血液：取红细胞，加等量四氯化碳，震荡后离心，上清液就是红细胞溶血液(也就是红细胞已经溶血，故称红细胞溶血液)。第 6 泳道为红细胞基质：在制备上述红细胞溶血液时，在上清液之下、四氯化碳之上有一膜样成分，它就是红细胞基质(实际上，它就是红细胞膜)。第 7 泳道为血浆：将血浆直接加在原点，电泳结果代表血浆中所有成分对电泳释放的反应。第 8 泳道=第 1 泳道。电泳后，先染丽春红，再染联苯胺。

2.5 等低渗全程电泳 取 10 支 0.5mlEP 管，按表 39-1 加入标本(全血或红细胞)和蒸馏水。此时第 1 管为等渗条件，第 2～9 管为不同程度的低渗条件。将此 10 支 EP 管内容物轻轻混匀，各取 6μl 加入 10 个滤纸条(4mm×1mm)，插入凝胶进行电泳(电压 5V/cm，电流 2mA/cm)。第一次电泳 2 小时，停电 5 分钟，再交替通电—停电造成多带释放。

表 39-1 两种红细胞的形态比较

管号	1	2	3	4	5	6	7	8	9	10
标本(μl)	20	18	16	14	12	10	8	6	4	2
蒸馏水(μl)	0	2	4	6	8	10	12	14	16	18
总体积(μl)	20	20	20	20	20	20	20	20	20	20

2.6 双向电泳对角线电泳

2.7 普泳时球形红细胞的一些特点

3 结果

3.1 两种红细胞的形态比较结果 见图 39-1。

由此图可以看出，左图中有多个球型红细胞，右图中有多个靶形红细胞。

3.2 两种红细胞的全血多组分电泳结果 见图 39-2。

由此图可以看出，靶形红细胞增多症患者的全血和红细胞都出现明显的多带释放，而球形红细胞增多症患者的全血和红细胞都显示多带释放。二者在基质(相当于红细胞膜残骸)方面也不相同，全血基质(泳道 2)是球形红细胞看不清、靶形红细胞则出现较多血红蛋白，红细胞基质(泳道 6)是球形红细胞有较多血红蛋白、靶形红细胞则只有一些非血红蛋白成分。

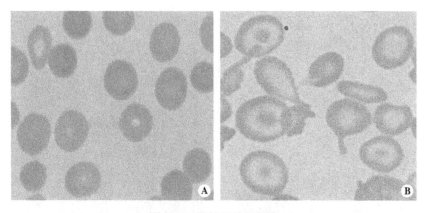

图 39-1 比较两种红细胞

注释：A 图来自球向红细胞增多症，B 图来自靶形红细胞增多症

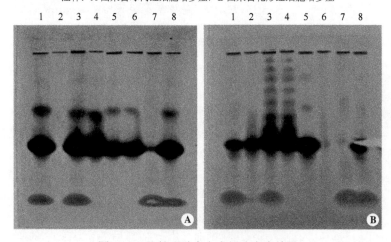

图 39-2 比较两种全血多组分电泳结果

注释：A 图为球型红细胞增多症患者的全血多组分电泳结果。B 图为靶型红细胞增多症患者的全血多组分电泳结果。泳道 1 为全血溶血液，泳道 2 为全血基质，泳道 3 为全血，泳道 4 为红细胞，泳道 5 为红细胞溶血液，泳道 6 为红细胞基质，泳道 7 为血浆，泳道 8=泳道 1

3.3 两种红细胞的等低渗全程(室温)电泳结果 见图 39-3。

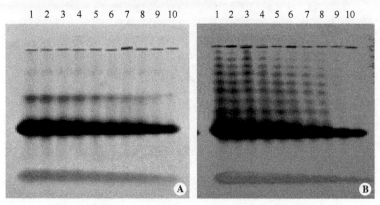

图 39-3 比较两种等低渗全程电泳室温结果

注释：A 图为球形红细胞增多症患者的全血结果，B 图为球形红细胞增多症患者的全血结果。泳道 1 为全血，泳道 2 为全血 9 份加蒸馏水 1 份，泳道 3 为全血 8 份加蒸馏水 2 份，泳道 4 为全血 7 份加蒸馏水 3 份，以此类推，泳道 9 为全血 2 份加蒸馏水 8 份，泳道 10 为全血 1 份加蒸馏水 9 份

由此图可以看出，靶形红细胞增多症患者的全血出现明显的多带释放，而球形红细胞增多症患者的全血则没有多带释放。

3.4 两种红细胞的等低渗全程(37℃)电泳结果 见图39-4。

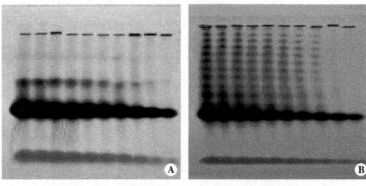

图39-4 比较两种等低渗全程电泳37℃结果

注释：A图为球形红细胞增多症患者的全血结果，B图为球形红细胞增多症患者的全血结果；各泳道的情况同图39-3

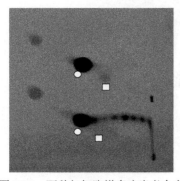

图39-5 两种红细胞增多症患者全血的双向对角线电泳结果

注释：上层为球形红细胞增多症患者全血，下层为靶形红细胞增多症患者全血，左上方的区带为血浆白蛋白；○上方的区带为血红蛋白 A_1，□上方的区带为血红蛋白 A_2

由此图可以看出，靶形红细胞增多症患者的全血出现明显的多带释放，而球形红细胞增多症患者的全血则没有多带释放。

3.5 两种红细胞的双向电对角线电泳结果 见图39-5。

由此图可以看出，靶形红细胞增多症患者的全血出现明显的多带释放，而球形红细胞增多症患者的全血则没有多带释放。球形红细胞增多症患者全血中的血红蛋白 A_2 基本处于对角线上。靶形红细胞增多症患者全血中的血红蛋白 A_2 出现于对角线的下方，而且血红蛋白 A_2 含量较少。上下对比，球形红细胞的血红蛋白 A_2 靠后，靶形红细胞的血红蛋白 A_2 靠前。

3.6 球形红细胞增多症患者红细胞普通电泳时的一些特点 见图39-6。

由此图可以看出，球形红细胞释放出来的血红蛋白 A_1 颜色鲜红，而且其前方没有血红蛋白 A_3。

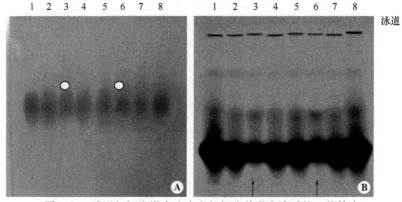

图39-6 球形红细胞增多症患者红细胞普通电泳时的一些特点

注释：泳道3、6为球形红细胞增多症红细胞，其余泳道为正常人红细胞，泳道1、8加样量增多；A图：电泳过程中直接拍照，球形红细胞释放出来的血红蛋白 A_1 呈鲜红色(见○下方的红带)，其余血红蛋白 A_1 呈暗红色；B图：电泳结束后丽春红-联苯胺染色，球形红细胞释放出来的血红蛋白 A_1 的前方缺乏血红蛋白 A_3(见箭头↑指向的位置)

4 讨论

红细胞内血红蛋白电泳释放的发现和研究已经经历了 30 多年。1981 年发现"初释放"，当时称之为"血红蛋白 A_2 现象"[15-18]，2007 年发现"再释放"[11, 12, 19]。在此基础上，我们发现再释放在多种疾病中显示不同程度的增强[20-25]，唯独遗传性球形红细胞增多症明显减弱甚至消失。与此同时，我们还注意到靶形红细胞增多症时情况恰恰相反，此时再释放明显增强，特别是地中海贫血时再释放更是显著增强，与球形红细胞增多症形成鲜明对比，本文结果就是二者的典型例子。

靶形红细胞增多症可以来自后天疾病，肝病特别是阻塞性黄疸时再释放也明显增强[8-10]，此时可能是血浆中黄疸有关物质影响到红细胞膜蛋白，从而影响到血红蛋白的再释放。地中海贫血的靶形红细胞增多症是遗传因素在起作用，中国海南黎族患者的结果[7]是红细胞膜上的收缩蛋白(spectrin)增多。遗传性球形红细胞增多症的情况复杂，不同国家涉及不同的红细胞膜蛋白[1-5]，中国人的研究结果[6]是收缩蛋白减少或缺乏。现在看来，对于中国人的球形红细胞增多症和靶形红细胞增多症来说，收缩蛋白可能对其红细胞的形态影响很大，从而造成红细胞内血红蛋白的电泳释放明显不同。但是，其他国家的情况如何，中国人的结果是否都是如此，还需今后进行更广泛的研究。

参 考 文 献

[1] Delaunay J. The molecular basis of hereditary red cell membrane disorders. Blood Rev, 2007, 21 (1) : 1-20.
[2] Perrota S, Gallagher P G, Mohandas N. Hereditary spherocytosis. The Lancet, 2008, 372 (9647) : 1411-1426.
[3] Gallagher P G, Forget P G. Hematologically important mutation: band 3 and protein 4. 2 variants in hereditary spherocytosis. Blood Cell Mol Dis, 1997, 23 (3) : 417-421.
[4] Stuya G, Gruhn B, Vogelasang H, et al. Flow cytometry as a diagnostic tool for hereditary spherocytosis. Acta hematol, 2006, 116 (3) : 186-191.
[5] King M J, Smethe J S, Mushens R. Eosin-5-maleimide binding to band 3 and Rh-related protein forms the basis of a screening test for hereditary spherocytosis. Br J Haematol, 2004, 124 (1) : 106-113.
[6] 赵新民, 潘华珍, 方芳, 等. 遗传性球星红细胞增多症临床和实验研究. 中华小儿血液, 1996, 1 (4) : 186-188.
[7] Yao HX, Chen ZB, Su QH, et al. Erythrocyte membrane protein abnormalities in β-thalassemia of the Li nationality in Hainan. Chin Med J, 2001, 114 (5) : 486-488.
[8] 凌柱三. 正常人及贫血疾患中的靶形红细胞. 输血及血液学, 1978, 2 (2) : 39.
[9] 吴教仁. 畸形红细胞与靶形红细胞在若干贫血疾患筛选中的意义. 临床检验杂志, 1988, 6 (3) : 159-160.
[10] 傅玉, 汤华. 病毒性肝炎患者靶形红细胞与肝损伤程度的研究. 国际检验医学杂志, 2009, 30 (2): 155-157.
[11] 秦文斌, 高丽君, 苏燕, 等. 血红蛋白释放试验与轻型 β-地中海贫血. 包头医学院学报, 2007, 23 (6): 561-563.
[12] Su Y, Shao G, Gao LJ, et al. RBC electrophoresis with discontinuous power supply - a newly established hemoglobin release test. Electrophoresis, 2009, 30: 3041-3043.
[13] 秦文斌. 活体红细胞内血红蛋白的电泳释放. 中国科学生命科学, 2011, 41 (8) : 597- 607.
[14] Su Y, Gao L J, Qin WB. Interactions of hemoglobin in lived red blood cells measured by the electrophoesis release test. Kurien BT, Scofield RHl. Protein Electrophoresis: Methods and Protocols. New York: Springer Science ＋ Busibess Media, 2012.
[15] 秦文斌, 梁友珍. 血红蛋白 A_2 现象 I A_2 现象的发现及其初步应用. 生物化学与生物物理学报, 1981, 13 (2) : 199-205.
[16] 秦文斌. 红细胞外血红蛋白 A 与血红蛋白 A_2 之间的相互作用. 生物化学杂志, 1991, 7 (5) : 583-587.

[17] 秦文斌. 血红蛋白的 A_2 现象发生机制的研究——"红细胞 HbA_2" 为 HbA_2 与 HbA 的结合产物. 生物化学与生物物理进展, 1991, 18 (4) : 286-288.

[18] Su Y, LJ Gao, Ma Q, et al. Interactions of hemoglobin in lived red blood cells measured by the electrophoesis release test. Electrophoresis, 2010, 31: 2913-2920.

[19] Su Y, Shen J, Gao LJ, et al. Molecular interaction of re-released proteins in electrophoesis of human erythrocytes. Electrophoresis, 2012, 33: 1042-1045.

[20] 韩丽红, 闫斌, 高雅琼, 等. 普通外科患者血红蛋白释放试验的比较研究. 临床和实验医学杂志, 2009, 8 (7) : 67-69.

[21] 张晓燕, 高丽君, 高雅琼, 等. 血糖浓度和血红蛋白释放试验的比对研究. 国际检验医学杂志, 2010, 31 (6) : 524-525.

[22] 高雅琼, 王彩丽, 高丽君, 等. 尿毒症患者低渗血红蛋白释放试验的初步结果. 临床和实验医学杂志, 2010, 9 (1) : 12-13.

[23] 王翠峰, 高丽君, 乌兰, 等. 血红蛋白释放试验与血液流变学检测结果相关性的研究. 中国医药导报, 2010, 7 (4) : 64-66.

[24] 张咏梅, 高丽君, 苏燕, 等. 陈旧血电泳出现快泳红带: 红细胞释放"高铁血红素？". 现代预防医学杂志, 2011, 38 (15) : 3040-3042.

[25] 宝勿仁必力格, 王翠峰, 高丽君, 等. 肝硬化患者血液标本的血红蛋白释放试验结果明显异常. 临床和实验医学杂志, 2011, 10 (24) : 1915-1917.

第四十章　α-地贫与球形红细胞增多症血红蛋白电泳释放的比较研究

苏燕[1#]　马宏杰[2#]　张宏汪[3#]　高丽君[1]　贾国荣[2]

秦文斌[1※]　贺其图[2※]

(1. 包头医学院　血红蛋白研究室，包头　014010；2. 包头医学院第一附属医院　血液科，包头　014010；3. 包钢集团第三职工医院　检验科，包头　014010)

摘　要

在游离红细胞释放血红蛋白方面，球形红细胞增多症几乎没有多带释放，而α-地贫则多带释放非常明显。全血中红细胞释放血红蛋白的情况也是球形红细胞增多症没有多带释放、α-地贫多带释放非常明显。以上是等渗条件下的实验结果，各种低渗条件下的全程实验结果继续显示这两种标本的明显不同。双向电泳也如此，球形红细胞增多症患者全血中没有多带释放，α-地贫患者全血中多带释放非常明显。

关键词　球形红细胞增多症；α地贫；血红蛋白释放；血红蛋白 A_3

1　前言

地中海贫血是一组因珠蛋白链合成障碍而导致的遗传性溶血性贫血，是一种常见的常染色体隐性遗传病，也是世界上最常见和发病率最高的遗传性疾病[1, 2]，由于早期病例都来自地中海，故称地中海贫血或海洋性贫血。根据合成障碍的肽链不同，将地中海贫血分为α-地中海贫血和β-地中海贫血等[1, 2]。α-地中海贫血的基因突变主要有两大类：点突变和缺失突变，本文患者属α-地中海贫血缺失型——SEA/αα[3, 4]。球形红细胞增多症几乎都是遗传性的，它不属于遗传性血红蛋白病[5-7]。

α-地中海贫血和球形红细胞增多症早有大量研究，只是无人涉及它们红细胞内血红蛋白的电泳释放，本文从这个角度进行研究，得到非常有趣的结果，详见正文。

2　材料与方法

2.1　标本来源　遗传性球形红细胞增多症血液标本来自包头医学院第一附属医院血液科，患者，女，45岁，溶血性贫血、黄疸、脾大，血涂片中球形红细胞大于20%，渗透脆性增高。α地贫血液标本来自包钢集团第三职工医院检验科，患者，女，35岁，溶血性贫血、脾大，血涂片中有大量靶形红细胞，渗透脆性降低。我们用PCR技术证明其为缺失型α地贫——SEA/αα[3, 4]。

\# 并列第一作者

※ 通讯作者：秦文斌，电子邮箱：qinwenbinbt@sohu.com；贺其图，电子邮箱：heqitu@163.com

2.2 血红蛋白释放试验方法学的总体情况 参见文献[8-10]。

2.3 全血多组分电泳 通常为 8 个泳道。第 1 泳道为全血溶血液：取全血，加等量四氯化碳，震荡后离心，上清液就是全血溶血液。第 2 泳道为全血基质：在制备上述全血溶血液时，在上清液之下、四氯化碳之上有一膜样成分，它就是全血基质。第 3 泳道为全血：将全血直接加在原点，电泳结果代表全血中所有成分对电泳释放的反应。第 4 泳道为红细胞：先由全血分离出红细胞，将此红细胞直接加在原点，电泳结果代表红细胞中所有成分对电泳释放的反应。第 5 泳道为红细胞溶血液：取红细胞，加等量四氯化碳，震荡后离心，上清液就是红细胞溶血液。第 6 泳道为红细胞基质：在制备上述红细胞溶血液时，在上清液之下、四氯化碳之上有一膜样成分，它就是红细胞基质。第 7 泳道为血浆：将血浆直接加在原点，电泳结果代表血浆中所有成分对电泳释放的反应。第 8 泳道=第 1 泳道。电泳后，先染丽春红，再染联苯胺。

2.4 等低渗全程电泳 参见文献[9]。

2.5 双向对角线电泳 参见文献[8-10]。

2.6 单向初释放电泳 参见文献[8-10]。

3 结果

3.1 两种标本的全血多组分电泳结果 见图 40-1。

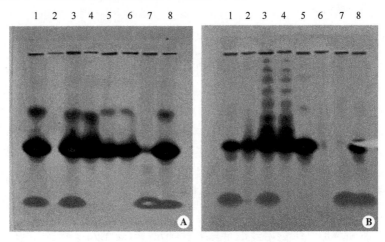

图 40-1 比较两种全血多组分电泳结果

注释：A 图为球型红细胞增多症患者的全血多组分电泳结果。B 图为 α-地贫患者的全血多组分电泳结果。泳道 1 为全血溶血液，泳道 2 为全血基质，泳道 3 为全血，泳道 4 为红细胞，泳道 5 为红细胞溶血液，泳道 6 为细胞基质，泳道 7 为血浆，泳道 8=泳道 1

由图 40-1 可以看出，α-地贫患者的全血和红细胞都出现明显的多带释放，而球形红细胞增多症患者的全血和红细胞都没有多带释放。全血基质方面(泳道 2)，球形红细胞增多症的全血基质中无成分，α-地贫则出现较多血红蛋白，红细胞基质(泳道 6)方面相反，球形红细胞有较多血红蛋白、α-地贫则只有一些非血红蛋白成分。

3.2 两种标本的等低渗全程(室温)电泳结果 见图 40-2。

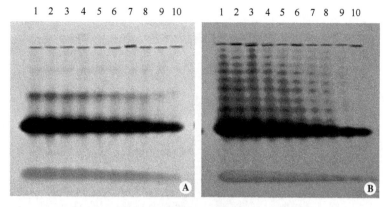

图 40-2 比较两种等低渗全血全程电泳室温结果

注释：A 图为球形红细胞增多症患者的全血全程结果，B 图为 α-地贫患者的全血全程结果。泳道 1 为全血，泳道 2 为全血 9 份加蒸馏水 1 份，泳道 3 为全血 8 份加蒸馏水 2 份，泳道 4 为全血 7 份加蒸馏水 3 份，以此类推，泳道 9 为全血 2 份加蒸馏水 8 份，泳道 10 为全血 1 份加蒸馏水 9 份

由图 40-2 可以看出，α-地贫患者的全血全程出现明显的多带释放，而球形红细胞增多症患者的全血全程则没有多带释放。

3.3 两种标本的等低渗全程(37℃)电泳结果　见图 40-3。

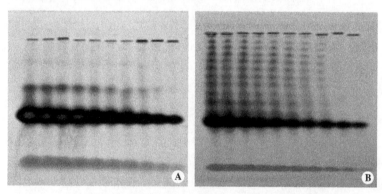

图 40-3 比较两种等低渗全血全程电泳 37℃结果

注释：A 图为球形红细胞增多症患者的全血全程结果，B 图为 α-地贫患者的全血全程结果各泳道的情况同图 41-2

由图 40-3 可以看出，α-地贫患者的全血全程出现明显的多带释放，而球形红细胞增多症患者的全血全程则没有多带释放。

3.4 两种标本的双向电泳结果　见图 40-4。

由图 40-4 可以看出，α-地贫患者的全血出现明显的多带释放，而球形红细胞增多症患者的全血则没有多带释放。球形红细胞增多症患者全血中的血红蛋白 A_2 基本处于对角线上。α-地贫患者全血中的血红蛋白 A_2 出现于对角线的下方，而且血红蛋白 A_2 含量较少。上下对比，球形红细胞的血红蛋白 A_2 靠后，α-地贫的血红蛋白 A_2 靠前。看来，此时球形红细胞增多症患者的溶血

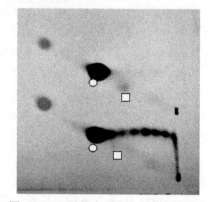

图 40-4 两种标本的全血双向电泳结果

注释：上层为球形红细胞增多症患者全血，下层为 α-地贫患者全血，左上方的区带为血浆白蛋白。○上方的区带为血红蛋白 A_1，□上方的区带为血红蛋白 A_2

程度大于α-地贫。

3.5 球形红细增多症患者红细胞单向初释放电泳时的一些特点　见图40-5。

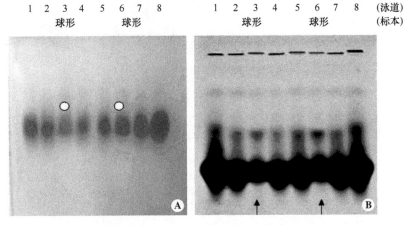

图40-5　球形红细增多症患者红细胞单向初释放电泳结果

注释：泳道3、6为球形红细胞增多症红细胞，其余泳道为正常人红细胞，泳道1、8加样量增多。A图：电泳过程中直接拍照，球形红细胞释放出来的血红蛋白A_1呈鲜红色(见○下方的红带)，其余泳道血红蛋白A呈暗红色。B图：电泳结束后丽春红-联苯胺染色，球形红细胞释放出来的血红蛋白A的前方缺乏血红蛋白A_3(见箭头↑指向的位置)

由图40-5可以看出，球形红细胞释放出来的血红蛋白A_1颜色鲜红，而且其前方没有血红蛋白A_3。由此可知，球形红细胞增多症红细胞内血红蛋白也不正常。

4　讨论

红细胞内血红蛋白电泳释放的发现和研究已经历了30多年。1981年发现"初释放"，当时称之为"血红蛋白A_2现象"[11-14]，2007年发现"再释放"[15-17]。在此基础上，我们发现再释放在多种疾病中显示不同程度的增强[18-23]，唯独遗传性球形红细胞增多症明显减弱甚至消失。与此同时，我们还注意到地中海贫血时情况恰恰相反，此时再释放显著增强，与球形红细胞增多症形成鲜明对比，本文结果就是二者的典型例子。关于地中海贫血分子机制，中国海南黎族患者的结果[24]是红细胞膜上的收缩蛋白(spectrin)增多。遗传性球形红细胞增多症的分子机制，全世界135例患者有61例为收缩蛋白缺乏，还有22例为收缩蛋白锚蛋白联合缺乏[25]。现在看来，对于球形红细胞增多症和地中海贫血来说，收缩蛋白可能对其红细胞的形态影响很大，从而造成红细胞内血红蛋白的电泳释放明显不同。

这两类疾病在血红蛋白电泳释放方面出现非常明显的差异。可以这样说，遗传性球形红细胞增多症的血红蛋白电泳释放几乎为零(全血中红细胞没有再释放，游离红细胞也没有)，而α-地贫的再释放接近百分百(全血中红细胞的再释放很强，游离红细胞者强度也非常强)，二者图像呈鲜明对比。

值得注意的是，球形红细胞增多症患者的红细胞做单向初释放电泳时，球形红细胞释放出来的HbA_1呈鲜红颜色，而且其前方没有HbA_3。众所周知，球形红细胞增多症不属于遗传性血红蛋白病，但从本文的血红蛋白释放结果来看，它也可被列入血红蛋白病范围，我们可以称之为"血红蛋白A_3缺乏症"。

参 考 文 献

[1] 曾溢滔. 人类血红蛋白. 北京: 科学出版社, 2002: 82-86.
[2] 张俊武, 龙桂芳. 血红蛋白与血红蛋白病. 南宁: 广西科学技术出版社, 2003: 204-244.
[3] 区采莹, 蒙晶, 郑诗华. 多重 PCR 筛查高发群体缺失型 α-地贫基因的研究. 中国热带医学, 2002, 2 (2): 129-131.
[4] 张俊武, 龙桂芳. 血红蛋白与血红蛋白病. 南宁: 广西科学技术出版社, 2003: 296.
[5] Perrota S, Gallagher P G, Mohandas N. Hereditary spherocytosis. The Lancet, 2008, 372 (9647): 1411-1426.
[6] Stuya G, Gruhn B, Vogelsang H, et al. Flow cytometry as a diagnostic tool for hereditary spherocytosis. Acta Hematol, 2006, 116 (3): 186-191.
[7] King M J. Using the eosin-5-maleimide binding test in the differential diagnosis of hereditary spherocytosis and hereditary pyropoikilocytosis. Cytometry B Clin Cytom, 2008, 74: 244-250.
[8] Su Y, Shao G, Gao LJ, et al. RBC electrophoresis with discontinuous power supply-a newly established hemoglobin release test. Electrophoresis, 2009, 30: 3041-3043.
[9] 秦文斌. 活体红细胞内血红蛋白的电泳释放. 中国科学生命科学, 2011, 41 (8): 597-607.
[10] Su Y, Shen J, Gao LJ, et al. Molecular interaction of re-released proteins in electrophoesis of human erythrocytes. Electrophoresis, 2012, 33: 1042-1045.
[11] 秦文斌, 梁友珍. 血红蛋白 A_2 现象 I. 此现象的发现及其初步应用. 生物化学与生物物理学报, 1981, 13 (2): 199-205.
[12] 秦文斌. 红细胞外血红蛋白 A 与血红蛋白 A_2 之间的相互作用. 生物化学杂志, 1991, 7 (5): 583-587.
[13] 秦文斌. 血红蛋白的 A_2 现象发生机制的研究——"红细胞 HbA_2" 为 HbA_2 与 HbA 的结合产物. 生物化学与生物物理进展, 1991, 18 (4): 286-288.
[14] Su Y, Gao LJ, Ma Q, et al. Interactions of hemoglobin in lived red blood cells measured by the electrophoesis release test. Electrophoresis, 2010, 31: 2913-2920.
[15] 秦文斌, 高丽君, 苏燕, 等. 血红蛋白释放试验与轻型 β-地中海贫血. 包头医学院学报, 2007, 23 (6): 561-563.
[16] Su Y, Shao G, Gao LJ, et al. RBC electrophoresis with discontinuous power supply – a newly established hemoglobin release test. Electrophoresis, 2009, 30: 3041-3043.
[17] Su Y, Shen J, Gao LJ, et al. Molecular interaction of re-released proteins in electrophoesis of human erythrocytes. Electrophoresis, 2012, 33: 1042-1045.
[18] 韩丽红, 闫斌, 高雅琼, 等. 普通外科患者血红蛋白释放试验的比较研究. 临床和实验医学杂志, 2009, 8 (7): 67-69.
[19] 张晓燕, 高丽君, 高雅琼, 等. 血糖浓度和血红蛋白释放试验的比对研究. 国际检验医学杂志, 2010, 31 (6): 524-525.
[20] 高雅琼, 王彩丽, 高丽君, 等. 尿毒症患者低渗血红蛋白释放试验的初步结果. 临床和实验医学杂志, 2010, 9 (1): 12-13.
[21] 王翠峰, 高丽君, 乌兰, 等. 血红蛋白释放试验与血液流变学检测结果相关性的研究. 中国医药导报, 2010, 7 (4): 64-66.
[22] 张咏梅, 高丽君, 苏燕, 等. 陈旧血电泳出现快泳红带: 红细胞释放"高铁血红素?". 现代预防医学杂志, 2011, 38 (15): 3040-3042.
[23] 宝勿仁必力格, 王翠峰, 高丽君, 等. 肝硬化患者血液标本的血红蛋白释放试验结果明显异常. 临床和实验医学杂志, 2011, 10 (24): 1915-1917.
[24] Yao HX, Chen ZB, Su QH, et al. Erythrocyte membrane protein abnormalities in β-thalassemia of the Li nationality in Hainan. Chin Med J, 2001, 114 (5): 486-488.
[25] 吕照江, 张之南. 遗传性球形红细胞增多症研究进展. 中华血液学杂志, 1995, 16 (4): 214-217.

附 依据研究发表的论文

Comparative Study of the Amount of Re-released Hemoglobin from α-Thalassemia and Hereditary Spherocytosis Erythrocytes

PUBLISHED BY

INTECH

open science | open minds

World's largest Science, Technology & Medicine Open Access book publisher

Comparative Study of the Amount of Re-released Hemoglobin from α-Thalassemia and Hereditary Spherocytosis Erythrocytes

Yan Su, Hongjie Ma, Hongwang Zhang, Lijun Gao, Guorong Jia, Wenbin Qin and Qitu He

Abstract

Hemoglobin release test (HRT), which is established by our lab, is a new experiment to observe the re-released hemoglobin (Hb) from erythrocytes. In this study, one-dimension HRT, double dimension HRT, and isotonic and hypotonic HRT were performed to observe the re-released Hb from the blood samples of normal adult, hereditary spherocytosis (HS), and α-thalassemia. The results showed that compared with normal adult, the re-released Hb from HS blood sample was decreased significantly; however, the re-released Hb from α-thalassemia blood sample was increased significantly. The mechanism of this phenomenon was speculated to have relation with the abnormal amount of membrane-binding Hb.

Keywords: hereditary spherocytosis; α-thalassemia; hemoglobin release test; eryth-rocyte, hemoglobin

1 Introduction

Erythrocytes, also called red blood cells (RBCs), are the most common type of blood cell. In humans, mature erythrocytes are flexible and oval biconcave disks. They lack cell nucleus and most organelles, in order to accommodate maximum space for hemoglobin (Hb), which has the

Additional information is available at the end of the chapter http: //dx.doi.org/10.5772/60947

important oxygen-transporting function. This protein makes up about 96% of the erythrocytes' dry content (by weight), and around 35% of the total content (including water) [1]. Hb is an assembly of two α-globin family chains (including α and ξ chains) and two β-globin family chains (including β, γ, δ, and ε chains). Each globin subunit has an embedded heme group and each heme group contains an iron atom that can bind one oxygen molecule through iron-induced dipole forces. These subunits are bound to each other by salt bridges, hydrogen bonds, and hydrophobic interactions. Three Hb variants exist in normal adult erythrocytes, that is HbA ($α_2β_2$, over 95%), HbF ($α_2γ_2$, <1%), and HbA_2 ($α_2δ_2$, 1.5%~3.5%) [2, 3].

The erythrocyte membrane plays many roles that aid in regulating erythrocytes' surface deformability, flexibility, adhesion to other cells, and immune recognition. These functions are highly dependent on the composition of the membrane, which includes 3 layers: the glycocalyx on the exterior, which is rich in carbohydrates; the lipid bilayer, which contains many transmembrane proteins, besides its lipidic main constituents; and the membrane skeleton, a structural network of proteins located on the inner surface of the lipid bilayer. The determinant of normal membrane cohesion is the system of "vertical" linkages between the phospholipid bilayer and membrane skeleton, formed by the interactions of the cytoplasmic domains of various membrane proteins with the spectrin- based skeletal network. Band 3 and Rh-associated glycoprotein (RhAG) provide such links by interacting with ankyrin, which in turn binds to β-spectrin. Protein 4.2 binds to both band 3 and ankyrin and can regulate the avidity of the interaction between band 3 and ankyrin. Glycophorin C, band 3, XK, Rh, and Duffy all bind to protein 4.1R, the third member of the ternary junctional complex with β-spectrin and actin[4].

Thalassemia is an inherited autosomal recessive blood disorder characterized by abnormal formation of Hb, which results in improper oxygen transport and destruction of erythrocytes. Normally, the majority of adult Hb (HbA), is composed of two α- and two β-globin chains, which are arranged into a heterotetramer. The β-globin chain is encoded by a single gene on chromosome 11[5], and α-globin chain is encoded by two closely linked genes on chromosome 16[6]. A normal person has two loci encoding the β-chain, and four loci encoding the α-chain. Thalassemia patients have defects in either the α- or β-globin chain. According to which chain is affected, thalassemias are classified into α-thalassemias and β-thalassemia. α-thalassemias result in decreased α-globin production, which result in an excess of β-chains in adults and excess γ-chains in newborns. The excess β-globin chains form unstable tetramers (HbH, $β_4$), which have abnormal oxygen dissociation curves. β-thalassemias are characterized as either βo or β-thalassemia major if formation of any β chains is prevented, the most severe form of β-thalassemia; as either β+ or β thalassemia intermedia if some β-globin chain formation are allowed; or as β-thalassemia minor if β-globin chain production is not terribly compromised[7-8]. In contrast to the β-thalassemias, which are usually caused by point mutations of the β- globin gene, the α-thalassemia syndromes are usually caused by the deletion of one or more α-globin genes and are subclassified according to the number of α-globin genes that are deleted (or mutated): one gene deleted (α+-thalassemia); two genes deleted on the same chromosome or in cis (α0-thalassemia); three genes deleted (HbH disease); or four genes deleted (hydrops fetalis with Hb Bart's) [8]

Hereditary spherocytosis (HS) is an autosomal dominant erythrocyte membranopathy[9], but does not belong to hereditary hemoglobinopathies. This disorder is caused by mutations in genes relating to membrane proteins. These proteins include spectrin (α and β), ankyrin[10], band 3, protein 4.2[11], and other erythrocyte membrane proteins that allow for the erythrocytes to change their shapes. The abnormal erythrocytes are sphere-shaped (spherocytosis) rather than being the normal biconcave disk shaped. This difference in shape not only interferes with the ability to be flexible to travel from the arteries to the smaller capillaries, but also makes the erythrocytes more prone to rupture.

Hemoglobin release test (HRT), also called electrophoresis release test (ERT), which is performed by electrophoresing live erythrocytes directly on a starch–agarose mixed gel with intermittent electric current, was established by our lab in 2007[3, 12]. Starch–agarose mixed gel electrophoresis is a routine method used to separate and analyze Hb in our lab since 1980. Hb within the erythrocytes can only be released once during routine starch–agarose mixed gel electrophoresis, which is performed with continuous power supply, and this phenomenon is named as "initial release" now. The difference in mobility of HbA_2 between erythrocyte and hemolysate sample (also called HbA_2 phenomenon) was found during an "initial release" experiment in 1981[12]. In 2007, a sudden power outage was encountered during the electrophoresis of erythrocytes, however, the experiment was not abandoned and electrophoresis was continued after the power was restored. To our surprise, another new Hb band was found to be released from the origin, which was named "single-band re-release" as opposed to the "initial release". When the power outages were simulated more than once, multiple Hb bands would appear between HbA and origin, and this phenomenon was named as multiple-band re-release or ladder-band re-release[13]. Based on these experiments, isotonic and hypotonic HRT and double-dimensional HRT were developed subsequently. Then the re-released Hb was observed in many patients' erythrocytes, and its amount varied in different patients[14, 15]. Some of the patients had increased Hb re-release, such as β-thalassemia, some general surgery patients, cirrhosis, and some gastro enteric tumor patients, but the specific screening experiment had not been done and the exact mechanism of this phenomenon had not been clear. The erythrocyte membrane or cytoskeleton binding Hb was speculated to have rela-tionship with this phenomenon. To further study the mechanism of Hb re-release, the effects of blood type, blood viscosity, different membrane-destroying methods, exogenous hydrogen peroxide, and glutaraldehyde treatments on the amount of re-released Hb were observed subsequently, and the re-released Hb was speculated to have relationship with the abnormality of erythrocyte membrane and Hb. In this study, re-released Hb from two hereditary hemolytic diseases, HS (erythrocyte membrane disorder) and α-thalassemia (Hb disorder), was observed with a variety of HRT experiments.

2 The comparative study of the re-released hemoglobin from α-thalassemia and hereditary spherocytosis erythrocytes

This study had been approved by the local ethics committee, and one HS patient (coming from the first hospital of Baotou Medical College) and one α- thalassemia patient (coming from the third Worker's Hospital of Baogang Group) were included. The HS patient, diagnosed as spectrin defect, is a 45-year-old female with hemolytic anemia, jaundice, and splenomegaly. The α-thalassemia patient, diagnosed as Southeast Asian deletion (SEA) by PCR, is a 35-year- old female with hemolytic anemia and splenomegaly. Before collecting their blood, patients were asked to sign the consent information. Venous blood samples were anticoagulated with EDTA, and then routine blood examination, osmotic fragility test and HRT were performed respectively within 24 h. The spherical erythrocytes of HS patient are more than 20% in peripheral blood smear, and the osmotic fragility is increased (max: 0.39% vs 0.40%—0.45%, min: 0.75% vs 0.55%—0.6%) . As to the α-thalassemia patient, a large number of target erythrocytes exist in peripheral blood smear and the osmotic fragility is decreased.

Whole blood was divided into two parts, one part was used to prepare blood samples, and the other part was used to prepare RBC samples. Blood samples were prepared by adding the same volume of CCl_4 into the anticoagulated blood. After turbulent mixing and centrifuging (12 000 rpm for 10 minutes), the upper layer was whole blood hemolysate, the middle layer between hemolysate and CCl_4 was whole blood stroma. The other part of whole blood was firstly made into packed RBCs by washing the RBCs with saline for 4-5 times until the supernatant was colorless[3, 12, 13]. Then the same volume of CCl_4 was added into the packed RBCs, and after mixing and centrifuging (3000 rpm for 10 minutes), the upper red solution was RBC hemolysate, and the middle layer between hemolysate and CCl_4 was RBC stroma.

The starch–agarose mixed gel was prepared by dissolving 0.24 g of agarose and 1.72 g of starch in 90 ml of TEB buffer (42.1 mmol/L Tris, 1.71 mmol/L EDTA, 6.47 mmol/L boric acid, pH 8.6) [3, 12, 13]. The solution was heated until the agarose melts, and then the gel was laid on a 17 cm×17 cm glass while hot. After solidification, about 8 μl of samples were applied on the cathodic side of the gel by using 3 mm filter paper. After adding blood samples on the starch–agarose mixed gel, electrophoresis was carried out in borate buffer (0.3 mol/L boracic acid, 0.06 mol/L NaOH, pH9.0) at 5 V/cm for 2 hours, then paused for 15 minutes and ran for 15 minutes by turns. It took about 6 hours for the entire electrophoresis. After electrophoresis, the red bands on the gel were firstly observed directly with eyes, and then the gel was sequentially stained with Ponceau Red (0.1% Ponceau S, 5% glacial acetic acid and 2% glycerol) and Benzidine (0.6 g of benzidine, 25 ml of glacial acetic acid, 10 ml of glycerol, add deionized water to 500 ml, then keep the solution in 75℃ water bath for 1 hour until the benzidine is dissolved completely. Sodium nitroprusside and 30% H_2O_2 should be added to this solution before use) for 4 hours, respectively. Finally, the gel was rinsed with rinsing solution (5% glacial acetic acid, 2% glycerin) until the background was clear.

Routine one-directional HRT was performed to compare the re-released Hb from normal, α-thalassemia, and HS patients' blood samples, which were prepared from whole blood, whole blood hemolysate, whole blood stroma, RBCs, RBCs hemolysate, RBCs stroma and plasma respectively. Comparing with the normal control, the re-released Hb from HS and α-thalas-semia erythrocytes had opposite changes (Figure 1). In normal control, there was nearly no HbA in the sample of whole blood stroma, but a small amount of HbA in the RBCs stroma; both whole blood and RBCs sample of normal control had re-released Hb; however re-released Hb did not appear in whole blood and RBCs hemolysate samples. As to HS, there was more HbA appearing in RBCs stroma sample, and no re-released Hb appeared in any of the blood samples. On the contrary, the HbA of

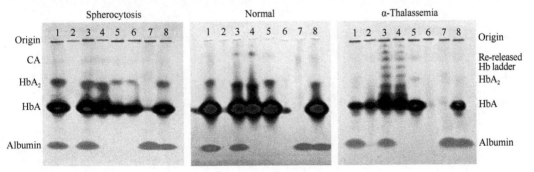

Figure 1 One dimension HRT of spherocytosis and α-thalassemia blood samples. Samples 1-8 were whole blood hemolysate, whole blood stroma, whole blood, RBCs, RBCs hemolysate, RBCs stroma, plasma, whole blood hemolysate

α-thalassemia whole blood stroma increased signifycantly, but that of α-thalassemia RBCs stroma was hard to see; in addition, the re-released Hb from whole blood and RBCs sample of α-thalassemia were increased significantly and Hb ladder was formed obviously.

To observe the effect of hypotonic treatment on the re-released Hb, isotonic and hypotonic HRT was performed with normal adult, HS patient, and α-thalassemia patient at room temperature or 37℃ for 1 hour. During the experiment, the whole blood or packed RBCs were firstly diluted with H_2O in the proportion from 10∶0 to 1∶9 (named as tube 1 to 10, respectively), and then kept at room temperature or 37℃ for 1 hour. Then one-direction HRT was performed as described above. The result of room temperature isotonic and hypotonic HRT showed that the whole blood sample (tube 1) of normal control had slight ladder of re-released Hb, when diluted with H_2O (tube 2 to tube 5), the re-released Hb was decreased, but increased from tube 6 to tube 8, and then decreased again from tube 9 (Figure 2). Compared with the normal control, the re-released Hb ladder decreased obviously in the HS patient, but increased significantly in the α-thalassemia patient. (Figure 2).

The result (Figure 3) of 37℃ isotonic and hypotonic HRT was similar with that at room temperature except for the disappearance of re-released Hb ladder from tube 1 of the normal control.

Double-direction HRT (or diagonal HRT) was performed to observe not only the re-released HbA but also the re-released HbA_2. Firstly, one-direction HRT was per formed as described above, then the direction of electric field was changed vertical to the original one, and another cycle of HRT was performed. The result showed that there was few re-released HbA in normal whole blood,

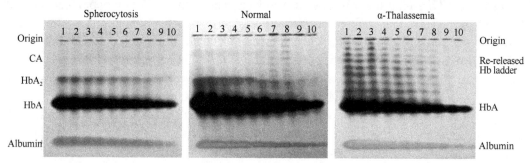

Figure 2 Whole blood isotonic and hypotonic HRT results of normal adult, spherocytosis, and α-thalassemia at room temperature. Whole blood was diluted with H_2O in the proportion from 10∶0 to 1∶9, respectively (tube 1 to 10), and kept at room temperature for 1 hour

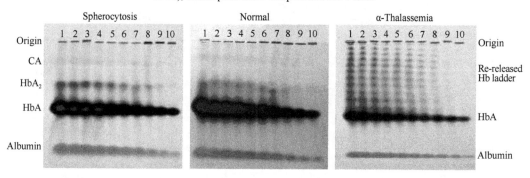

Figure 3 Whole blood isotonic and hypotonic HRT results of normal adult, spherocytosis, and α-thalassemias at 37℃. Whole blood was diluted with H_2O in the proportion from 10∶0 to 1∶9, respectively (tube 1 to 10), and kept at 37℃ for 1 hour

but the re-released HbA_2 was difficult to observe (Figure 4A) . Compared with normal, the re-released HbA from HS whole blood was decreased, but that from α-thalassemia whole blood was increased significantly. The amount of HbA_2 in α-thalassemia whole blood was less than the normal control obviously, and the re-released HbA_2 could not be detected in our experiment (Figure 4B) .

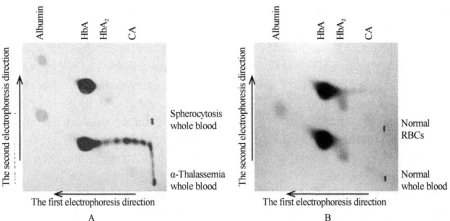

Figure 4 Double-direction HRT of normal adult, spherocytosis, and α-thalassemias blood sample. A was the double-direction HRT of spherocytosis and α-thalassemia whole blood sample; B was the double-direction HRT of the normal whole blood and RBCs sample

3 Discussion

Both HS and thalassemia belong to hereditary hemolytic disorders, which include hemoglobinopathies, erythrocyte membranopathy, and erythrocyte enzymopathy[16]. HS is the representative erythrocyte membranopathy[17], and thalassemia is the classic hemoglobinopathy[18-20]. In this study, Clinical tests showed that anemia, splenomegaly, and jaundice were the common clinical signs and symptoms of these two patients[16]. Some of the erythrocytes of the HS patient were spherical, but that of α-thalassemia were target-shaped. The osmotic fragility of erythrocytes increased in HS, but decreased in α-thalassemia. The morphology and osmotic fragility changes were caused by the defects of these two disorders. The abnormalities in HS erythrocyte membrane proteins, particularly ankyrin, α- and β-spectrin, band 3 and protein 4.2, result in the loss of membrane surface area relative to intracellular volume, which leads to spherically shaped erythrocytes with decreased deformability and increased fragility. Increased erythrocyte fragility leads to vesiculation and further membrane loss[21], so HS erythrocytes are unable to withstand the introduction of small amounts of free water that occurs when they are placed in increasingly hypotonic saline solutions. As a consequence, HS erythrocytes hemolyze more readily than normal erythrocytes at any saline concentration[17].

The thalassemia erythrocyte membranes exhibit morphological, biochemical, and mechanical abnormalities due to oxidative damage induced by binding of unmatched globin chains to the cytoplasmic surface of the membrane. So both α- and β- thalassemia erythrocytes become renitent and are less deformable than normal erythrocytes. The morphology and mechanical properties of the erythrocytes membrane are controlled by the cytoskeletal network underlying the lipid bilayer. Spectrin is the principal structural element of the erythrocyte cytoskeleton, regulating membrane cytoskeletal functions[22].

The re-released Hb was compared between these two kinds of hereditary hemolytic disorders by HRT. The results showed that comparing with the normal control, the re-released Hb from HS whole blood or erythrocytes was decreased, but increased distinctively from that of α- thalassemia erythrocytes during routine, two-directional, and isotonic and hypotonic HRT. The re-released Hb is speculated to have relationship with membrane-binding Hb, and the abnormal membrane-binding Hb will lead to abnormal Hb re-release. As known, most of the Hb exists in cytoplasm; only small amount of Hb binds with the membrane through interaction with the cytoskeletal proteins or membrane lipids. The abnormality of both membrane and Hb will change the amount of membrane-binding Hb, and will further lead to the variation of re-released Hb during HRT. HRT was established by our lab in 2007, and in the previous studies, the re-released Hb usually increased from some patients' erythrocytes during HRT, such as β-thalassemia patients, some general surgery patients, cirrhosis, and some gastroenteric tumor patients. In our study, the re-released Hb from α-thalassemia erythrocytes was increased significantly like before[12], but the re-released Hb from HS erythrocytes was decreased a lot. The abnormal membrane-binding Hb was speculated to be the reason.

It is well known that in vivo and under normal physiological conditions, intraerythrocytic hemoglobin may exist in three different forms represented by oxygenated, deoxygenated and partially oxidized Hb. Apart from the first two derivatives whose relative proportions arecontinuously changing during the oxygenation deoxygenation cycle, met-hemoglobin (MetHb) is normally present at a steady-state level of about 1%[23]. MetHb usually binds with membrane, and the re-released Hb from normal erythrocytes is speculated to be the membrane-binding MetHb. Oxidative damage can lead to the oxidative membrane damage and increased proportion of MetHb. The oxidization of band 3 leads to dissociation of ankyrin from band 3, and then tetrameric MetHb cross-link with the cytoplasmic domain of oxidized band 3 dimer[24]. In addition to MetHb, the abnormal Hb in all kinds of hemoglobinopathies is speculated to be the other main source of re-released Hb. α-thalassemia has the defect in α- globin syntheses, the relative excess of β-globin increases and the abnormal HbH ($β_4$) forms, which can bind with the membrane and lead to the increased Hb re-release.

Hb usually has interaction with spectrin, and the spectrin defect in HS patient interfere the binding of Hb with membrane, so the membrane-binding Hb and re-released Hb decreased obviously. There are five main kinds of erythrocyte skeleton proteins; defect of different cytoskeletal protein might leads to different results.

In conclusion, the change of re-released Hb is only an experimental phenomenon of HRT, and the mechanism of HRT has not been clear very much. In the future, more and more studies are needed to clarify these.

Acknowledgements

This work was supported by grants from Natural Science Foundation of China (81160214), Major Projects of Higher Education Scientific Research in the Inner Mongolia Autonomous Region (NJ09157), Key Science and Technology Research Project of the Ministry of Education, Natural Science Foundation of Inner Mongolia (2010BS1101) . We also especially acknowledge all of the people who donated their blood samples for our research.

Author details

Yan Su[1], Hongjie Ma[2], Hongwang Zhang[3], Lijun Gao[1], Guorong Jia[2], Wenbin Qin[1*] and Qitu He[2*]

*Address all correspondence to: qinwenbinbt@sohu.com; Heqitu@163.com
[1] Laboratory of Hemoglobin, Baotou Medical College, Baotou, China
[2] Department of Hematology, the First Affiliated Hospital of Baotou Medical College, Baotou, China Clinical Laboratory, the third Worker's Hospital of Baogang Group, Baotou, China Yan Su and Hongjie Ma contribute equally to this work.

References

[1] Weed RI, Reed CF, Berg G. Is hemoglobin an essential structural component of hu-man erythrocyte membranes? The Journal of Clinical Investigation. 1963; 42: 581-588.

[2] Ribeil JA, Arlet JB, Dussiot M, Moura IC, Courtois G, Hermine O. Ineffective erythro-poiesis in β-thalassemia. Scientific World Journal. 2013; 2013: 394295. DOI: 10.1155/2013/394295.

[3] Su Y, Gao L, Ma Q, Zhou L, Qin L, Han L, Qin W. Interactions of hemoglobin in live red blood cells measured by the electrophoresis release test.Electrophoresis.2010; 31 (17) : 2913-2920. DOI: 10.1002/elps.201000034.

[4] Barcellini W, Bianchi P, Fermo E, Imperiali FG, Marcello AP, Vercellati C, Zaninoni A, Zanella A. Hereditary red cell membrane defects: diagnostic and clinical aspects. Blood Transfusion. 2011; 9 (3) : 274-277.

[5] Schwartz E, Cohen A, Surrey S. Overview of the beta thalassemias: genetic and clini-cal aspects. Hemoglobin. 1988; 12 (5-6) : 551-564.

[6] Bernini LF, Harteveld CL. Alpha-thalassaemia. Bailliere's Clinical Haematology.1998; 11 (1) : 53-90.

[7] Clarke GM, Higgins TN. Laboratory investigation of hemoglobinopathies and thalassemias: review and update. Clinical Chemistry. 2000; 46 (8 Pt 2) : 1284-1290.

[8] Forget BG, Bunn HF. Classification of the Disorders of Hemoglobin. Cold Spring Harb Perspect Med. 2013; 3: a011684. DOI: 10.1101/cshperspect.a011684.

[9] Guitton C, Garçon L, Cynober T, Gauthier F, Tchernia G, Delaunay J, Leblanc T, Thuret I, Bader-Meunier B. Hereditary spherocytosis: guidelines for the diagnosis and management in children. Archives de Pediatrie. 2008; 15 (9) : 1464-1473. DOI: 10.1016/j.

[10] Gallagher PG, Forget BG. Hematologically important mutations: spectrin and ankyr-in variants in hereditary spherocytosis. Blood Cells, Molecules and Diseases. 1998; 24 (4) : 539-543.

[11] Perrotta S, Gallagher PG, Mohandas N. Hereditary spherocytosis. Lancet.2008; 372 (9647) : 1411-1426. DOI: 10.1016/S0140-6736 (08) 61588-3.

[12] Su Y, Shao G, Gao L, Zhou L, Qin L, Qin W. RBC electrophoresis with discontinuous power supply-a newly established hemoglobin release test. Electrophoresis. 2009; 30 (17) : 3041-3043. DOI: 10.1002/elps.200900176.

[13] Su Y, Shen J, Gao L, Tian Z, Tian H, Qin L, Qin W. Molecular Interactions of Re-re-leased proteins in Electrophoresis of Human Erythrocytes. Electrophoresis. 2012; 33 (9-10) : 1402-1405. DOI: 10.1002/elps.201100644.

[14] Qin W. Electrophoresis release of hemoglobin from living red blood cells. Scientia Sinica (Vitae) . 2014; 41 (8) : 597-607.

[15] Li JX, Su Y. Clinical research progress of red cell hemoglobin release test. Progress in Veterinary Medicine. 2014; 35 (10) : 104-107.

[16] Dhaliwal G, Cornett PA, Tierney LM Jr. Hemolytic anemia. American Academy of Family Physicians. 2004; 69 (11) : 2599-2606.

[17] Gallagher PG. Abnormalities of the erythrocyte membrane. Pediatric Clinic of North America. 2013; 60 (6) : 1349-1362. DOI: 10.1016/j.pcl.2013.09.001.

[18] De Franceschi L, Bertoldi M, Matte A, Santos Franco S, Pantaleo A, Ferru E, Turrini F. Oxidative stress and β-thalassemic erythroid cells ehind the molecular defect. Oxi-dative Medicine and Cellular Longevity. 2013;

2013: 985210. DOI: 10.1155/2013/985210.
[19] Modell B, Darlison M. Global epidemiology of haemoglobin disorders and derived service indicators. Bulletin of the World Health Organization. 2008; 86 (6) : 480-487.
[20] Weatherall DJ. The global problem of genetic disease. Annals of Human Biology.2005; 32 (2) : 117-122.
[21] Alaarg A, Schiffelers RM, van Solinge WW, van Wijk R. Red blood cell vesiculation in hereditary hemolytic anemia. Frontiers of Physiology. 2013; 4 (365) : 1-82. DOI: 10.3389/fphys.
[22] Rutaiwan T, Pornpimol M, Prapon W. Status of red cell membrane protein phosphor-ylation in thalassemia. ScienceAsia-Journal of the Science Society of Thailand. 2002; 28: 313-317.
[23] Giardina B, Scatena R, Clementi ME, Ramacci MT, Maccari F, Cerroni L, Condò SG. Selective binding of met-hemoglobin to erythrocytic membrane: a possible involve-ment in red blood cell aging. Advances in Experimental Medicine and Biology. 1991; 307: 75-84.
[24] Arashiki N, Kimata N, Manno S, Mohandas N, Takakuwa Y. Membrane peroxidation and methemoglobin formation are both necessary for band 3 clustering: mechanistic insights into human erythrocyte senescence. Biochemistry. 2013; 52 (34) : 5760-5769. DOI: 10.1021/ bi400405p.

附：反馈外界对本文的好评

Dear Prof. Su,

We are pleased to inform you that your paper "Comparative Study of the Amount of Re-released Hemoglobin from α-Thalassemia and Hereditary Spherocytosis Erythrocytes" has achieved impressive readership results. The chapter you have published with InTech in the book "Inherited Hemoglobin Disorders" has so far been accessed 100 times. Congratulations on the significant impact that your work has achieved to date.

The top downloads of your paper are from the following five countries: United States of America, China, India, France, Romania.

Dear Prof. Qin，

We are pleased to inform you that your chapter has been downloaded 500 times to date.

CHAPTER TITLE：Comparative Study of the Amount of Re-released Hemoglobin from α-Thalassemia and Hereditary Spherocytosis Erythrocytes

BOOK TITLE：Inherited Hemoglobin Disorders

We congratulate you on this achievement.

Such readership results demonstrate some very important factors about the reach and usage of your InTechOpen published research.

-Visibility—more than 500 researchers worldwide read，downloaded and interacted with your published content.

-Impact—this achievement demonstrates the influence your research has had within the scientific community.

-Connectivity—researchers from all over the world have been able to connect with your research to obtain relevant information to further develop their own research projects.

Attached you will find your Chapter Performance Metrics Report. For a more detailed report，visit your Author Panel：http：//www.intechopen.com/account/login.

We thank you for choosing InTechOpen as your Open Access publisher，and congratulate you once again on your success.

InTechOpen Metrics

InTech Open Access Publisher
Janeza Trdine 9,
51000 Rijeka, Croatia
T/F +385 (51) 770 447
E info@intechopen.com

www.intechopen.com

YOUR CHAPTER BY STATS AND NUMBERS

JUL 30, 2016

Chapter metrics for "Comparative Study of the Amount of Re-released Hemoglobin from α-Thalassemia and Hereditary Spherocytosis Erythrocytes", published in the book:

Inherited Hemoglobin Disorders

Edited by: Anjana Munshi

ISBN 978-953-51-2198-5
Publisher: InTech
Publication date: November 2015

This document outlines some of the major factors influencing your paper statistics. The following are taken into consideration:

1. Cumulative downloads by countries/time frame
2. Cumulative downloads by countries
3. Cumulative downloads by time frame

INTECH
open science | open minds

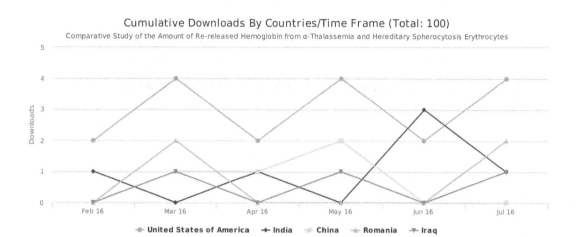

The first graph shows the number of chapter downloads in the last six months by country. The countries represented in the graph are the TOP 5 countries from which your paper was accessed.

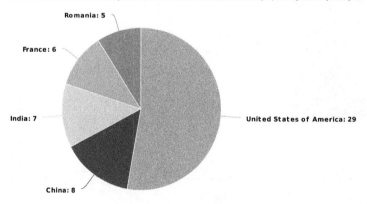

The pie chart above shows the download share by country. Again, the countries represented in the graph are the TOP 5 countries from which your paper was accessed.

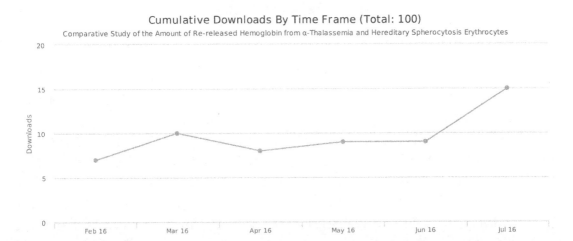

The last graph presents the accumulated download share in the last six months. The time period is defined from the publication date up to today.

For more information about your chapter statistics log in into your author's profile by following the link below

http://www.intechopen.com/account/login

Once logged in, you can choose one of the options in the Charts selection. By defining the time frame, a summary of the total number of downloads by countries within the selected time period will be displayed. You can also view specific charts referring only to countries, or to a time frame of your choice.

Thank you for choosing InTech as your Open Access publisher.

第四十一章 缺氧对小鼠红细胞再释放血红蛋白的影响

我们在 1981 年发现人类红细胞通过电泳来释放血红蛋白，当时是只通一次电，称为"初释放"[1, 2]，后来再通电，又有血红蛋白释放出来，称为"再释放"[1, 3, 4]。再释放与临床关系密切[1, 5-7]，这里我们介绍一个动物模型——缺氧对小鼠红细胞再释放血红蛋白的影响。取 8 只小鼠做缺氧处理，具体如下：1 号小鼠放入缺氧装置后 15 分钟取出，2 号小鼠放入约 20 分钟死亡(心脏取血，其他都是眼球取血)，3 号小鼠放入约 20 分钟，4 号小鼠放入约 25 分钟，5 号小鼠放入约 45 分钟(缺氧 2 次)，6 号小鼠放入约 70 分钟(缺氧 3 次)，7 号小鼠放入约 90 分钟(缺氧 3 次)，8 号小鼠放入约 120 分钟(缺氧 4 次)。将 8 只缺氧处理后小鼠血，分离红细胞，取一部分做在等渗条件下的血红蛋白再释放实验，结果见图 41-1，由此图可以看出，随着缺氧程度的加深，小鼠红细胞在等渗条件下的再释放带呈逐渐减弱趋势。再用另一部分红细胞加等量蒸馏水，造成低渗，重复上诉再释放实验，结果见图 41-2，由此图可以看出，与等渗结果不同，随着缺氧程度的加深，小鼠红细胞在低渗条件下的再释放带呈逐渐增强趋势。在这里，我们看到缺氧对小鼠红细胞释放血红蛋白有影响，而且等渗与低渗结果相反，是一个非常有趣的现象。这说明红细胞结构的复杂性：完整的红细胞(等渗)缺氧时再释放减弱，部分破坏的红细胞(加等量蒸馏水，低渗)再释放增强。看来，血红蛋白电泳释放与红细胞膜有关，初释放的血红蛋白与膜成分疏松结合，而再释放的血红蛋白与膜成分结合牢固，需要第二次通电才能释放出来。这种牢固结合在缺氧时阻碍血红蛋白的释放，加等量水破坏红细胞膜改变了牢固结合状态，缺氧时助长血红蛋白的释放。膜成分很多，哪种膜蛋白结合何种血红蛋白、谁与谁结合疏松或牢固，现有文献[8-11]都没有解答这个问题。总之，血红蛋白释放研究开辟了红细胞膜研究的信领域，科研无止境，永远在路上。

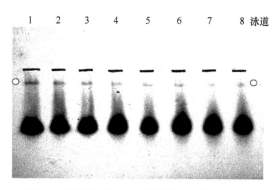

图 41-1 小鼠缺氧时红细胞在等渗条件下的再释放结果
注释：缺氧情况，1. 正常对照(放入 15 分钟后取出)；2. 缺氧 1 次(放入约 20 分钟)(死亡，心脏取血，其他都是眼球取血)；3. 缺氧 1 次(放入约 20 分钟)；4. 缺氧 1 次(放入约 25 分钟)；5. 缺氧 2 次(放入约 45 分钟)；6. 缺氧 3 次(放入约 70 分钟)；7. 缺氧 3 次(放入约 90 分钟)；8. 缺氧 4 次(放入约 120 分钟)。两个○之间的区带为再释放带

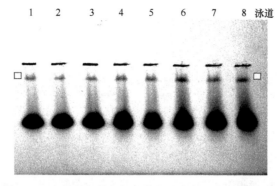

图 41-2 小鼠缺氧时红细胞在低渗条件下的再释放结果

注释：缺氧情况，1. 正常对照(放入 15 分钟后取出)；2. 缺氧 1 次(放入约 20 分钟)(死亡，心脏取血，其他都是眼球取血)；3. 缺氧 1 次(放入约 20 分钟)；4. 缺氧 1 次(放入约 25 分钟)；5. 缺氧 2 次(放入约 45 分钟)；6. 缺氧 3 次(放入约 70 分钟)；7. 缺氧 3 次(放入约 90 分钟)；8. 缺氧 4 次(放入约 120 分钟)；两个□之间的区带为再释放带

参 考 文 献

[1] 秦文斌. 活体红细胞内血红蛋白的电泳释放. 中国科学生命科学, 2011, 41 (8)：597-607.
[2] Su Y, Gao LJ, Ma Q, et al. Interactions of hemoglobin in lived red blood cells measured by the electrophoesis release test. Electrophoresis, 2010, 31: 2913-2920.
[3] Su Y, Shao G, Gao LJ, et al. RBC electrophoresis with discontinuous power supply - a newly established hemoglobin release test. Electrophoresis, 2009, 30: 3041-3043.
[4] Su Y, Shen J, Gao LJ, et al. Molecular interaction of re-released proteins in electrophoesis of human erythrocytes. Electrophoresis, 2012, 33: 1042-1045.
[5] 张晓燕, 高丽君, 高雅琼, 等. 血糖浓度和血红蛋白释放试验的比对研究. 国际检验医学杂志, 2010, 31 (6)：524-525.
[6] 高雅琼, 王彩丽, 高丽君, 等. 尿毒症患者低渗血红蛋白释放试验的初步结果. 临床和实验医学杂志, 2010, 9 (1)：12-13.
[7] 宝勿仁必力格, 王翠峰, 高丽君, 等. 肝硬化患者血液标本的血红蛋白释放试验结果明显异常. 临床和实验医学杂志, 2011, 10 (24)：1915-1917.
[8] 张莉, 秦桂秀, 张小莉, 等. 红细胞膜流动性在新生儿缺氧缺血性脑病中的变化与意义. 中国药物与临床, 2010, 10 (10)：1154-1155.
[9] 司本辉, 张庆成, 李积荣, 等. 久居高原地区缺氧环境对红细胞膜蛋白结构和血液流变学的影响. 高原医学杂志, 2001, 11 (3)：26-27.
[10] 杨晓静, 毛宝龄, 钱桂生. 慢性常压缺氧和缺氧伴高二氧化碳大鼠带 3 蛋白与红细胞内酸碱调节的实研究. 第三军医大学学报, 1995, 17 (1)：1-6.
[11] 高钰琪, 彭鹰, 孙秉庸. 慢性缺氧对大鼠红细胞压积和脆性以及右心室重量的影响. 第三军医大学学报, 1990, 12 (6)：539-540.

第四十二章 贲门癌与血红蛋白释放试验

郭俊[1#] 闫斌[2#] 高丽君[3] 孙晓荣[4] 韩丽红[3] 韩丽莎[4] 秦文斌[3※]

(1. 包头市肿瘤医院 介入科,包头 014010;2. 包头医学院第一附属医院 普外科,包头 014010;3. 包头医学院 血红蛋白研究室,包头 014010;4. 包头医学院 病理生理教研室,包头 014010)

摘 要

我们对包头市肿瘤医院多种肿瘤标本进行过血红蛋白释放研究,首先发现肝内胆管癌的血红蛋白释放试验结果特殊,与此同时,在两例贲门癌患者的全血中检出明显的HP-HB(结合珠蛋白-血红蛋白)。是否所有贲门癌都有HP-HB?为了弄清这个问题,我们研究了第三例贲门癌,结果是否定的,此患者全血里没有查到HP-HB。我们还比较研究了贲门癌与甲状腺癌的双向电泳释放情况,结果发现二者差异明显。对于第三例标本我们进一步研究了介入对血红蛋白释放的影响,通过比较介入前后的释放结果,发现介入后的后退再释放减弱,说明介入处理能够改善血红蛋白的电泳释放。

关键词 贲门癌;HP-HB;介入;血红蛋白释放试验

1 前言

癌症(cancer),也称恶性肿瘤(malignant neoplasm),为由控制细胞生长增殖机制失常而引起的疾病。癌细胞除了生长失控外,还会局部侵入周遭正常组织甚至经由体内循环系统或淋巴系统转移到身体其他部分。由于癌症对人类的危险最大,所以人类对它下的功夫也最多,有关癌症的文章铺天盖地,机制研究、治疗层出不穷,而且也有了很大进步。例如,人乳头瘤病毒是宫颈癌的病因[1-5],干细胞移植在癌症治疗中起到越来越大的作用[6-9]。

如上所述,有关癌症研究的报道很多,但是没有关于电泳释放方面的资料。

这里边,首先是血红蛋白电泳释放,然后是非血红素蛋白的都有释放,它们与癌症的机制和治疗有什么关系,都是我们的研究目标。本文是这方面研究的开端。

2 材料和方法

2.1 材料 贲门癌患者血液标本来自包头市肿瘤医院介入科,甲状腺癌患者血液标本来自包头医学院第一附属医院普外科。

2.2 方法 常规操作参见文献[10-15]。双向全程释放电泳如下所示。

2.2.1 标本处理

取10支试管,第1管只加全血,不加蒸馏水,第2~9管里加蒸馏水由少到多、加全

\# 并列第一作者
※ 通讯作者:秦文斌,电子邮箱 qwb5991309.atom.com

血由多到少，构成一个连续的、完整的等低渗条件，具体操作参见表 42-1。

表 42-1 全血的等低渗全程处理

管号	1	2	3	4	5	6	7	8	9	10
全血(μl)	20	18	16	14	12	10	8	6	4	2
蒸馏水(μl)	0	2	4	6	8	10	12	14	16	18

注释：第 1 管为等渗，原来的全血，没有蒸馏水。第 2~10 管为低渗。水量逐渐增加，全血相应减少。第 10 管中全血占 10%，蒸馏水占 90%

2.2.2 双向电泳 用以上一系列标本直接做电泳。

(1) 第一向电泳：先普泳，再做前进再释放和后退再释放。普泳=电势梯度 6V/cm，泳 2 小时 15 分钟左右，停电 15 分钟。前进再释放=电势梯度 6V/cm，再通电半小时。后退再释放=电势梯度 6V/cm，倒极再通电 15 分钟。

(2) 第二向电泳：普泳=电势梯度 6V/cm，倒极转向再泳 1 小时 15 分钟左右。

2.2.3 染色

(1) 染丽春红：将凝胶板直接放入丽春红染液中过夜，取出、照相，再晾干或烤干。

(2) 染联苯胺：将凝胶板直接放入联苯胺染液中，加 3% H_2O_2 直到血红蛋白变成蓝黑颜色，转入漂洗液(5%乙酸、1%甘油)换洗两次，每次 5 分钟，取出晾干。

2.2.4 结果保存 晾干凝胶与玻璃板结合，可长期保存。

3 结果

3.1 两例贲门癌的实验结果 见图 42-1。

由图 42-1 可以看出，这两例贲门癌都有结合珠蛋白 HP，其他标本都没有。此次筛查中发现异常的有肝内胆管癌(见□处)和贲门癌(见◇处)，本文涉及贲门癌。这批 7 个标本中有 2 个贲门癌，它们都在全血里出现快泳联苯胺阳性成分(见◇处)，但红细胞里并没有这种成分。这种成分只出现于全血，而且位于 HbA_1 前方(阳极侧)说明它不是 HbA_3，而是结合珠蛋白(haptoglobin, HP)。为什么贲门癌有 HP 而其他标本没有，是不是所有贲门癌都有，这是我们要研究的问题。

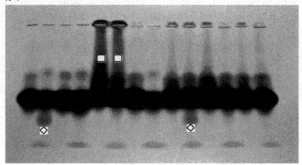

图 42-1 两例贲门癌都有结合珠蛋白

注释：泳道 1、2=贲门癌，3、4=肝癌，5、6 =肝内胆，7、8 =肝癌，9、10=贲门癌，11、12=肝癌，13、14=肝癌；标本 R=红细胞，B=全血

3.2 第三例贲门癌的单向电泳实验结果　见图42-2。

由图42-2可以看出，全血中未出现结合珠蛋白HP。看来，并非所有贲门癌都有HP。

3.3 比较贲门癌与甲状腺癌的双向全程电泳结果　见图42-3。

由图42-3可以看出，贲门癌的前进释放带和后退释放带都比较弱，而甲状腺癌的这两种释放带都相当强，二者差别明显。

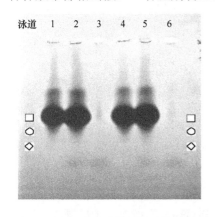

图42-2　第三例贲门癌没有结合珠蛋白
注释：泳道1=红细胞，2=全血，3=血浆，4=1，5=2，6=3；两个□之间的成分是HbA_1，两个○之间的成分是HbA_3，两个◇之间的位置是结合珠蛋白应该出现的地方，参见图42-1

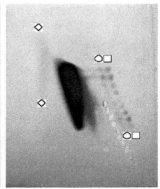

图42-3　贲门癌与甲状腺癌的双向全程电泳图
注释：左图为贲门癌，右图为甲状腺癌；两个◇处为白蛋白，两个○处是前进带，两个□处是后退带

3.4 第三例贲门癌介入前后的双向全程电泳实验结果　见图42-4。

由图42-4可以看出，介入前，前进带有1和8、9、10，后退带没有1，有8、9、10；介入后，前进带有1和8、9、10(看不清)，后退带全无。介入前后再释放结果不同，这说明它对红细胞释放血红蛋白有影响。根据我们的经验，再释放增强表示病情加重，再释放减弱表示病情减轻，现在结果从电泳释放角度支持介入操作的治疗效果。

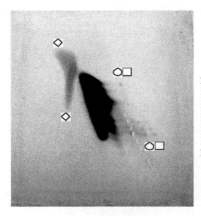 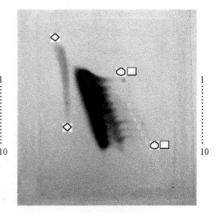

图42-4　贲门癌介入前后的双向全程电泳图
注释：左图为介入前，右图为介入后；两个◇处为白蛋白，两个○处是前进带，两个□处是后退带

4　讨论

贲门是一个特殊的解剖部位，是食管通向胃的开口，组织学上食管的鳞状上皮与胃的柱

状上皮在此截然分界，其远方 0.5～4cm 的一圈环形区内有不规则分布的贲门腺体，或呈管状，或呈分支状。此区域大多宽约 2cm。近 30 年来，胃癌总的发病率在持续下降，但是与之相反的是贲门癌发病相对地有所增加。这也提示贲门癌的病因学与胃其他部位的癌有所不同。应该作为一个独立的疾病进行分析研究。

胃癌的组织发生学中，胃溃疡、胃息肉(腺瘤)及慢性萎缩性胃炎过去皆被认为是胃癌的癌前期病变。近年的研究发现上述几种情况发生癌变的机会很小，贲门癌作为胃癌的特殊类型，上述病变与贲门癌的组织发生关系不大。目前比较多认为贲门癌是起源于有多方向分化潜能的贲门腺的颈部干细胞，干细胞可以形成具有贲门或腺上皮特点的腺癌。光镜、电镜和组化研究发现贲门癌是混合型，有力支持该观点。不典型增生是贲门癌的癌前病变，它也是上述与贲门癌发病有关的溃疡、息肉、萎缩性胃炎共有的关键病理过程。当它们发生不典型增生的改变时才可能癌变，其中结肠型发生多数具有不典型增生的性质。

从释放角度来看，首先要了解贲门癌的释放特点，所以一开始就注意到 HP，以为它是贲门癌的独特之处，但在后来的病例里没有得到证实。贲门癌还有什么特殊释放情况？目前资料很少，这里有甲状腺癌的双向全程电泳结果，属于配对资料，可供比较。我们由图 42-4 可以看出，贲门癌的前进释放带和后退释放带都比较弱，而甲状腺癌的这两种释放带都相当强，二者差别明显。是不是所有癌症的进退带都增强唯独贲门癌减弱呢？有待今后的大量研究。

参 考 文 献

[1] Hausen HZ. Papillomaviruses and cancer: from basic studies to clinical application. Nat Rev Cancer, 2002, 2 (5) : 342-350.
[2] Hausen HZ. A synthetic E7 gene of human papillomavirus type 16 that yields enhanced expression of the protein in mammalian cells and is useful for DNA immunization studies. J Virol, 2003, 77 (8) : 4928-4937.
[3] Hausen HZ. Classification of papillomaviruses. Virology, 2004, 324 (1) : 17-27.
[4] Hausen HZ. Recognition of conserved amino acid motifs of common viruses and its role in autoimmunity. PLoS Pathog, 2005, 1 (4) : e41.
[5] Hausen HZ. Perspectives of contemporary papillomavirus research. Vaccine, 2006, 24 Suppl 3: S3/iii-iv.
[6] 徐岚, 胡炯, 吴文, 等. 造血干细胞移植治疗血液恶性肿瘤的生存分析. 肿瘤, 2007, 27 (4) : 312-315.
[7] 曹慧敏. 再生障碍性贫血无关供者造血干细胞移植治疗进展. 白血病, 2010, 19 (5) : 316-319.
[8] 李旭东, 何易, 王东宁, 等. 非亲缘供者异基因造血干细胞移植治疗恶性血液病15例. 中国组织工程研究, 2012, 16 (6) : 1071-1074.
[9] 莫晓冬, 黄晓军. 替代供者造血干细胞移植在成人血液恶性肿瘤治疗中的作用. 中华血液学杂志, 2014, 35 (3) : 262-265.
[10] 秦文斌. 红细胞内血红蛋白的电泳释放——发现和研究. 北京: 科学出版社, 2015, 196-199.
[11] 秦文斌. 活体红细胞内血红蛋白的电泳释放. 中国科学生命科学, 2011, 41 (8) : 597-607.
[12] Su Y, Shao G, Gao LJ, et al. RBC electrophoresis with discontinuous power supply-a newly established hemoglobin release test. Electrophoresis, 2009, 30: 3041-3043.
[13] Su Y, Gao LJ, Ma Q, et al. Interactions of hemoglobin in lived red blood cells measured by the electrophoesis release test. Electrophoresis, 2010, 31: 2913-2920.
[14] Su Y, Shen J, Gao LJ, et al. Molecular interaction of re-released proteins in electrophoesis of human erythrocytes. Electrophoresis, 2012, 33: 1042-1045.
[15] Su Y, Gao LJ, Qin WB. Interactions of hemoglobin in lived red blood cells measured by the electrophoesis release test. Protein Electrophoresis, 2012, 869: 393-402.

第五篇 交叉互作电泳

前　言

1981 年我们发现"血红蛋白 A_2 现象"[1]，主要内容就是"活体红细胞内血红蛋白 A_2 与 A_1 之间的相互作用"[2,3]，这是发生在红细胞内部的事件。在红细胞外这两种血红蛋白是否也能发生相互作用？经过一系列努力，1991 年我们发现在红细胞外血红蛋白 A_2 与 A_1 之间也能够进行相互作用[4]，只是它出现于凝胶电场里，当我们让快泳的血红蛋白 A_1 穿过(追过)慢泳的血红蛋白 A_2 时，后者的区带变形，我们称之为"交叉互作"。

这种交叉互作的结构基础来源于血红蛋白的分子杂交，最早的知识是异常血红蛋白 Richmond，它能与其他血红蛋白形成不对称杂交产物[5]。后来这一概念得到广泛证实[6-8]，质谱分析技术也证实不对称杂交产物的存在[9]。

根据上述知识，我们做如下说明：血红蛋白多为四聚体，由两个 α 链和两个其他链组成，电场中一个血红蛋白四聚体(HbA_1 或 HbA_2)，可以分解成两个二聚体并构成平衡，两种泳速不同的二聚体相遇时生成一过性交叉互作产物，此时它的泳速位于二者之间，继续前行，二者逐渐分开，就出现了"A_2 的电泳位置前移，A_1 的电泳速度变慢"现象，此时长条区带中间变形，机制见式(1)。

$$\left.\begin{array}{l}(\alpha^A\beta^A)_2 \longleftrightarrow 2(\alpha^A\beta^A) \\ HbA_1 \\ (\alpha^A\beta^{A_2})_2 \longleftrightarrow 2(\alpha^A\beta^{A_2}) \\ HbA_2\end{array}\right\} \longleftrightarrow \begin{array}{c}2(\alpha^A\beta^A\alpha^A\delta^{A_2}) \\ HbA_2 \text{ 与 } HbA_1 \text{ 交叉互作产物}\end{array} \qquad (1)$$

上面说的是成人血红蛋白之间的交叉互作，后来我们又试验了胎儿血红蛋白 F 与血红蛋白 A_2 之间的相互作用。实验证明，当快泳的血红蛋白 F 追过慢泳的血红蛋白 A_2，也出现了类似于血红蛋白 A_2 与血红蛋白 A_1 之间的交叉互作，机制见式(2)。

$$\left.\begin{array}{l}(\alpha^A\gamma^F)_2 \longleftrightarrow 2(\alpha^A\gamma^F) \\ HbF \\ (\alpha^A\delta^{A_2})_2 \longleftrightarrow 2(\alpha^A\delta^{A_2}) \\ HbA_2\end{array}\right\} \longleftrightarrow \begin{array}{c}2(\alpha^A\gamma^F\alpha^A\delta^{A_2}) \\ HbA_2 \text{ 与 } HbF \text{ 交叉互作产物}\end{array} \qquad (2)$$

成人血红蛋白里还有一个血红蛋白 A_3，其泳速比血红蛋白 A_1 还快，它追过血红蛋白 A_2 时会怎么样呢？ 出乎意料，它们之间没有发生相互作用。都是成人血红蛋白，HbA_1 能与 HbA_2 互作，而 HbA_3 就不行，这怎么理解？看来血红蛋白 A_3 的结构有特殊情况。众所周知，HbA_3 与 HbA_1 不同之处在于，前者的 β 链 N 端缬氨酸的游离氨基结合有葡萄糖等[10,11]，其机制参见式(3)，可能是这些成分影响到它与 HbA_2 的交叉互作。在这里我们看到蛋白质结构与功能关系的又一有趣的实例。

$$\left.\begin{array}{c}(\alpha^A\beta^{糖})_2 \longleftrightarrow\!\!\!\times\!\!\!\longrightarrow 2(\alpha^A\beta^{糖}) \\ \text{HbA}_3 \\ (\alpha^A\delta^{A_2})_2 \longleftrightarrow 2(\alpha^A\delta^{A_2}) \\ \text{HbA}_2\end{array}\right\} \longleftrightarrow \begin{array}{c}2(\alpha^A\beta^{糖}\alpha^A\delta^{A_2}) \\ \text{一过性交叉互作产物}\end{array} \quad (3)$$

人类各种血红蛋白之间的交叉互作如上，动物不同血红蛋白之间是否也有交叉互作？这里最好的例子就是大鼠的血红蛋白。大鼠有四种血红蛋白：HbA、HbB、HbC 和 HbD，我们的实验证明它们之间都可以发生交叉互作，以 HbA 与 HbB 互作为例，机制见式(4)。

$$\left.\begin{array}{c}(\alpha^A\beta^A)_2 \longleftrightarrow 2(\alpha^A\beta^A) \\ \text{大鼠 HbA} \\ (\alpha^A\beta^B)_2 \longleftrightarrow 2(\alpha^A\beta^B) \\ \text{大鼠 HbB}\end{array}\right\} \longleftrightarrow \begin{array}{c}2(\alpha^A\beta^A\alpha^A\beta^B) \\ \text{HbA 与 HbB 交叉互作产物}\end{array} \quad (4)$$

人类自身血红蛋白的交叉互作如式(3)，大鼠自身血红蛋白的交叉互作如上式(4)，下边讨论人血红蛋白与动物血红蛋白之间是否也存在类似行为？在这方面工作量更大。我们对脊椎动物中各类动物(哺乳类、鸟类、爬行类、两栖类、硬骨鱼、软骨鱼及无颌纲动物)都进行了研究，结果发现，前三类动物的血红蛋白都能与人血红蛋白发生交叉互作，而后几类动物血红蛋白就没有这种作用。众所周知，哺乳类、鸟类、爬行类共属于羊膜动物，而两栖类、硬骨鱼、软骨鱼及无鄂纲都属于非羊膜动物。看来，我们从血红蛋白交叉互作角度发现了脊椎动物进化的分子机制，羊膜动物和非羊膜动物成为血红蛋白分子进化的分水岭。

现在的问题是，为什么羊膜动物血红蛋白能与人类血红蛋白交叉互作而非羊膜动物的就不能呢？具体机制尚不完全清楚，我们的初步意见如下：羊膜动物血红蛋白与人类血红蛋白 A_2、A_1 类似，能够形成四聚体与二聚体之间的平衡，从而能与人类血红蛋白发生类似式(1)的交叉互作，机制见式(5)。

$$\left.\begin{array}{c}(\alpha^A\beta^X)_2 \longleftrightarrow 2(\alpha^A\beta^X) \\ \text{Hb}^※ \\ (\alpha^A\delta^{A_2})_2 \longleftrightarrow 2(\alpha^A\delta^{A_2}) \\ \text{HbA}_2\end{array}\right\} \longleftrightarrow \begin{array}{c}2(\alpha^A\beta^X\alpha^A\delta^{A_2}) \\ \text{HbA}_2 \text{ 与 HbX 交叉互作产物}\end{array} \quad (5)$$

式中，※，HbX=羊膜动物血红蛋白。

非羊膜动物的血红蛋白情况特殊，实验结果是它们不能与人类血红蛋白发生交叉互作。从上述反应原理来看，非羊膜动物的血红蛋白可能无法形成血红蛋白二聚体，从而不能进行交叉互作，机制见式(6)。至于它们为什么不能与成人血红蛋白互作，目前尚找不到统一解释，但有文献[12]报道，无颌纲动物的盲鳗在一般情况下以单体形式存在，只有在 pH 低于 7 的环境中才可以相互聚合，我们的电泳缓冲液是 pH 9 左右，不发生互作是可以理解的。另外，一些低等脊椎动物，如软骨鱼、两栖动物、爬行动物的血红蛋白容易发生聚合而形成多聚体[13]，而人、哺乳动物及鸟类血红蛋白不容易发生聚合。此聚合通常在溶血后发生，而我们做血红蛋白互作的实验时首先要将两种血红蛋白制成溶血液。如果其中一种血红蛋白在制备时已经形成多聚体，反应倾向于生成多聚物的方向，就不易于另一血红蛋白发生相互作用。根据上述资料，我们推测，非羊膜动物中软骨鱼、两栖动物血红蛋白可能因形成多聚体而无法与成人血红蛋白发生互作。另外，爬行动物的血红蛋白也容易形成多聚体，它却能够与成人血红蛋白发生互作，看来，还有其他机制在起作用，有待继续研究。可以这样说，非羊膜动物血红蛋白不与人血红蛋白互作，源于前者没有稳定的四聚体，因为有的非羊膜动物血红蛋白是单体，另一些非羊膜动物血红蛋白是多聚体。

$$\left. \begin{array}{c} (\alpha^A\beta^Y)_2 \longleftrightarrow 2(\alpha^A\beta^Y) \\ HbY^{※} \\ (\alpha^A\delta^{A_2})_2 \longleftrightarrow 2(\alpha^A\delta^{A_2}) \\ HbA_2 \end{array} \right\} \longleftrightarrow\!\!\!\!\!\!\times\!\!\!\!\!\!\longrightarrow \begin{array}{c} 2(\alpha^A\beta^Y\delta^{A_2}) \\ HbA_2 与 HbY 交叉互作产物 \end{array} \quad (6)$$

式中，※，HbY=非羊膜动物血红蛋白。

总之，我们通过具体实验发现了红细胞外血红蛋白的交叉互作，明确了它的方法和应用，这些都是客观存在的。至于对其机制的探讨是否正确，请国内外学者批评指正。

参 考 文 献

[1] 秦文斌, 梁友珍. 血红蛋白 A_2 现象 I. 此现象的发现及其初步应用. 生物化学与生物物理学报, 1981, 13 (2)：199-205.

[2] Su Y, Gao LJ, Ma Q, et al. Interactions of hemoglobin in lived red blood cells measured by the electrophoresis release test. Electrophoresis, 2010, 31: 2913-2920.

[3] 秦文斌. 活体红细胞内血红蛋白的电泳释放. 中国科学生命科学, 2011, 41 (8)：597-607.

[4] 秦文斌. 红细胞外血红蛋白 A 与血红蛋白 A_2 之间的相互作用. 生物化学杂志, 1991, 7 (5)：583-587.

[5] Efremov GD, Huisman T HJ, Smith LL, et al. Hemoglobin richmond, a human hemoglobin which forms asymmetric hybrids with other hemoglobins. J Biol Chem, 1969, 244 (22)：6105-6116.

[6] Marden MC, Griffon N, Poyart C. Asymmetric hemoglobin hybrids. J Mol Biol, 1996, 263 (1)：90-97.

[7] Marden MC, Kiger L, Poyart C, et al. Identifying the conformational state of bi-liganded haemoglobin. Cell Mol Life Sci, 1998, 54 (12)：1365-1384.

[8] Kiger L, Marden MC. Asymmetric (deoxy dimer/azido-met dimer) hemoglobin hybrids dissociate within seconds. J Mol Biol, 1999, 291 (1)：227-236.

[9] Oforiacquah SF, Green BN, Davies SC, et al. Mass spectral analysis of asymmetric hemoglobin hybrids: demonstration of Hb FS ($\alpha_2\gamma\beta^S$) in sickle cell disease. Analytical Biochemistry, 2001, 298 (1)：76-82.

[10] Bookchin RM, Gallop PM. Structure of hemoglobin A1c: nature of the N-terminal beta chain blocking group. Biochem Biophys Res Commun, 1968, 32 (1)：86-93.

[11] Koenig J, Blobstein S H, Cerami A. Structure of carbohydrate of hemoglobin AIc. J Biol Chem, 1977, 252: 2992-2997.

[12] Fago A, Weber R E. The hemoglobin system of hagfish, Myxine glutinosa, aggregation state and functional properties. Biochimica Biophysica, 1955, 249: 109-115.

[13] Goodman M, Moore GW. Darwinian evolution in the genealogy of heamoglobin. Nature, 1975, 253: 603-608.

第四十三章　淀粉-琼脂糖混合凝胶交叉互作电泳

秦文斌

(包头医学院　血红蛋白研究室，包头　014010)

1　材料与方法

1.1　红细胞溶血液的制备　取人或动物的抗凝血液，离心去血浆，用生理盐水洗涤红细胞5次。向红细胞层加入1.5倍体积蒸馏水及等体积四氯化碳，激烈震荡后离心，取上层红色溶血液备用。

1.2　电泳纯血红蛋白的制备　用红细胞溶血液做淀粉-琼脂糖混合凝胶电泳，详见文献[1-3]。电泳分离后不染色，连同凝胶抠出所需的血红蛋白(红色)，放入1.5ml EP管中，后者转入-20℃冰箱过夜，次日取出，室温融化，高速离心，上层红色液体就是电泳纯的血红蛋白。还有一种简单办法：将两个交叉后者血红蛋白连同凝胶直接转到一个新玻璃板上，摆好位置(快泳穿过慢泳)，周围倒上新胶，然后再行电泳。

1.3　交叉互作电泳技术　用淀粉-琼脂糖混合凝胶做交叉互作电泳，具体如下。制胶：淀粉1g，琼脂糖0.15g，TEB pH 8.6 的缓冲液55ml，煮胶。玻板17cm×10cm，水平倒胶。加样：将溶血液加入在新华3号滤纸条，此滤纸条再插入凝胶。每次实验有互作双方的两种溶血液，先用普通电泳判定其中血红蛋白的泳速快慢。交叉互作电泳时，按电泳方向将泳速快的标本放在后方(阴极侧)、慢的放在前方(阳极侧)，让泳速快的标本穿过或兜过泳速慢者。含标本的滤纸条有长短两种。穿过实验时，慢泳标本用长条，放在阳极侧、快泳标本用短条，放在阴极侧，兜过实验时，二者关系反过来，慢泳标本用短条，放在阳极侧、快泳标本用长条，放在阴极侧，参见图43-1、图43-2。电泳结果可直接观察红色区带的变形，必要时染氨基黑10B或丽春红-联苯胺。

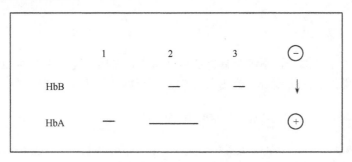

图43-1　穿过试验(HbB 穿过 HbA)示意

注释：泳道1、3为对照，HbA、HbB 单独电泳，互不干扰；泳道2为交叉互作电泳；HbB 的泳速比 HbA 快，电泳过程中 HbB 短区带(—)穿过 HbA 长区带(——)中部。如有互作，HbA 的长区带中间变形，如无互作，则不变形

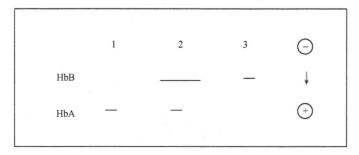

图 43-2　兜过试验(HbB 兜过 HbA)示意

注释：基本同上示意图 43-1，只是长短区带位置对调。1、3 仍为对照。2 为交叉互作电泳；HbB 的泳速比 HbA 快，电泳过程中 HbB 长区带(——)兜过 HbA 短区带(—)。如有互作，HbB 的长区带中间变形，如无互作，则不变形

2　结果

交叉互作电泳结果判定：电泳后看长条(被穿过或兜过)标本的区带中部是否变形(参见箭头↓所示处)，变形者为发生互作，无变形时，未发生互作。

2.1　穿过试验结果的判定　参见图 43-3。

2.2　兜过试验结果的判定　参见图 43-4。

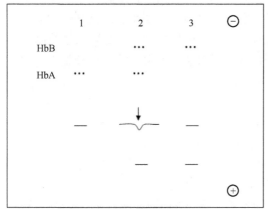

图 43-3　穿过试验结果的判定示意

注释：泳道 1、3 为对照，泳道 2 为交叉互作。HbB 的泳速比 HbA 快，电泳过程中 HbB 穿过 HbA 中部。如有互作，HbA 的长区带中间变形(如本图)，如无互作，则不变形

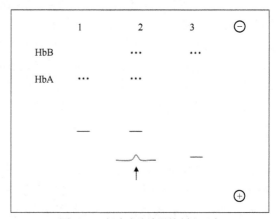

图 43-4　兜过试验结果的判定示意

注释：基本同示意图 43-3，只是长短区带位置对调。1、3 仍为对照。2 为交叉互作，HbB 的泳速比 HbA 快，电流过程中 HbB 长区带兜赤 HbA 短区带。如有互作，HbB 的长区带中间变形(如本图)，如无互作，则不变形

3　讨论

这里是本篇的总论部分，不涉及具体的血红蛋白，之后各论里的各种血红蛋白，都按上述规律来判定结果，长区带变形者为交叉互作性，否则就是没有发生互作。

参 考 文 献

[1] 秦文斌. 红细胞外血红蛋白 A 与血红蛋白 A_2 之间的相互作用. 生物化学杂志, 1991, 7 (5)：583-587.

[2] Yu H, Wang ZL, Qin WB, et al. Structural basis for the specific interaction of chicken haemoglobin with bromophenol blue: a computational analysis. Molecular Physics, 2010, 108 (2)：215-220.

[3] Su Y, Gao LJ, Ma Q, et al. Interactions of hemoglobin in lived red blood cells measured by the electrophoresis release test. Electrophoresis, 2010, 31: 2913-2920.

第四十四章　血红蛋白F与血红蛋白A_2之间的交叉互作

韩丽红[#]　闫　斌[#]　高丽君　苏　燕　秦文斌[※]

(包头医学院　血红蛋白研究室，包头 014010)

摘　要

我们曾经用淀粉-琼脂糖混合凝胶电泳证明成人血红蛋白A_1与A_2发生交叉互作，当时未对胎儿血红蛋白F进行研究。现在，再用这种方法研究血红蛋白F与血红蛋白A_2是否也有相互作用。结果表明，血红蛋白F与血红蛋白A_2也能发生交叉互作。

关键词　淀粉-琼脂糖混合凝胶电泳；胎儿血红蛋白；血红蛋白F；血红蛋白A_2；交叉互作

1　前言

本文所使用的技术属于交叉电泳范畴，在这方面日本学者中村正二郎曾专门著书[1]，介绍他们利用纸上电泳和琼脂电泳研究血清蛋白及免疫球蛋白的研究成果。我们所使用的支持体为淀粉-琼脂糖混合凝胶，不同于纸上电泳和琼脂电泳，研究对象也不一样，这里研究的是血红蛋白。我们用本法发现成人血红蛋白A_1与A_2之间发生交叉互作[2-4]。本文的研究内容是胎儿血红蛋白F与成人血红蛋白A_2之间是否存在交叉互作问题。

2　材料和方法

2.1　材料　正常成人血液来自本研究室人员，脐带血来自包钢集团第三职工医院妇产科。

2.2　方法　先由成人血液分离出来红细胞并制备红细胞溶血液。由脐带血分离红细胞，再用淀粉-琼脂糖混合凝胶电泳制备电泳的血红蛋白F，然后让血红蛋白F穿过成人红细胞溶血液，观察穿过溶血液后血红蛋白A_2区带变形情况，变形者(区带中部弯向阳极呈"V"状)为发生交叉互作，未变形为未发生相互作用。

详细操作参见文献[2-4]。

3　结果

血红蛋白F穿过成人红细胞溶血液后的结果见图44-1。由此图可以看出，血红蛋白F穿过成人红细胞溶血液中的血红蛋白A_2时，后者变形(参见箭头↓处)，与游离血红蛋白F相

[#]并列第一作者
[※]通信作者：秦文斌；电子邮箱：qwb5991309@tom.com

比，进入溶血液中的血红蛋白 F 泳速稍慢。

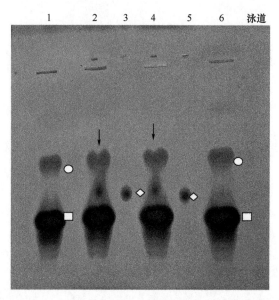

图 44-1 血红蛋白 F 穿过成人红细胞溶血液后的结果

注释：泳道 1=溶血液(○为 HbA$_2$，□为 HbA$_1$)，泳道 2=HbF 穿过溶血液，泳道 3=HbF(见◇处)，泳道 4=HbF 穿过溶血液，泳道 5=HbF(见◇处)，泳道 6=溶血液(○为 HbA$_2$，□为 HbA$_1$)

4 讨论

血红蛋白 F 穿过血红蛋白 A$_2$ 时，A$_2$ 区带也变形，说明二者也发生交叉互作。此时，血红蛋白 F 泳速减慢，说明穿过时二者曾一度结合并很快分开，它们的反应式如下所示。

$$\left.\begin{array}{l}(\alpha^A\gamma^F)_2 \longleftrightarrow 2(\alpha^A\gamma^F) \\ \text{HbF} \\ (\alpha^A\delta^{A_2})_2 \longleftrightarrow 2(\alpha^A\delta^{A_2}) \\ \text{HbA}_2\end{array}\right\} \longleftrightarrow 2(\alpha^A\gamma^F\alpha^A\delta^{A_2}) \atop \text{一过性交叉互作产物} \quad (1)$$

穿过实验的安排是，将慢泳成分(血红蛋白 A$_2$)放在前面，快泳成分(血红蛋白 F)放在后边，电泳过程中 F 赶上并进入 A$_2$ 时反应向右进行，出现一过性交叉互作产物。继续电泳，F 超过 A$_2$ 时反应向左进行，又恢复为 HbF 和 HbA$_2$，但留下了血红蛋白 A$_2$ 变形的痕迹。

我们以前发现的成人 HbA$_1$ 和 HbA$_2$ 交叉互作[2-4]也是同样道理，那时的反应式如下所示。

$$\left.\begin{array}{l}(\alpha^A\beta^A)_2 \longleftrightarrow 2(\alpha^A\beta^A) \\ \text{HbA}_1 \\ (\alpha^A\delta^{A_2})_2 \longleftrightarrow 2(\alpha^A\delta^{A_2}) \\ \text{HbA}_2\end{array}\right\} \longleftrightarrow 2(\alpha^A\beta^A\alpha^A\delta^{A_2}) \atop \text{一过性交叉互作产物} \quad (2)$$

现在看来，这里是两个血红蛋白二聚体一过性结合成四聚体，然后再分开，留下了结合过的痕迹，让我们发现了它们的交叉互作。

以上是人体内部几种血红蛋白之间的交叉互作，人体血红蛋白与各种动物血红蛋白之间有无交叉互作，各种动物血红蛋白之间有无交叉互作，都是需要进一步研究的有趣课题。在这些方面，我们已经做了一些工作[5, 6]，有待再深入研究。

参 考 文 献

[1] Nakamura S. Cross Electrophoresis. Its Principle and Applications. New York: Elsevier.
[2] 秦文斌. 红细胞外血红蛋白 A_2 与 A_1 间的相互作用. 生物化学杂志, 1991, 7 (5) : 583-587.
[3] Su Y, Gao LJ, Ma Q, et al. Interactions of hemoglobin in lived red blood cells measured by, the electrophoesis release test. Electrophoresis, 2010, 31: 2913-2920.
[4] 秦文斌. 活体红细胞内血红蛋白的电泳释放. 中国科学, 2011, 41 (8) : 597-607.
[5] 邵国, 睢天林, 秦文斌. 几种鸟类蛋白与人血红蛋白之间的相互作用的研究. 包头医学院学报, 2000, 16 (2) : 148-152.
[6] 苏燕, 秦文斌, 睢天林. 几种鱼类血红蛋白不与人血红蛋白相互作用及其进化意义. 包头医学院学报, 2000, 16 (2) : 1149-1151.

第四十五章 血红蛋白 A_3 与血红蛋白 A_2 之间没有交叉互作

韩丽红[#] 苏 燕[#] 高丽君 秦文斌[※]

(包头医学院 血红蛋白研究室，包头 014010)

摘 要

我们曾经用淀粉-琼脂混合凝胶电泳证明成人血红蛋白 A_1 与 A_2 发生交叉互作，后来还发现胎儿血红蛋白 F 也能与血红蛋白 A_2 发生交叉互作。当时未对快泳血红蛋白 A_3 进行研究。现在，再用这种方法研究血红蛋白 A_3 与血红蛋白 A_2 是否也有相互作用。结果表明，与血红蛋白 A_1 和血红蛋白 F 不同，血红蛋白 A_3 并不与血红蛋白 A_2 发生交叉互作。

关键词 血红蛋白 A_1；血红蛋白 A_2；血红蛋白 A_3；血红蛋白 F；交叉互作

1 前言

本文所使用的技术属于交叉电泳范畴，在这方面日本学者中村正二郎曾专门著书[1]介绍他们利用纸上电泳和琼脂电泳研究血清蛋白及免疫球蛋白的研究成果。我们所使用的支持体为淀粉-琼脂糖混合凝胶，不同于纸上电泳和琼脂电泳，研究对象也不一样，这里研究的是血红蛋白。我们用本法发现成人血红蛋白 A_1 与 A_2 之间发生交叉互作[2-4]，还有血红蛋白 F 与血红蛋白 A_2 之间的交叉互作(待发表)。本文的研究内容是快泳血红蛋白 A_3 与成人血红蛋白 A_2 之间是否存在交叉互作问题。

2 材料和方法

2.1 材料 正常成人血液来自本研究室人员，脐带血来自包钢集团第三职工医院妇产科。

2.2 方法 由成人血液分离出来红细胞并制备红细胞溶血液，由脐带血分离红细胞并制备胎儿红细胞溶血液，再用淀粉-琼脂糖混合凝胶电泳由成人溶血液制备电泳的血红蛋白 A_1、A_2 和 A_3，由胎儿溶血液制备电泳的血红蛋白 F，然后让血红蛋白 A_1、A_3、F 穿过血红蛋白 A_2，观察穿过后血红蛋白 A_2 区带变形情况，变形者(区带中部弯向阳极呈"V"状)为发生交叉互作，未变形为未发生相互作用。详细操作参见文献[2-4]。

3 结果

血红蛋白 A_1、A_3 和 F 穿过血红蛋白 A_2 后的结果见图 45-1。

[#] 并列第一作者
[※] 通信作者：秦文斌，电子邮箱：qwb5991309@tom.com

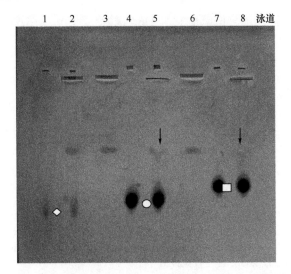

图 45-1 血红蛋白 A_1、A_3、F 穿过血红蛋白 A_2 的结果

注释：泳道 1=血红蛋白 A_3(见◇处)，泳道 2= A_3 穿过 HbA_2，泳道 3=Hb HbA_2，泳道 4= HbA_1(见○处)，泳道 5= HbA_1 穿过 HbA_2，泳道 6= HbA_2，泳道 7= HbF(见□处)，泳道 8= HbF 穿过 HbA_2

由此图可以看出，血红蛋白 A_1 穿过血红蛋白 A_2 时，后者变形(见泳道 5 箭头↓处)，血红蛋白 F 穿过血红蛋白 A_2 时，后者也变形(见泳道 8 箭头↓处)，但血红蛋白 A_3 穿过血红蛋白 A_2 时，后者不变形(见泳道 2)。

4 讨论

我们以前发现的成人 HbA_1 和 HbA_2 交叉互作[2-4]的反应式如下所示。

$$\left.\begin{aligned} &(\alpha^A\beta^A)_2 \longleftrightarrow 2(\alpha^A\beta^A) \\ &HbA_1 \\ &(\alpha^A\delta^{A_2})_2 \longleftrightarrow 2(\alpha^A\delta^{A_2}) \\ &HbA_2 \end{aligned}\right\} \longleftrightarrow 2(\alpha^A\beta^A\alpha^A\delta^{A_2}) \quad (1)$$

一过性交叉互作产物

血红蛋白 F 穿过血红蛋白 A_2 时，A_2 区带也变形，说明二者也发生交叉互作[5]。此时，血红蛋白 F 泳速减慢，说明穿过时二者曾一度结合并很快分开，它们的反应式如下所示。

$$\left.\begin{aligned} &(\alpha^A\gamma^F)_2 \longleftrightarrow 2(\alpha^A\gamma^F) \\ &HbF \\ &(\alpha^A\delta^{A_2})_2 \longleftrightarrow 2(\alpha^A\delta^{A_2}) \\ &HbA_2 \end{aligned}\right\} \longleftrightarrow 2(\alpha^A\gamma^F\alpha^A\delta^{A_2}) \quad (2)$$

一过性交叉互作产物

穿过实验的安排是将慢泳成分(血红蛋白 A_2)放在前面，快泳成分(血红蛋白 F)放在后边，电泳过程中 F 赶上并进入 A_2 时反应向右进行，出现一过性交叉互作产物。继续电泳，F 超过 A_2 时反应向左进行，又恢复为 HbF 和 HbA_2，但留下了血红蛋白 A_2 变形的痕迹。

HbA_3 和 HbA_2 不发生交叉互作，怎么理解，参见反应式(3)，问题应当在于快泳 HbA_3 的粗结构，它可能不同于 HbA_1 和 HbF。快泳血红蛋白也称糖化血红蛋白(另一种命名法称 HbA_1)，层析法可分成 HbA_1a、HbA_1b、HbA_1c、…其中主要成分是 HbA_1c，它的特点是在β珠蛋白链 N 端缬氨酸的游离氨基结合有葡萄糖等糖类化合物[6,7]，它的粗结构是$(\alpha^A\beta^{糖})_2$，代入反应式(3)，可能是这些糖化成分影响到它与 HbA_2 的交叉互作。在这里我们看到蛋白质结构与功能关系的又一有趣的实例。

$$\left.\begin{array}{c}(\alpha^A\beta^{糖})_2 \longleftrightarrow\!\!\!\times\!\!\!\longrightarrow 2(\alpha^A\beta^{糖}) \\ HbA_3 \\ (\alpha^A\delta^{A_2})_2 \longleftrightarrow 2(\alpha^A\delta^{A_2}) \\ HbA_2\end{array}\right\} \longleftrightarrow\!\!\!\times\!\!\!\longrightarrow (\alpha^A\beta^{糖}\alpha^A\delta^{A_2}) \atop 一过性交叉互作产物 \quad (3)$$

参 考 文 献

[1] Nakamura S. Cross Electrophoresis. Its Principle and Applications. New York: Elsevier, 1967.
[2] 秦文斌. 红细胞外血红蛋白 A_2 与 A_1 间的相互作用. 生物化学杂志, 1991, 7 (5)：583-587.
[3] Su Y, Gao LJ, Ma Q, et al. Interactions of hemoglobin in lived red blood cells measured by the electrophoresis release test. Electrophoresis, 2010, 31: 2913-2920.
[4] 秦文斌. 活体红细胞内血红蛋白的电泳释放. 中国科学, 2011, 41 (8)：597- 607.
[5] 韩丽红，闫斌，高丽君，等. 血红蛋白 F 与血红蛋白 A_2 之间的交叉互作 (待发表).
[6] Bookchin RM, Gallop PM.Structure of hemoglobin A1c: nature of the N-terminal beta chain blocking group. Biochem. Biophys. Res. Commun. 1968, 32 (1)：86-93.
[7] Koenig R J, Blobstein S H, Cerami A. Structure of carbohydrate of hemoglobin. AIc. J. Biol. Chem, 1977, 252: 2992-2997.

第四十六章　大鼠血红蛋白之间的交叉互作
——大鼠红细胞结晶的可能机制

裴娟慧[1#]　高丽君[2#]　秦文斌[2※]　韩丽莎[3※]　高雅琼[2]　韩丽红[2]　苏　燕[2]

(1. 北京航天总医院　心内科，北京　100076；2. 包头医学院　血红蛋白研究室，包头　014010；3. 包头医学院　病理生理教研室，包头　014010)

摘　要

目的： 我们早已知道大鼠红细胞容易形成结晶，本文利用双向电泳及交叉电泳来研究其可能机制。

方法： 用显微镜观察大鼠红细胞形成的结晶，双向对角线电泳观察大鼠红细胞及其全血内血红蛋白与对角线的关系，再用交叉电泳研究血红蛋白之间的相互作用。

结果： 显微镜下看到大鼠红细胞有结晶形成，红细胞双向对角线电泳时出现多个脱离对角线的血红蛋白成分。交叉电泳证明这些血红蛋白在快泳成分穿过慢泳成分时出现区带变形。

结论： 大鼠红细胞容易形成结晶与其血红蛋白之间的相互作用有关。

关键词　大鼠；红细胞；血红蛋白；结晶；相互作用

1　前言

大鼠红细胞内血红蛋白容易形成结晶，1968年法国学者Lessin L.S.对此进行了深入研究[1]，此后Lessin L.S.等在研究血红蛋白C病时再次提到大鼠红细胞内血红蛋白形成结晶与血红蛋白C病红细胞的类似性[2]。John M. E. 研究了大鼠血红蛋白的结构、功能等问题[3]，Frolov E. N. 等人研究了大鼠红细胞内血红蛋白的动力学[4]。但是这些文章都未涉及红细胞内血红蛋白之间的相互作用。本文的研究目的是利用我研究室建立的血红蛋白释放试验HRT及血红蛋白交叉电泳[5-12]，研究大鼠红细胞内血红蛋白的交叉互作。我们曾经发现轻型β-地中海贫血时HRT异常，也想研究另一种血红蛋白病——血红蛋白C病的情况，但实验室没有这种标本，选择具有类似红细胞的大鼠也是研究目的之一。

2　材料及方法

2.1　材料　大鼠为SD雄性大鼠，体重250～300g，由北京维通利华实验动物技术有限公司提供。肝素腹腔注射抗凝，主动脉采血。

\# 并列第一作者

※ 并列通信作者：秦文斌，电子邮箱：qwb5991309@tom.com；韩丽莎，电子邮箱：lishahan@sina.com

基金项目：包头医学院秦文斌基金会基金(200704)

2.2 方法

(1) 显微镜观察：载物片上加一滴盐水，再用 tip 头蘸少许大鼠全血与盐水混匀，40×10 倍镜下观察结果。

(2) 双向双层对角线电泳，操作参见文献[6，10-12]，主要操作是在淀粉-琼脂糖混合凝胶上，加入两个样品，上边是红细胞溶血液，下边是红细胞，第一向普通电泳后转 90°进入第二向，电泳后观察有无成分脱离对角线。

(3) 双向三层对角线电泳，基本同双向双层对角线电泳，只是标本为大鼠全血，上层为全血溶血液，中层为全血，下层为全血基质。

(4) 交叉电泳，操作参见文献[7，10，11]，主要操作是从电泳后的淀粉-琼脂糖混合凝胶中抠取血红蛋白部分(包括凝胶)，重新布局(长条的慢泳血红蛋白放在阳极侧，短条的快泳血红蛋白放在阴极侧，以便通电后让快泳血红蛋白穿过慢泳血红蛋白)，然后再铺胶，进行交叉电泳。

3 结果

3.1 显微镜结果　见图 46-1。

由图 46-1 可以看出显微镜下大鼠血液出现大量针状结晶。

3.2 双向双层对角线电泳　见图 46-2。

由图 46-2 可以看出，大鼠红细胞溶血液中主要有四种血红蛋白(染色前都为红色)，由左上方向右下方，依次命名为：HbA，HbB，HbC 和 HbD。按含量多少排列为 HbA＞HbB＞HbC＞HbD，它们都位于对角线上。红细胞释放的血红蛋白，在上述四种血红蛋白的基础上，出现多个离开对角线成分：在 HbA 右侧、HbB 右侧、HbC 左侧、HbD 左侧都有横向移动成分(参见相应的↓或↑)。

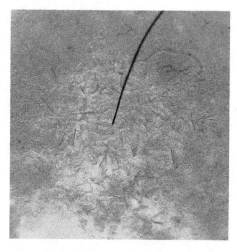

图 46-1　大鼠血液的显微镜下结晶
注释：40×10 倍镜下观察

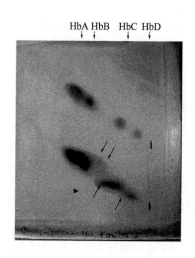

图 46-2　大鼠红细胞及其溶血液的双向电泳
注释：上层为溶血液，下层为红细胞，考马斯亮蓝染色

3.3 双向三层对角线电泳　见图 46-3。

由图 46-3 可以看出，全血溶血液中的血红蛋白(染色前显红色)都在对角线上，在它们的左上方还有一个白蛋白(Alb)。全血中血红蛋白则仍然脱离对角线，与图 46-2

的红细胞结果类似。全血基质中的血红蛋白脱离对角线情况更明显，而且原点有大量血红蛋白残留，第二向电泳时又有一部分释放出来，形成上升的垂直线(参见图中↓处)。

3.4 血红蛋白交叉电泳 大鼠 HbB、HbC 穿过 HbC、HbD，结果见图 46-4。

由图 46-4 可以看出，HbB 穿过 HbC 时 HbC 区带变形、中部下凸，HbB 穿过 HbD 时 HbD 区带变形、中部下凸，HbC 超过 HbD 时 HbD 区带变形、中部下凸。上述结果中区带变形者见"○"处。

3.5 血红蛋白交叉电泳 大鼠 HbA、HbB、HbC 穿过 HbB、HbC、HbD，全部四种血红蛋白成分的交叉电泳结果见图 46-5。

由图 46-5 可以看出，泳道 2 HbB 带出现变形、中部下凸；泳道 3 HbC 带出现变形、中部下凸；泳道 4 HbD 带出现变形、中部下凸；泳道 6 HbC 带出现变形、中部下凸；泳道 7 HbD 带出现变形、中部下凸；泳道 8 HbD 带出现变形、中部下凸；泳道 9 为大鼠 HbC 对照。泳道 9 加样太靠边，看不清楚，可再参见图 46-4 之泳道 4、5。上述区带变形者见"○"处。

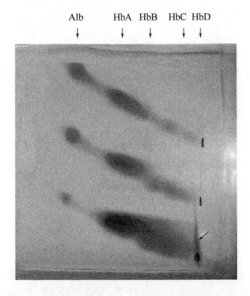

图 46-3 大鼠全血溶血液、全血及全血基质的双向电泳
注释：上层为全血溶血液，中层为全血，下层为全血基质，考马斯亮蓝染色

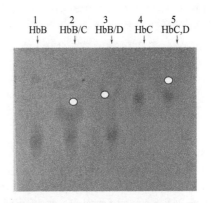

图 46-4 大鼠 HbB、HbC 穿过 HbC、HbD 的结果
注释：泳道由左向右，1. HbB，2. HbB 穿过 HbC，3. HbB 穿过 HbD，4. HbC，5. HbC 穿过 HbD

4 讨论

2006 年我们在研究哺乳动物自身血红蛋白相互作用时已经发现大鼠血红蛋白之间的相互作用[8]，但那时尚未进行双向对角线电泳，没有看到这方面的特点。本文将大鼠红细胞及其溶血液同时进行双向双层对角线电泳，可以清楚看到溶血液出现四种血红蛋白成分(HbA、HbB、HbC、HbD)，而且它们都在对角线上。与溶血液不同，红细胞的四种血红蛋白成分界限不清，有多个离开对角线成分。根据我们对血红蛋白 A_2 现象的研究[10]，两种相互作用的血红蛋白(HbA_2 与 HbA_1)在双向电泳中都脱离对角线，而且互相对应。

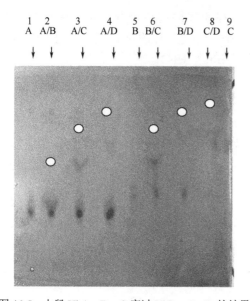

图 46-5 大鼠 HbA、B、C 穿过 HbB、C、D 的结果
注释：上方为负极，下方为正极。泳道由左向右 1. 大鼠 HbA 对照，2. 大鼠 HbA 穿过 HbB，3. 大鼠 HbA 穿过 HbC，4. 大鼠 HbA 穿过 HbD，5. 大鼠 HbB 对照，6. 大鼠 HbB 穿过 HbC，7. 大鼠 HbB 穿过 HbD，8. 大鼠 HbC 穿过 HbD，9. 大鼠 HbC 对照

这些结果说明，大鼠红细胞内血红蛋白之间存在相互作用迹象，需要用交叉电泳来证实。血红蛋白交叉电泳结果进一步证明，上述四种血红蛋白之间都有交叉互作：HbA 与 HbB、HbA 与 HbC、HbA 与 HbD、HbB 与 HbC、HbB 与 HbD、HbC 与 HbD 都有互作。由此可见，大鼠红细胞生成结晶与其中血红蛋白存在相互作用有关。现在看来，红细胞释放型对角线电泳与血红蛋白交叉电泳联合运用，可以帮助我们发现和证实细胞内血红蛋白之间的交叉互作。早在 30 多年前，我们用双向电泳发现了正常人红细胞的血红蛋白 A_2 脱离对角线，再用交叉电泳发现了血红蛋白 A_2 与血红蛋白 A_1 之间的相互作用[5, 10]。现在，我们在大鼠红细胞看到了类似情况。看来，双向电泳与交叉电泳相互配合，能够发现红细胞内血红蛋白之间的相互作用。对于其他细胞，如血小板[13]等，我们也看到其双向电泳中蛋白质有脱离对角线成分，推测存在细胞内蛋白质之间的相互作用，但尚未做交叉电泳等其他分析，有待进一步深入研究。

参 考 文 献

[1] Lessin LS. Structure mol'eculaire de l'hémoglobine cristallisée d'érythrocytes de rat, étudiée par cryodécapage. Nouv Rev Fr Hematol, 1968, 8 (4) : 423-436.

[2] Lessin L S, Jensen W N, Ponder E. Molecular mecha-nism of hemolytic anemia in homozygous hemoglobin C disease. Electron micros copic study by the freeze-etchig technique. J Exp Med, 1969, 130 (3) : 443-466.

[3] John M E. Structural, functional and conformational properties of rat hemoglobins. Eur 1 Biochem, 1982, 124: 105-310.

[4] Frolov E N, Fischer M, Graffweg E, et al. Hemoglobin dynamics in rat erythrocytes investigated by Mössbauer spectroscopy. European Biophysics Journal, 1991, 19 (5) : 253-256.

[5] 秦文斌, 梁友珍. 血红蛋白 A_2 现象 I . 此现象的发现及其初步应用. 生物化学与生物物理学报, 1981, 13 (2) : 199-205.

[6] 秦文斌. 红细胞外血红蛋白 A 与血红蛋白 A_2 之间的相互作用. 生物化学杂志, 1991, 7 (5) : 583-587.

[7] 秦文斌. "红细胞 HbA_2" 为 HbA_2 与 HbA 的结合产物. 生物化学与生物物理进展, 1991, 18 (4) : 286-288.

[8] 裴娟慧, 秦文斌, 韩丽莎. 哺乳纲动物绵羊和大白鼠自身血红蛋白相互作用的研究和意义. 包头医学院学报, 2006, 22 (3) : 241-243.

[9] Su Y, Shao G, Gao L J, et al. RBC electrophoresis with discontinuous power supply–a newly established hemoglobin release test. Electrophoresis, 2009, 30: 3041-3043.

[10] Su Y, Gao L J, Ma Q, et al. Interactions of hemoglobin in lived red blood cells measured by the electrophoesis release test. Electrophoresis, 2010, 31: 2913-2920.

[11] 秦文斌. 活体红细胞内血红蛋白的电泳释放. 中国科学: 生命科学, 2011, 41 (8) : 597-607.

[12] Su Y, Gao L J, Qin W B. Protein electrophoresis, methods and protocols. Methods in Molecular Biology, 2012, 869 (17) : 393-402.

[13] 乔姝, 沈木生, 韩丽红, 等. 血小板成分电泳释放的初步研究. 现代预防医学, 2011, 38 (4) : 685-688.

第四十七章　羊膜动物与非羊膜动物血红蛋白的交叉互作行为不同
——脊椎动物血红蛋白的分子进化

邵　国[#]　苏　燕[#]　王占黎[#]　武莎莎　孟　俊　杨　东
邢晓燕　折志刚　秦文斌[※]

(包头医学院　血红蛋白研究室，包头　014010)

摘　要

背景与目的： 在用淀粉-琼脂糖凝胶交叉电泳研究"血红蛋白 A_2 现象"的机制过程中发现成人血红蛋白 HbA_2 与 HbA 存在相互作用。在此基础上，研究成人血红蛋白与脊椎动物红蛋白之间是否也有相互作用。

方法： 利用淀粉-琼脂糖混合凝胶进行交叉电泳，根据穿过、兜过时区带的变形情况，来判定成人 Hb 与各种脊椎动物 Hb 之间是否发生相互作用。

结果： 与成人 Hb 发生互作的有哺乳动物 Hb、鸟类 Hb、爬行动物 Hb；不发生互作的有两栖动物 Hb、硬骨鱼 Hb、软骨鱼 Hb、圆口类 Hb。

结论： 从交叉电泳结果来看，脊椎动物血红蛋白的分子进化存在明显差异。哺乳动物、鸟类及爬行动物的血红蛋白能与成人血红蛋白发生相互作用；两栖类、硬骨鱼、软骨鱼及无颌纲动物血红蛋白与成人血红蛋白不发生相互作用。这说明，在脊椎动物进化过程中，其血红蛋白在结构上发生变化，造成与成人血红蛋白互作上的差异。由此看来，脊椎动物血红蛋白结构差异可能存在一条分水岭，即羊膜动物血红蛋白与无羊膜动物血红蛋白有所不同。

关键词　血红蛋白；交叉电泳；脊椎动物；羊膜动物；无羊膜动物

1　前言

1991 年我们用交叉电泳技术证明，在红细胞外，成人血红蛋白 A_2 与血红蛋白 A 可以发生相互作用[1]，后来用小鼠、大鼠及家兔等哺乳动物血红蛋白与成人血红蛋白进行交叉电泳，发现都有相互作用[2, 3]，此时我们推测，这种发生在电场中的相互作用可能是各种血红蛋白之间的普遍规律。接下来进行鸟类血红蛋白与成人血红蛋白之间的交叉电泳，情况也是如此[4]。但是，后来用鱼类血红蛋白与成人血红蛋白做的实验，没有出现相互作用的迹象[5, 6]。继续实验，爬行动物血红蛋白[7]与成人血红蛋白有互作，两栖动物血红蛋白[8]与成人血红蛋白没有互作。最后，用无颌纲动物血红蛋白[9]做实验，也是没有互作。上述

[#] 并列第一作者
[※] 通信作者：秦文斌，电子邮箱：qwb5991309@tom.com

结果表明,在脊椎动物血红蛋白的进化过程中,互作与否的分界线位于两栖动物血红蛋白与爬行动物血红蛋白之间,即羊膜动物血红蛋白与成人血红蛋白有互作,非羊膜动物血红蛋白与成人血红蛋白没有互作。由此看来,在进化过程中,羊膜动物血红蛋白与非羊膜动物血红蛋白的分子结构出现明显差异,造成与成人血红蛋白互作不同,爬行动物血红蛋白成为进化的分水岭。

2 材料与方法

2.1 血液来源 正常成人血液来自包头医学院第一附属医院检验科血库。各种动物血来自本校动物室或动物市场。盲鳗红细胞由丹麦 Aarhus 大学动物生理学系 Roy Weber 教授惠赠,再次表示感谢。

2.2 红细胞溶血液的制备 取人或动物的抗凝血,离心去血浆,用生理盐水洗涤红细胞 5 次。向红细胞层加入 1.5 倍体积的蒸馏水及等体积的四氯化碳,激烈震荡后离心,取上层红色溶血液备用。

2.3 交叉电泳技术 淀粉-琼脂糖混合凝胶电泳,按我室常规,具体如下。制胶:淀粉 1g,琼脂糖 0.2g,TEB pH 8.6 的缓冲液 55ml,煮胶。玻璃板 17cm×10cm,水平倒胶。加样:将溶血液加入到新华 3 号滤纸条,此滤纸条再插入凝胶。每次实验有互作双方的两种溶血液,先用普通电泳判定其中血红蛋白的泳速快慢。交叉电泳时,按电泳方向,将泳速快的标本放在后方(阴极侧)、慢的放在前方(阳极侧),让泳速快的标本穿过或兜过泳速慢者。含标本的滤纸条有长短两种。穿过实验时,慢泳标本用长条,放在阳极侧、快泳标本用短条,放在阴极侧。兜过实验时,二者关系反过来,慢泳标本用短条,放在阳极侧、快泳标本用长条,放在阴极侧,参见图47-1、图47-2。电泳结果可直接观察红色区带的变形,必要时染氨基黑10B。

2.4 交叉电泳结果判定 电泳后看长条(被穿过或兜过)标本的区带中部是否变形(长条区带的中段出现弯曲,如⌒或‿)变形者为发生互作,无变形(仍是直线)时,未发生互作。参见图47-1、图47-2。

3 结果

3.1 交叉电泳实例 包括互作阳性与阴性两种类型。

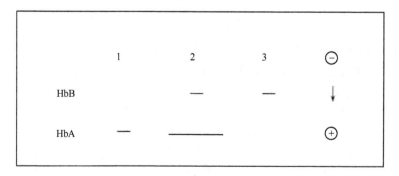

图 47-1 穿过试验(HbB 穿过 HbA)

注释:1、3 为对照,HbA、HbB 单独电泳,互不干扰。2 为交叉电泳,穿过试验:HbB 的泳速比 HbA 快,电泳过程中 HbB 短区带(—)穿过 HbA 长区带(——)中部。如有互作,HbA 的长区带中间变形,出现⌒状,此时⌒朝下;如无互作,则仍为——,不变形

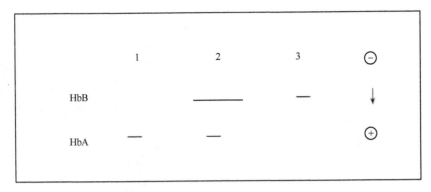

图 47-2 兜过试验(HbB 兜过 HbA)

注释：基本同上图 47-1，只是长短区带位置对调。1、3 仍为对照。2 为交叉电泳，兜过试验：HbB 的泳速比 HbA 快，电泳过程中 HbB 长区带(——)兜过 HbA 短区带(—)。如有互作，HbB 的长区带中间变形，成为 ⌒，此时 ⌒ 朝上；如无互作，则仍为——状，不变形

(1) 互作阳性结果：参见图 47-3～图 47-10。
(2) 互作阴性结果：参见图 47-11～图 47-22。

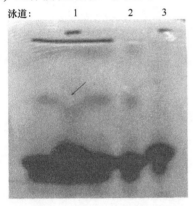

图 47-3 成人 Hb 与小鼠 Hb 互作

注释：泳道 1 交叉电泳：小鼠 Hb 穿过成人 Hb，发生互作(见箭头处)；2 对照：成人血红蛋白溶液；3 对照：小鼠血红蛋白溶液

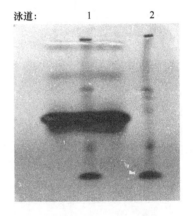

图 47-4 成人 Hb 与小鼠血清蛋白不互作

注释：泳道 1 交叉电泳：小鼠血清白蛋白等穿过成人 Hb，未发生互作；2 对照：小鼠血清

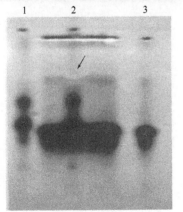

图 47-5 成人 Hb 与大鼠 Hb 互作

注释：泳道 1 对照：大鼠血红蛋白溶液；2 交叉电泳：大鼠 Hb 穿过成人 Hb，发生互作(见箭头处)；3 对照：成人血红蛋白溶液

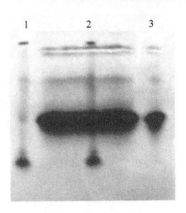

图 47-6 成人 Hb 与大鼠血清蛋白不互作

注释：泳道 1 对照：大鼠血红蛋白溶液；2 交叉电泳：大鼠血清白蛋白等穿过成人 Hb，未发生互作；3 对照：成人血红蛋白溶液

图 47-7 纯 HbA2 与大鼠 Hb 互作
注释：上边的短加样框为含大鼠 Hb 淀琼胶；下边的长加样框为含成人 HbA2 淀琼胶；将此二胶转到另一玻璃板，再倒胶，进行交叉电泳：大鼠 Hb 穿过成人 HbA2，发生互作（见箭头处）

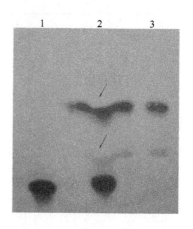

图 47-8 成人 Hb 与鸭 Hb 互作
注释：泳道 1 对照，成人 HbA；2 交叉电泳，成人 HbA 穿过鸭 Hb，发生互作（见箭头处）；3 对照，鸭血红蛋白溶液

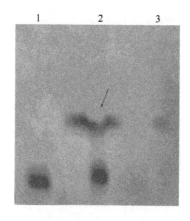

图 47-9 中华鳖穿过成人血红蛋白 A2
注释：泳道 1 对照，中华鳖 Hb；2 交叉电泳，中华鳖 Hb 穿过成人 HbA2，发生互作（见箭头处）；3 对照，成人 HbA2

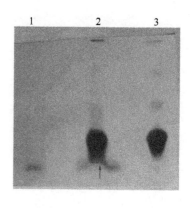

图 47-10 中华鳖兜过成人血红蛋白 A
注释：泳道 1 对照，中华鳖 Hb；2 交叉电泳，中华鳖 Hb 兜过成人 Hb，发生互作（见箭头处）；3 对照，成人 Hb

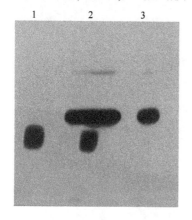

图 47-11 牛蛙血红蛋白与成人血红蛋白不互作
注释：泳道 1 对照，牛蛙 Hb；2 交叉电泳，牛蛙 Hb 穿过成人 Hb，未发生互作；3 对照，成人血红蛋白溶液

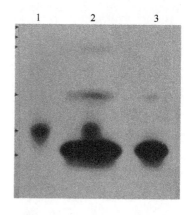

图 47-12 花背蟾蜍血红蛋白与成人血红蛋白不互作
注释：泳道 1 对照，花背蟾蜍 Hb；2 交叉电泳，花背蟾蜍 Hb 穿过成人 HbA2，未发生互作；3 对照，成人血红蛋白溶液

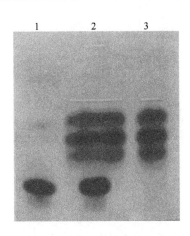

图 47-13 成人 Hb 与鲤鱼 Hb 不互作

注释：泳道 1 对照，成人血红蛋白溶液；2 交叉电泳，成人 HbA 穿过鲤鱼 Hb，未发生互作；3 对照，鲤鱼血红蛋白溶液

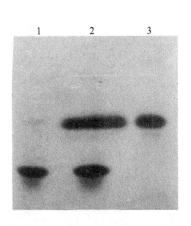

图 47-14 成人 Hb 与鲫鱼 Hb 不互作

注释：泳道 1 对照，成人血红蛋白溶液；2 交叉电泳，成人 HbA 穿过鲫鱼 Hb，未发生互作；3 对照，鲫鱼血红蛋白溶液

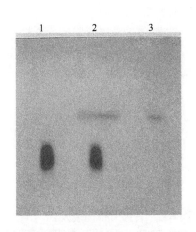

图 47-15 条纹斑竹鲨与 A_2 不互作

注释：泳道 1 对照，条纹斑竹鲨血红蛋白溶液；2 交叉电泳，条纹斑竹鲨穿过成人 HbA_2，未发生互作；3 对照，成人 HbA_2

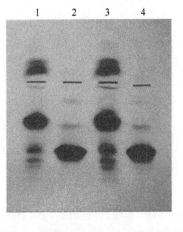

图 47-16 盲鳗溶血液与成人溶血液比较

注释：1 盲鳗血红蛋白溶液(由阳极到阴极：A、B、C、D、E、F 六个区带)；2 成人血红蛋白溶液(由阳极到阴极：A、A_2、CA 三个区带)；3 =1，4 =2

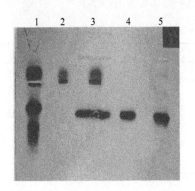

图 47-17 盲鳗 HbA 与成人 Hb 不互作

注释：1 对照，盲鳗血红蛋白溶液；2 对照，盲鳗 HbA；3 交叉电泳，成人 HbA 穿过盲鳗 HbA，未发生互作；4 对照，成人 HbA；5 对照，成人血红蛋白溶液

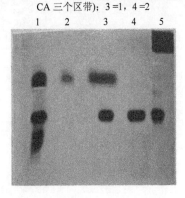

图 47-18 盲鳗 HbB 与成人 Hb 不互作

注释：1 对照，盲鳗血红蛋白溶液；2 对照，盲鳗 HbB；3 交叉电泳，成人 HbA 穿过盲鳗 HbB，未发生互作 4 对照，成人 HbA；5 对照，成人血红蛋白溶液

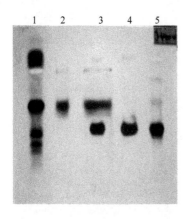

图 47-19　盲鳗 HbC 与成人 Hb 不互作

注释：1 对照，盲鳗血红蛋白溶液；2 对照，盲鳗 HbC；3 交叉电泳，成人 HbA 穿过盲鳗 HbC，未发生互作；4 对照，成人 HbA；5 对照，成人血红蛋白溶液

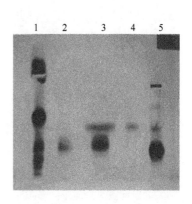

图 47-20　盲鳗 HbD 与成人 Hb 不互作

注释：1 对照，盲鳗血红蛋白溶液；2 对照，盲鳗 HbD；3 交叉电泳，盲鳗 HbD 穿过成人 HbA_2，未发生互作；4 对照，成人 HbA_2；5 对照，成人血红蛋白溶液

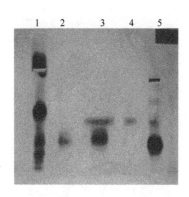

图 47-21　盲鳗 HbE 与成人 Hb 不互作

注释：1 对照，盲鳗血红蛋白溶液；2 对照，盲鳗 HbE；3 交叉电泳，盲鳗 HbE 穿过成人 HbA_2，未发生互作；4 对照，成人 HbA_2；5 对照，成人血红蛋白溶液

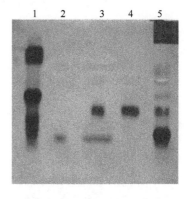

图 47-22　盲鳗 HbF 与成人 Hb 不互作

注释：1 对照，盲鳗血红蛋白溶液；2 对照，盲鳗 HbF；3 交叉电泳，盲鳗 HbF 兜过成人 HbA_2，未发生互作；4 对照，成人 HbA_2；5 对照，成人血红蛋白溶液

3.2　脊椎动物 Hb 与成人 Hb 之间的相互作用综合表　参见表 47-1。

表 47-1　脊椎动物血红蛋白与成人血红蛋白[※1]之间的相互作用表

动物纲名	动物名	与成人血红蛋白互作与否	文献
哺乳动物	小鼠	(+)[※2]	[1]
	大鼠	(+)[※2]	[1]
	家兔	(+)[※3]	[2]
鸟类	家鸽	(+)	[3]
	家鸡	(+)	[3]
	家鸭	(+)	[3]
	暗绿柳莺	(+)	[3]
	山雀	(+)	[3]
爬行类	沙蜥	(+)	[4]
	中华鳖	(+)	[4]
	蝮蛇	(?)[※4]	[4]

续表

动物纲名	动物名	与成人血红蛋白互作与否	文献
两栖类	牛蛙	(−)※5	[5]
	黑斑蛙	(−)※5	[5]
	花背蟾蜍	(−)※5	[5]
硬骨鱼类	金鱼	(−)※6	[6]
	鲤鱼	(−)※6	[6]
	鲫鱼	(−)※6	[6]
	泥鳅	(−)※6	[6]
软骨鱼类	条文斑竹鲨	(−)	[7]
无颌纲	盲鳗	(−)	[8]

注释：※1 多数情况下是动物的快泳血红蛋白与成人血红蛋白 A_2 发生互作，也有成人血红蛋白 A_1 与动物慢泳血红蛋白发生互作；※2 还看到动物血红蛋白与异常血红蛋白 HbE 发生互作；※3 还看到动物血红蛋白与异常血红蛋白 HbG 发生互作；※4 蝮蛇血红蛋白与成人血红蛋白互作不明确，需要再研究；※5 这几种动物血红蛋白之间也无相互作用；※6 这几种动物血红蛋白之间也无相互作用，同一动物的不同血红蛋白组分之间也没有互作。

3.3 脊椎动物 Hb 与成人 Hb 之间的相互作用综合图
参见图 47-23。

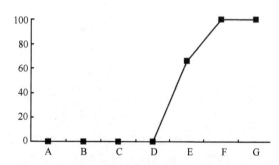

图 47-23 成人 Hb 与脊椎动物 Hb 间相互作用图
注释：A=无颌纲 Hb；B=软骨鱼类 Hb；C=硬骨鱼类 Hb；D=两栖动物 Hb；E=爬行动物 Hb；F=鸟类 Hb；G=哺乳类 Hb

4 讨论

血红蛋白分子进化的研究由来已久，Morris Godman 等[10]曾对 55 种不同种属动物血红蛋白的氨基酸序列进行比较，推导出进化树和血红蛋白的进化规律，为物种进化的相关研究奠定了基础。Dickerson RE 和 Geis I.出版了一本涉及血红蛋白分子进化的书[11]，主要是比较不同物种血红蛋白的氨基酸序列(一级结构)，同时比较它们的静态四级结构。近来，Griffith WP.和 Kaltashov I A. 提出一种方法，用质谱技术，研究血红蛋白组装时珠蛋白间相互作用的高度不对称性[12]，在此基础上提出血红蛋白进化涉及其构象与珠蛋白间结合的关系[13]。Csaba Pál 等人提出蛋白进化的综合观点[14]，他们认为，系统生物学及基因组学方面的进展，使进化研究由个别蛋白质的结构与功能转向总体细胞因子，其中包括编码基因在染色体上的位置、它们的表达模式、它们在生物网络中的位置等。Eric A. Ortlund 等[15]提出构象的异常抑制(conformational epistasis)对蛋白进化的影响。Jamie T 等[16]强调，分子修补(molecular tinkering)在蛋白进化中的作用。

本文所使用的技术属于交叉电泳范畴，日本学者中村正二郎[17]利用纸上电泳和琼脂电泳研究血清蛋白及免疫球蛋白。我们所使用的支持体为淀粉-琼脂糖混合凝胶，研究对象是血红蛋白之间的相互作用。我们是研究血红蛋白 A_2 现象的机制时发现了成人血红蛋白 A_2 与血红蛋白 A_1 能够相互作用，从而想到人血红蛋白与动物血红蛋白的互作问题[1]。血红蛋白相互作用也涉及蛋白质的立体结构问题。

这里的血红蛋白相互作用，是用交叉电泳方法进行研究的，是血红蛋白在电泳过程中(或者说是在电场中)发生的相互作用。此种互作机制尚不清楚。不过，根据实验结果可以看出，互作是一过性的，两种血红蛋白能够结合，不久又分开，遗留下来曾经一过性结合的痕迹(互作部位的区带变形)。成人血红蛋白 A 与血红蛋白 A_2 之间的互作机制,我们的推测是,在电场中两种血红蛋白相遇时发生一过性分子杂交,临时交换亚单位,生成杂交分子,具体如式(1);关于成人血红蛋白与羊膜动物血红蛋白中间的相互作用,可能与上述成人血红蛋白内部互作相同或相似,如式(2);非羊膜动物血红蛋白与成人血红蛋白在交叉电泳中没有反应,暂列成式(3),以便讨论。

$$\left.\begin{array}{l}(\alpha^A\beta^A)_2 \longleftrightarrow 2(\alpha^A\beta^A) \\ \text{HbA} \\ (\alpha^A\delta^{A_2})_2 \longleftrightarrow 2(\alpha^A\delta^{A_2}) \\ \text{HbA}_2\end{array}\right\} \longleftrightarrow \begin{array}{l}2(\alpha^A\beta^A\alpha^A\delta^{A_2}) \\ \text{一过性杂交分子}\end{array} \quad (1)$$

$$\left.\begin{array}{l}(\alpha^A\beta^A)_2 \longleftrightarrow 2(\alpha^A\beta^A) \\ \text{HbA} \\ (\alpha^B\beta^B)_2 \longleftrightarrow 2(\alpha^B\beta^B) \\ \text{羊膜动物Hb}\end{array}\right\} \longleftrightarrow \begin{array}{l}2(\alpha^A\beta^A\alpha^B\beta^B) \\ \text{一过性杂交分子}\end{array} \quad (2)$$

$$\left.\begin{array}{l}(\alpha^A\beta^A)_2 \longleftrightarrow 2(\alpha^A\beta^A) \\ \text{HbA} \\ (\alpha^C\beta^C)_2 \longleftrightarrow ?(\alpha^C\beta^C) \\ \text{非羊膜动物Hb}\end{array}\right\} \longleftrightarrow \begin{array}{l}?(\alpha^A\beta^A\alpha^C\beta^C)? \\ \text{一过性杂交分子?}\end{array} \quad (3)$$

关于无羊膜动物血红蛋白不与成人血红蛋白互作的机制,目前还没有一个明确解释,但根据文献和我们研究室所做的系列实验,可做出以下推测。

4.1 解离再结合问题 两种血红蛋白能够发生相互作用,是指两种血红蛋白分别解聚生成的二聚体可以通过离子键及疏水基团间的 van der Waals(范德瓦耳斯)力相互连接形成新的杂交四聚体分子。而形成这种新的杂交四聚体分子的基础,是两种血红蛋白一、二、三、四级结构的相似性。只有两种血红蛋白二聚体氨基酸组成、立体结构相似,两对血红蛋白二聚体才可能以与原四聚体相似的方式进行连接,组合成新的杂交四聚体分子。两种血红蛋白越相似,就越可能形成新杂交分子;两种血红蛋白不能够发生相互作用,可能是由于两种血红蛋白一、二、三、四级结构的差别很大,两种血红蛋白的二聚体不容易通过离子键及疏水基团间的 Van ker waals 力相互连接而形成新的杂交四聚体分子,或者这种新的杂交四聚体分子数量很少,不易被检出,因此被认为两种血红蛋白不可以发生相互作用。从血红蛋白分子本身的进化角度来看,依据 Dickerson 和 Geis 提出的血红蛋白分子的进化规律[11],各种血红蛋白在氨基酸排列顺序上具有同源性,但在脊椎动物的进化过程中,两物种相差越远,同种蛋白质(血红蛋白)的氨基酸排列顺序差异越大。例如,人与大猩猩β链之间只有 1 个氨基酸残基不同,人与兔β链之间有 14 个氨基酸残基不同(9.6%),猪与小鼠β链有 32 个氨基酸残基不同(22%),说明人与哺乳动物(较高等的羊膜动物)及哺乳动物之间的同源性较大。随着脊椎动物与人的进化距离的加大,有些氨基酸残基发生了置换,血红蛋白一级结构的差异即开始逐步变大,这些置换的氨基酸残基大多数位于对血红蛋白结构和功能影响不大的位置,但由于置换的存在,必然会引起一些改变,这其中就包括血红蛋白二、三、四级结构的改变。这方面,最突出的例子是无颌纲的盲鳗。这种血红蛋白在一般情况下以单体形式存在,只有在 pH 低于 7 的环境中才可以相互聚合[9, 18]。我们的电泳缓冲液是 pH 9 左右,不发生互作是可以理解的。

4.2 反应动力学问题

(1) 上述血红蛋白相互作用如一般的化学反应一样存在反应平衡，即四聚体解聚生成二聚体、二聚体再结合成四聚体、两对不同来源的二聚体结合成新杂交四聚体、重新解聚生成两对二聚体的各个反应之间存在着动态平衡，各个反应的平衡常数不同。像成人血红蛋白这样在生理条件及电泳时均以稳定的四聚体形式存在的血红蛋白，其反应方向倾向于生成四聚体，反应体系中二聚体含量较少。当反应体系中存在有与之相似的另一血红蛋白二聚体分子时，由于可以生成新杂交四聚体，反应产物被消耗，向二聚体解聚方向的反应加强，反应体系中二聚体含量增多，反应有利于新杂交四聚体的生成，此时我们认为发生了血红蛋白相互作用；而当反应体系中不存在有与之相似的另一血红蛋白二聚体分子时，不能生成新杂交四聚体，反应方向倾向于生成四聚体，此时认为两种血红蛋白不可以发生血红蛋白相互作用。

(2) 不同血红蛋白的四聚体——二聚体解离平衡常数($K_{4,2}$)不同。非常大或者非常小的$K_{4,2}$均可极大地改变其电泳及色谱特性[19, 20]。人类 HbKansas 和 HbHirose 就是极大增加解离的两个例子[其一氧化碳衍生物(CO-derivative)Hb 的 $K_{4,2}$ 分别为 10^{-4}mol/L 及 1mol/L]。人 Hb $K_{4,2}$ 值为 10^{-6}mol/L，哺乳动物 Hb 与人相似，而鱼类 Hb 的 $K_{4,2}$ 为 10^{-8}mol/L(鲨 Hb 的 $K_{4,2}$ 为 10^{-8}mol/L 或更低)，比人 Hb 低至少 100 倍，相差如此之大的两种血红蛋白，其四聚体——二聚体解离平衡在同一实验条件下是分别进行的[21, 22]，使得两种血红蛋白的二聚物是分离着的，因此无法生成杂交四聚体或者其生成的量太少而不易被检出。我们推测，非羊膜动物的硬骨鱼类血红蛋白可能属于此类，因为它自身的血红蛋白可以保温互作[23, 24]，与成人血红蛋白却无互作，可能属于反应动力学差异或在构象上与成人血红蛋白相距较远。

4.3 多聚体问题 一些低等脊椎动物，如软骨鱼、两栖动物、爬行动物(硬骨鱼例外)的血红蛋白容易发生聚合形成多聚体[10]，而人、哺乳动物及鸟类血红蛋白不容易发生聚合。此聚合通常在溶血后发生，而我们做血红蛋白相互作用的实验时首先要将两种血红蛋白制成溶血液。如果其中一种血红蛋白在制备时已经形成多聚体，反应倾向于生成多聚物的方向，就不易与另一血红蛋白发生相互作用。根据上述资料，我们推测，非羊膜动物中软骨鱼、两栖动物血红蛋白可能因形成多聚体而无法与成人血红蛋白发生互作。另外，爬行动物血红蛋白也容易形成多聚体，它却能够与成人血红蛋白发生互作，这如何解释；三种爬行动物中蝮蛇血红蛋白结果又特殊，这又如何解释，都是需要继续研究的问题。

参 考 文 献

[1] 秦文斌. 红细胞外血红蛋白 A 与血红蛋白 A_2 之间的相互作用. 生物化学杂志, 1991, 7: 583.
[2] 武莎莎, 睢天林, 秦文斌. 大、小鼠血红蛋白与人正常、异常血红蛋白之间相互作用的研究. 生物化学杂志, 1997, (专刊): 37.
[3] 孟峻, 睢天林, 秦文斌. 家兔血红蛋白与人正常、异常血红蛋白之间相互作用的研究. 生物化学杂志, 1997, (专刊): 36.
[4] 邵国, 睢天林, 秦文斌. 几种鸟类蛋白与人血红蛋白之间的相互作用的研究. 包头医学院学报, 2000, 16(2): 148.
[5] 王占黎, 睢天林, 秦文斌. 几种爬行动物血红蛋白与人血红蛋白之间相互作用的研究. 上海: 中国生物化学与分子生物学会第八届会员代表大会暨全国学术会议论文摘要, 2001: 311.
[6] 杨冬, 秦文斌, 睢天林. 几种两栖动物血红蛋白不与人血红蛋白相互作用及其进化意义. 上海: 中国生物化学与分子生物学会第八届会员代表大会暨全国学术会议论文摘要, 2001: 311.

[7] 苏燕,秦文斌,睢天林. 几种鱼类血红蛋白不与人血红蛋白相互作用及其进化意义. 包头医学院学报, 2000, 16 (2) : 1149.

[8] 邢晓燕,秦文斌,睢天林. 一种软骨鱼(条纹斑竹鲨)血红蛋白不与人血红蛋白相互作用及其进化意义. 上海:中国生物化学与分子生物学学会第八届会员代表大会暨全国学术会议论文摘要, 2001: 311.

[9] 折志刚,睢天林,秦文斌. 无鄂纲动物盲幔血红蛋白不与人血红蛋白相互作用及其进化意义. 上海:中国生物化学与分子生物学学会第八届会员代表大会暨全国学术会议论文摘要, 2001: 311.

[10] Goodman M, Moore G W. Darwinian evolution in the genealogy of haemoglobin. Nature, 1975, 253: 603-608.

[11] Dickerson R E, Geis I. Hemoglobin: Structure, Function, Evolution, and Pathology. Benjamin: Cummings Pub. Co. , 1983.

[12] Griffith W P, Kaltashov I A. Highly asymmetric interactions between globin chains during hemoglobin assembly revealed by electrospray ionization mass spectrometry. Biochemistry, 2003, 42: 10024-10033.

[13] Griffith W P, Kaltashov I A. Phenotypic analysis of Atlantic cod hemoglobin chains using a combination of top-down and bottom-up mass spectrometric approaches. International Journal of Mass Apectrometry, 2008, 278(2-3): 114-121.

[14] Pál C, Papp B, Lercher M J. An integrated view of protein sevolutions. Nature Reviews Genetics, 2006, 7: 337-348.

[15] Ortlund E A, Bridgham J T, Redinbol M R. et al. Crystal structure of an ancient protein: evolution by conformational epistasis. Science, 2007, 317 (5844) : 1544-1548.

[16] Bridgham J T, Eick G N, Larroux C, et al. Protein evolution by molecular tinkering: diversification of the nuclear receptor superfamily from a ligand -dependent ancestor 2010: PLoS Biology.

[17] Nakamura S. Cross Electrophoresis. Its Principle and Applications. New York: Elsevier, 1967.

[18] Fago A, Weber R E. The hemoglobin system of hagfish, Myxine glutinosa, aggregation state and functional properties. Biochimica Biophysica, 1955, 249: 109-115.

[19] Fyhn UE, Sullivan B. Elasmobranch hemoglobin: dimerization and polymerization in various species. Comp biochem Physiol, 1975, 50 (1) : 119-129.

[20] Dafre AL, Reischl E. High hemoglobin mixed disulfide content in hemolysates from stressed shark. Comp Biochem Physiol, 1990, 96 (2) : 215-219.

[21] Guidotty G, Konigsberg W, Craig L C. On the dissociation of normal adult human hemoglobin. Proc Natl Acad Sci U S A, 1963, 50: 774-782.

[22] Macleod R M, Hill RJ. Demonstration of the hybrid hemoglobin 2 A A S. J Bio Chem, 1973, 248 (1) : 100-103.

[23] Parkhurst LJ, Goss DJ. Ligand binding kinetic studies on the hybrid hemoglobin alpha (human) : beta (carp) : a hemoglobin with mixed conformations and sequential conformational changes. Biochemistry, 1984, 23 (10) : 2180-2186.

[24] 苏燕,邵国,睢天林,等. 淀粉-琼脂糖凝胶混合保温法分析鲤鱼血红蛋白的粗结构. 包头医学院学报, 2007, 23 (4) : 343-344.

第四十八章 几种鱼类血红蛋白不与人血红蛋白相互作用及其进化意义

苏 燕 邵 国 睢天林 秦文斌

(包头医学院 生物化学与分子生物学教研室，包头 014010)

摘 要

目的：观察鱼类血红蛋白与人血红蛋白及其相互之间的相互作用。

方法：利用淀粉-琼脂糖混合凝胶交叉电泳的方法。

结果：人血红蛋白穿过或兜过四种鱼类血红蛋白时并没有出现通常相互作用(简称互作)所观察到的"峰"或"谷"的改变。

结论：鱼类血红蛋白与人血红蛋白之间没有互作，这一结果对了解血红蛋白的四级结构及分子进化都具有一定的意义。

关键词 鱼类；血红蛋白；相互作用；进化

1 前言

以前，我们利用交叉电泳证明，在红细胞外，人血红蛋白 A_2 与血红蛋白 A 可以发生相互作用[1]。后来推测，这种相互作用可能是各种血红蛋白之间的普遍规律。目前通过实验已经证实哺乳类中的家兔、大小鼠及几种鸟类的血红蛋白均可同人血红蛋白发生相互作用[2-5]。为了进一步检验这种设想是否正确，我们选择市场上常见的四种鱼类：鲤鱼、鲫鱼、金鱼和泥鳅，用它们的红细胞制备溶血液，然后与人的红细胞溶血液进行交叉电泳实验，观察它们之间是否发生相互作用。

2 材料与方法

2.1 材料 硼酸缓冲液(pH 9.0)，1×TEB 缓冲液(pH 8.6)，氨基黑 10B 染色液，5%冰醋酸溶液。正常成人血样来自附属医院血库，各种鱼类血样来自水产品市场。

2.2 方法

(1) 新鲜鱼抗凝血的采集：采用心脏穿刺取血[1]，肝素抗凝。

(2) 溶血液制备：取人或鱼的抗凝血，离心除去血浆，然后用 0.9%生理盐水(泥鳅用 0.7%生理盐水)洗涤红细胞 3~4 次，最后向红细胞层加入 1~1.5 倍体积的蒸馏水及 0.5 倍体积的四氯化碳，激烈振荡后离心，取上层红色溶血液备用。

(3) 电泳分析：利用我室传统的淀粉-琼脂糖混合凝胶电泳。加样方法根据需要分为两种：单排加样，即在板的一端距边缘 1.5~2cm 处或中央并排加样，使加样线处于同一水平。双排加样，适用于快泳血红蛋白与相对慢泳血红蛋白之间的相互作用。具体又分为以下两种：穿过实验，前排(阳极侧)为一条长加样线，加慢泳血红蛋白，后排(阴极侧)为一短加样线，加快泳血红蛋白，相互拉开一定距离并在两边配上不交叉对照样品；兜过实验，前排(阳极侧)为一条短加样线，加慢泳血红蛋白，后排(阴极侧)为一长加样线，加快泳血红蛋白，相互拉开一定距离，并在两边配上不交叉对照样品。

(4) 染色方法及电泳结果的保存：将电泳完毕的胶板置于氨基黑 10B 染色液中染色 1 小时后转入 5%的冰醋酸中反复漂洗，直至胶板底色清亮透明，然后取出，观察结果、拍照、阴干保存。

3 结果

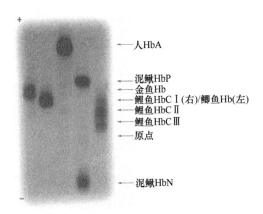

图 48-1 人血红蛋白与四种鱼类血红蛋白电泳位置的相对比较

3.1 人血红蛋白与四种鱼类，即鲫鱼(crucian carp)、鲤鱼(common carp)、泥鳅(loach)及金鱼(goldfish)血红蛋白电泳位置的相对比较。使用淀粉-琼脂糖混合凝胶电泳，采用单排加样法，加样线位于板中央，样品均为溶血液，电压为 6～8V/cm，电泳为 3～4 小时，电泳结果见图 48-1。

由此图可以看到鱼类血红蛋白与人血红蛋白相比具有如下特点：①四种鱼的血红蛋白组分及电泳行为各不相同，鲤鱼具有三种主要的血红蛋白组分，由阳极侧向阴极侧分别命名为 HbCⅠ、HbCⅡ及HbCⅢ。泥鳅有两种主要的血红蛋白组分，它们对称地分布于加样线的阴阳极两侧，分别命名为HbP(positive)和HbN(nagative)。金鱼和鲫鱼均只有一个主要的血红蛋白组分。②四种鱼的血红蛋白与人血红蛋白 A 相比均为相对慢泳血红蛋白，电泳速度比较为人 HbA＞泥鳅 HbP＞金鱼 Hb＞鲫鱼 Hb=鲤鱼 HbCⅠ＞鲤鱼 HbCⅡ＞鲤鱼 HbCⅢ＞泥鳅 HbN。根据上述特点我们便可以选用适当的交叉电泳方法来分析各种血红蛋白间的相互作用。

3.2 鱼类血红蛋白与人血红蛋白之间的相互作用

(1) 鲤鱼血红蛋白与人血红蛋白之间的相互作用。分别采用双排穿过和兜过实验，让快泳人血红蛋白穿过或兜过相对慢泳鲤鱼血红蛋白。电压为 6～8V/cm，通电 5 小时左右，结果见图 48-2。

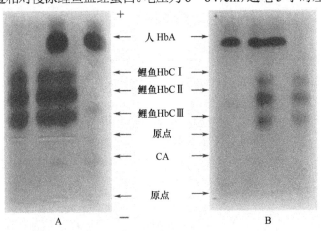

图 48-2 鲤鱼血红蛋白与人血红蛋白之间的相互作用
A. 穿过实验；B. 兜过实验

由上述两图可见,人血红蛋白 A 在穿过和兜过鲤鱼血红蛋白后,区带位置及形状均未发生变化,说明两者之间可能没有发生相互作用。

(2) 泥鳅血红蛋白与人血红蛋白之间的相互作用。分别采用双排穿过和兜过实验,加样线位于板中央,让快泳人血红蛋白穿过或兜过相对慢泳泥鳅血红蛋白。电压为 6～8V/cm,通电 5 小时左右,结果见图 48-3。

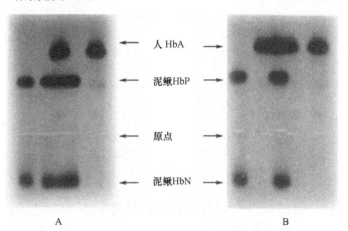

图 48-3 泥鳅血红蛋白与人血红蛋白之间的相互作用
A. 穿过实验;B. 兜过实验

可见,人血红蛋白在穿过和兜过泥鳅的两个血红蛋白组分后,二者的血红蛋白电泳区带与对照相比均未发生改变,说明两者之间可能未发生相互作用。

(3) 鲫鱼血红蛋白与人血红蛋白间的相互作用。分别采用双排穿过和兜过实验,让快泳人血红蛋白穿过或兜过相对慢泳鲫鱼血红蛋白。电压为 6～8V/cm,通电 5 小时左右,结果见图 48-4。

由图 48-4 可知,人血红蛋白在穿过和兜过鲫鱼血红蛋白后,二者的血红蛋白电泳区带与对照相比均未发生改变,说明人血红蛋白与鲫鱼血红蛋白之间可能也不存在相互作用。

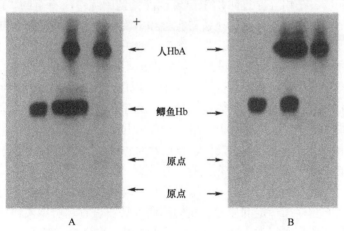

图 48-4 鲫鱼血红蛋白与人血红蛋白之间的相互作用
A. 穿过实验;B. 兜过实验

(4) 金鱼血红蛋白与人血红蛋白间的相互作用。分别采用双排穿过和兜过实验,让快泳人血红蛋白穿过或兜过相对慢泳金鱼血红蛋白,结果见图 48-5。

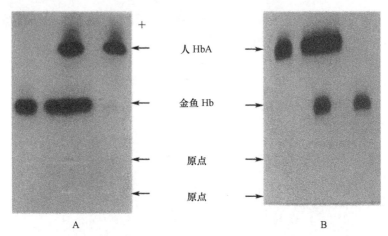

图 48-5 金鱼血红蛋白与人血红蛋白之间的相互作用
A. 穿过实验；B. 兜过实验

由图 48-5 可知，人血红蛋白在穿过和兜过金鱼血红蛋白后，二者的血红蛋白电泳区带与对照相比均未发生改变，说明金鱼血红蛋白与人血红蛋白之间可能也不存在相互作用。

4　讨论

在我们以前的实验中已经明确证明，在电场中人血红蛋白 A 和 A_2 可以发生相互作用，很可能是一种"一过性对称杂交"[1]。在此基础上，我们又发现人血红蛋白与一些哺乳动物如大小鼠、家兔等血红蛋白之间也有这种相互作用[2-4]。因此，我们曾设想，这种相互作用也许是各种血红蛋白间相互作用的普遍规律。但在按照原来计划进行鱼类血红蛋白与人血红蛋白相互作用的实验时，出现了与过去设想迥然不同的情况。在完全相同的实验条件下，同时比较研究了鸟类血红蛋白或鱼类血红蛋白与人血红蛋白的相互作用。实验结果表明，鱼类血红蛋白与人血红蛋白在相互穿过时均未出现以前见到的互作所特有的"波浪形区带"，说明两者之间可能没有发生相互作用。一系列实验和多次重复的结果，都一再证明二者在相互作用方面存在明显差异[5]。此项结果不支持作者以前关于"各种血红蛋白间都可能发生相互作用，即血红蛋白互作的普遍性"的设想，从而提出不同种类动物间血红蛋白相互作用可以有差异。在这种情况下，我们有可能提出一个血红蛋白分子进化方面的新问题。

过去谈血红蛋白的分子进化，多数是讲各种血红蛋白在氨基酸序列方面的同源性。进化过程相近者，同源性较大。这些都属于一级结构或者化学结构范畴。现在，我们是从立体结构方面，特别是从相互作用、交换二聚体、一过性对称杂交方面进行比较研究，并且发现哺乳类、鸟类血红蛋白与鱼类血红蛋白之间存在区别。

现在可以认为，在生物进化过程中，特别是脊椎动物的进化过程中，鸟类和鱼类之间在血红蛋白分子进化方面存在一定的差异。我们的研究还在进行，初步实验证明，这个差异可能还要再深入到两栖类与爬行类之间。因此，我们推断，血红蛋白的互作可能在羊膜动物和非羊膜动物之间存在分水岭，这将给血红蛋白分子进化提供新的知识。

参 考 文 献

[1] 秦文斌. 红细胞外血红蛋白 A 与血红蛋白 A_2 之间的相互作用. 生物化学杂志, 1991, 7 (5) : 583-585.
[2] 秦良谊, 秦文斌. 几种哺乳动物血红蛋白与人血红蛋白 A_2 之间的相互作用. 生物化学杂志, 1997, (专刊) : 37.
[3] 孟峻, 睢天林, 秦文斌. 家兔血红蛋白与人正常、异常血红蛋白之间相互作用的研究. 生物化学杂志, 1997, (专刊) : 36.
[4] 武莎莎, 秦文斌, 睢天林. 大小鼠血红蛋白与人正常、异常血红蛋白之间相互作用的研究. 生物化学杂志, 1997, (专刊) : 37.
[5] 邵国, 睢天林, 秦文斌. 几种鸟类血红蛋白与人血红蛋白之间相互作用的研究. 包头医学院学报, 2000, 16 (2) : 148.

第四十九章　美国锦龟血红蛋白与人血红蛋白相互作用的生物信息学和计算机模拟研究

于　慧　王占黎　秦文斌　睢天林

(包头医学院　血红蛋白研究室，包头　014010)

摘　要

目的： 研究美国锦龟(western painted turtle)血红蛋白与人血红蛋白在淀粉-琼脂糖交叉电泳行为中发生相互作用的机制。

方法： 采用 Accelrys 公司 Discovery Studio(DS)Modeling1.1 和 Insight Ⅱ2000 软件，利用生物信息学和计算机模拟的方法，对美国锦龟血红蛋白的序列进行分析，预测其二级结构，对其空间结构进行同源模建，并计算表面静电势分布，同时将美国锦龟血红蛋白与人血红蛋白的一级结构、二级结构、三级结构及静电势分布进行比较分析。

结果： 美国锦龟血红蛋白 B 链与人血红蛋白有 65.8% 的同源序列，而 AD 和 AA 链与人的血红蛋白 A 链的同源性分别为 61.0%和 67.4%。二者具有相似的空间结构和静电势分布特点。

结论： 美国锦龟血红蛋白与人血红蛋白在淀粉-琼脂糖交叉电泳中发生相互作用是基于结构的同源性。

关键词　生物信息学；计算机模拟；美国锦龟；血红蛋白；相互作用

1　前言

21 世纪是生命科学的时代，也是信息时代，生物信息学和计算机模拟是在数学、计算机科学和生命科学的基础上形成的一门新型交叉学科，是对生物信息的收集、加工、传播、分析和解析的科学[1]，能够对蛋白质的结构进行预测和模拟，它的发展为分子生物学的研究提供了新的研究思路和强有力的工具。美国锦龟是低等的羊膜类动物，它的血红蛋白能够与人的血红蛋白在交叉电泳中发生相互作用[2]。目前，美国锦龟血红蛋白的一级结构已经确定，至今还没有得到晶体结构，要阐述蛋白质相互作用的分子基础，只有氨基酸顺序的知识是不够的，必须知道它们的三维结构。本文运用生物信息学和计算机模拟的方法，对美国锦龟血红蛋白的结构和功能及与人血红蛋白的相互作用和机制进行分析。

2　材料和方法

在本文的各个研究过程中都用到了 Discovery Studio(DS)Modeling1.1 和 Insight Ⅱ 2000。

2.1　序列比对和二级结构预测　美国锦龟血红蛋白的成分有主要成分 HbA 和次要成分

HbD 两种[3]。从 PDB 数据库中获取人血红蛋白(PDB 号：2HHB)的序列和结构信息，从 NCBI 数据库中获取美国锦龟血红蛋白α^A、α^D 和β链的序列信息(编号分别为 HATTAP、HATTDP、HBTTP)。用 Discovery Studio(DS)Modeling1.1 中的 Align123 对人和美国锦龟血红蛋白序列进行比对。对人的血红蛋白α链和β链及美国锦龟血红蛋白α^A、α^D 和β链的二级结构分别进行预测。

2.2 蛋白质准备 几种动物及人的血红蛋白的三维结构都已通过 X-射线晶体衍射技术得到解析。本文选择 2HHB，其分辨率为 1.74Å。按照 PDB 文件中的说明，调整蛋白质分子的键级，加氢，设定环境 pH 为 8.6。用 DS CHARMm 对氢原子进行能量最小化，并对该蛋白质的其他部分做和谐约束，接着解除对蛋白质的约束，再进行一次相似的最小化过程。结果显示蛋白质和血红素之间产生了微小的位置变化(RMSD 小于 1Å)，从而优化了它们彼此之间及与蛋白间的相互作用。

2.3 同源模建 本文采用 InsightⅡ分子模拟软件包中的 MODELER 模块，以人血红蛋白的晶体结构(2HHB)为参考蛋白，通过同源蛋白搭建得到美国锦龟血红蛋白的结构。初始模型首先用最陡下降法优化 2000 步，然后采用共轭梯度法优化，直到能量 RMS 偏差小于 0.001 kcal/mol·Å。在 298K 下用分子动力学平衡 10ps，模拟步长为 1fs。动力学优化得到的构象用分子力学优化 5000 步后得到美国锦龟血红蛋白的最终构象。

2.4 静电势计算 静电势的计算采用 InsightⅡ分子模拟软件包中的 Delphi 模块。通过有限差分方法求解 Poisson-Boltzmann 方程来得到分子的静电势分布。Delphi 可以把蛋白质和它周围的空间分为不同的介电区域，同时能够有效考虑溶液中离子强度对蛋白质静电分布的影响。

3 结果

3.1 序列比对及氨基酸组成、性质和二级结构分析 对美国锦龟血红蛋白与人血红蛋白的序列进行比对(图 49-1，图 49-2，图 49-3)，其中棕色代表两条相比较的链的氨基酸完全相同，灰色代表氨基酸的性质相似。由图可知，如果仅考虑完全相同的氨基酸，美国锦龟血红蛋白β链与人血红蛋白有 65.8%的同源序列，而α^D和α^A链与人的血红蛋白α链相比较，同源性分别为 61.0%和 67.4%；如果把性质相似的氨基酸也考虑在内，则美国锦龟血红蛋白β链与人血红蛋白有 83.6%的同源性，而α^D和α^A链与人的血红蛋白α链相比较，同源性分别为 76.6%和 79.4%。同时针对美国锦龟血红蛋白各肽链的组成及性质进行分析，并对其二级结构分别进行预测(表 49-1，表 49-2)，发现美国锦龟血红蛋白各肽链的理论等电点、极性氨基酸的数目、分子量及二级结构等因素与人的血红蛋白有一定区别。

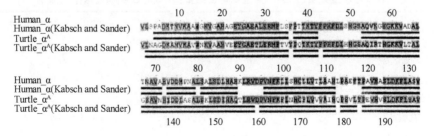

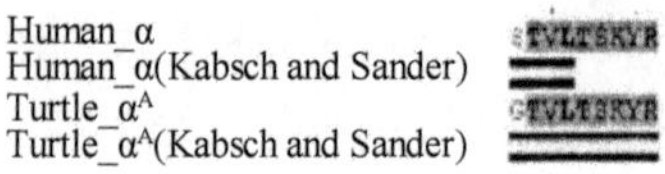

```
Human_α
Human_α(Kabsch and Sander)
Turtle_αᴬ
Turtle_αᴬ(Kabsch and Sander)
```

图 49-1　人血红蛋白α链与美国锦龟血红蛋白αᴬ链序列比对和各自的二级结构特征

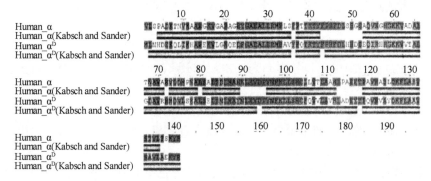

图 49-2　人血红蛋白α链与美国锦龟血红蛋白αᴰ链序列比对和各自的二级结构特征

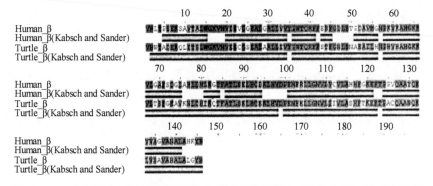

图 49-3　人血红蛋白β链与美国锦龟血红蛋白β链序列比对和各自的二级结构特征

表 49-1　人血红蛋白α链与美国锦龟血红蛋白αᴬ链、αᴰ链的性质和二级结构分析

项目	人血红蛋白α链	美国锦龟血红蛋白αᴬ链	美国锦龟血红蛋白αᴰ链
分子量	15 184	15 641	16 241
酸性氨基酸数目	12	13	19
碱性氨基酸数目	14	14	14
α链螺旋数目	7	6	5
β链折叠数目	0	0	0
理论等电点	8.07	8.07	6.20

表 49-2　人血红蛋白β链与美国锦龟血红蛋白β链的性质和二级结构分析

项目	人血红蛋白β链	美国锦龟血红蛋白β链
分子量	15 998	16 303
酸性氨基酸数目	16	12
碱性氨基酸数目	15	13
α链螺旋数目	10	5
β链折叠数目	0	0
理论等电点	6.74	7.95

3.2 结构模建 利用 MODELER 软件包，以人血红蛋白为模板，自动构建美国锦龟血红蛋白的模型，模建获得的初始结构利用 DISCOVER 程序包，经过分子力学优化、分子动力学模拟退火得到其优势构象(图 49-4，图 49-5)。由图可以直观地比较两种血红蛋白空间结构的差异。宏观构象表明，美国锦龟血红蛋白与人血红蛋白有相似的结构，但把美国锦龟血红蛋白的$\alpha^A\beta$、$\alpha^D\beta$二聚体分别与人的$\alpha\beta$二聚体分别进行比较，它们之间的结构还是有少许的差别，包括美国锦龟血红蛋白的α^A、$\alpha^D\beta$二聚体之间也有部分差异，这种结构的差别是由各自的氨基酸组成的差异所决定的。

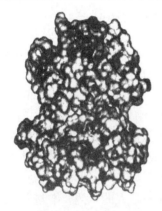 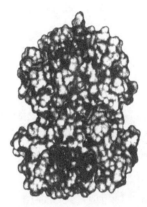

图 49-4　人血红蛋白$\alpha_1\beta_1$二聚体模型及静电势分布　　图 49-5　美国锦龟血红蛋白$\alpha^A\beta$二聚体模型及静电势分布

3.3 静电势分布 除了保守残基外，在每个二聚体中，静电势的分布还会受到非保守残基的影响，同时在形成二聚体时分子构象的变化对分子表面的静电分布也会有一定的影响。图 49-4、图 49-5、图 49-6 分别为人血红蛋白$\alpha\beta$二聚体、美国锦龟血红蛋白$\alpha^A\beta$二聚体和美国锦龟血红蛋白$\alpha^D\beta$二聚体在 pH 8.6 情况下的静电势分布。整体的静电势分布和保守残基的静电势分布比较具有一定的相似性。但在二聚体接触界面上，人血红蛋白和美国锦龟血红蛋白在一些区域存在一定的差异。

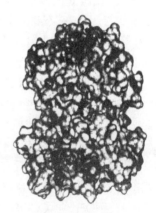

图 49-6　美国锦龟血红蛋白$\alpha^D\beta$二聚体模型及静电势分布

4 讨论

美国锦龟血红蛋白与人的血红蛋白相比较，对于具有重要功能的保守区，美国锦龟血红蛋白与人血红蛋白有高度的一致性。例如，高度保守的几个氨基酸中，疏水氨基酸有 B10-Leu、FG5-Val、CD1-Phe、F4-Leu 和 HC2-Tyr 等，极性或两性的氨基酸有 B6-Gly、B12-Arg、F8-His 和 G1-Asp 等，大家知道，血红蛋白 B6 和 E8 位置都是 Gly，这是由于它的体积小，大于氢原子的侧链将不允许螺旋彼此接近地像原来那么紧密。B10 的 Leu 与血红素右侧形成疏水基团，而 F4-Leu 在左侧形成疏水基团，B12-Arg 与邻位亚基的 GH5 和 H5 形成 CO 键，与同一亚基的 B8 形成盐桥，CD1-Phe 在血红素周围，与其他疏水残基形成疏水的血红素口袋，防止水分子进入。HC2-Tyr 与 FG5-Val 形成氢键稳定的 H 螺旋。对于氨基酸的组成，从结果可以看到，美国锦龟血红蛋白α^D链与人的α链及其他链的差别比较大，它的理论等电点只有6.20，而它的血红蛋白β链和α^A链的理论等电点分别为7.95和8.07，从Asp和Glu及Lys和Arg的数目来看，其α^A链、β链与人的基本相近。例如，美国锦龟血红蛋白β链与人的进行比较，即使有氨基酸的不同，也是性质相似的Asp和Glu互相替代，对功能没有太大的影响。例如，B6Glu-Asp、B7Asp-Glu、B23Asp-Glu、B74Asp-Glu、B94Asp-Glu，但α^D链的酸性氨基酸数目明显多于其他二者，为19个。

血红蛋白α1C：β2FG接触点有两种可以区别的构象，此处相当于构象跃迁的"开关"，而α1FG：β2C接触点的位移很小，发现前者的大多数氨基酸恒定不变。从结果可以看到，美国锦龟β链的43位氨基酸是Ser，而不是Asp，相当于少了1个带负电的残基，因此，与人血红蛋白比较，推测其易于解离成二聚体。参与正常波尔效应的氨基酸为β94-Asp、β146-His、α1-Val，美国锦龟血红蛋白与人的基本相同，区别在于β94位上是Glu，而不是人血红蛋白Asp，由于这两种氨基酸性质相似，因此很少影响其功能。但美国锦龟α^D链的第一位氨基酸是Met，不是Val，研究已经发现，鸟类、几种龟及有鳞目动物的α珠蛋白有α^D和α^A两种，而且证明HbD较HbA有较高的氧亲和力和弱的协同效应，Met替代Val也许就是原因之一[4]。2，3-DPG在人的血红蛋白亲和力方面具有别构效应，2，3-DPG分子与每个β链的三个带正电荷基团的氨基酸NA1-Val，EF6-Lys，H21-His互相吸引，形成六个盐桥，通过一级结构比较我们知道，美国锦龟血红蛋白的这些氨基酸残基与人的一致，推测它的这个结合部位与人血红蛋白有相似的结构特征。

对于美国锦龟血红蛋白与人的血红蛋白在电泳过程中的相互作用问题，我们从几个方面进行讨论。假设血红蛋白是四聚体之间的作用。首先，疏水相互作用通常认为是蛋白质折叠的主要动力之一[5]，它在维持蛋白质三级结构方面占有突出的地位，蛋白质折叠总是倾向于把疏水残基埋藏在分子内部，对于血红蛋白在溶液中聚合物的形成过程，疏水相互作用无疑也是维持多聚体稳定性的因素之一，但我们对疏水性相互作用还缺乏足够的理论基础，还很难从定量的角度直接衡量或认知疏水性相互作用，但一般认为蛋白质中或蛋白质间的疏水相互作用常常与非极性基团密切相关。评价蛋白质的疏水相互作用的强弱当然要考虑非极性基团在接触表面上所占有的比例和疏水区域能否形成好的表面接触。从以上结果我们可知道，血红蛋白包含很大比例的疏水氨基酸残基，但主要折叠在分子内部，用以形成疏水的血红素口袋，因此亚基之间疏水相互作用并不是很强。如果单纯依赖疏水相互作用，很难形成稳定的聚合物。另外，静电作用也参与蛋白质的相互作用，蛋白质表面可发生化学反应的基团一般是—NH$_2$、—OH、—COOH和—SH等[6]。在美国锦龟血红蛋白与人血红蛋白的比较中，大部分氨基酸属于保守氨基酸，这部分氨基酸保守序列会产生基本相似的静电势分布，而血红蛋白之间静电势分布的差别则主要是由少量非保守残基体现出来的。一般而言，pH和蛋白的静电分布直接相关，通过考察聚合物中静电势分布可以得到血红蛋白之间静电相互作用的重要信息，设想把人的血红蛋白与美国锦龟的血红蛋白分别在酸碱度不同的缓冲体系中作用，它们还有无相互作用呢？由于这种相互作用是在电泳过程中发生的，而且表现一过性行为，所用的电泳TEB缓冲液的pH为8.6，根据理论等电点，美国锦龟血红蛋白此时与人血红蛋白一样，都带有负电荷，因此两种血红蛋白之间应该存在静电斥力而不是相互作用。

所以，我们还必须从血红蛋白结构方面来讨论。已经知道，尽管人的血红蛋白在生理条件下以稳定的四聚体形式存在，但当血红蛋白不在红细胞内而出现在血浆时，则有明显的二聚体存在，即血红蛋白的$\alpha_1\beta_1$结合力要比$\alpha_1\alpha_2$亚基之间的结合能力强，因此可以推测，我们在进行血红蛋白溶血液的制备过程中，很大程度上人为破坏了四级结构。已知蛋白分子间存在空间及电荷的互补性是分子之间发生相互作用的关键因素。研究发现，美国锦龟血红蛋白与人血红蛋白具有相似的结构特点和静电势分布，因此，二聚体之间有发生相互作用的结构基础。因为影响电泳速度的因素有很多，包括电荷、分子量、蛋白质形状等，正是由于人的血红蛋白与美国锦龟的血红蛋白带电荷数目及分子量的差别，在电泳过程中形成了这种结合力与电泳力的较量，而电泳力占优势的情况下，表现出一过性行为。但美国锦龟血红蛋白与人血红蛋白二聚体之间依靠什么力进行相互作用呢？可以肯定，如果美国锦龟血红蛋白与人血红蛋白 $\alpha\beta$ 二聚体之间确实是基于结构同源而发生的相互作用，那么，这种作用方式与人血红蛋白二聚体之间的作用方式一定是相似的。非羊膜动物血红蛋白不与人血红蛋白发生相互作用，可能就是由于结构的不相吻合，这也许就是进化的结果。

参 考 文 献

[1] 祁美艳，李伟. 生物信息学的现状与展望. 日本医学介绍, 1996, (8)：371.
[2] 王占黎，睢天林，秦文斌. 几种爬行动物血红蛋白与人类血红蛋白之间相互作用的研究. 上海：中国生物化学与分子生物学学会第八届委员代表大会暨全国学术会议论文摘要, 2000: 311.
[3] Rucknagel K P, Braunitzer G. Hemoglobins of reptiles. The primary structure of the major and minor hemoglobin component of adult Western Painted Turtle (Chrysemys picta bellii). Biol Chem Hoppe Seyler, 1988, 369 (2)：123.
[4] Matsuura MS, Fushitani K, Riggs AF. The amino acid sequence of alpha and beta chains of hemoglobin from the snake, Liophis miliaris. Biol Chem, 1989, 264: 5515.
[5] Tsai CJ, Lin SL, Wolfson JH, et al. Studies of protein-protein interfaces: a statistical analysis of the hydrophobic effect. Protein Sci, 1997, 6: 53.
[6] Blundell TL, Carney D, Gardner S, et al. Knowledge-based protein modeling and design. Eur Biochem, 1988, 172: 513.

第五十章 生物信息学方法推测脊椎动物血红蛋白相互作用机制

王程[1] 隋春红[1] 秦文斌[2] 睢天林[2] 吕士杰[1]

(1. 吉林医药学院，吉林 132013；2. 内蒙古科技大学包头医学院，包头 014010)

摘 要

目的：通过分析脊椎动物血红蛋白(Hb)分子进化过程，解释 Hb 相互作用的现象并推测作用机制。

方法：采用 NCBI、PDB 等在线生物信息学网站及 SMS、ANTH EPROT 5.0、Clustalx 2 0、MEGA 4、Vector NTI 9 等软件包对比脊椎动物门各纲羊膜和非羊膜动物 Hb 氨基酸多序列的同源相似性，查找保守位点，构建分子进化树，预测二级结构，对比三级结构模型，推测 Hb 相互作用现象的发生机制。

结果：氨基酸多序列对比显示，非羊膜动物α链没有保守氨基酸(cAA)，β链有 3 个 cAA，羊膜动物仅α链、β链分别有 27 个和 68 个 cAA；分子进化树显示非羊膜动物 Hb 氨基酸每位点替代值(SpS)远大于羊膜动物($P<0.01$)；预测二级结构分析 Hb 作用发现非羊膜动物α链 40 位氨基酸主要参与无规则卷曲结构，羊膜动物α链 40 位和β链 94 位氨基酸均主要参与α螺旋结构；对比空间结构模型发现羊膜动物α链均有一个苏氨酸(41 位)与β链的组氨酸(98 位)形成氢键，而非羊膜动物不能形成此氢键。

结论：脊椎动物 Hb 相互作用发生的关键可能在于是否存在α链苏氨酸(41 位)与β链组氨酸(98 位)形成的氢键。

关键词 生物信息学；血红蛋白；相互作用；氢键

1 前言

自从发现人血红蛋白(hemoglobin, Hb)A 与 HbA$_2$ 可以发生相互作用[1]，即推测这种相互作用可能是各种血红蛋白之间的普遍规律。但通过实验证实，人类、哺乳纲、鸟纲及大部分爬行纲动物的 Hb 之间可发生相互作用(简称互作)，但不能与小部分爬行纲、两栖纲、鱼纲动物 Hb 发生互作，小部分爬行纲、两栖纲、鱼纲动物 Hb 之间也不能发生互作[2-6]。因此推断 Hb 互作发生的分水岭可能存在于羊膜动物和非羊膜动物之间，互作的机制在于 Hb 分子进化过程中珠蛋白的氨基酸不断发生置换来改变空间结构，满足物种对环境的适应有关。本研究采用生物信息学方法，通过分析 Hb 分子进化过程，解释 Hb 互作的现象并推测作用机制。

2 材料与方法

2.1 材料 随机选择脊椎动物门各纲代表动物氨基酸序列及空间结构图像，材料来源于 NCBI 数据库(http://www.ncbi.nlm.nh.gov/)和 PDB 数据库(http://www.rcsb.org/)。如下所列：七鳃鳗(lamprey, Hb I: BAF47284, Hb II: BAF47285, PDB ID: 1UC3), 盲鳗(hagfish, HbA: 1IT2_A, HbB: 1IT2_B), 鲨(shark, Hbα: P07408, Hbβ: P07409, PDB ID: 1GCV), 电鳐(torpedinifomies, Hbα-1: P20244, Hbβ-1P20246), 肺鱼(lungfish, Hbα: P02020, Hbβ: P02138), 空棘鱼(coelacanth, Hbα: P23740, Hbβ: P23741), 金枪鱼(tuna, PDB ID: 1V4U), 鳟鱼(trout, PDB ID: 1OUT), 鲤鱼(car, Hbα: P02016, Hbβ-A/B: P02139), 大黄鱼(crocea, Hbα: AAV52697, Hbβ: AAV91971, PDB ID: 1SPG), 蝾螈(salamander, Hbα-1: P06640), 蟾蜍(bufonidae, Hbα: AAN41264, Hbβ-3: P02011), 蛙(rana, Hbα-3: P02011, Hbβ: P02135), 蛇(snake, Hbα: P41331, Hbβ: P41332), 蜥蜴(lizard, Hbα: AAB20248, Hbβ: AAB20247), 龟(turtle, Hbα-A: AAB33014, Hbβ: AAB33015, PDB ID: 1V75), 鳄(crocodile, Hbα: P02131, Hbβ: P02130), 鹦鹉(parrot, PDB ID: 2ZFB), 鸡(chicken, Hbα-1: NP_001004376, Hbβ: NP_001075173, PDB ID: 1HBR), 鸵鸟(ostrich, Hbα-A: P01981, Hbβ: P02123), 鹅(goose, Hbα-A: P689454, Hbβ: P01991, PDB ID: 1A4F), 鸭嘴兽(platypus, Hbα: P01979, Hbβ: P02111), 袋鼠(kangaroo, Hbα: P01975, Hbβ: P02106), 蝙蝠(bat, Hbα: AAB24575, Hbβ: AAB24574), 兔(rabbit, Hbα: 711680A, Hbβ: 660912A), 小鼠(mouse, Hbα: PI 1757, Hbβ: PI 1758), 大鼠(rat, Hbα: PO1943, Hbβ: P02090), 猫(cat, Hbα: P07405, Hbβ: P68871), 犬(dog, Hbα: P07405, Hbβ: P04244, PDB ID: 2B7H), 狼(wolf, PDB ID: 1FHJ), 鲸(whale, Hbα: P18978, Hbβ: PI8984), 马(horse, Hbα: P01958, Hbβ: P02062, PDB ID: 1GOB), 驴(donkey, PDB ID: 1SOH), 猪(pig, PDBED: 1QPW), 牛(cow, Hbα: P01966, Hbβ: P02070, PDBID: 1G09), 黑猩猩(chimpanzee, Hbα: P69907, Hbβ: P68873), 猴(monkey, Hbα: P63108, Hbβ: P02030), 人(human, Hbα: P69905, Hbβ: P68871, PDB ID: 1A3N), 人-小鼠互作(PDB ID: 1JEB)。

2.2 方法 应用 SMS 软件包，以 100%相似率，亮氨酸(leucine, L)、异亮氨酸(isoleucine, I)、缬氨酸(valine, V), 色氨酸(tryptophan, W)、苯丙氨酸(phenylalanine, F)、酪氨酸(tyrosine, Y), 赖氨酸(lysine, K)、精氨酸(arginine, R)、组氨酸(histidine, H), 天冬氨酸(asparticacid, D)、谷氨酸(glutamic acid, E)、甘氨酸(glycine, G)、丙氨酸(alanine, A)、丝氨酸(serine, S), 脯氨酸(praline, P), 半胱氨酸(cysteine, C), 苏氨酸(threonine, T)、天冬酰胺(asparagine, N)、谷氨酰胺(glutamine, Q)、甲硫氨酸(methionine, M), 对以上八个类似性氨基酸组，对比羊膜和非羊膜动物 Hb 氨基酸多序列的同源相似性，查找保守位点，构建分子进化树。应用 Clustalx 2.0、MEGA 4 软件包构建进化树，SPSS 11.5 统计羊膜和非羊膜动物氨基酸每位点替代值(substitutions per site, SpS)的差异, $P<0.01$，有意义。应用 ANTHEPROT 5.0 软件包预测 Hb 二级结构，比较 Hb 亚基接触面上 $\alpha_{N_{首位}}$、α_{40}、α_{126}、$\alpha_{140/C_{末位}}$、β_{94}、$\beta_{146/C_{末位}}$ 8 个氨基酸参与形成的二级结构。应用 Vector NTI 9 软件包分析 α_{39-42}、β_{94-98} 氨基酸支链的氢键空间结构。

3 结果

3.1 同源性分析 氨基酸多序列对比显示，非羊膜动物以蛙为参照，α 链没有保守氨基酸(conservative amino acid，CAA)，β链有3个CAA：131(V)、135(L)、139。羊膜动物以人为参照，α链有27个CAA：40(L)、56(V)、59(H)、60(G)、62(K)、63(V)、66(A)、70(A)、75(D)、82(S)、84(L)、85(H)、87(L)、88(H)、93(R)、94(V)、95(D)、98(N)、99102(L)、111(A)、113(H)、125(S)、128(K)、133(V)、137(L)、141(Y)、142(R)，β链68个CAA：8(E)、9(K)、12(V)、16(W)、18(K)、19(V)、25(G)、26(G)、29(L)、30(G)、33(L)、34(V)、35(V)、36(Y)、37(P)、38(W)、41(R)、43(F)、46(F)、47(G)、55(V)、58(N)、61(V)、64(H)、65(G)、67(K)、68(V)、69(L)、71(A)、72(F)、73(S)、75(G)、79(L)、80(D)、82(L)、83(K)、86(F)、87(A)、89(L)、90(S)、92(L)、93(H)、94(C)、96(K)、98(H)、99(V)、100(D)、101(P)、104(F)、107(L)、108(G)、111(L)、112(V)、115(L)、116(A)、122(E)、123(F)、131(Y)、132(Q)、133(K)、134(V)、138(V)、139(A)、141(A)、142(L)、143(A)、146(Y)、147(H)。

3.2 分子进化树的构建与分析 对羊膜动物和非羊膜动物的Hbα、β亚基构建分子进化树，如图50-1~图50-4所示。分别对Hbα、Hbβ的氨基酸SpS进行独立小样本t检验，非羊膜动物Hbα、Hbβ氨基酸SpS均显著大于羊膜动物($P<0.01$)，即非羊膜动物Hbα、Hbβ氨基酸变异较羊膜动物大。

3.3 二级结构预测与分析 羊膜动物和非羊膜动物Hb二级结构预测，如图50-5~图50-8所示。Hb亚基之间主要通过8对盐键连接，涉及α_N首位、α_{40}、α_{126}、$\alpha_{140/C}$末位、β_{94}、$\beta_{146/C}$末位8个氨基酸[7]，比较这些氨基酸参与形成的二级结构，发现α_N首位、α_{126}、$\alpha_{140/C}$末位和$\beta_{146/C}$末位5个氨基酸均较保守结构，羊膜动物和非羊膜动物无区别，α_N首位为无规则卷曲(non-regular curl，C)结构或β折叠(β-sheet，S)结构，α_{126}为α螺旋(α-helix，H)结构，$\alpha_{140/末位}$多为β转角结构(β-turn，T)或C结构，$\beta_{146/末位}$多为C结构。α_{40}、β_{94}参与形成的二级结构有了明显变化，但保守于羊膜动物和非羊膜动物内部，非羊膜动物α_{40}主要参与C结构，非羊膜动物参与的二级结构多样化，包括H、C、S结构；羊膜动物α_{40}和β_{94}均参与H结构。

3.4 空间结构分析 根据羊膜动物和非羊膜动物二级结构的变化，观察空间结构模型Hbα$_{35-45}$氨基酸支链，发现羊膜动物Hbα$_{41}$ T与Hb β$_{98}$H形成氢键(T-H氢键)，而非羊膜动物Hb没有，如图50-9。

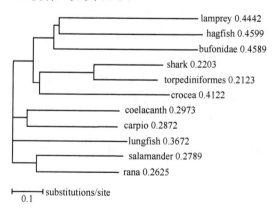

图50-1 非羊膜动物Hbα亚分子进化树

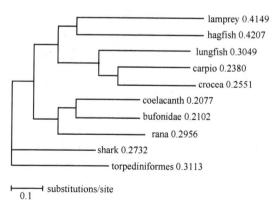

图50-2 非羊膜动物Hbβ亚分子进化树

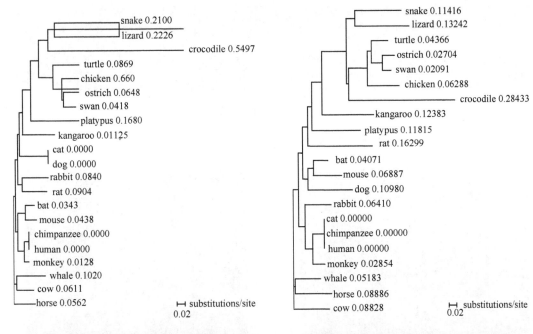

图 50-3　羊膜动物 Hbα 亚分子进化树

图 50-4　羊膜动物 Hbβ 亚分子进化树

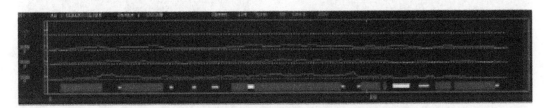

图 50-5　非羊膜动物(鲤鱼)Hbα 亚基二级结构

图 50-6　羊膜动物(黑猩猩)Hbα 亚基二级结构

图 50-7　非羊膜动物(鲨)Hbβ 亚基二级结构

图 50-8 羊膜动物(犬)Hbβ 亚基二级结构

羊膜动物 Hb 互作时，也存在此氢键，如图 50-10。

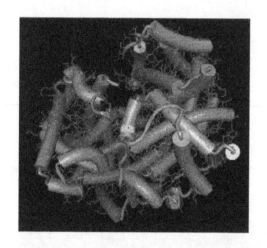 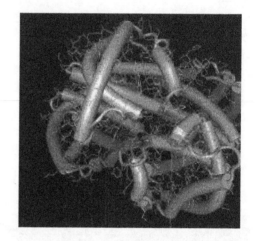

图 50-9 鸡 Hb 空间结构及 T—H 氢键位置　　图 50-10 人-小鼠互作 Hb 空间结构及 T—H 氢键位置

T-H 氢键结构由 T 的羟基与 H 的羰基连接，如图 50-11。

图 50-11 T—H 氢键结构

T—H 氢键具有 A—H … B(O—H … O)形式，键角约 165°，键长为(0.359 ± 0.053)nm，根据 DesirajuGR[8]氢键判定方法为弱氢键或羟基氢桥。

4 讨论

Hb 在体内分布十分广泛,是研究各种动物之间亲缘关系及系统进化上的良好材料。Morris Godman[9]通过比较55种不同种属动物Hb的氨基酸序列后推导出了进化树和Hb的进化规律。本研究室长期以来一直进行Hb的研究,发现了脊椎动物Hb互作的现象,但互作机制一直没有明确的解释。一般认为,Hb 两种能发生互作,这两种Hb须分别解聚成二聚体,并通过离子键及范德华力相互连接,组成新的杂交四聚体分子[10]。裴娟慧[11]推测互作两种Hbα、β亚基接触面上必须同时含有一个"互作结构域"。刘小舟[12]等对已知结构的哺乳纲、鸟纲和鱼纲Hb晶体结构数据进行了比较,发现鸟纲和鱼纲Hb各亚基间的距离比哺乳纲大。α_1-β_1亚基间距离:鱼纲>鸟纲>哺乳纲,而且一级结构比较发现,鸟纲、鱼纲和哺乳纲Hb的残基选择性不同,鱼纲Hb分子表面和亚基之间的界面倾向于选择侧链更大的残基,这可能与非羊膜动物不发生互作有关。Tsai[13]认为蛋白质折叠的主要动力是疏水作用,但单纯依靠疏水作用不能单独使蛋白质发生互作。而静电作用可使蛋白质发生互作,参与的化学基团一般是—NH_3,—COOH,—OH,—SH 和 =O 等[14],所以Hb互作必然有氢键的参与。于慧等[15]计算锦龟与人的静电势分布,发现保守残基处有相似性。用生物信息学方法做假设:①检查多种动物Hb氨基酸同源相似性,发现Hbα、β亚基前段比后段保守氨基酸多,进化树分析非羊膜动物Hbα、β氨基酸变异较羊膜动物大,推测"互作结构域"会存在于羊膜动物Hbα、β的前段保守氨基酸序列中;②Hb 互作应该发生在亚基接触面上,Hb亚基之间主要通过8对由支链氨基(—NH_3)和支链羧基(—COOH)构成的盐键连接,这些连接与其二级结构基础有密切联系,比较连接Hbα、β亚基的氨基酸所参与形成的二级结构,发现非羊膜动物α_{40}主要参与C,羊膜动物α_{40}和β_{94}均主要参与H,推测"互作结构域"会存在于羊膜动物α_{40}和β_{94}所构成的二级结构中。取以上两点推测的交集,观察空间结构模型$Hb\alpha_{35-45}$氨基酸支链,发现羊膜动物$Hb\alpha_{41}$ T 与 $Hb\beta_{98}$ H 形成 T—H 氢键,而非羊膜动物Hb没有。羊膜动物Hb发生互作时,也存在此氢键,因此推测Hb互作与此氢键有关,即Hb互作发生的关键可能在于是否存在 T—H 氢键。

利用 T—H 氢键的推断可以很好地解释一些Hb 互作时的现象。由于只有羊膜动物有 T—H 氢键,故Hb互作只发生在羊膜动物之间。而羊膜动物与非羊膜动物之间也不能发生互作,是因为互作的两种Hb必须同时具有 T—H 氢键,Hb互作实验在淀粉-琼脂糖凝胶电泳中进行。两种Hb互作之后,互作时形成的杂交四聚体分子不能牢固结合,其原因可能是在互作时形成的 T—H 氢键不牢固,同种Hbα、β亚基能很快恢复自身连接的氢键。同种Hbα、β亚基自身连接氢键多,连接力大,故在凝胶中不易分离。Hb互作现象是一个复杂的过程,T—H 氢键作用只是基于结构同源性的一种推测,对Hb互作现象机制的研究还需要大量的先进实验来证明。

参 考 文 献

[1] 秦文斌. 红细胞外血红蛋白A与血红蛋白A_2之间的相互作用. 中国生物化学与分子生物学报, 1991, 7 (5): 583-585.

[2] 秦良谊, 秦文斌. 几种哺乳动物血红蛋白与人血红蛋白A_2之间的相互作用. 生物化学杂志, 1997, 13(专刊): 37.

[3] 邵国, 睢天林, 秦文斌. 几种鸟类血红蛋白与人血红蛋白之间相互作用的研究. 包头医学院学报, 2000,

16 (2) : 148.
[4] 苏燕, 邵国, 眭天林, 等. 几种鱼类血红蛋白与人血红蛋白之间相互作用的研究. 包头医学院学报, 2007, 23 (5) : 452-454.
[5] 焦勇钢, 秦文斌, 眭天林. 软骨鱼系动物与爬行纲动物血红蛋白 A_2 现象的研究. 包头医学院学报, 2005, 21 (2) : 106-108.
[6] 王程, 秦文斌, 眭天林. 鱼纲动物鲤鱼和两栖纲动物牛蛙血红蛋白自身不相互作用的研究及其进化意义. 包头医学院学报, 2006, 22 (3) : 244-247.
[7] 查锡良, 周爱如. 生物化学. 6版. 北京: 人民卫生出版社, 2004: 22-23.
[8] Desiraju G R. Hydrogen bridges in crystal engineering: interactions without borders. Acc Chem Res, 2002, 35 (7) : 565-573.
[9] Goodman M, Moore G W, Matsuda G. Darwinian evolution in the genealogy of haemoglobin. Nature, 1975, 253 (5493) : 603-608.
[10] 秦文斌. 血红蛋白的分子杂交. 生物化学与生物物理学进展, 1975, 2 (3) : 54.
[11] 裴娟慧, 秦文斌, 韩丽莎. 哺乳纲动物绵羊和大白鼠自身血红蛋白相互作用的研究及意义. 包头医学院学报, 2006, 22 (3) : 241-243.
[12] 刘小舟, 李松林, 梁宇和, 等. 鸟类血红蛋白的四级结构和残基选择性特征. 生物物理学报, 2000, 16 (3) : 459-467.
[13] Tsai C J, Lin S L, Wolfson H J, et al. Studies of protein-protein interfaces: a statistical analysis of the hydrophobic effect. Protein Sci, 1997, 6 (1) : 53-64.
[14] Blunde II T, Carney D, Gardner S, et al. Knowledge-based protein modelling and design. Eur J Biochem, 1988, 172 (3) : 513-520.
[15] 于慧, 王占黎, 秦文斌, 等. 美国锦龟血红蛋白与人血红蛋白相互作用的生物信息学和计算机模拟研究. 包头医学院学报, 2005, 21 (3) : 205-209.

第五十一章 血红蛋白的聚丙烯酰胺凝胶交叉互作电泳

高丽君[#] 苏 燕[#] 秦文斌[※]

(包头医学院 血红蛋白研究室,包头 014010)

摘 要

1991年我们利用淀粉-琼脂糖混合凝胶电泳发现血红蛋白(Hb)A_2与HbA_1之间的交叉互作,后来用这种电泳在动物血红蛋白交叉互作方面也做了不少研究。但是,众所周知,淀粉-琼脂糖混合凝胶电泳的分辨率远不如聚丙烯酰胺凝胶电泳,迄今尚无人用它进行交叉电泳。为此,本文用聚丙烯酰胺凝胶进行交叉电泳,观察血红蛋白之间、血红蛋白与白蛋白之间及血红蛋白与PRX之间是否发生交叉互作。结果表明,交叉互作只发生在HbA_2与HbA_1之间。

关键词 血红蛋白;交叉电泳;聚丙烯酰胺凝胶电泳;白蛋白;PRX

1 前言

交叉电泳是日本学者Nakamura S所创建的[1],他的电泳支持体为琼脂,所用标本是血清,研究内容是血浆蛋白之间的交叉互作。我们在研究血红蛋白A_2现象的机制时,发现血红蛋白A_1穿过血红蛋白A_2时后者区带变形,说明二者发生相互作用[2-4]。查阅文献后认为它属于交叉电泳范畴。必须指出,我们的交叉电泳与日本学者不同:支持体为淀粉-琼脂糖混合凝胶,标本是红细胞溶血液,研究内容是血红蛋白之间的交叉互作。众所周知,琼脂电泳的分辨率较低,淀琼电泳稍好,聚丙烯酰胺凝胶电泳的分辨率明则显高于前二者。本文就是利用聚丙烯酰胺凝胶电泳来研究血红蛋白之间的交叉互作,以及血红蛋白与PRX(peroxiredoxin 过氧化物还原酶)之间是否存在交叉互作问题。

2 材料和方法

2.1 材料 正常成人血液来自我研究室成员,PRX购自北京友谊中联生物科技有限公司(下属于Abcam公司),货号ab79947。

2.2 方法

(1) 准备实验 由正常成人红细胞制备溶血液,再用常规淀琼电泳分离HbA_1和HbA_2,见图51-1、图51-2,抠胶后冷冻再融化,高速离心后得到电泳纯的血红蛋白等,然后用于聚丙烯酰胺凝胶交叉电泳(图51-3,图51-4)。

[#] 并列第一作者
[※] 通讯作者:秦文斌,电子邮箱:qwb5991309@tom.com

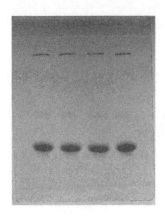 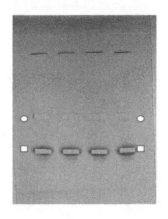

图 51-1　溶血液的单向电泳　　图 51-2　电泳后抠出血红蛋白胶

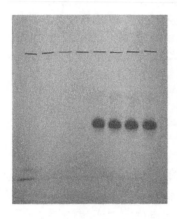 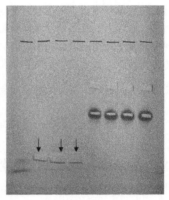

图 51-3　溶血液及血浆的电泳　　图 51-4　电泳后抠出血红蛋白及白蛋白

(2) 交叉电泳

1) 血红蛋白 A_2 与血红蛋白 A_1 之间的交叉电泳。用淀琼电泳制备的血红蛋白进行聚丙烯酰胺凝胶交叉电泳的实验安排，见图 51-5。

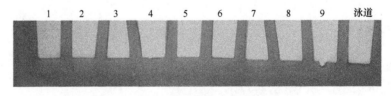

图 51-5　第一次聚丙烯酰胺凝胶加样安排

注释：加样安排如下所示。泳道 1 对照组，先加入 HbA_2。泳道 2 实验组，先加 HbA_2，后补加 HbA_1，让 HbA_1 追过 HbA_2，等待结果。泳道 3 对照组，先加入 HbA_2。泳道 4 对照组，补加入 HbA_1。泳道 5 对照组，补加入 HbA_1+HbA_2。泳道 6 对照组，先加入 HbA_2。泳道 7 实验组，先加 HbA_2，后补加 HbA_1，让 HbA_1 穿过(追过)HbA_2，等待结果。泳道 8 对照组，先加入 HbA_2。泳道 9 对照组，补加入 HbA_1+HbA_2。

2) 血红蛋白 A_2、A_1 与白蛋白之间的交叉电泳。用淀琼电泳制备的血红蛋白及白蛋白进行聚丙烯酰胺凝胶交叉电泳的实验安排，见图 51-6。

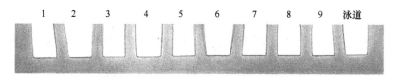

图 51-6 第二次聚丙烯酰胺凝胶加样安排

注释：加样安排如下所示。泳道 1 对照组，先加入 HbA_2。泳道 2 实验组，先加入 HbA_2，后补加 HbA_1，让 HbA_1 追过 HbA_2，等待结果。泳道 3 对照组，先加入 HbA_2。 泳道 4 对照组，先不加，后补加 HbA_1。 泳道 5 对照组，先加入 HbA_2。泳道 6 实验组，先加入 HbA_2，后补加白蛋白，让白蛋白追过 HbA_2，等待结果。泳道 7 对照组，先加入 HbA_2。泳道 8 对照组，先不加，后补加白蛋白。 泳道 9 预备组，加入 PRX，看电泳位置，给第三轮实验做准备

3) 血红蛋白 A_2、A_1、白蛋白与 PRX 之间的交叉电泳。用淀琼电泳制备的血红蛋白、白蛋白与购来的 PRX 进行聚丙烯酰胺凝胶交叉电泳的实验安排，见图 51-7。

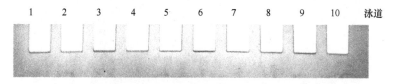

图 51-7 第三次聚丙烯酰胺凝胶加样安排

注释：加样安排如下所示。泳道 1 对照组，先加入 PRX。泳道 2 对照组，后加入 HbA_1。泳道 3 实验组，先加入 PRX，后补加 HbA_1，让 HbA_1 追过 PRX，等待结果。泳道 4 对照组，后加入 HbA_1。泳道 5 对照组，后加入 HbA_2。泳道 6 实验组，先加入 PRX，后补加 HbA_2，让 HbA_2 追过 PRX，等待结果。泳道 7 对照组，后加入 HbA_2。泳道 8 对照组，后加入白蛋白。泳道 9 实验组，先加入 PRX，后补加白蛋白，让白蛋白追过 PRX，等待结果。泳道 10 对照组，后加入白蛋白

3 结果

3.1 血红蛋白 A_2 与血红蛋白 A_1 之间的交叉电泳结果

用淀琼电泳制备的血红蛋白进行聚丙烯酰胺凝胶交叉电泳，结果见图 51-8。

由图 51-8 可以看出，与泳道 1、3、6、8 先加的 HbA_2 相比，泳道 2、7 先加的 HbA_2 因

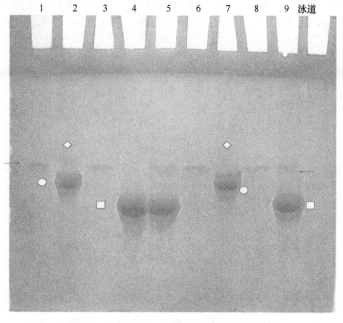

图 51-8 血红蛋白 A_2 与 A_1 之间的交叉电泳

注释：两个箭头之间的是 HbA_2，两个○之间的是 HbA_2 与 HbA_1 互作产物，两个□之间的是 HbA_1

HbA$_1$的穿过而向前(阳极)移动(见◇处)。与此对应，泳道 4 的 HbA$_1$，它没有穿过先加的 HbA$_2$，独自泳向前方(阳极)。同样，泳道 5、9 加的是混合物(HbA$_1$+HbA$_2$)，彼此分开，互不影响。

3.2 血红蛋白 A$_2$、血红蛋白 A$_1$ 与白蛋白之间的交叉电泳结果 用淀琼电泳制备的血红蛋白及白蛋白进行聚丙烯酰胺凝胶交叉电泳的实验，结果见图 51-9。

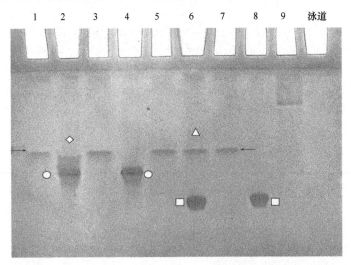

图 51-9 血红蛋白 A$_2$、血红蛋白 A$_1$ 与白蛋白之间的交叉电泳
注释：两个箭头之间的是 HbA$_2$，两个〇之间的是 HbA$_2$ 与 HbA$_1$ 互作产物，两个□之间的是 HbA$_1$

由图 51-9 可以看出，与泳道 1、3、5、7 先加的 HbA$_2$ 相比，泳道 2 先加的 HbA$_2$ 因 HbA$_1$ 的穿过而向前(阳极)移动(见◇处)。与此对应，泳道 4 的 HbA$_1$ 没有穿过先加的 HbA$_2$，独自泳向前方(阳极)。泳道 6，白蛋白穿过 HbA$_2$ 后，彼此分开，互不影响(见△处)。与泳道 8 相比，白蛋白的电泳位置不变。泳道 9 显示 PRX 的电泳位置，供下次实验参考。

3.3 血红蛋白 A$_2$、血红蛋白 A$_1$、白蛋白与 PRX 之间的交叉电泳结果 用淀琼电泳制备的血红蛋白、白蛋白及购买来的 PRX 进行聚丙烯酰胺凝胶交叉电泳，结果见图 51-10。

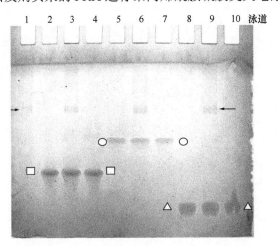

图 51-10 第三次聚丙交叉电泳结果
注释：两个箭头之间的是 PRX，两个〇之间的是 HbA$_2$，两个□之间的是 HbA$_1$，两个△之间的是白蛋白

由图 51-10 可以看出，与泳道 1、6、9 先加的 PRX 和泳道 2、4 后加的 HbA$_1$ 相比，泳

道3中HbA$_1$追过PRX后PRX电泳位置未变。与泳道1、3、9先加的PRX和泳道5、7后加的HbA$_2$相比，泳道6中HbA$_2$追过PRX后PRX电泳位置未变。与泳道1、3、6先加的PRX和泳道8、10后加的白蛋白相比，泳道9中白蛋白追过PRX后PRX电泳位置未变。

4 讨论

1981年我们发现血红蛋白A$_2$现象[5]，这种现象的主要内容是红细胞电泳时释放出来的"血红蛋白A$_2$"与溶血液(红细胞裂解液)电泳时泳出来的血红蛋白A$_2$，二者电泳位置不同。也就是说，完整红细胞里血红蛋白A$_2$的存在状态，不同于红细胞裂解后释放出来的游离血红蛋白A$_2$。推测活体红细胞内血红蛋白A$_2$可能与另外一种物质结合存在。后来我们用双向电泳来分析红细胞"血红蛋白A$_2$"，证明它是血红蛋白A$_2$与血红蛋白A$_1$的结合产物[6]，也就是说，它们之间存在相互作用。红细胞裂解液里也有血红蛋白A$_1$和血红蛋白A$_2$，为什么它们之间没有相互作用？原因不明。但是，我们发现，在淀粉-琼脂糖混合凝胶电场中让快泳的血红蛋白A$_1$穿过慢泳的血红蛋白A$_2$时，后者区带变形，说明在这种情况下二者可以发生相互作用[2]。根据血红蛋白粗结构的知识，我们认为，在电场里成人血红蛋白存在四聚体与2×二聚体之间的平衡，当HbA$_1$二聚体穿过HbA$_2$二聚体时，在二者结合的瞬间形成一种"一过性杂交四聚体"，它将HbA$_2$带向前方。后来，在电场作用下二者又逐渐分开，此时被穿过的HbA$_2$区带变形，反应式如下所示。

$$\left.\begin{array}{l}(\alpha^A\beta^A)_2 \longleftrightarrow 2(\alpha^A\beta^A) \\ \text{HbA}_1 \\ (\alpha^A\delta^{A_2})_2 \longleftrightarrow 2(\alpha^A\delta^{A_2}) \\ \text{HbA}_2\end{array}\right\} \longleftrightarrow \underbrace{2(\alpha^A\beta^A\alpha^A\delta^{A_2})}_{\text{一过性交叉互作产物}} \quad (1)$$

后来，我们用这种电泳又发现HbA$_2$与HbF之间也能交叉互作[7]，反应式如下所示。

$$\left.\begin{array}{l}(\alpha^A\gamma^F)_2 \longleftrightarrow 2(\alpha^A\gamma^F) \\ \text{HbF} \\ (\alpha^A\delta^{A_2})_2 \longleftrightarrow 2(\alpha^A\delta^{A_2}) \\ \text{HbA}_2\end{array}\right\} \longleftrightarrow \underbrace{2(\alpha^A\gamma^F\alpha^A\delta^{A_2})}_{\text{一过性交叉互作产物}} \quad (2)$$

用这种电泳，我们还发现大鼠的四种血红蛋白(HbA、HbB、HbC、HbD)之间能够发生交叉互作。

同样，在脊椎动物，我们发现，羊膜动物血红蛋白可以与人血红蛋白发生交叉互作，而非羊膜动物血红蛋白则不能与人血红蛋白发生交叉互作。众所周知，羊膜动物包括哺乳动物、鸟类和爬行动物，非羊膜动物包括两栖动物、鱼类和无颌纲动物。我们认为，包括人类在内的羊膜动物，它们的血红蛋白按反应式(1)、(2)进行交叉互作；非羊膜动物血红蛋白不与人类血红蛋白交叉互作，由于它们的血红蛋白(HbX)处于单体、多聚体等状态，不能解离为可供交叉互作的二聚体，反应式如(3)所示。

$$\left.\begin{array}{l}(\alpha^X\beta^X)_n \leftrightarrow\!\!\!\times\!\!\!\leftrightarrow 2(\alpha^X\beta^X) \\ \text{HbX} \\ (\alpha^A\delta^{A_2})_2 \longleftrightarrow 2(\alpha^A\delta^{A_2}) \\ \text{HbA}_2\end{array}\right\} \longleftrightarrow 2(\alpha^X\beta^X\alpha^A\delta^{A_2}) \quad (3)$$

一过性交叉互作产物

以上实验和理论都是来自淀粉-琼脂糖混合凝胶电泳。众所周知，这种电泳的分辨率远不如聚丙烯酰胺凝胶电泳，但是，迄今尚无人用后者进行交叉电泳。本文用聚丙烯酰胺凝胶进行交叉电泳，看它有无特殊之处。实验结果表明，HbA_2 与 HbA_1 之间发生交叉互作，白蛋白不与血红蛋白交叉互作。同样，PRX 也不与血红蛋白交叉互作。关于白蛋白不与血红蛋白交叉互作，以前已经用淀粉-琼脂糖混合凝胶电泳证明过[2]，PRX 是在对"红细胞 HbA_2"做质谱分析时证明它与血红蛋白存在相互作用[3]，但在本项研究中未发现它与血红蛋白之间有交叉互作。没有交叉互作不等于没有其他互作，蛋白质之间的相互作用有多种[7]，PRX 与血红蛋白之间的相互作用属于何种类型有待继续研究。

参 考 文 献

[1] Nakamura S. Cross Electrophoresis. New York: Elsevier Amsterdam, 1967.
[2] 秦文斌. 红细胞外 HbA_2 与 HbA 间的相互作用. 生物化学杂志, 1991, 7 (5)：583-587.
[3] Su Y, Gao L J, Ma Q, et al.Interactions of hemoglobin in lived red blood cells measured by the electrophoresis release test. Electrophoresis, 2010, 31: 2913-2920.
[4] 秦文斌. 活体红细胞内血红蛋白的电泳释放. 中国科学生命科学, 2011, 41 (8)：597-607.
[5] 秦文斌, 梁友珍. 血红蛋白 A_2 现象Ⅰ. 此现象的发现及其初步应用. 生物化学与生物物理学报, 1981, 13 (2)：199-205.
[6] 秦文斌. 血红蛋白的 A_2 现象发生机制的研究——"红细胞 HbA_2" 为 HbA_2 与 HbA 的结合产物. 生物化学与生物物理进展, 1991, 18 (4)：286-288.
[7] Fu H A. Protein-Protein Interactions: Methods and Applications. Clifton: Humana Press, 2004.

第六篇　混合互作电泳

前　言

1. "混合互作"，实际就是"相互作用"，为了区别于"交叉互作"而产生。

2. 与交叉互作相比，混合互作更为普遍存在。

3. 首先说一般性观察，我们常常做"双释放"实验，就是比较红细胞和全血的再释放，看二者有什么关系。

4. 双释放可以有多种情况，但大多数结果是全血的再释放不同于红细胞的再释放。

5. 二者不同的原因在哪里？血浆和血浆成分，它们影响着全血里的红细胞，这就是"浆胞互作"，血浆成分与红细胞之间的相互作用，混合互作。

6. 血浆成分是怎样与红细胞互作的呢？血浆蛋白、氨基酸、血糖、无机盐等，都有各自的作用，这就是本篇的内容。

第五十二章　蒿甲醚抗疟机制的研究
——蒿甲醚与红细胞成分的混合互作

韩丽红[#]　高丽君[#]　周立社　苏　燕　高雅琼　宝勿仁　秦文斌[※]

(包头医学院　血红蛋白研究室，包头　014010)

摘　要

蒿甲醚为青蒿素的衍生物，有强大且快速的杀灭作用，它的抗疟活性较青蒿素大6倍，主要用于凶险型恶性疟的急救。此药为注射液，便于直接进行研究。

众所周知，疟原虫寄生于红细胞，从而造成疾病——疟疾。疟原虫繁殖需要红细胞内的血红蛋白，并产生一系列不良后果。蒿甲醚等青蒿素类药物能够特效地治好疟疾，一定有一套反制措施，消除对疟原虫有利的环境，最终消灭疟原虫。

对于红细胞的内环境，我们有过一些研究[1, 2]。通过红细胞内血红蛋白的电泳释放，我们知道红细胞内的血红蛋白有两种存在状态：游离状态和结合状态。结合状态又分两种：疏松结合状态和牢固结合状态。

蒿甲醚治疗疟疾时，对红细胞内上述血红蛋白状态有何影响，这是本文研究的主要内容。实验结果表明，蒿甲醚对红细胞内游离及结合状态的血红蛋白都有影响。

关键词　青蒿素类药物；蒿甲醚；红细胞内血红蛋白的电泳释放；游离状态的血红蛋白；结合状态的血红蛋白

1　前言

蒿甲醚是中国自主研制的抗疟良药，主要用于凶险型恶性疟的急救。但是，青蒿素类药物作用机制一直未被彻底破解[3]。现在有一些假说，如铁参与青蒿素的激活[4-13]、血红素参与青蒿素的激活[14-15]、线粒体参与青蒿素的激活[16-17]、血红蛋白参与青蒿素的激活[18]等。文献里还涉及青蒿素类药物的作用靶点问题，靶点之一是血红素的烷基化[19-24]，还有蛋白靶点[25]、线粒体模型[26-28]等。

上述各种假说和观点表明，青蒿素类药物的作用机制相当复杂，可能涉及不同的作用分子，这些作用可能协同，也可能竞争或拮抗。随着研究手段的不断发展，应当寻找易于操作的、接近于体内环境下的疟原虫生物模型，从而减少体内外实验结果不一致的现象。

对于红细胞的内环境，我们有过一些研究[1, 2]。通过红细胞内血红蛋白的电泳释放，我们知道红细胞内的血红蛋白有两种存在状态：游离状态和结合状态。结合状态又分两种：疏松结合状态和牢固结合状态。青蒿素类药物治疗疟疾时，对红细胞内上述血红蛋白状态有何影响，这是本文研究的主要内容。实验结果表明，这类药物对红细胞内两种状态的血红蛋白

[#] 并列第一作者
[※] 通讯作者．秦文斌，电子邮箱：qwb5991309@tom.com

都有影响，对牢固结合状态的血红蛋白影响更明显，这说明，青蒿素类药物能够全面、深入地干扰红细胞内血红蛋白的存在状态，从而达到治疗疟疾的目的。

2 材料与方法

2.1 材料

2.1.1 蒿甲醚注射液 购自昆明制药集团股份有限公司，批号14HM20 1-11。

2.1.2 对照用的花生油 鲁花5S压榨一级花生油，购自超市，因为蒿甲醚注射液的辅料为花生油，所以用它做对照。

2.1.3 正常人血液来源于包头医学院第一附属医院体检科。

2.2 方法
本文用淀粉-琼脂糖混合凝胶双向电泳，分析蒿甲醚对全血及红细胞成分的影响。总体实验参见文献[1，2]，具体操作如下所示。

2.2.1 蒿甲醚与红细胞成分的相互作用

(1) 准备标本：①正常人全血300µl，加生理盐水至1ml，混匀后，低速离心3分钟，弃上清液,留沉淀。重复4次,再加等量的生理盐水即为实验用的RBC；②取制备好的RBC 100µl加20µl花生油混合均匀为对照组；③取制备好的RBC 100µl加20µl蒿甲醚混合均匀为实验组；将上述2管放置室温24小时后使用。

(2) 制淀粉-琼脂糖混合凝胶：17cm×17cm的大胶板。

(3) 上样：10µl。上层为对照，下层为实验。

(4) 电泳：按6V/cm的电势梯度，先普通电泳1小时，停电15分钟，再电泳1小时15分钟，换电极再停电20分钟后；转向普通电泳1小时15分钟。

(5) 染色

1) 丽春红：电泳结束后将胶板放入丽春红染色液中染色24小时后，拍照留图。后将胶板烤干或室温放置待干。

2) 联苯胺：将干燥好的胶板按联苯胺染色方法染色后拍照留图。

2.2.2 蒿甲醚与全血成分的相互作用

(1) 准备标本：准备正常人全血300µl。①取全血100µl加花生油20µl混合均匀为对照组；②取全血100µl加蒿甲醚20µl混合均匀为实验组；将上述2管放置室温24小时后使用。

(2) 制淀粉-琼脂糖混合凝胶：17cm×17cm的大胶板。

(3) 上样：10µl。上层为对照，下层为实验。

(4) 电泳：按6V/cm的电势梯度，先普通电泳1小时，停电15分钟，再电泳1小时15分钟，换电极再停电15分钟后；转向普通电泳1小时15分钟。

(5) 染色

1) 丽春红：电泳结束后将胶板放入丽春红染色液中染色24小时后，拍照留图。后将胶板烤干或室温放置待干。

2) 联苯胺：将干燥好的胶板按联苯胺染色方法染色后拍照留图。

2.2.3 蒿甲醚与红细胞成分相互作用指纹图

(1) 准备标本

1) 取正常人全血300µl，加生理盐水至1ml，混匀后低速离心3分钟，弃上清液，留沉淀。重复4次，再加等量的生理盐水即为实验用的RBC。

2) 取 200μl 全血加 40μl 花生油混合均匀备用。

3) 稀释样品：准备 10 支 0.5μl 的小 EP 管编号 0～9。混合好后将上述 10 管均放于室温 24 小时后使用。

管号	0	1	2	3	4	5	6	7	8	9
盐水	0	2	4	6	8	10	12	14	16	18
RBC	20	18	16	14	12	10	8	6	4	2

(2) 制淀粉-琼脂糖混合凝胶：17cm×17cm 的大胶板。

(3) 上样：6μl。从左上到右下按 1～9 顺序上样。

(4) 电泳：按 6V/cm 的电势梯度，先普通电泳 1 小时 45 分钟，停电 15 分钟，再电泳 30 分钟，换电极再停电 15 分钟后；转向普通电泳 1 小时 15 分钟。

(5) 染色

1) 丽春红：电泳结束后将胶板放入丽春红染色液中染色 24 小时后，拍照留图。后将胶板烤干或室温放置待干。

2) 联苯胺：将干燥好的胶板按联苯胺染色方法染色后拍照留图。

2.2.4 蒿甲醚与全血成分相互作用指纹图

(1) 准备标本

1) 取正常人全血 200μl，加 40μl 花生油混合均匀备用。

2) 稀释样品：准备 10 支 0.5μl 的小 EP 管编号 0～9。混合好后将上述 10 管均放置于室温 24 小时后使用。

管号	0	1	2	3	4	5	6	7	8	9
盐水	0	2	4	6	8	10	12	14	16	18
全血	20	18	16	14	12	10	8	6	4	2

(2) 制淀粉-琼脂糖混合凝胶：17cm×17cm 的大胶板。

(3) 上样：6μl。从左上到右下按 1～9 顺序上样。

(4) 电泳：按 6V/cm 的电势梯度，先普通电泳 1 小时 45 分钟，停电 15 分钟，再电泳 30 分钟，换电极再停电 15 分钟后；转向普通电泳 1 小时 15 分钟。

(5) 染色

1) 丽春红：电泳结束后将胶板放入丽春红染色液中染色 24 小时后，拍照留图。后将胶板烤干或室温放置待干。

2) 联苯胺：将干燥好的胶板按联苯胺染色方法染色后拍照留图。

3 结果

3.1 蒿甲醚与红细胞成分的相互作用　结果见图 52-1。

由图 52-1 可以看出，红细胞成分与花生油(对照)作用时，HbA_2 现象(初释放现象)基本正常，红细胞成分与蒿甲醚作用时，HbA_2 现象不正常，游离血红蛋白向后(阴极侧)延伸。

3.2 蒿甲醚与全血成分的相互作用　结果见图 52-2。

由图 52-2 可以看出,全血成分与花生油(对照)作用时,HbA$_2$ 现象(初释放现象)基本正常,全血成分与蒿甲醚作用时,HbA$_2$ 现象消失。白蛋白方面,上层白蛋白里有一些 MHA,下层白蛋白里没有或很少。

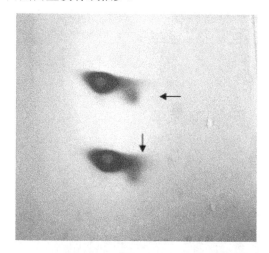

图 52-1 蒿甲醚与红细胞成分的相互作用双向双层电泳　定退普

注释：上层为红细胞+花生油(对照)；下层为红细胞+蒿甲醚；HbA$_2$ 现象见←所指处,游离血红蛋白后退见↓所指处

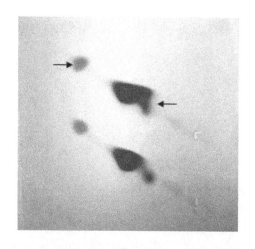

图 52-2 蒿甲醚与全血成分的相互作用双向双层电泳　定退普

注释：上层为全血+花生油(对照)，下层为全血+蒿甲醚；HbA$_2$ 现象见←所指处,白蛋白见→所指处

3.3 花生油(对照)与红细胞成分的相互作用指纹图　结果见图 52-3。

由图 52-3 可以看出，对照标本与红细胞成分相互作用时，定释带明显，从上往下都明显，后退带也有一些，可看到下边有一点。

3.4 蒿甲醚与红细胞成分的相互作用指纹图　结果见图 52-4。

由图 52-4 可以看出，在蒿甲醚作用下，定释带明显减弱，后退带消失。

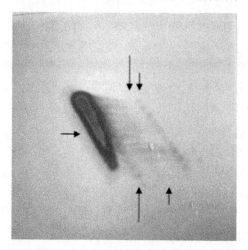

图 52-3 红细胞成分与花生油相互作用指纹图　定退普

注释：游离血红蛋白见→所指处,红细胞 HbA$_2$ 见长↑所指处,碳酸酐酶见短↑所指处(红色处),定释带见长↑所指处,后退带见短↓所指处

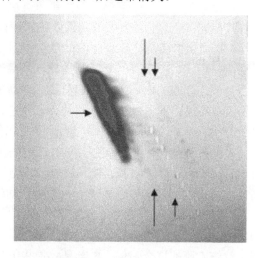

图 52-4 蒿甲醚与红细胞成分相互作用指纹图　定退普

注释：游离血红蛋白见→所指处,红细胞 HbA$_2$ 见长↑所指处,碳酸酐酶见短↑所指处(红色处),定释带见长↓所指处,后退带见短↓所指处

3.5 花生油(对照)与全血成分的相互作用指纹图 结果见图 52-5。

由图 52-5 可以看出，对照标本与全血成分相互作用时，白蛋白里有 MHA，有定释带，上下两头较明显，后退带看不清。

3.6 蒿甲醚与全血成分的相互作用指纹图 结果见图 52-6。

由图 52-6 可以看出，在蒿甲醚作用下，白蛋白里 MHA 稍微减少，定释带基本消失。

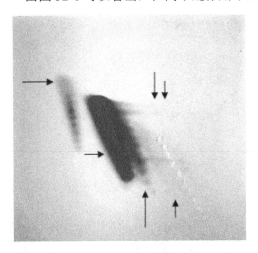
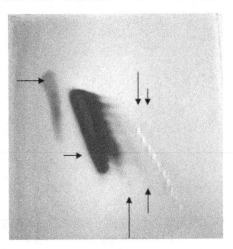

图 52-5 全血成分与花生油相互作用指纹图 定退普
注释：白蛋白见长→所指处，游离血红蛋白见短→所指处，红细胞 HbA_2 见长↑所指处，碳酸酐酶及纤维蛋白原见短↑所指处(红色处)，定释带见长↓所指处，后退带见短↓所指处

图 52-6 全血成分与蒿甲醚相互作用指纹图 定退普
注释：白蛋白见长→所指处，游离血红蛋白见短→所指处，红细胞 HbA_2 见长↑所指处，碳酸酐酶及纤维蛋白原见短↑所指处(红色处)，定释带见长↓所指处，后退带见短↓所指处

4 讨论

青蒿素类药物治疗疟疾的效果举世闻名，它们作用机制的研究一直在进行，虽然尚未完全解决问题，但还是得到大量成果[3-28]。众所周知，疟原虫寄生于红细胞，从而造成疾病——疟疾。疟原虫繁殖需要红细胞内的血红蛋白，并产生一系列不良后果。青蒿素类药物能够特效地治好疟疾，一定有一套反制措施，消除对疟原虫有利的环境，最终消灭疟原虫。

对于红细胞的内环境，我们有过一些研究[1, 2]。通过红细胞内血红蛋白的电泳释放，我们知道红细胞内的血红蛋白有两种存在状态：游离状态和结合状态。结合状态又分两种：疏松结合状态(初释放现象、HbA_2 现象)和牢固结合状态(再释放现象)。

青蒿素类药物治疗疟疾时，对红细胞内上述血红蛋白状态有何影响，这是本文研究的主要内容。实验结果表明，这类药物对红细胞内游离血红蛋白及结合血红蛋白都有影响。在游离血红蛋白方面，蒿甲醚能使它向阴极移动(参见图 52-1)，机制不明。对疏松结合血红蛋白来说，它表现为 HbA_2 现象、初释放现象，是红细胞内 HbA_2 与 HbA_1 结合、再与红细胞膜疏松结合，第一次通电就能脱落下来[1, 2]，蒿甲醚能使 A_2 现象消失(参见图 52-2)，说明它能作用于红细胞膜的这个部位，影响到疟原虫的生存。还有牢固结合血红蛋白部分，它表现为再释放现象，这类血红蛋白在第一次通电时释放不出来，只有再次(第二次或二次以上)通电时才能释放出来，所以称为再释放现象[1, 2]。再释放还可分两种：①等渗条件下的再释放；②等低渗全程条件下的再释放。前者见图 52-1、图 52-2，后者见图 52-3～图 52-6。等低渗全程条件下的再释放,是让红细胞处于等渗和一系列(9 个等级)低渗状态,观察此条件下的再释放情况。此时,信息量大增,有可能反映个体差异,用来检测药物作用,也应当效果不错。

在本项研究中,蒿甲醚使红细胞成分指纹图里的 10 个定释带明显减弱和减少、使少量后退带完全消失(参见图 52-7)。全血指纹图增加了血浆因素,信息量又多了一些。此时,除了定释带和后退带外,又出现了连成一片的 10 个白蛋白区带,这里边可以显示有没有 MHA(高铁血红素白蛋白)和数量多少。在本项研究中,蒿甲醚使全血成分指纹图里的 10 个定释带全部消失,使白蛋白里的 MHA 减弱、变少(参见图 52-8)。MHA 是溶血指标,此时的溶血来自花生油,花生油是蒿甲醚注射液的辅料,所以我们拿它做对照。在这里,我没看到花生油与全血成分互作时出现溶血,蒿甲醚与全血成分互作时溶血减弱、变少,再一次显示出蒿甲醚的治疗作用。

通过上述实验研究,可以认为,青蒿素类药物蒿甲醚,能够比较全面、深入地干扰红细胞和全血内血红蛋白的多种存在状态,从而达到治疗疟疾的目的。

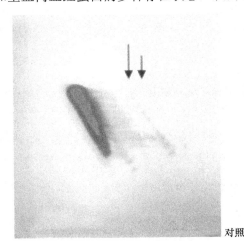

图 52-7　比较对照及蒿甲醚与红细胞成分的相互作用蒿甲醚使再释放减弱

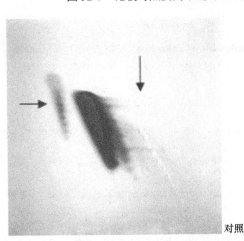
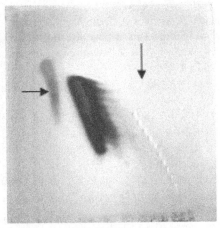

图 52-8　比较对照及蒿甲醚与全血成分的相互作用蒿甲醚使再释放减弱

参 考 文 献

[1] 秦文斌. 活体红细胞内血红蛋白的电泳释放. 中国科学: 生命科学, 2011, 41 (8): 597-603.
[2] 秦文斌. 红细胞内血红蛋白的电泳释放——发现和研究. 北京: 科学出版社, 2015: 70.
[3] 孙辰, 李坚, 周兵. 青蒿素类药物的作用机制: 一个长期未决的基础研究挑战. 中国科学: 生命科学, 2012, 42 (5): 345-354.
[4] Meshnick S R, Yang Y Z, Lima V, et al. Iron-dependent free radical generation from the antimalarial agent artemisinin (qinghaosu). Antimicrob Agents Chemother, 1993, 37: 1108-1114.
[5] Eckstein-Ludwig U, Webb R J, Van Goethem I D, et al. Artemisinins target the SERCA of Plasmodium falciparum. Nature, 2003, 424: 957-961.
[6] Wu Y, Yue Z Y, Wu Y L. Interaction of qinghaosu (artemisinin) with cysteine sulfhydryl mediated by traces of non-heme iron. Angew Chem Int Ed Engl, 1999, 38: 2580-2582.
[7] Posner G H, Oh C H, Wang D, et al. Mechanism-based design, synthesis, and in vitro antimalarial testing of new 4-methylated trioxanes structurally related to artemisinin: the importance of a carbon-centered radical for antimalarial activity. J Med Chem, 1994, 37: 1256-1258.
[8] Butler A R, Gilbert B C, Hulme P, et al. EPR evidence for the involvement of free radicals in the iron-catalysed decomposition of qinghaosu (artemisinin) and some derivatives; antimalarial action of some polycyclic endoperoxides. Free Radical Res, 1998, 28: 471-476.
[9] Jefford C W, Vicente M G H, Jacquier Y, et al. The Deoxygenation and isomerization of artemisinin and artemether and their relevance to antimalarial action. Helv Chim Acta, 1996, 79: 1475-1487.
[10] Haynes R K, Chan W C, Lung C M, et al. The Fe^{2+}-mediated decomposition, PfATP6 binding, and antimalarial activities of artemisone and other artemisinins: the unlikelihood of C-centered radicals as bioactive intermediates. Chem Med Chem, 2007, 2: 1480-1497.
[11] O'Neill P M, Bishop L P, Searle N L, et al. Biomimetic Fe (II) -mediated degradation of arteflene (Ro-42-1611). The first EPR spin-trapping evidence for the previously postulated secondary carbon-centered cyclohexyl radical. J Org Chem, 2000, 65: 1578-1582.
[12] Haynes R K, Vonwiller S C. The behaviour of qinghaosu (artemisinin) in the presence of non-heme iron (II) and (III). Tetrahedron Lett, 1996, 37: 257-260.
[13] Wu W M, Wu Y k, Wu Y L, et al. Unified mechanistic framework for the Fe (II) -induced cleavage of Qinghaosu and derivatives/analogues. The first spin-trapping evidence for the previously postulated secondary C-4 radical. J Am Chem Soc, 1998, 120: 3316-3325.
[14] Hong Y L, Yang Y Z, Meshnick S R. The interaction of artemisinin with malarial hemozoin. Mol Biochem Parasitol, 1994, 63: 121-128.
[15] Meshnick S R, Thomas A, Ranz A, et al. Artemisinin (qinghaosu): the role of intracellular hemin in its mechanism of antimalarial action. Mol Biochem Parasitol, 1991, 49: 181-189.
[16] Li W, Mo W, Shen D, et al. Yeast model uncovers dual roles of mitochondria in action of artemisinin. PLoS Genet, 2005, 1: e36.
[17] 王娟, 周兵. 线粒体呼吸链抑制剂对青蒿素代谢速率的影响. 清华大学学报 (自然科学版), 2010, 50: 944-946.
[18] Klonis N, Crespo-Ortiz M P, Bottova I, et al. Artemisinin activity against Plasmodium falciparum requires hemoglobin uptake and digestion. Proc Natl Acad Sci USA, 2011, 108: 11405-11410.
[19] Robert A, Coppel Y, Meunier B. Alkylation of heme by the antimalarial drug artemisinin. Chem Commun (Camb), 2002: 414-415.
[20] Pandey A V, Tekwani B L, Singh R L, et al. Artemisinin, an endoperoxide antimalarial, disrupts the hemoglobin catabolism and heme detoxification systems in malarial parasite. J Biol Chem, 1999, 274: 19383-19388.
[21] Robert A, Benoit-Vical F, Claparols C, et al. The antimalarial drug artemisinin alkylates heme in infected mice. Proc Natl Acad Sci USA, 2005, 102: 13676-13680.
[22] Meunier B, Robert A. Heme as trigger and target for trioxane-containing antimalarial drugs. Acc Chem Res, 2010, 43: 1444-1451.
[23] Kannan R, Sahal D, Chauhan V S. Heme-artemisinin adducts are crucial mediators of the ability of artemis-

inin to inhibit heme polymerization. Chem Biol, 2002, 9: 321-332.
[24] Loup C, Lelievre J, Benoit-Vical F, et al. Trioxaquines and heme-artemisinin adducts inhibit the in vitro formation of hemozoin better than chloroquine. Antimicrob Agents Chemother, 2007, 51: 3768-3770.
[25] Uhlemann A C, Cameron A, Eckstein-Ludwig U, et al. A single amino acid residue can determine the sensitivity of SERCAs to artemisinins. Nat Struct Mol Biol, 2005, 12: 628-629.
[26] Srivastava I K, Rottenberg H, Vaidya A B. Atovaquone, a broad spectrum antiparasitic drug, collapses mitochondrial membrane potential in a malarial parasite. J Biol Chem, 1997, 272: 3961-3966.
[27] Wang J, Huang L, Li J, et al. Artemisinin directly targets malarial mitochondria through its specific mitochondrial activation. PLos One, 2010, 5: e9582.
[28] 王娟, 黄丽英, 龙伊成, 等. 线粒体通透性转移孔与青蒿素抗疟机制研究. 现代生物医学进展, 2009, 9: 4006-4009.

第五十三章 青蒿素类药物与一些物质的混合互作
——表现为凝集反应

韩丽红 [1#] 王满元 [2#] 高丽君 [1#] 周立社 [1] 苏燕 [1] 高雅琼 [1]
宝勿仁 [1] 秦文斌 [1※]

(1. 包头医学院　血红蛋白研究室，包头 014010；2. 首都医科大学　中医药学院，北京 100069)

摘　要

在用红细胞电泳技术研究青蒿素类药物的抗疟机制过程中，我们注意到药物与血液接触时出现凝集现象。进一步实验，发现青蒿素类药物能与多种物质发生凝集作用，如红细胞、红细胞溶血液、氯化血红素、硫酸亚铁、三氯化铁、氯化钠等。这是一种现象，说明有相互作用，它的机制和意义有待深入研究。

关键词　青蒿素；双氢青蒿素；凝集作用；氯化血红素；硫酸亚铁；三氯化铁；氯化钠

1　前言

青蒿素类药物治疗疟疾的效果举世闻名，这类药物作用机制的研究一直在进行，虽然尚未完全解决问题，但是已经得到大量成果[1-19]。目前提出的青蒿素作用机制假说基本涉及两个方面：青蒿素的激活和青蒿素的作用靶点。在青蒿素的激活方面，有人认为铁参与青蒿素的激活[2-11]，有人主张血红素参与青蒿素的激活[2,3,12,14,15]，但还有争议[2,12]。还有人提出线粒体参与青蒿素的激活[16,17]，另外，对于铁参与青蒿素的激活，也有不同意见[8,18]。还有人认为，青蒿素的抗疟活性与血红蛋白有关[19]。在青蒿素的作用靶点方面，有人认为血红素的烷基化是重要靶点[20-26]，但也有反对意见[27-30]，还有线粒体模型[31-33]及其不同观点[34-37]。

上述各种观点和假说表明，青蒿素类药物的作用机制相当复杂，可能涉及不同的相互作用。本文作者在实验过程中发现青蒿素类药物能与一些物质发生凝集，这也是一种相互作用，它与上述多项研究的关系尚不清楚。

2　材料与方法

2.1　材料

2.1.1　青蒿素类药物　由王满元博士(屠呦呦的博士生)提供，青蒿素粉末状，双氢青蒿素粉末状、避光保存。蒿甲醚注射液购自中国昆明制药集团股份有限公司,批号 14HM201-11。氯化血红素购置于合肥博美生物科技有限责任公司,进口产品,高纯级,批号 180137。硫酸亚铁购置于新乡市化学试剂厂,分析纯,批号 840918。DMSO(二甲基亚砜)来自石瑞丽教授。

2.1.2　正常人血液　来源与包钢集团第三职工医院检验科,由宝勿仁提供。

2.1.3　ABO 血型正定型检定卡　购置于长春博德生物技术有限责任公司。

2.2　方法

2.2.1　全血与青蒿素类药物的凝集作用

(1) 准备好 ABO 血型正定型检定卡。

(2) 制备所需试剂:①DMSO 原液为第一液;②将 0.055g 的青蒿素溶解于 1ml DMSO 溶液中为第二液;③将 0.035g 的双氢青蒿素溶解于 1ml DMSO 溶液中为第三液。

(3) 加样:①检定卡左侧三孔均为全血 50μl;②检定卡右侧三孔,上孔加第一液 50μl,中孔加第二液 50μl,下孔加第三液 50μl。留图 53-1。

(4) 用竹签拉混左右两孔液体:①从左向右拉 1 下,留图 53-2;②从左向右拉 3 下,留图 53-3。

2.2.2　红细胞与青蒿素类药物的凝集作用

(1) 准备好 ABO 血型正定型检定卡。

(2) 制备所需试剂:同上述全血实验。

(3) 加样:①检定卡左侧三孔均为带有等量盐水的红细胞 50μl;②检定卡右侧三孔,同上述全血实验,留图 53-4。

(4) 用竹签拉混两孔液体:①从左向右拉 1 下,留图 53-5;②从左向右拉 3 下,留图 53-6。

2.2.3　溶血液与青蒿素类药物的凝集作用

(1) 准备好 ABO 血型正定型检定卡

(2) 制备所需试剂:同上述全血实验。

(3) 加样:①检定卡左侧三孔均为红细胞溶血液 50μl;②检定卡右侧三孔,同上述全血实验,留图 53-7。

(4) 用竹签拉混两孔液体:①从左向右拉 1 下,留图 53-8;②从左向右拉 3 下,留图 53-9。

2.2.4　氯化血红素与青蒿素类药物的凝集作用

(1) 备好 ABO 血型正定型检定卡。

(2) 制备所需试剂:同上述全血实验。氯化血红素液,用 1%的氢氧化钠 1ml 溶解 0.01g 氯化血红素。

(3) 加样:①检定卡左侧三孔均为氯化血红素液 50μl;②检定卡右侧三孔,同上述全血实验,留图 53-10。

(4) 用竹签拉混两孔液体:①从左向右拉 1 下,留图 53-11;②从左向右拉 3 下,留图 53-12;③从左向右分叉拉,留图 53-13。

2.2.5　硫酸亚铁与青蒿素类药物的凝集作用

(1) 准备好 ABO 血型正定型检定卡。

(2) 制备所需试剂:同上述全血实验。硫酸亚铁染液,用双蒸水 1ml 溶解 0.025g 硫酸亚铁。

(3) 加样:①检定卡左侧三孔均为硫酸亚铁溶液 50μl;②检定卡右侧三孔,同上述全血实验,留图 53-14。

(4) 用竹签拉混两孔液体:①从左向右拉 1 下,留图 53-15;②从左向右拉 3 下,留图 53-16;

③从左向右分叉拉，留图 53-17；④待 10 分钟后抬高纸片右侧，使液体从右侧流向左侧，留图 53-18。

2.2.6 三氯化铁与青蒿素类药物的凝集作用

(1) 准备好 ABO 血型正定型检定卡。

(2) 制备所需试剂：同上述全血实验。溶解，用 1ml 双蒸水，加 0.1g 三氯化铁，约 10 分钟完全溶解，呈黄色。后又稀释 3 倍，呈淡黄色备用。

(3) 加样：①检定卡左侧三孔均为三氯化铁溶液 50μl；②检定卡右侧三孔：同上述全血实验，留图 53-19；

(4) 用竹签拉混两孔液体：①从左向右拉 1 下，留图 53-20；②从左向右拉 3 下，留图 53-21；③从左向右分叉拉 1 下，留图 53-22；④从左向右分叉拉 3 下，留图 53-23；⑤待 10 分钟后抬高纸片右侧，使液体从右侧流向左侧，正照，留图 53-24；⑥斜照，留图 53-25。

2.2.7 氯化钠与青蒿素类药物的凝集作用

(1) 准备好 ABO 血型正定型检定卡。

(2) 制备所需试剂：同上述全血实验。0.9%氯化钠生理盐水。

(3) 加样：①检定卡左侧三孔均为氯化钠生理盐水 50μl；②检定卡右侧三孔，同上述全血实验，留图 53-26。

(4) 用竹签拉混两孔液体：①从左向右拉 1 下，留图 53-27；②从左向右分叉拉 3 下，留图 53-28；③从左向右分叉拉 3 下，待 10 分钟后抬高纸片右侧，使液体从右侧流向左侧，斜照，留图 53-29，正照，留图 53-30。

2.2.8 DMSO 与青蒿素类药物的凝集作用

(1) 准备好 ABO 血型正定型检定卡。

(2) 制备所需试剂：同上述全血实验。DMSO 溶液。

(3) 加样：①检定卡左侧三孔均为 DMSO 50μl；②检定卡右侧三孔，同上述全血实验，留图 53-31。

(4) 用竹签拉混两孔液体：①从左向右拉 1 下，留图 53-32；②从左向右分叉拉 3 下，留图 53-33；③待 10 分钟后抬高纸片右侧，使液体从右侧流向左侧，留图 53-34。

2.2.9 蒿甲醚与青蒿素类药物的凝集作用

(1) 准备好 ABO 血型正定型检定卡。

(2) 制备所需试剂：同上述全血实验。蒿甲醚注射液。

(3) 加样：①检定卡左侧三孔均为液体蒿甲醚 50μl；②检定卡右侧三孔，同上述全血实验，留图 53-35。

(4) 用竹签拉混两孔液体：①从左向右拉 1 下，留图 53-36；②从左向右分叉拉 3 下，留图 53-37；③待 10 分钟后抬高纸片右侧，使液体从右侧流向左侧，留图 53-38。

3 结果

3.1 全血与青蒿素类药物的凝集作用 结果见图 53-1～图 53-3。

由图 53-3 可以看出，右侧中孔和下孔出现明显的凝集颗粒。

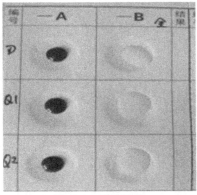

图 53-1　准备阶段

注释：左侧三孔，都是全血；右侧三孔，上孔为 DMSO，中孔为 DMSO 溶解青蒿素，下孔为 DMSO 溶解双氢青蒿素

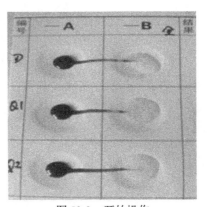

图 53-2　开始操作

注释：用竹签将全血由左孔拉到右孔，只一次

3.2　红细胞与青蒿素类药物的凝集作用　结果见图 53-4～图 53-6。

由图 53-6 可以看出，右侧中孔和下孔出现明显的凝集颗粒，下孔更明显。

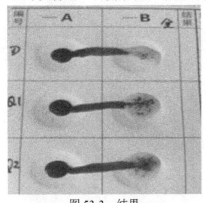

图 53-3　结果

注释：用竹签将全血由左孔拉到右孔，拉三次

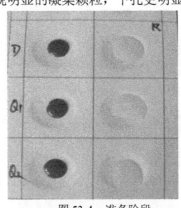

图 53-4　准备阶段

注释：左侧三孔，都是红细胞；右侧三孔，上孔为 DMSO，中孔为 DMSO 溶解青蒿素，下孔为 DMSO 溶解双氢青蒿素

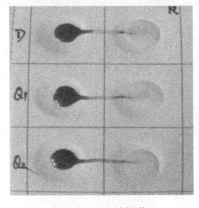

图 53-5　开始操作

注释：用竹签将全血由左孔拉到右孔，只一次

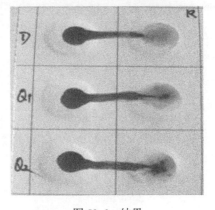

图 53-6　结果

注释：用竹签将全血由左孔拉到右孔，拉三次

3.3　溶血液与青蒿素类药物的凝集作用　结果见图 53-7～图 53-9。

由图 53-9 可以看出，右侧中孔和下孔出现明显的凝集颗粒，中孔不如下孔明显。

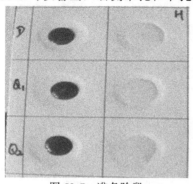

图 53-7　准备阶段

注释：左侧三孔，都是红细胞溶血液；右侧三孔，上孔为 DMSO，中孔为 DMSO 溶解青蒿素，下孔为 DMSO 溶解双氢青蒿素

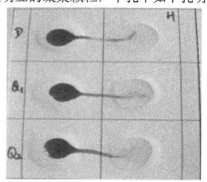

图 53-8　开始操作

注释：用竹签将全血由左孔拉到右孔，只一次

3.4　氯化血红素与青蒿素类药物的凝集作用　结果见图 53-10～图 53-13。

由图 53-12 可以看出，右侧中孔和下孔出现凝集，中孔更明显。由图 53-13 可以看出，右侧中孔和下孔出现"所指"，显示凝固趋势。

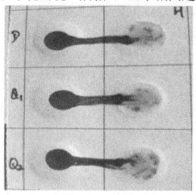

图 53-9　结果

注释：用竹签将全血由左孔拉到右孔，拉三次

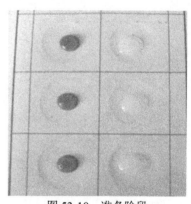

图 53-10　准备阶段

注释：左侧三孔，都是氯化血红素溶液；右侧三孔，上孔为 DMSO，中孔为 DMSO 溶解青蒿素，下孔为 DMSO 溶解双氢青蒿素

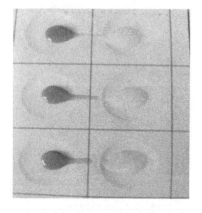

图 53-11　开始操作

注释：用竹签将全血由左孔拉到右孔，只一次

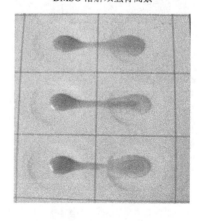

图 53-12　结果

注释：用竹签将全血由左孔拉到右孔，拉三次

3.5 硫酸亚铁与青蒿素类药物的凝集作用 结果见图 53-14～图 53-18。由图 53-18 可看出，右侧中孔和下孔出现"手指"，显示凝固趋势，上孔不凝而流动。

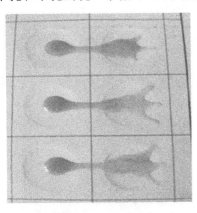

图 53-13 结果
注释：用竹签将全血由左孔拉到右孔，再往右拉出三个"手指"；注意，右上孔"手指"拉不开

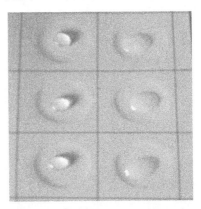

图 53-14 准备阶段
注释：左侧三孔，都是硫酸亚铁溶液；右侧三孔，上孔为 DMSO，中孔为 DMSO 溶解青蒿素，下孔为 DMSO 溶解双氢青蒿素

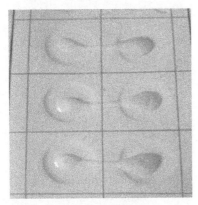

图 53-15 开始操作
注释：用竹签将全血由左孔拉到右孔，只一次

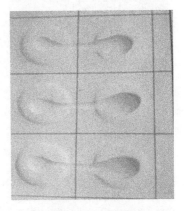

图 53-16 结果
注释：用竹签将全血由左孔拉到右孔，拉三次

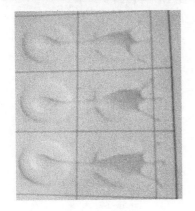

图 53-17 结果
注释：用竹签将全血由左孔拉到右孔，再往右拉出三个"手指"；注意，右上孔"手指"拉不开

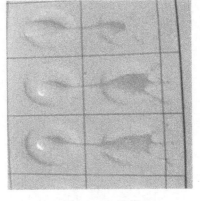

图 53-18 结果
注释：抬高纸片右侧，使液体从右侧流向左侧；注意，右上孔"手指"消失，液体流入左孔

3.6 三氯化铁与青蒿素类药物的凝集作用　　结果见图 53-19～图 53-25。

由图 53-24 和图 53-25 可以看出，右侧中孔和下孔出现"手指"不变形，显示凝固，上孔不凝而流动。

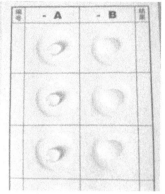

图 53-19　准备阶段

注释：左侧三孔，都是三氯化铁溶液；右侧三孔，上孔为 DMSO，中孔为 DMSO 溶解青蒿素，下孔为 DMSO 溶解双氢青蒿素

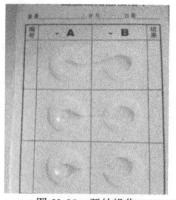

图 53-20　开始操作

注释：用竹签将全血由左孔拉到右孔，只一次

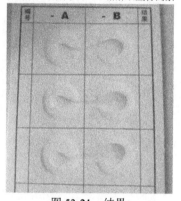

图 53-21　结果

注释：用竹签将全血由左孔拉到右孔，拉三次

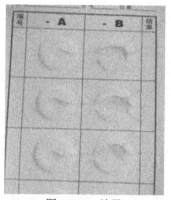

图 53-22　结果

注释：用竹签分叉拉 1 下

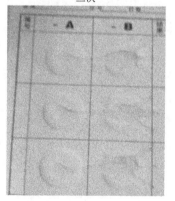

图 53-23　结果

注释：用竹签分叉拉 3 下

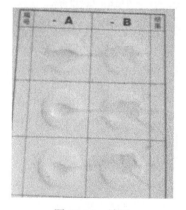

图 53-24　结果

注释：抬高右侧使液体从右侧自然流入左侧，正照

3.7 氯化钠与青蒿素类药物的凝集作用 结果见图 53-26～图 53-30。

由图 53-29 和图 53-30 可以看出，右侧中孔和下孔出现"手指"不变形，显示凝固，上孔不凝而流动。

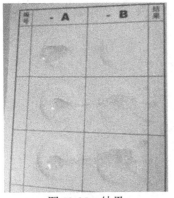

图 53-25 结果
注释：抬高右侧使液体从右侧自然流入左侧，斜照

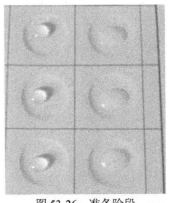

图 53-26 准备阶段
注释：左侧三孔，都是氯化钠溶液；右侧三孔，上孔为 DMSO，中孔为 DMSO 溶解青蒿素，下孔为 DMSO 溶解双氢青蒿素

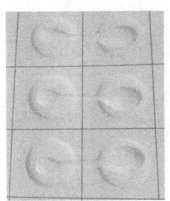

图 53-27 开始操作
注释：用竹签将全血由左孔拉到右孔，只一次

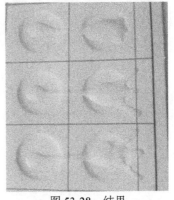

图 53-28 结果
注释：用竹签从左向右拉分叉观察

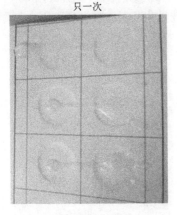

图 53-29 结果
注释：抬高纸片右侧，使液体从右侧流向左侧，斜照

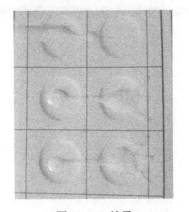

图 53-30 结果
注释：抬高纸片右侧，使液体从右侧流向左侧

3.8 DMSO 与青蒿素类药物的凝集作用 结果见图 53-31～图 53-34。

由图 53-34 可以看出，右侧中孔和下孔与上孔结果没有差别。

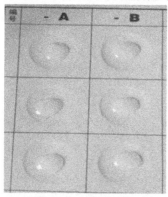

图 53-31　准备阶段

注释：左侧三孔，都是 DMSO 溶液；右侧三孔，上孔为 DMSO，中孔为 DMSO 溶解青蒿素，下孔为 DMSO 溶解双氢青蒿素

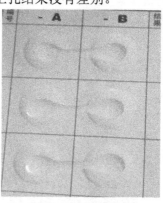

图 53-32　开始操作

注释：用竹签将全血由左孔拉到右孔，只一次

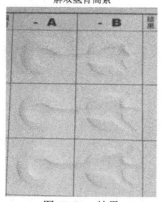

图 53-33　结果

注释：用竹签从左向右拉分叉观察

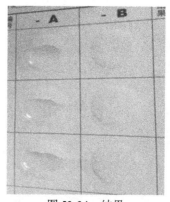

图 53-34　结果

注释：抬高纸片右侧，使液体从右侧流向左侧

3.9　蒿甲醚与青蒿素类药物的凝集作用　结果见图 53-35～图 53-38。

由图 53-38 可以看出，右侧中孔和下孔与上孔结果相反，即上孔凝固、中孔和下孔不凝固。

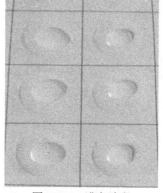

图 53-35　准备阶段

注释：左侧三孔，都是蒿甲醚溶液；右侧三孔，上孔为 DMSO，中孔为 DMSO 溶解青蒿素，下孔为 DMSO 溶解双氢青蒿素

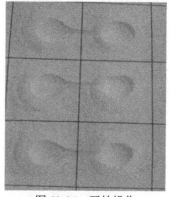

图 53-36　开始操作

注释：用竹签将全血由左孔拉到右孔，只一次

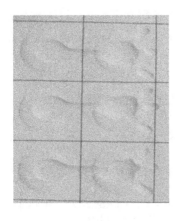
图53-37 结果
注释：用竹签从左向右拉分叉观察

图53-38 结果
注释：抬高纸片右侧，使液体从右侧流向左侧

4 讨论

凝集反应是一种现象，是两种物质相遇而发生相互作用的一种表现。我们是在青蒿素类药物与红细胞溶血液混合时发现了凝集反应，参见图53-39。结果表明，双氢青蒿素与溶血液混合时凝集颗粒最多，青蒿素次之，对照管最少。在试管里观察凝集反应界限不清，我们决定试用ABO血型正定型检定卡，得到上面的一系列结果。

图53-39 红细胞溶血液与青蒿素类药物混合时的凝集现象
注释：左侧管，溶血液+DMSO；中间管，溶血液+DMSO(溶有青蒿素)；右侧管，溶血液+DMSO(溶有双氢青蒿素)

凝集反应是指细菌或红细胞等颗粒形抗原与相应抗体特异性结合后，在适当电解质存在下，出现肉眼可见的凝集现象。凝集反应在临床上的应用主要是对于抗原和抗体的检测，对于抗原的检测，临床上常用反向间接凝集试验，如乳胶凝集抑制试验检测绒毛膜促性腺激素等。对于抗体的检测，血型鉴定及交叉配血是临床上常用的凝集反应。

本文中的凝集反应与抗原抗体无关，是青蒿素类药物与各种物质相互作用的结果。这里有青蒿素类药物与全血的凝集反应、与红细胞的凝集反应、与溶血液的凝集反应……与氯化钠的凝集反应等。全血、红细胞和溶血液含有生物大分子成分，氯化血红素属于低分子有机化合物，其余就是简单的无机化合物。青蒿素类药物能与这么大范围的物质都发生互作而凝集，说明它们作用的强大。此强大作用的机制是什么？目前尚不明确。推测与青蒿素分子内"过氧桥"有关，但作用细节如何、怎样发生凝集，都不得而知。有趣的是，蒿甲醚与青蒿素及双氢青蒿素相互作用时，不发生凝集，对照反而有凝集趋势(参见图53-35～图53-38)，与其他结果相反，这又是为什么？

我们发现了青蒿素类药物与多种物质相互作用而发生凝集,过去文献未见记载。这种现象可见于抗原-抗体反应,但本项发现似乎不属于这类反应。我们曾经发现血红蛋白 A_2 现象,即红细胞内 HbA_2 与 HbA_1 之间的相互作用,还有大鼠红细胞内 HbA HbB HbC HbD 之间的相互作用[38, 39],这些都不是抗原-抗体反应,能否发生凝集?我们想进一步试验研究。

总之,这是一种简单而有趣的现象,现在发表出来,抛砖引玉,供大家批评和研究。

参 考 文 献

[1] 孙辰, 李坚, 周兵. 青蒿素类药物的作用机制: 一个长期未决的基础研究挑战. 中国科学: 生命科学, 2012, 42 (5) : 345-354.

[2] Meshnick S R, Yang Y Z, Lima V, et al. Iron-dependent free radical generation from the antimalarial agent artemisinin (qinghaosu). Antimicrob Agents Chemother, 1993, 37: 1108-1114.

[3] Eckstein-Ludwig U, Webb R J, Van Goethem I D, et al. Artemisinins target the SERCA of Plasmodium falciparum. Nature, 2003, 424: 957-961.

[4] Wu Y, Yue Z Y, Wu Y L. Interaction of qinghaosu (artemisinin) with cysteine sulfhydryl mediated by traces of non-heme iron. Angew Chem Int Ed Engl, 1999, 38: 2580-2582.

[5] Posner G H, Oh C H, Wang D, et al. Mechanism-based design, synthesis, and in vitro antimalarial testing of new 4-methylated trioxanes structurally related to artemisinin: the importance of a carbon-centered radical for antimalarial activity. J Med Chem, 1994, 37: 1256-1258.

[6] Butler A R, Gilbert B C, Hulme P, et al. EPR evidence for the involvement of free radicals in the iron-catalysed decomposition of qinghaosu (artemisinin) and some derivatives; antimalarial action of some polycyclic endoperoxides. Free Radical Res, 1998, 28: 471-476.

[7] Jefford C W, Vicente M G H, Jacquier Y, et al. The Deoxygenation and isomerization of artemisinin and artemether and their relevance to antimalarial action. Helv Chim Acta, 1996, 79: 1475-1487.

[8] Haynes R K, Chan W C, Lung C M, et al. The Fe^{2+}-mediated decomposition, PfATP6 binding, and antimalarial activities of artemisone and other artemisinins: the unlikelihood of C-centered radicals as bioactive intermediates. Chem Med Chem, 2007, 2: 1480-1497.

[9] O'Neill P M, Bishop L P, Searle N L, et al. Biomimetic Fe(Ⅱ)-mediated degradation of arteflene (Ro-42-1611). The first EPR spin-trapping evidence for the previously postulated secondary carbon-centered cyclohexyl radical. J Org Chem, 2000, 65: 1578-1582.

[10] Haynes R K, Vonwiller S C. The behaviour of qinghaosu (artemisinin) in the presence of non-heme iron (Ⅱ) and (Ⅲ). Tetrahedron Lett, 1996, 37: 257-260.

[11] Wu W M, Wu Y k, Wu Y L, et al. Unified mechanistic framework for the Fe(Ⅱ)-induced cleavage of Qinghaosu and derivatives/analogues. The first spin-trapping evidence for the previously postulated secondary C-4 radical. J Am Chem Soc, 1998, 120: 3316-3325.

[12] Golenser J, Domb A, Leshem B, et al. Iron chelators as drugs against malaria pose a potential risk. Redox Rep, 2003, 8: 268-271.

[13] Hong Y L, Yang Y Z, Meshnick S R. The interaction of artemisinin with malarial hemozoin. Mol Biochem Parasitol, 1994, 63: 121-128.

[14] Efferth T. Willmar schwabe award 2006: antiplasmodial and antitumor activity of artemisinin—from bench to bedside. Planta Med, 2007, 73: 299-309.

[15] Meshnick S R, Thomas A, Ranz A, et al. Artemisinin (qinghaosu) : the role of intracellular hemin in its mechanism of antimalarial action. Mol Biochem Parasitol, 1991, 49: 181-189.

[16] Li W, Mo W, Shen D, et al. Yeast model uncovers dual roles of mitochondria in action of artemisinin. PLoS Genet, 2005, 1: e36.

[17] 王娟, 周兵. 线粒体呼吸链抑制剂对青蒿素代谢速率的影响. 清华大学学报 (自然科学版), 2010, 50: 944-946.

[18] Haynes R K, Ho W Y, Chan H W, et al. Highly antimalaria-active artemisinin derivatives: biological activity does not correlate with chemical reactivity. Angew Chem Int Ed Engl, 2004, 43: 1381-1385.

[19] Robert A, Coppel Y, Meunier B. Alkylation of heme by the antimalarial drug artemisinin. Chem Commun

(Camb), 2002, 5 (5) : 414-415.
[20] Pandey A V, Tekwani B L, Singh R L, et al. Artemisinin, an endoperoxide antimalarial, disrupts the hemoglobin catabolism and heme detoxification systems in malarial parasite. J Biol Chem, 1999, 274: 19383-19388.
[21] Robert A, Benoit-Vical F, Claparols C, et al. The antimalarial drug artemisinin alkylates heme in infected mice. Proc Natl Acad Sci USA, 2005, 102: 13676-13680.
[22] Kannan R, Kumar K, Sahal D, et al. Reaction of artemisinin with haemoglobin: implications for antimalarial activity. Biochem J, 2005, 385: 409-418.
[23] Meunier B, Robert A. Heme as trigger and target for trioxane-containing antimalarial drugs. Acc Chem Res, 2010, 43: 1444-1451.
[24] Kannan R, Sahal D, Chauhan V S. Heme-artemisinin adducts are crucial mediators of the ability of artemisinin to inhibit heme polymerization. Chem Biol, 2002, 9: 321-332.
[25] Loup C, Lelievre J, Benoit-Vical F, et al. Trioxaquines and heme-artemisinin adducts inhibit the in vitro formation of hemozoin better than chloroquine. Antimicrob Agents Chemother, 2007, 51: 3768-3770.
[26] Cazelles J, Robert A, Meunier B. Alkylating capacity and reaction products of antimalarial trioxanes after activation by a heme model. J Org Chem, 2002, 67: 609-619.
[27] Asawamahasakda W, Ittarat I, Chang C C, et al. Effects of antimalarials and protease inhibitors on plasmodial hemozoin production. Mol Biochem Parasitol, 1994, 67: 183-191.
[28] Haynes R K, Monti D, Taramelli D, et al. Artemisinin antimalarials do not inhibit hemozoin formation. Antimicrob Agents Chemother, 2003, 47: 1175.
[29] Coghi P, Basilico N, Taramelli D, et al. Interaction of artemisinins with oxyhemoglobin Hb-FeII, Hb-FeII, carboxyHb-FeII, heme-FeII, and carboxyheme FeII: significance for mode of action and implications for therapy of cerebral malaria. Chem Med Chem, 2009, 4: 2045-2053.
[30] Meshnick S R. Artemisinin and heme. Antimicrob Agents Chemother, 2003, 47: 2712-2713.
[31] Srivastava I K, Rottenberg H, Vaidya A B. Atovaquone, a broad spectrum antiparasitic drug, collapses mitochondrial membrane potential in malarial parasite. J Biol Chem, 1997, 272: 3961-3966.
[32] Wang J, Huang L, Li J, et al. Artemisinin directly targets malarial mitochondria through its specific mitochondrial activation. PLoS One, 2010, 5: e9582.
[33] 王娟, 黄丽英, 龙伊成, 等. 线粒体通透性转移孔与青蒿素抗疟机制研究. 现代生物医学进展, 2009, 9: 4006-4009.
[34] del Pilar Crespo M, Avery T D, Hanssen E, et al. Artemisinin and a series of novel endoperoxide antimalarials exert early effects on digestive vacuole morphology. Antimicrob Agents Chemother, 2008, 52: 98-109.
[35] Afonso A, Hunt P, Cheesman S, et al. Malaria parasites can develop stable resistance to artemisinin but lack mutations in candidate genes atp6 (encoding the sarcoplasmic and endoplasmic reticulum Ca^{2+} ATPase), tctp, mdr1, and cg10. Antimicrob Agents Chemother, 2006, 50: 480-489.
[36] Ellis D S, Li Z L, Gu H M, et al. The chemotherapy of rodent malaria, XXXIX. Ultrastructural changes following treatment with artemisinine of Plasmodium berghei infection in mice, with observations of the localization of [3H]-dihydroartemisinine in P. falciparum in vitro. Ann Trop Med Parasit, 1985, 79: 367-374.
[37] Kawai S, Kano S, Suzuki M. Morphologic effects of artemether on Plasmodium falciparum in Aotus trivirgatus. Am J Trop Med Hyg, 1993, 49: 812-818.
[38] 秦文斌. 活体红细胞内血红蛋白的电泳释放. 中国科学生命科学, 2011, 41 (8) : 597-607.
[39] 秦文斌. 红细胞内血红蛋白的电泳释放——发现和研究. 北京: 科学出版社, 2015.

第五十四章 偏重亚硫酸钠与红细胞相互作用

——将偏重亚硫酸钠加入标本 无氧与有氧同在一胶板

高丽君[1] 宝勿仁必力格[1,2] 秦文斌[1※]

(1. 包头医学院 血红蛋白研究室，包头 014010；2. 包钢集团第三职工医院 检验科，包头 014010)

摘 要

用普通缓冲液制胶，无氧标本是将偏重亚硫酸钠加入血液或红细胞里，有氧标本不加此物质。然后，在同一胶板做二层双向电泳，丽春红-联苯胺染色，观察结果。本文共检测四种标本：正常分娩、脑出血、胃癌和乳腺癌患者的红细胞标本。结果表明，普通红细胞都有 HbA_2 现象，即有 HbA_2 与 HbA_1 之间的相互作用。但是，含偏重亚硫酸钠的红细胞，都没有 HbA_2 现象（A_2 现象消失），即 HbA_2 与 HbA_1 之间没有发生相互作用。

关键词 正常分娩；脑出血；胃癌；乳腺癌；胃癌；一个胶板；两种标本；红细胞加入偏重亚硫酸钠；双向二层电泳

1 前言

偏重亚硫酸钠处理，能够促进血红蛋白 S 病患者的红细胞镰化，我们把它加入制胶缓冲液，用于梅花鹿的红细胞，发现它能改善梅花鹿红细胞内血红蛋白的电泳释放。在此基础上我们建立起无氧条件下的血红蛋白释放实验。

用普通缓冲液制胶，无氧标本是将偏重亚硫酸钠加入血液或红细胞里，有氧标本不加此物质。然后，同一胶板做二层双向电泳，丽春红-联苯胺染色，观察结果。详见正文。

2 材料与方法

2.1 材料 血液标本来自包钢集团第三职工医院检验科，分别为正常分娩、脑出血、胃癌和乳腺癌患者的血液标本。

2.2 方法 患者红细胞的双向双层电泳，标本处理方法如下所示。

(1) 洗 RBC 方法：取 200μl 全血于 1ml EP 管中，加生理盐水至 1ml，上下混合后，低速离心 3 分钟，弃上清液留下层沉淀。再加生理盐水至 1ml，再上下混合后低速离心 3 分钟，

※通讯作者：秦斌，电子邮箱：qwb5991309@tom.com

弃上清液留沉淀。重复洗 3 遍。下层的沉淀即为加样用的 RBC。上层标本制备是 20μlRBC +2μl 饱和偏重亚硫酸钠，下层标本制备是 20μlRBC +2μl 生理盐水，上样 10μl。

(2) 电泳条件：定退普 普通电泳 2.5 小时后，停电 15 分钟再泳 30 分钟。倒极再泳 15 分钟。将胶板转向 90°，再普泳 1.5 小时。取下胶板。

(3) 染色：先用丽春红染色过夜拍照留图，再烤干后染联苯胺。

3 结果

结果见图 54-1。

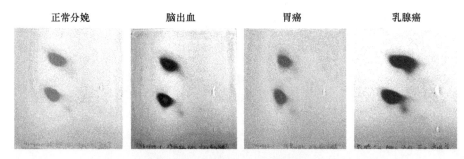

图 54-1 四种情况的综合结果

注释：图中都是红细胞的双向二层电泳结果，由左向右依次为正常分娩、脑出血、胃癌、乳腺癌的血液标本结果；上层都是偏重亚硫酸钠红细胞，下层都是无偏重亚硫酸钠红细胞

结果：下层都有 A_2 现象，即都有 HbA_2 与 A_2 之间的相互作用；上层都是 A_2 现象消失，即都没有 HbA_2 与 A_1 之间的相互作用。

4 讨论

现在看来，无氧条件时红细胞内部发生了很大变化，最明显的就是 HbA_2 现象消失。A_2 现象[1-4]是红细胞内 HbA_2 与 HbA_1 之间的相互作用，在红细胞内二者是结合存在的。这种结合只有在严重贫血时才遭到破坏。在这里，我们看到无氧条件下这种结合也遭到破坏，可见这种条件能够影响红细胞的内部结构，且作用深刻。

HbA_2 现象是发生在红细胞内部的事情，后来我们发现在红细胞外 HbA_2 与 HbA_1 可以发生交叉互作[3-14]。现在的问题是，无氧条件下 HbA_2 现象消失，此条件下的血红蛋白交叉互作如何？也会消失吗？下一章里我们就来讨论这个问题。

参 考 文 献

[1] 秦文斌，梁友珍. 血红蛋白 A_2 现象 I . 此现象的发现及其初步应用. 生物化学与生物物理学报, 1981, 13 (2)：199-205.
[2] 秦文斌. 血红蛋白的 A_2 现象发生机制的研究"红细胞 HbA_2"为 HbA_2 与 HbA 的结合产物. 生物化学与生物物理进展, 1991, 18 (4)：286-288.
[3] 秦文斌. 红细胞外血红蛋白 A 与血红蛋白 A_2 之间的相互作用. 生物化学杂志, 1991, 7 (5)：583-587.
[4] Su Y, Gao L J, Ma Q, et al. Interactions of hemoglobin in lived red blood cells measured by the electrophoresis release test. Humana Press, 2012, 869 (17)：393-402.
[5] 秦文斌. 红细胞内血红蛋白的电泳释放——发现和研究. 北京：科学出版社, 2015：66-148.

[6] 韩丽红, 闫斌, 高丽君, 等. 血红蛋白F与血红蛋白A_2之间的交叉互作. 中国医药指南, 2015 (5) : 62-63.
[7] 秦文斌. 红细胞内血红蛋白的电泳释放——发现和研究. 北京: 科学出版社, 2015: 98-100.
[8] 秦文斌. 红细胞内血红蛋白的电泳释放——发现和研究. 北京: 科学出版社, 2015: 101-105.
[9] 秦文斌. 红细胞内血红蛋白的电泳释放——发现和研究. 北京: 科学出版社, 2015: 106-115.
[10] 苏燕, 邵国, 睢天林, 等. 几种鱼类血红蛋白不与人血红蛋白相互作用及其进化意义. 包头医学院学报, 2000, 23 (2) : 452-454.
[11] 于慧, 王占黎, 秦文斌, 等. 美国锦龟血红蛋白与人血红蛋白相互作用的生物信息学和计算机模拟研究. 包头医学院学报, 2005, 21 (3) : 205-209.
[12] Yu H, Wang Z L, Qin W B et al. Structural basis for the specific interaction of chicken hemoglobin with bromphenolblue: a computational analysis. Molecular Physics, 2010, 108 (2) : 215–220.
[13] 王程, 隋春红, 秦文斌, 等. 生物信息学方法推测脊椎动物血红蛋白相互作用机制. 吉林医药学院学报, 2008, 29 (5) : 249-253
[14] 秦文斌. 红细胞内血红蛋白的电泳释放——发现和研究. 北京: 科学出版社, 2015: 143-148.

第五十五章 肝素与DNA之间的相互作用

王占黎[#] 高丽君[#] 周立社[#] 王步云 邵 国 苏 燕 秦良谊 秦文斌[※]

(包头医学院 血红蛋白研究室,包头 014010)

摘 要

目的:研究肝素与DNA的相互作用。

方法:将肝素与DNA混合后进行常规琼脂糖凝胶(含溴乙锭)电泳,观察DNA区带的颜色、亮度、电泳速度。再做琼脂糖凝胶穿过电泳,让肝素穿过DNA,继续观察DNA区带的颜色、亮度、电泳速度。

结果:穿过实验中,DNA区带显橙红色,亮度不变、泳速加快。混合实验结果与肝素量多少有关。肝素量大时,DNA区带显白色,亮度下降,泳速变慢。

结论:肝素能与DNA互作,两种实验的结果不同,可能来自不同的互作机制。

关键词 肝素 DNA;互作

1 前言

肝素是一类结构复杂的糖胺聚糖,它在人体内具有多重生物活性,除了经典的抗凝血作用外,肝素还能与生长因子和细胞因子多种蛋白质发生相互作用。所以,肝素-蛋白质相互作用成为当前的一项重要课题[1-7]。我们在研究肝素与各种物质(包括血液蛋白质和各种染料)相互作用过程中,就肝素是否能与DNA发生互作进行了研究。结果发现,肝素能与DNA互作,而且有其自身的特点,特报告如下。

2 材料与方法

2.1 材料 肝素钠注射液来自江苏万邦生化医药股份有限公司,2ml含12 500U,辅料为氯化钠。DNA为我研究室常规PCR产物。三重PCR产物为淋球菌(630bp)、衣原体(241bp)和β-肌动蛋白(内对照 318bp),单重PCR产物只有β-肌动蛋白(318bp)。

2.2 混合互作实验 向十个琼脂糖凝胶(含溴乙锭)孔中各加入三重PCR产物10μl,第1孔及第10孔为对照,再向2~9孔加入不同数量的生理盐水或肝素注射液,具体如表55-1所示。

表55-1 混合互作实验1~10孔应加的物质及体积

孔号	1	2	3	4	5	6	7	8	9	10
加入物质	无	盐水	肝素	盐水	肝素	盐水	肝素	盐水	肝素	无
体积(μl)	0	2	2	8	8	16	16	32	32	0

电泳条件:电势梯度10V/cm,电泳时间30分钟,置于透射式紫外灯上,数码相机拍照留图。

[#] 并列第一作者
[※] 通讯作者:秦文斌,电子邮箱:qwb5991309@tom.com

2.3 交叉互作实验 向五个琼脂糖凝胶(含溴乙锭)孔中各加入三重 PCR 产物 10μl，电泳分离出来 3 个 DNA 区带后，再向第 1、3、5 孔中加入盐水 10μl，第 2、4 孔中加入肝素注射液 10μl，继续电泳，看到肝素孔与盐水孔有差别后，数码相机拍照留图。

3 结果

3.1 混合结果 如图 55-1 所示。

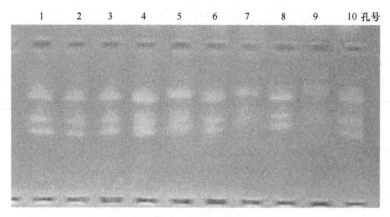

图 55-1 混合互作结果

由图 55-1 可以看出，量少时(如 2μl)肝素与盐水结果差别不大。随着数量增多，差别越来越明显。32μl 时，盐水孔 DNA 亮度基本未变，肝素孔 DNA 泳速变慢、亮度明显下降。

3.2 先分离 DNA，再使肝素穿过。见图 55-2。

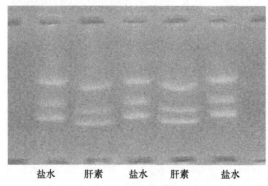

图 55-2 交叉互作结果

由图 55-2 可以看出，肝素穿过 DNA 时，DNA 泳速加快、区带变长、变弯，但亮度未变。

4 讨论

文献中提到过肝素与病毒发生互作，但具体内容都是肝素与病毒的外壳蛋白发生互作，没有涉及肝素与病毒中核酸部分的内容[8, 9]。有一篇文章提到了肝素对 DNA 的作用[10]，提到肝素能够改变 DNA 的沉降系数。

(1) 在我们的混合互作实验里，肝素与 DNA 混合后，DNA 泳速变慢、量减少。这说明肝素与 DNA 发生互作。其 DNA 泳速变慢的粗略机制推测如下所示。

$$DNA^+\text{-}^-\text{肝素}^+ = (DNA\text{-}\text{肝素})^+ \text{（泳向阴极）}$$

肝素量大时 DNA 量减少的粗略机制推测如下：大量肝素与 DNA 互作，占据了 DNA 与溴乙锭结合的部位，使 DNA 无色，看似减少。

$$DNA\text{-}EB + HEP \longrightarrow DNA\text{-}HEP + EB$$

(2) 在我们的穿过互作实验里，肝素穿过 DNA 时，DNA 泳速加快、区带变长、变弯。这说明肝素穿过 DNA 时，二者发生互作。其 DNA 泳速加快的粗略机制推测如下所示。

$$DNA^+\text{-}^-\text{肝素}^- = (DNA\text{-}\text{肝素})^- \quad \text{（泳向阳极）}$$

(3) 为什么混合互作与穿过互作结果不同？推测是由于来自肝素注射液的成分不均一。即此时肝素至少含有两种电泳成分(快泳和慢泳)。

$$HEP = HEPf + HEPs$$

混合互作时起作用的肝素成分，与穿过互作时起作用的成分不同，穿过时起作用的是快泳成分，它不占据 DNA 与溴乙锭的结合部位；混合时起作用的是慢泳成分，它占据 DNA 与溴乙锭的结合部位。

$$HEPf + DNA\text{-}EB \longrightarrow EB\text{-}DNA\text{-}HEPf \tag{1}$$

$$HEPs + DNA\text{-}EB \longrightarrow DNA\text{-}HEPs + EB \tag{2}$$

(4) 先分离 DNA，再与肝素混合的实验中，可以看到泳速变慢"白色的" DNA，这是支持反应(2)的结果。

(5) 已知肝素类物质的结构复杂、成分多样性[11, 12]，将肝素分为快泳和慢泳两种显然过于简单化。我们准备用双向电泳将肝素成分全部展开，然后再与 DNA 互作，观察其"互作多样性"。在此基础上，探讨与此相对应的肝素结构。

参 考 文 献

[1] Conrad H E. Heparin-Binding Proteins. San Diego: Academic Press, 1998.
[2] Capila I, linhard RJ. Heparin-protein interactions. Angew Chem Int Ed Engl, 2002, 14(3): 385-412.
[3] Coombe DR, Kettle WC. Heparin sulfate-protein interactions: therapeutic potential through structure-function insights. Cell mol life sci, 2005, 62(2): 410-424.
[4] Kreuger J, Spillmann D, Li JP, et al. Interactions between heparan sulfate and proteins: the concept of specificity. J Cell Biol, 2006, 174(3): 323-327.
[5] Ding SS, Cui YL, Gong Z, et al. Heparin for growth factor delivery systems. Chemistry, 2008, 20(12): 1998-2011.
[6] Free O, A, Courty P, Wilkinson J, et al. Identification of heparin-binding sites in proteins by selective labeling. Mol Cell Proteomics, 2009, 8(10): 2256-2265.
[7] Xiao SJ, Hu PP, Tan KJ, et al. Spectra characterization of the conformational changes of human cellular prion protein induced by heparin. Acta Chimica Sinica, 2010, 68(10): 1032-1036.
[8] Rux A H, Lambris J D, Friedman HM, et al. Kinetic analysis of glycoprotein C of herpes simplex virus types 1and 2 binding to heparin, heparan sulfate, and complement component C3b. Virology, 2002, 294, 324-332.
[9] Kalia M, Chandra V, Rahman S A, et al. Heparan sufate proteoglycan are required for cellular binding of the hepatitis E virus ORF2 capsid protein and for viral infection. J Viol, 83(24): 12714-12724.
[10] Paponova V D, Ruzga B, SamoilovaV K, et al. Effect of heparin on DNA. Bulletin of Experimental Biology and Medicine, 99(3): 228-291.
[11] Esko JD, Lindahl U. Molecular diversity heparan sulfate. J Clin Invest, 2001, 108: 169-173.
[12] Casu B, Lindahl U. Structure and biological interactions of heparin and heparan sulfate. Adv Carbohydr Chem Biochem, 57: 157-206.

第五十六章 肝素与染料之间的相互作用——用凝胶电泳发现肝素与染料之间的相互作用

杨文杰[1] 韩丽红[1] 高丽君[2] 王步云[1] 秦文斌[2]

(包头医学院 1. 生物化学教研室；2. 包头医学院血红蛋白研究室，包头 014060)

摘 要

目的：观察肝素与23种染料之间的相互作用。

方法：利用淀粉-琼脂糖凝胶电泳及聚丙烯酰胺凝胶电泳，观察染料加肝素后电泳行为的变化。

结果：淀粉-琼脂糖凝胶电泳中染料加肝素后多数结果是泳速变慢，有一些留在原点，也有前移、后退者。二氯荧光素、荧光红、荧光素、甲酚红、苯胺蓝、溴酚兰、茜素红S、胭脂红、亮绿、氨基黑10B、丽春红G、曙红、苯胺黑、氯酚红加肝素后泳速变慢。亚甲蓝、灿烂甲酚蓝、甲基紫加肝素后电泳留在原点。刚果红、洋红加肝素后电泳变化不明显。四氯荧光素、氯化硝基四氮唑加肝素后电泳荧光看不清。聚丙烯酰胺凝胶电泳中溴酚兰加入肝素后泳速明显变慢，随着肝素量增加溴酚兰的泳速加快。与淀粉-琼脂糖凝胶电泳结果一致，但它的电泳结果更明显。

结论：本次所研究的23种染料，几乎都能与肝素发生相互作用，为进一步深入研究打下基础。

关键词 电泳；肝素；染料；相互作用

Using gel Electrophoresis Found the Interaction between heparin and Dyes

Yang Wen-jie[1]　Han Li-hong[1]　Gao Li-jun[2]　Wang Bu-yun[1]　Qin Wen-bin[2]

(1. Biochemistry Department; 2. Hemoglobin Laboratory of Baotou Medical College, 014060, Baotou, China)

ABSTRACT　Objective: To investigate the relationship between heparin and 23 kinds of dyes. **Methods:** Starch agarose gel electrophoresis and polyacrylamide gel electrophoresis were used to detect the changes of electrophoresis behavior after dye plus heparin. **Results:** Most electrophoresis results were slower than before after plus heparin, there were some stay in origin, also had moved forward, backward by starch agarose gel electrophoresis. The electrophoresis speed of dichlorofluorescein, fluorescent red, fluorescein, cresol red, aniline blue, bromophenol blue, the Sin purple S, carmine, light green, amido black 10B, Ponceau G, eosin, aniline black, chlorophenol red after plus heparin became slower. Methylene blue, brilliant cresyl blue, methyl violet after plus heparin electrophoresis stay at the origin. The electrophoresis of congo red, magenta after plus heparin did not change significantly. There were no electrophoresis fluorescence for tetrachloro-fluorescein, nitro tetrazolium blue after plus heparin. In polyacrylamide gel electrophoresis of bromophenol blue after adding heparin electrophoresis speed slowed down than before obviously, with the amount of heparin increased bromophenol blue electrophoresis speed up. Results of bromophenol blue were consistent with the starch agarose gel electrophoresis, but its electrophoresis results are more obvious. **Conclusion:** The study of 23 kinds of dyes, almost can interact with heparin.

Key words: Electrophoresis; Heparindye; Interactions
Chinese Library Classification: R96   Document code: A
Article ID: 1671-6273(2012)

1 前言

肝素是一类结构复杂的糖胺聚糖，它在人体内具有多重生物活性，除了经典的抗凝血作用外，肝素还能与生长因子和细胞因子多种蛋白质发生相互作用。所以，肝素-蛋白质相互作用成为当前的一项重要课题[1-6]。肝素和硫酸类肝素在体内具有抗凝血、调血脂、抗炎、抗动脉粥样硬化、调节血管生成等多方面的生物活性。这些生物活性是通过与多种蛋白质的相互作用来发挥的。在肝素与染料互作方面，已经有过光谱学研究[7]和电化学分析[8,9]，本文则从电泳角度研究肝素与染料的相互作用。我们利用淀粉-琼脂糖凝胶电泳及聚丙烯酰胺凝胶电泳，观察染料加肝素后电泳行为的变化。本次所研究的23种染料几乎都能与肝素发生相互作用，但是肝素与染料互作时的结构与功能关系还需进一步研究。

2 材料与方法

2.1 试剂

2.1.1 肝素   肝素钠注射液(江苏万邦生化医药股份有限公司)，主要成分为硫酸氨基葡聚糖的钠盐，属黏多糖类物质，平均分子量 12 000，辅料为氯化钠、注射用水。

2.1.2 染料   共23种，包括四碘荧光素、二氯荧光素、荧光红、荧光素、亚甲蓝、甲基绿、甲酚红、刚果红、苯胺蓝、溴酚蓝、茜素红S、胭脂红、亮绿、氨基黑10B、洋红、灿烂甲酚蓝、丽春红G、溴酚蓝(?)(标签上字看不清)、曙红、苯胺黑、甲基紫、氯酚红、氯化硝基四氮唑蓝。这些染料，大部分是进口产品，级别为试剂级或生物染料级。

2.2 实验方法和步骤

2.2.1 淀粉-琼脂糖混合凝胶电泳检测肝素与染料之间的相互作用   电泳方法同前[11,12]。

2.2.2 染料选择及其与肝素的互作实验   对准备好的30多种染料，做水溶型试验：取1.5ml EP管，加入蒸馏水1ml、染料1耳勺，振荡后染料全溶解者留下，不溶解者剔除。30多种染料中有23种溶于水。以下用水溶性染料做实验。每种染料用2个1.5ml EP管，各加入染料8μl，其中之一管加入生理盐水8μl，另一管加入肝素氯化钠注射液8μl。混匀后上电泳，每种染料占两个泳道，前边是染料加盐水，后边是染料加肝素，然后通电，电势梯度6V/cm，电泳时间45分钟。对于电泳结果，有以下几种处理：用数码相机直接照相、置于紫外光灯上照相(透射和反射)。实验分两批进行。

2.2.3 第一批染料   四碘荧光素、二氯荧光素、荧光红、荧光素、亚甲蓝、甲基绿、甲酚红、刚果红、苯胺蓝、溴酚蓝，共10种。

2.2.4 第二批染料 茜素红 S、胭脂红、亮绿、氨基黑 10B、洋红、灿烂甲酚蓝、丽春红 G、溴酚蓝(?)、曙红、苯胺黑、甲基紫、氯酚红、氯化硝基四氮唑蓝，共 13 种。

2.2.5 PAGE 测肝素与溴酚蓝之间的相互作用 PAGE 浓度为 10%。比较溴酚蓝与加入不同量肝素后的电泳行为，由电泳仪上直接照相观察。

3 结果

3.1 第一批染料与肝素互作的电泳结果

由图 56-1-A 可以看出，有一些染料泳向阳极，另一些泳向阴极，加肝素者多数有变化。其中泳道 1、2 四碘荧光素看不清，泳道 3、4 为二氯荧光素也不明显，泳道 5 荧光红泳向阳极，呈黄绿色，泳道 6 荧光红加肝素后稍后退，泳道 7、8 荧光素类似荧光红，泳道 9 亚甲蓝泳向阴极，呈蓝色，泳道 10 亚甲蓝加肝素后变成原点(蓝色)不动，泳道 11 为甲基绿泳向阴极更远，呈浅蓝绿色，泳道 12 甲基绿加肝素后阴极成分减弱、阳极出现新成分，泳道 13 甲酚红泳向阳极，泳道 14 甲酚红加肝素后泳速变慢，泳道 15 刚果红留在原点，呈深红色，泳道 16 刚果红加肝素后电泳位置和颜色都不变，泳道 17 苯胺蓝泳向阳极，呈浅紫色，泳道 18 苯胺蓝加肝素后泳速稍变慢，色更浅，泳道 19 溴酚蓝泳向阳极，呈蓝色，泳道 20 加肝素后泳速变慢、颜色变浅。图 56-1-B 基本同图 56-1-A，泳道 5、6(荧光红)泳道 7、8(荧光素) 结果更明显，泳道 17、18(苯胺蓝)显深蓝色。图 56-1-C 泳道 3、4(二氯荧光素)结果开始明显，泳道 5、6(荧光红)泳道 7、8(荧光素) 结果更明显，泳道 17、18(苯胺蓝)出现多带，差别有三处，相对最明显者位于中间处。

3.2 第二批染料与肝素互作的电泳结果

由图 56-2-A 可以看出，多数染料泳向阳极，少数泳向阴极，加肝素者多数有变化。图 56-2-B 基本同图 56-2-A。图 56-2-C 泳道 17、18 显示明显荧光。

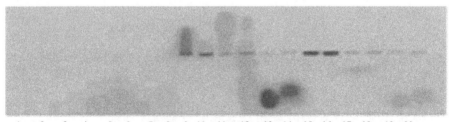

图 56-1-A 第一批染料电泳后直接照相结果

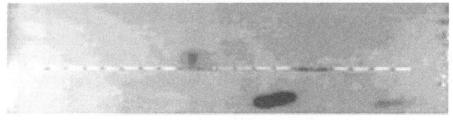

图 56-1-B UV 灯上(透射)照相结果

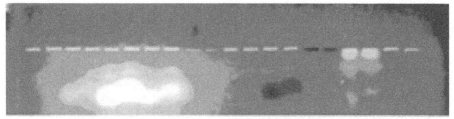

图 56-1-C UV 灯下(反射)照相结果

注释：电泳图中上方为负极，下方为正极。由左向右 20 个泳道，单数为染料，双数为染料加肝素。泳道 1、2 为四碘荧光素，泳道 3、4 为二氯荧光素，泳道 5、6 为荧光红，泳道 7、8 为荧光素，泳道 9、10 为亚甲蓝，泳道 11、12 为甲基绿，泳道 13、14 为甲酚红，泳道 15、16 为刚果红，泳道 17、18 为苯胺蓝，泳道 19、20 为溴酚蓝。

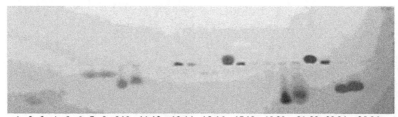

图 56-2-A 第二批染料电泳后直接照相结果

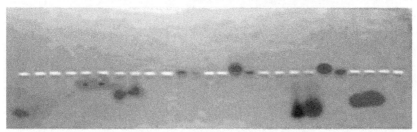

图 56-2-B UV 灯上(透射)照相结果

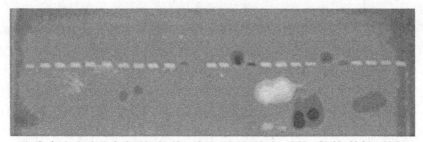

图 56-2-C UV 灯下(反射)照相结果

注释：电泳图中上方为负极，下方为正极。由左向右 26 个泳道，单数为染料，双数为染料加肝素。泳道 1、2 为茜素红 S，泳道 3、4 为胭脂红，泳道 5、6 为亮绿，泳道 7、8 为氨基黑 10B，泳道 9、10 为洋红，泳道 11、12 为灿烂甲酚蓝，泳道 13、14 为丽春红 G，泳道 15、16 为溴酚蓝，泳道 17、18 为曙红，泳道 19、20 为苯胺黑，泳道 21、22 为甲基紫，泳道 23、24 为氯酚红，泳道 25、26 为氯化硝基四氮唑蓝，共 13 种。

3.3 淀琼电泳综合结果

淀粉-琼脂糖凝胶电泳中染料加肝素后多数结果是泳速变慢，有一些留在原点，也有前移、后退者。二氯荧光素、荧光红、荧光素、甲酚红、苯胺蓝、溴酚蓝、茜素红S、胭脂红、亮绿、氨基黑10B、丽春红G、曙红、苯胺黑、氯酚红淀琼电泳原染料泳向阳极,加肝素后泳速变慢。亚甲蓝、灿烂甲酚蓝、甲基紫淀琼电泳原染料退向阴极,加肝素后电泳留在原点。刚果红、洋红淀琼电泳加肝素后电泳变化不明显。四氯荧光素、氯化硝基四氮唑蓝淀琼电泳加肝素后电泳荧光看不清。

3.4 PAGE检测肝素与溴酚蓝互作结果

由图56-3-B可以看出,溴酚蓝加入肝素后泳速明显变慢,随着肝素量增加溴酚蓝的泳速加快(图56-3-A,图56-3-B)。

图56-3-A 聚丙烯酰胺凝胶电泳仪(电泳在进行中)

图56-3-B 电泳仪的直接照相结果

泳道由左向右,1、3、5为溴酚蓝,2、4、6为溴酚兰加肝素(第2泳道肝素量与溴酚蓝相等,第4泳道肝素量增加一倍,第6泳道肝素量增加两倍)

4 讨论

肝素与染料互作的研究种类不多,每一项目只研究一两种染料,而且用的是光谱学和电化学分析方法[7-9],本文用电泳方法可同时研究多种染料与肝素的相互作用,效率较高。本文所研究的多种染料大多数都与肝素有相互作用,从电泳行为角度可分成两类:泳向正极成分和泳向负极成分。泳向正极成分中有一部分属于荧光染料,泳向负极成分则几乎不显荧光。荧光染料中苯胺蓝的情况比较特殊:电泳后直接照相时,只有一个白点被暗蓝色区带包围,加肝素者也差不多,稍靠后一点;UV灯上透射照相时,见到两个蓝带,苯胺蓝色深,加肝素者色浅,也是稍靠后一点;UV灯上反射照相时,苯胺蓝出现6~7条浅蓝绿色荧光区带,加肝素者有三条带泳速变慢,也就是说有一部分成分与肝素发生相互作用。荧光红与荧光素的情况与苯胺蓝有一些类似,UV灯上透射照相时,出现3~4条浅蓝绿色荧光区带,加肝素者有一条带泳速变慢。二氯荧光素的荧光区带比荧光素弱、区带减少,四碘荧光素的荧光区带更弱、甚至消失,可见氯化或碘化处理对荧光素的影响较大。第二批染料中曙红的荧光明显,快泳的次要成分不受肝素影响,慢泳的主要成分与肝素互作而变得更慢。PAGE检测肝素与溴酚蓝互作结果,也是溴酚蓝加肝素时泳速变慢,与淀琼电泳结果一致,但它比淀琼电泳结果更明显。淀琼电泳可以观察泳向正极和负极的两类染料,PAGE无法检测泳向负极的染料。总之,我们用两种电泳方法都证明了多种染料能与肝素发生互作,为进一步深入研究，特别是结构分析打下基础。肝素的结构与功能关系的文章不少[10-20]，都是肝素与各种蛋白互作时的结构与功能关系，没有涉及肝素与染料互作时的结构与功能关系。肝素是一类结构复杂的糖胺聚糖，它的哪个部位与染料结合、有多少个结合部位、正负电荷的分布情况如何等都是需要深入研究的内容。

参 考 文 献

[1] Capila I, linhard RJ. Heparin-protein interactions. Angew Chem Int Ed Engl, 2002, 14(3): 385-412.

[2] Coombe DR, Kettle WC. Heparin sulfate-protein interactions; therapeutic potential through structure-function insights. Cell Mol Life Sci, 2005, 62(2) : 410-424.

[3] Kreuger J, Spillmann D. Interactions between heparan sulfate and proteins; the concept of specificity. J Cell Biol, 2006, 174(3) : 323-327.

[4] Ding S, Cui YL.Heparin for growth factor delivery systems. Progress in Chemistry, 2008, 20(12): 1998-2011.

[5] Ori A, Free PC. Identification of heparin-binding sites in proteins by selective labeling. Mol Cell Proteomics, 2009, 8(10): 2256-2265.

[6] Xiao SJ, Hu PP. Spectra characterization of the conformational changes of human cellular prion protein induced by heparin. Acta Chimica Sinica, 2010, 68(10): 1032-1036.

[7] Nandinia R, Vishalakshib B.A spectroscopic study of interaction of cationic dyes with heparin. Orbital, 2009, 1(4): 255-272.

[8] Sun W, Ding YQ, Jiao K. Electrochemical studies on the interaction of heparin with crystal violet and its analytical application. J   Analytical Chemistry, 2006, 61(4): 357-364.

[9] Hui N, Sun W, Ding YQ, et al. Electrochemical studies on the interaction of heparin with phenosafranine and its analytical application. Journal of the Chinese Chemical Society, 2009, 56: 269-278.

[10] Su Y, Shao G, Gai L, et al. RBC electrophoresis with discontinuous power supply –a newly established hemoglobin release test. Electrophoresis, 2009, 30: 3041-3043.

[11] Su Y, Gao LJ, Ma Q, et al. Interactions of hemoglobin in lived red blood cells measured by the electrophoresis release test. Electrophoresis, 2010, 31: 2913-2920.

[12] 高宁国，程秀兰，杨敬，等. 肝素结构与功能的研究进展. 生物工程进展, 1999, 19(5): 4-13.

Gao NG, Cheng XL, Yang J, et al. Heparin structure and function of research progress. Progress of Bio-Engineering, 1999, 19(5): 4-13.

[13] Mulloy B, Forste MJ. Conformation and dynamics of heparin and heparan sulfate. Glycobiology, 2000, 10(11): 1147-1156.

[14] 孙自才，魏峥，魏可镁. 硫酸类肝素的结构、功能、修饰与合成. 化学进展, 2008, 20(7): 1136-1142.

Sun ZC, Wei Z, Wei KM. Heparan sulfate structure, function, modification and synthesis. Progress in Chemistry, 2008, 20(7): 1136-1142.

[15] Seki Y, Mizukura M, Ichimiya T, et al.O-sulfate groups of heparin are critical for inhibition of ecotropic murine leukemia virus infection by heparin. Virology, 2012, 424(1): 56-66.

[16] Hung SC, Lu XA, Lee JC, et al. Synthesis of heparin oligosaccharides and their interaction with eosinophil-derived neurotoxin. Org Biomol Chem, 2012, 10(4): 758-772.

[17] Krauel K, Hackbarth C, Fürll B, et al. Heparin-induced thrombocytopenia: in vitro studies on the interaction of dabigatran, rivaroxaban, and low-sulfated heparin, with platelet factor 4 and anti-PF4/heparin antibodies. Blood, 2012, 119(5): 1248-1255.

[18] Bhaskar U, Sterner E, Hickey AM, et al.Engineering of routes to heparin and related polysaccharides. Appl Microbiol Biotechnol, 2012, 93(1): 1-16.

[19] Rajabi M, Struble E, Zhou Z, et al. Potentiation of C1-esterase inhibitor by heparin and interactions with C1s protease as assessed by surface plasmon resonance. Biochim Biophys Acta, 2012, 1820(1): 56-63.

[20] Sankhala RS, Damai RS, Anbazhagan V, et al. Biophysical investigations on the interaction of the major bovine seminal plasma protein, PDC-109, with heparin. J Phys Chem B, 2011, 115(44): 12954-12962.

[原文发表于"现代生物医学进展, 2012, 12(27): 5259-5264"]

第五十七章 计算机模拟法分析鸡血红蛋白与溴酚蓝特异性互作的结构基础

Structural basis for the specific interaction of chicken hemoglobin with bromophenol blue: a computational analysis

Hui Yu[a], Zhanli Wang[b], Wenbin Qin[c] and Liangren Zhang[d]

([a] Department of *Laboratory Medicine, the Affiliated Tenth People's Hospital, Tongji University, Shanghai* 200072, *China*; [b] *College of pharmaceutical Science, Zhejiang University of Technology, Hangzhou* 310014, *China*; [c] *Laboratory of Hemoglobin, Baotou Medical College, Baotou* 014010, *China*; [d] *School of Pharmaceutical Science, Peking University, Beijing* 100083, *China*)

Abstract

It has been observed that bromophenol blue interacted specifically with chicken hemoglobin but not with carp hemoglobin during electrophoresis, but the mechanism of interaction is still not well understood. In this computational study, the binding of bromophenol blue to chicken hemoglobin has investigated using sequence alignment, homology modeling, electrostatic potential distribution and flexible docking methods. Molecular modelling studies reveal that bromophenol blue-binding site, formed by residues Val1α, Leu2α, Ala131α, Thr134α, Ala 138α and Arg141α, is located between two α chains of chicken hemoglobin, and the binding is dominated by hydrophobic interactions. Moreover, comparison of chicken and carp hemoglobin structural models provides a structural rationale for the recognition of bromophenol blue by chicken hemoglobin. These principles can in turn be used to study the molecular recognition mechanism and design a mimic of bromophenol blue for the development of new hemoglobin binders.

Key words hemoglobin; bromophenol blue; electrophoresis; molecular modeling

1 Introduction

Hemoglobins (Hbs) are members of the globin superfamily responsible for oxygen transport. This molecule, being one of the most well-studied protein systems, has a long evolutionary history [1,2]. They are universally present in higher vertebrates, and are found in some bacteria and plants [3]. In vertebrates, Hbs are typically tetrameric proteins consisting of two pairs of identical chains. The blood of adult chickens, as in most other birds, is composed of two Hb components that have identical chains but differ in the sequences of subunits. Hb A ($\alpha^A_2\beta_2$) and Hb D($\alpha^D_2\beta_2$) are expressed in a 3∶1 ratio in adult chickens[4,5]. Chromatography on DEAE-Sephadex resolved three components of carp Hb which were designated Ⅰ, Ⅱ, and Ⅲ in the order of their elution. Previous

---

(Received 6 October 2009; final version received 12 January 2010)
Corresponding author: Email: qwb@public.hh.nm.cn; liangren@bjmu.edu.cn

results indicate that the three major components of carp Hb are both structurally and functionally very similar [6]. The respiratory system of chicken differs from that of carp. Although their Hbs appear to be functionally similar, some qualitative differences do exist which could represent the functional basis of their possibly different physiological roles.

Bromophenol blue (BPB) is a common dye used as a component of loading buffers to monitor the progress of DNA and protein electrophoresis. It has been demonstrated that BPB can bind to some proteins including cytochrome c' from Chromatium vinosum, bovine serum albumin and human serum albumin [7-9]. The dye BPB is shown to bind to hydrophobic sites on the surface of these proteins and hydrophobic interactions are important.

In this study, we showed for the first time that BPB interacted with chicken Hb but not with carp Hb. It was a surprise to discover this phenomenon during electrophoresis studies of the Hbs. There is, therefore, a growing interest in investigation of interaction mechanism. Here we focus on the analysis of structural basis for the specific interaction of chicken Hb with BPB using computational approaches, and give evidence for the role of Val1α, Ala 131α and Ala 138α in the α-subunit's N- and C-termini regions. The structural differences between chicken Hb and carp Hb are also discussed. Information on the binding domain of chicken Hb is important for explaining the binding mechanism of chicken Hb and BPB and can further help us to design the selective binders of Hbs.

2 Experimental

BPB and agarose gel were purchased from Shanghai Yito Enterprise (Shanghai, China). Carbon tetrachloride, Tris (hydroxymethyl) aminomethane, Ethylene diamine tetraacetic acid (EDTA) and boric acid were from Beijing Chemical Works (Beijing, China). All the chemicals used were of analytical grade. Blood was collected by caudal vein puncture from specimens. Hb preparation was carried out at 4 ℃.The heparinized blood samples were washed by centrifugation four times with 0.9% NaCl. RBC were collected and haemolysis was accomplished by adding 1.5 volume of ultrapure water and 0.4 volume of carbon tetrachloride, followed by clarification by centrifugation at 10000 rpm for 30 min. Subsequently, 5% (w/v) NaCl was added to the sample, and the solution was centrifuged again at 18 000 rpm for 15 min. Finally, Hb was obtained in the haemolysate form. After isolation of Hb, additional purification was carried out by gel electrophoresis experiment. In this study we used only chicken Hb A and carp Hb I to examine the BPB-binding properties. A 1% separating agarose gel was prepared using Tris–HCl buffer. Electrophoresis was run in 1ÂTris- EDTA-borate (TEB, pH 8.6) buffer.

3 Computational details

All molecular modelling studies were performed using Insight II software developed by Accelrys, San Diego [10].

The α and β goblin chains of chicken Hb A(accession number: P01994 and P02112) and carp Hb I(accession number: P02016 and P02139) were obtained from the databank in the National Center for Biotechnology Information. The three-dimensional structures of Hbs were generated using the program Modeler. The crystal structures of chicken Hb D (PDB code: 1HBR) and Antarctic fish Hb (PDB code: 1S5Y) were used as reference proteins for chicken Hb A and carp Hb I, respectively. After the above steps, energy minimisation and molecular dynamics calculations were performed as described in previous study [11]. The obtained structures were evaluated with the

ProStat and Profile-3D programs [12].

The electrostatic potential of the binding site surfaces was calculated using DelPhi module [13]. DelPhi is a program for calculating electrostatic properties, including the effects of bulk solvent and ionic strength thereby providing crucial data for rationalising differences in the activity of macromolecules. The boundary between the protein and the water was defined by rolling a ball r=1.4 Å over the van der Waals boundary of the protein [14]. The dielectric constants of the proteins and of the surrounding medium were 4 and 80, respectively.

Affinity program was used for docking of BPB into the chicken Hb A binding site [15]. CVFF force filed was selected prior to perform docking calculations. To explore the effects of salvation implicitly, a distance-dependent dielectric constant ($\varepsilon=4$ rij) was used. The nonbonded cell multipole method was used for SA docking with input energy parameters. The initial position of BPB within the chicken Hb A binding site was found using a Monte Carlo type procedure to search both conformational and Cartesian space. The resulting structure was accepted on the basis of an energy check. Second, a simulated-annealing phase optimised BPB placement and the structures were then subjected to energy minimisation based on molecular dynamics. The final conformations were obtained through a simulated annealing procedure from 500 to 300 K, and then 1000 rounds of energy minimisation were performed to reach convergence. The lowest global structure obtained was used for computing intermolecular binding energies.

4 Results and discussion

4.1 Experimental results During gel shift studies of Hbs, it was observed that BPB was retarded on the gel in the presence of chicken Hb A with molar ratio of 1:1. Figure 57-1 shows the mobility of BPB and chicken Hb A alone and together during electrophoresis. Figure 57-1 also shows the mobility of carp Hb I for comparison with chicken Hb A. One can observe that the BPB band (lane 1) is depleted in the lane containing chicken Hb A relative to migration of BPB alone (lane 2). The results provide clear evidence that chicken Hb A can interact with BPB. We assume that the decrease of the electrophoretic mobility of BPB is partly due to the change in overall charge of BPB resulting from the binding of chicken HB A. However, no corresponding change in mobility of BPB band is observed in the presence of carp Hb I (lane 3). This result seems to indicate that the significant interaction between carp Hb I and BPB does not occur.

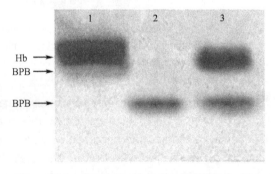

Figure 57-1 Gel electrophoresis of BPB in the presence and absence of Hbs. Lanes: (1) chicken Hb A + BPB; (2) BPB; (3) carp Hb I + BPB

4.2 Theoretical results In order to investigate the potential BPB-binding site in chicken Hb A, sequence alignments of chicken Hb A and carp Hb I were carried out(see supplementary data). The sequence identity between β chains of two Hbs is 57%. The regions of chicken Hb A which contains the highest number of mutated residues when compared with carp Hb I are the α-subunit's N- and C-termini. The replacement of Ser1α, Asp131α and Glu138α in carp Hb with Valα1, Ala 131α and Ala138α in chicken Hb increases the hydrophobic effect of these regions (numbering is according to chicken Hb A). The latter feature is very important, as BPB binding to proteins is essentially hydrophobic. Moreover, multiple sequence alignment of several known avian and teleost Hbs was performed. A comparison of the α-chains also revealed that all of these hydrophobic amino acids (Val11, Ala131 and Ala138) were fully conserved among the avian Hbs and were not found in the teleost Hbs (Figure 57-2). We therefore suppose that the binding of BPB to chicken Hb A may due to the α-subunit's N- and C-termini. Three hydrophobic residues, Val1, Ala131 and Ala138 which are conserved in α chains of chicken Hb A but not in carp Hb I might contribute to the interaction.

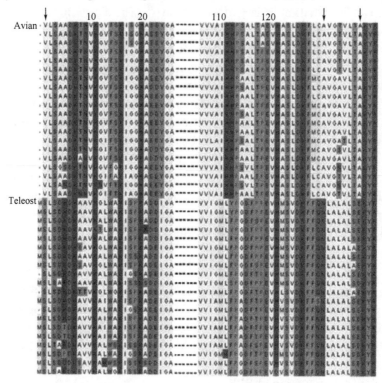

Figure 57-2 Multiple sequence alignment of avian and teleost Hb (Chain α) are shown, where the amino acids are colored according to their functions: red, light blue, pink, green, yellow and white stand for acidic, amido, aromatic, basic, hydroxyl, praline, sulfur, hydrophobic and other, respectively. Numbering is according to avian Hbs. The arrow indicates α1, α131 and α138, respectively

To further confirm the above hypothesis, the homology models of chicken Hb A and carp Hb I were built. The stereochemical quality of the modelled structures was checked by the ProStat. The bond lengths and the bond angles were not significantly different from the average values calculated from the known proteins. Using Profiles-3D, the overall self-compatibility scores for the predicted structures were all above the expected values (supplementary data, Figure 57-2). Our results indicate that the predicted models are reliable. These homology models would help us to understand the spatial arrangement and physicochemical properties of Hbs. First, the predicted structure of chicken Hb A was compared with that of carp Hb I with the aim of verifying if some substitutions are able to produce conformational changes. It can be seen that α-subunit's N- and C-termini of Hbs form the central cavity of the tetrameric molecules (Figure 57-3). The superposition of Cα atoms from two αchains of carp Hb I onto the corresponding atoms from chicken Hb A gives an RMS deviation of 1.30 Å. Visual inspection of the two structures indicated a generally good superimposition. The largest differences were observed at the N terminal and the C terminal regions (Figure 57-3). Second, the Connolly molecular surface was calculated using the DelPhi program. The surface is colour-coded by electrostatic potential as shown in Figure 57-4. The spectrum shows that negative and positive regions are opaque red and blue, respectively. The central cavity ormed by α-subunit's N- and C-termini of chicken Hb A shows a large hydrophobic region (white), surrounded by areas of negative charge (red), and areas of positive charge (blue) [Figure 57-4(A)]. By contrast, the corresponding region from carp Hb I is strongly negatively charged (red) [Figure 57-4(B)]. It has been proposed that the presence of hydrophobic patches is a necessary condition for BPB to be recognised by proteins. It is interesting to observe that the central cavity formed by α-subunit's N- and C-termini of chicken Hb

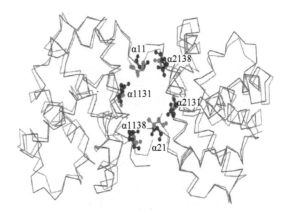

Figure 57-3 Comparison of the chicken Hb A model (green) and the carp Hb I model (blue).

α1, α131 and α138 are depicted in ball-stick representation

A shows a hydrophobic region and thus supporting the hypothesis that this site may be responsible for the binding of chicken Hb A to BPB. Carp Hb I represent loss of the BPB-binding property due to the absence of hydrophobic patches.

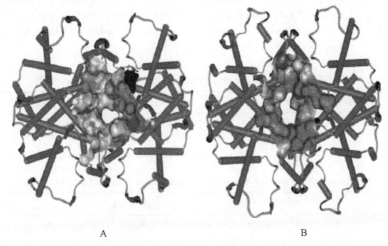

Figure 57-4 The central cavities formed by α-subunit's N- and C-termini of chicken Hb A (A) and carp Hb I (B) are represented as a molecular surface colored by electrostatic potential. Red represents negative charges, blue represents positive charge and white represents hydrophobic region

For the purpose of obtaining additional evidences that BPB binds to chicken Hb A in the α-subunit's N- and C-termini regions, computational docking studies were performed using chicken Hb A model and the energy-minimised structure of BPB as starting docking conformation. The predicted BPB binding mode is shown in Figure 57-5. In this mode, we find that the majority of BPB-binding site is located in the central cavity formed by α-subunit's N- and C-termini of chicken Hb A. Inspecting the model structure, it became obvious that a series of hydrophobic residues involved in complex formation: Val1α, Leu2α, Ala131α, Thr134α, and Ala138α. Atomic contacts also include putative hydrogen bonds with Thr134α and Arg141α(Figure 57-6). These interactions lead to a large stabilisation of BPB in this region. The molecular energy and intermolecular energy values for the formation of chicken Hb A-BPB complex are summarised in Table

57-1. The total interaction energy was calculated to be –55.94 kcal/mol. The van der Waals and electrostatic energies were –40.80 and –15.14 kcal/mol, respectively. The results show that the binding of BPB to chicken Hb A would be energetically favoured. Moreover, other possible interaction sites were not observed within the docking experiments. To determine the key residues that comprise the binding pocket of the model, the values of solvent accessible surface (SAS) area of each individual amino acid in the binding site were also calculated (Table 57-2).

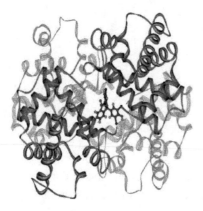

Figure 57-5 Proposed docking model of chicken Hb A and BPB. The backbones of α chains and β chains are rendered as cyan solid and purple line ribbons, respectively

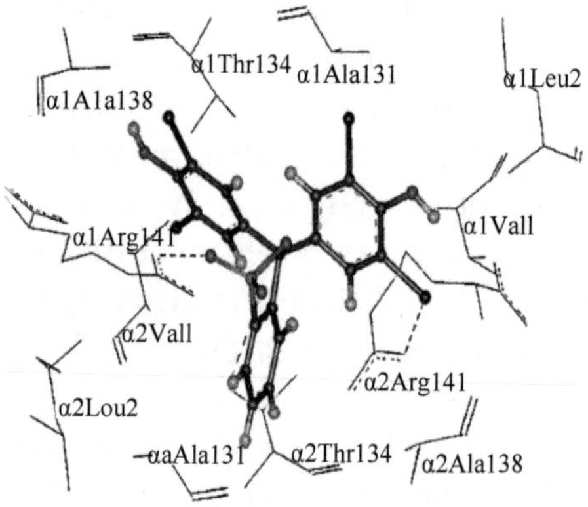

Figure 57-6 Illustration of the key residues involved in the binding of chicken Hb A. The hydrogen bonds are labeled by black lines

Table 57-1 The docking energy between chicken Hb A and BPB

Intermolecular energy(kcal/mol)			Molecular energy(kcal/mol)			
Vdw	Elect	Total	Vdw-repulsive	Vdw-dispersive	Elect	Total
–39.31	–17.49	–56.80	1840.32	–1784.55	–347.74	–291.97

Table 57-2 Computed solvent accessible surface (SAS) of amino acids in binding sites

	SAS area(Å^2)	Polar(Å^2)	Nonpolar(Å^2)
α1Val 1	143.457	27.491	115.966
α1Leu 2	6.509	3.035	3.474
α1Ala131	23.280	0.475	22.805
α1Thr134	40.429	12.186	28.243
α1Ala138	21.838	4.436	17.402
α1Arg141	49.554	45.618	3.936
α2Val 1	129.834	15.177	114.657
α2Leu 2	10.66	10.274	0.386
α2Ala131	21.159	0.242	20.917
α2Thr134	25.365	13.131	12.234

	SAS area(Å^2)	Polar(Å^2)	Continued Nonpolar(Å^2)
α2Ala138	20.205	2.568	17.637
α2Arg141	65.38	32.797	32.583

The results are very important for quantitatively determining the interaction tendencies of all residues in the binding site. Consistent with previous data, Val1α, Ala131α, Ala138α are the key residues involved in the binding of BPB. Furthermore, BPB was docked into the Carp Hb I model under the similar conditions. Our results indicated that BPB could not interact with carp Hb I, which were evaluated using the interaction energy function. The replacement of Val1α, Ala131α and Ala138α in chicken Hb with Ser1α, Asp131α and Glu138α in carp Hb decreases the hydrophobic effect of the central cavity formed by α-subunit's N- and C-termini of carp Hb, explaining why carp Hb could not participate in BPB binding.

5 Conclusion

In this paper, the sequence and structural characteristics, as well as the molecular electrostatic potential surface properties of chicken Hb A and carp Hb I have been investigated. A significant dissimilarity of the α-subunit's N- and C-termini regions of two Hbs was found. In particular, the central cavity formed by the α-subunit's N- and C-termini of chicken Hb A shows a large hydrophobic region, whereas the corresponding region from carp Hb I is strongly negatively charged. Furthermore, we have obtained a reasonable structural model of chicken Hb A in complex with BPB, and an interaction mechanism is proposed. Our results suggest that three hydrophobic residues, Val1α, Ala131α and Ala138α which are conserved in α chains of chicken Hb A but not in carp Hb I play critical roles in BPB binding. This is the first report demonstrating that key amino acid substitutions in the central cavity formed by the α-subunit's N- and C-termini of Hbs create a dramatic shift in the binding of BPB.

Acknowledgements

This work was supported by Science Foundation for The Excellent Youth Scholars of Tongji University, China(No. 2008kj0670) and Science Foundation for Youths of Shanghai Health Bureau, China (No. 2008Y020).

Note

Supplementary material can be viewed online.

References

[1] A. Efstratiadis, J.W. Posakony, T. Maniatis, R. M. Lawn, C. O'Connell, R. A. Spritz, J. K. DeRiel, B.G. Forget, S.M. Weissman, J. L. Slightom, A.E. Blechl, O. Smithies, F. E. Baralle, C.C. Shoulders and N. J. Proudfoot, Cell, 21, 653 (1980).
[2] E.W. Fisher, A. Rojnuckarin and S. Kim, J. Mol. Struct., Theochem, 592, 37 (2002).
[3] C.R. Anderson, E.O. Jensen, D.J. LLewellyn, E.S. Dennis and W.J. Peacock, Proc. Natl. Acad. Sci. USA, 93, 5682 (1996).
[4] J.E. Knapp, M.A. Oliveira, Q. Xie, S.R. Ernst, A.F. Riggs and M.L. Hackert, J. Biol. Chem, 274, 6411 (1999).
[5] B. A. Moss and E. O. P. Thompson, Aust. J. Biol. Sci, 22, 1455 (1969).

[6] R. G. Gillen and A. Riggs, J. Biol. Chem, 247, 6039(1972).
[7] A. L. Mayburd, Y. Tan and R.J. Kassner, Arch. Biochem. Biophys, 378, 40 (2000).
[8] S. Tayyab and M.A. Qasim, Int. J. Biol. Macromol, 12, 55 (1990).
[9] V. B. Gavrilov, V.P. Nikol'skaia, G.V. Kaler and S.V. Konev, Mol. Biol, 24, 1211 (1990).
[10] InsightII 2005, Accelrys Inc., San Diego, CA, USA.
[11] J. D. Madura, E. A. Salter, A. Wierzbicki, P. Dalal and J.P. Harrington, J. Mol. Struct., Theochem, 592, 173(2002).
[12] R. Luthy, J. U. Bowie and D. Eisenberg, Nature, 356, 83(1992).
[13] B. Honig and A. Nicholls, Science, 268, 1144 (1995).
[14] M. L. Connolly, Science, 221, 709 (1983).
[15] Z. L. Wang, L.R. Zhang, J. F. Lu and L. H. Zhang, J. Mol. Model, 11, 80 (2005).

[*Molecular Physis*. 2010,108(2): 215-220]

第七篇　其他细胞内成分的电泳释放

前　言

1　本书前述各篇都是以红细胞为研究对象。

2　本篇是其他细胞(红细胞以外的细胞)的问题，都是有氧条件下蛋白质的电泳释放。这些细胞包括血小板、粒细胞、淋巴细胞、胃癌细胞和小鼠胚胎成纤维细胞NIH3T3。

3　血小板标本来自包头医学院第一附属医院血液科和鄂尔多斯中心血站检验科[初步研究结果发表于现代预防医学　2011，38(4)：678-686]。

4　粒细胞标本来自包头医学院第一附属医院血液科。慢性粒细胞白血病患者血液标本于室温放置过夜，此时血液分为三层：上层为浅黄色透明的血浆，下层为红色的红细胞，中间层是乳白色的粒细胞。取中间层进行实验研究。

5　淋巴细胞标本来自包头医学院第一附属医院血液科。急性淋巴细胞白血病患者血液标本于室温放置过夜，此时血液分为三层：上层为浅黄色透明的血浆，下层为红色的红细胞，中间层是乳白色的粒细胞。取中间层进行实验研究。

6　胃癌细胞来自包头医学院第一附属医院消化科。取胃癌细胞培养物及其培养基，比较二者的电泳结果，观察胃癌细胞中蛋白质的电泳释放情况。

7　小鼠胚胎成纤维细胞NIH3T3由邵国老师教研室提供，没有培养基，单独做实验。

第五十八章 血小板内成分电泳释放的初步研究

乔姝[1] 沈木生[2] 韩丽红[3] 高丽君[3] 苏燕[3] 周立社[3] 秦良谊[3] 秦文斌[3]※

(1. 包头医学院 第一附属医院输血科, 包头 014010; 2. 鄂尔多斯中心血站 检验科, 鄂尔多斯 017000; 3. 包头医学院 血红蛋白研究室, 包头 014010)

摘 要

背景: 大约30年前, 我们发现"血红蛋白A_2现象", 近年来又提出"血红蛋白释放试验HRT", 这些都是一种"细胞成分的电泳释放", 只是这里的细胞是红细胞。现在我们要扩大范围, 进入其他细胞领域, 由比较容易获得的血小板入手。

目的: 观察血小板成分电泳释放情况, 给深入研究打下基础。

方法: 将来自血站的血小板液转入大EP管, 低速离心10分钟, 沉淀与上层液(血浆)分别转到不同滤纸条, 并排进行实验室常规的淀粉-琼脂糖混合凝胶电泳, 电泳条件: 普泳2小时15分钟, 然后定释(停15分钟再泳15分钟), 丽春红-联苯胺复合染色。必要时, 并排加入红细胞溶血液, 以便比较明确血小板成分的电泳位置。再做双向电泳, 观察两次释放的详细情况。

结果: 与血浆成分比较, 血小板显示: 原点残留明显、全程拖尾, 由原点一直拖到白蛋白。差别之处在于, 血小板的纤维蛋白原界限欠清晰或者稍后退, 白蛋白与$α_1$球蛋白之间有一模糊的区带。双向电泳结果表明, 血小板与血浆的差别更大, 有数个脱离对角线的成分。

结论: 单向电泳时, 血小板与血浆的差别在于白蛋白稍后退和纤维蛋白原附近等处, 双向电泳时, 这些地方都有横向带。由此看来, 血小板电泳释放产物中可能有蛋白复合物存在。它的详细机制有待进一步研究。

关键词 细胞成分电泳释放(EPRCC); 血小板; 血浆; 细胞; 电泳释放; 白蛋白; α球蛋白; 纤维蛋白原

Initial study on electrophoretic release of platelet component

QIAO Shu, SHEN Mu-sheng, HAN Li-hong, et al.

Blood Bank of the first affiliated hospital of Baotou Medical College, 014010

Abstract

Objective To observe the electrophoretic release of platelet component and do the groundwork for deep investigation.

Methods The platelet solution which obtained from blood station was added to EP tube of 1.5ml, centrifuge 10 minutes at 3000rpm, the sediment and supernatant(blood plasma)were separated and transfered to different slip of filter paper. Routine starch-agarose gel electrophoresis was carried

※通讯作者: 秦文斌, 电子信箱: qwb5991309@tom.com

out: first to do common electrophoresis and then timed-release electrophoresis, complex stain with ponceau red and benzidine. Two-dimensional electrophoresis was carried out to observe the detailed condition of twice release.

Results Comparing with plasma the platelet showed origin resid obviously and whole range tailing(from albumin to origin). Difference was at that the band of "fibrinogen" of platelet was not clear and there was a fuzzy band between albumin and α1 globulin. The results of two-dimensional electrophoresis showed that the difference between platelet and plasma was more obvious and there were several components separate oneself from diagonal.

Conclusion One-dimensional electrophoresis show that there is difference between platelet and plasma at post-albumin and nearby "fibrinogen" etc, two-dimensional electrophoresis show several trans.-bands at these region. It looks like that protein complex may exist in the electrophoretic release product of platelet. Its detailed mechanism needs further investigation .

Key words EPRCC(electrophoretic release of cell component); Platelet; Plasma; Cell; Electrophoretic release; Albumin; A-globulin; Fibrinogen

1 前言

1981年以来，我们一直在研究"血红蛋白 A_2 现象"[1~3]，近年来又提出"血红蛋白释放试验 HRT"[4~6]。这些都是一种"细胞成分的电泳释放"，只是以前的研究对象是红细胞。现在我们要扩大范围，进入其他细胞领域。先从比较容易获得的血小板入手，观察其细胞成分电泳释放的初步情况，为深入分析做好准备工作。

2 材料及方法

2.1 标本来源 血小板液来自鄂尔多斯中心血站和包头医学院第一附属医院检验科血库。

2.2 标本处理 将来自血站的血小板液转入大EP管，低速离心(3000rpm)10分钟，分成沉淀与上层液(血浆)两部分。

2.3 电泳分析

2.3.1 单向电泳 将离心后的沉淀与上层液(血浆)分别转到不同滤纸条，并排进行我室常规的淀粉-琼脂糖混合凝胶电泳，电泳条件：普泳2小时15分钟，然后定释(停15分钟再泳15分钟)，丽春红-联苯胺复合染色。

2.3.2 同上电泳 并排加入红细胞溶血液，以便比较明确血小板成分的电泳位置。

2.3.3 双向叠层电泳 上层为血浆，下层为血小板。

3 结果

3.1 血小板(沉淀物)与其上层血浆的电泳比较 血小板泳道的结果：原点残留明显；原点滤纸稍前移；由原点到白蛋白全程拖尾；定释，有带不集中；白蛋白与 $α_1$ 球蛋白之间有蛋白成分；纤维蛋白原的电泳位置也与血浆者有些不同。见图58-1。

3.2 血小板(沉淀物)与其上层血浆的电泳比较 增加成人红细胞溶血液，血小板泳道的结果图58-2与图58-1基本相同，主要差别是原点残留更加明显；纤维蛋白原靠后。与溶血液比较：血小板纤维蛋白原的电泳位置与溶血液中的碳酸酐酶相近。

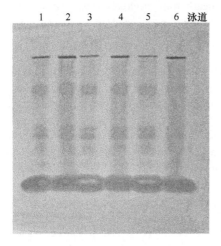

图 58-1　血小板与其血浆的比较电泳
注释：泳道由左向右，1 血浆，2 血小板，3 血浆，4 血小板，5 血浆，6 血小板

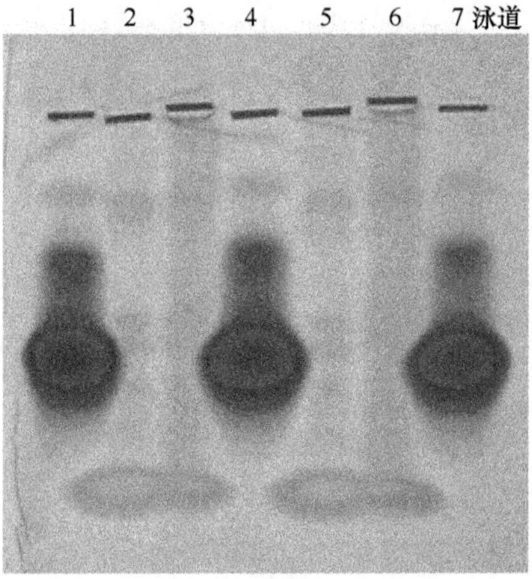

图 58-2　血小板与其血浆的比较电泳
注释：泳道由左向右，1 溶血液，2 血浆，3 血小板，4 溶血液，5 血浆，6 血小板，7 溶血液

3.3　血小板(沉淀物)与其上层血浆的双向电泳比较　血小板成分的电泳释放，第一次释放(对角线上)和第二次释放(垂直线上)都与血浆有很多不同之处。离开对角线的成分，通常是第一次释放时为"复合物"，而第二次释放时解离所致。见图 58-3。

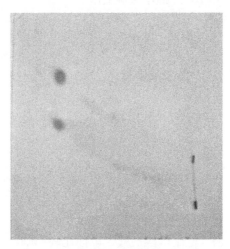

图 58-3　血小板与其血浆的双向电泳
注释：上层为血浆，下层为血小板。血浆方面，能看到白蛋白、α_1 球蛋白、α_2 球蛋白、β 球蛋白和纤维蛋白原，都在对角线上；原点没有上升成分。血小板成分的对角线方向，其在对角线上成分与血浆不完全对应，而且出现多个离开对角线成分——白蛋白、α 球蛋白、纤维蛋白原。第一向电泳时紧跟白蛋白的成分，第二向电泳时随白蛋白水平上升。α 球蛋白、纤维蛋白原位置也有类似情况。血小板成分的垂直线方向，原点二次释放现象明显，释放物顶部与白蛋白平齐

4　讨论

血小板(blood platelet)是哺乳动物血液中的有形成分之一，它具有特定的形态结构和生化组成，在正常血液中有较恒定的数量[如人的血小板数为$(100～300)\times 10^9$/L]，在止血、伤口愈合、炎症反应、血栓形成及器官移植排斥等生理和病理过程中有重要作用。蛋白质组学技术为血小板成分的分析提供了许多详细信息。此时，血小板需要先裂解，然后进行分析，所

得结果是血小板裂解液中的各种成分[7~11]。本文的分析角度不同,血小板不裂解,靠"电泳"来释放,研究释放出来的各种成分,这就是我们所提出"细胞成分的电泳释放"。思路来自"血红蛋白 A_2 现象",那是"红细胞成分的电泳释放",我们发现电泳释放出来的血红蛋白 A_2 与红细胞裂解出来的血红蛋白 A_2 不同,来自红细胞的 HbA_2,实际是溶血液 HbA_2 与溶血液 HbA 的结合物,而且还有第三种成分 $X^{[1~3]}$。近来,经过 SDS-PAGE 和 ESI-QUAD-TOF-MS (electrospray ionization quadrupole time-of-flight mass spectrometry)分析,证明这个 X 很可能是"peroxiredosin 或 Thioredoxin Peroxidase B"(待发表)。

如上所述,红细胞成分电泳释放时出现"血红蛋白 A_2 现象",证明在红细胞内或电泳释放过程中血红蛋白 A_2 与血红蛋白 A 等互作而形成复合物。血小板成分电泳释放时情况如何?血小板分泌的蛋白质多种多样,经过通常的蛋白质组学研究,已经分离和鉴定出 300 多种,有 82 种蛋白质通过重复实验证实,其中包括人血清白蛋白。我们在单向对比电泳中,看到血小板与其上清液(血浆)内白蛋白电泳行为相同成分,而且在双向电泳结果中脱离"对角线"的首先是白蛋白,它的右侧(阴极侧)有平行成分。这说明此右侧平行成分是白蛋白的一部分,在第一向电泳时它曾与慢泳成分(正电荷大于白蛋白的成分)结合存在。第二向电泳时,在电场作用下二者分离、慢泳成分脱落,这部分白蛋白恢复了自己的电泳速度,但出现在白蛋白的平行右侧。这就相当于红细胞成分电泳释放时出现的"血红蛋白 A_2 现象"。我们把这种情况叫作"大 A_2 现象",表示它与"血红蛋白 A_2 现象"类似,也是细胞成分电泳释放过程中出现的蛋白质的结合或互作。不过,它的"大 A_2 现象"不止一个,还有 α 球蛋白的右侧平移成分、纤维蛋白原的右侧平移(参见双向电泳图)。这些都说明,血小板成分的电泳释放比红细胞成分的电泳释放复杂,深入研究的内容很多。当然,本文是初步研究,下一步也要做 SDS-PAGE 和质谱分析,寻找互作的详细机制。我们认为,本课题的内容及"血红蛋白 A_2 现象"的系列研究,都是"电泳释放蛋白质组学研究"的雏形,深入研究有可能揭示出细胞内部的一些动态情况。

参 考 文 献

[1] 秦文斌, 梁友珍. 血红蛋白 A_2 现象 I. 此现象的发现及其初步意义. 生物化学与生物物理学报, 1981, 13: 199.
[2] 秦文斌. "血红蛋白 A_2 现象"专辑. 包头医学院学报, 1990, 7(3): 1-76.
[3] 秦文斌. 红细胞外血红蛋白 A 与血红蛋白 A_2 之间的相互作用. 生物化学杂志, 1991, 7: 577-587.
[4] 秦文斌, 高丽君, 苏燕, 等. 血红蛋白释放试验与轻型 β-地中海贫血. 包头医学院学报, 2007, 23(6): 559-563.
[5] 韩丽红, 闫斌, 高雅琼, 等. 一些外科患者血红蛋白释放试验的比较研究. 临床和实验医学杂志, 2009, 8(7): 67-69.
[6] Su Y, Shao G, Gao LJ, et al. RBC electrophoresis with discontinuous power supply a newly established hemoglobin release test. Electrophoresis, 2009, 30(17): 3041-3043.
[7] Maguire P B, Fitsgerald D J. Pletelete proteomics. J Thromb Haemost, 2003, 1(7): 1589-1601.
[8] Coppinger J A, Cagney G, Toomey S, et al. Characterization of the proteins released from activated platelets leads to localization of novel platelet proteins in human atherosclerotic lesions. Blood, 2004, 103(6): 2090-2104.
[9] Maguire P B, Moran N, Cagney G, et al. Application of proteomics to the study of platelet regulatory mechanisms. Trends Cardiovasc Med, 2004, 14(6): 207-220.
[10] Perrotta P L, Bahou WF. Proteomics in platelet science. Curr Hematol Rep, 2004, 3(6): 460-469.
[11] 邱宗荫, 尹一兵. 临床蛋白质组学. 北京: 科学出版社, 2008: 229-231.

[原文发表于"现代预防医学, 2011, 38(4): 678-686"]

第五十九章　粒细胞内蛋白质电泳释放的初步研究

贾国荣[1]　高丽君[2#]　马宏杰[1#]　苏　燕[2]　周立社[2]　卢　燕[1]
李　喆[1]　李　静[1]　贺其图[1※]　秦文斌[2※]

(1　包头医学院　第一附属医院　血液科，包头　014010；2　包头医学院　血红蛋白研究室，包头　014010)

摘　要

目的：我们曾长期研究红细胞内血红蛋白的电泳释放，现在要扩大范围，进入其他细胞领域，本文是研究粒细胞内蛋白质的电泳释放。

方法：由慢性粒细胞白血病患者血中分离出粒细胞成分，再用电泳释放技术观察其中蛋白质的释放情况。

结果：与血浆成分比较，粒细胞显示在白蛋白与 $α_1$ 球蛋白之间出现大量蛋白质，原点稍前方也有蛋白成分，可能分子量较大。

结论：粒细胞内可能存在自己特有的蛋白质，其详细情况有待深入研究。

关键词　细胞成分电泳释放；粒细胞；血浆；细胞；电泳释放；α 球蛋白

1　前言

慢性粒细胞白血病(CML)是一种影响血液及骨髓的恶性肿瘤，它的特点是产生大量不成熟的白细胞，这些白细胞在骨髓内聚集，抑制骨髓的正常造血；并且能够通过血液在全身扩散，导致患者出现贫血、容易出血、感染及器官浸润。慢性粒细胞白血病是一种相对少见的癌症，大约占所有癌症的 0.3%，占成人白血病的 20%；一般人群中，大约每 10 万有 1~2 个人患有该病。慢性粒细胞白血病可以发生于任何年龄的人群，但以 50 岁以上的人群最常见，平均发病年龄为 65 岁，男性比女性更常见。慢性粒细胞白血病进展缓慢，根据骨髓中白血病细胞的数量和症状的严重程度，分为三期：慢性期、加速期和急变期。其中，大约有 90%的患者诊断时为慢性期，每年 3%~4%慢性期进展为急变期。姚峰[1]做了慢性粒细胞白血病急变机制的比较蛋白质组学研究，他采用比较蛋白质组学技术对 CML 急变的差异蛋白表达分析，获得一组可能参与 CML 急变的相关蛋白质，它们可能是急变的可能标志物。这里涉及许多蛋白质，作者认为对这些蛋白质的深入研究将加深对 CML 急变机制的认识。

必须指出，当今的蛋白质组学研究，必须先破坏细胞，释放出来蛋白质，然后对蛋白质进行各种实验分析，它的结果与完整细胞里边蛋白质的真实情况相差多少不得而知。完整细

\# 并列第一作者
※通讯作者：贺其图，电子信箱：heqitu@136.com；秦文斌，电子信箱：qwb5991309@tom.com

胞内各种蛋白质的存在状态非常复杂，其中包括多种形式的蛋白质相互作用，破坏细胞可能使这些相互作用减弱或消失。我们用红细胞做的实验证明了这一点：完整红细胞有血红蛋白A_2现象(HbA_2与HbA_1之间的相互作用)，红细胞破坏后这种互作完全消失[3,4,8]。本文作者的研究手段是不破坏细胞，将完整细胞直接加入凝胶电场进行电泳[2-8]，此时电泳释放出来的蛋白质应该与传统蛋白质组学的结果不大相同，也许能够查出这种完整细胞的特异蛋白质。详情参见正文。

2 材料及方法

2.1 标本来源 慢性粒细胞白血病患者血液标本来自包头医学院第一附属医院血液科。

2.2 方法

2.2.1 分层试验 慢性粒细胞白血病患者血液标本于室温放置过夜，此时血液分为三层：上层为浅黄色透明的血浆，下层为红色的红细胞，中间层是乳白色的粒细胞。由上往下，逐一吸出各层成分，转入相应的3支EP管，进行以下实验。

2.2.2 三层成分的单向电泳比较(单带释放) 将血浆、粒细胞、红细胞并排加入淀粉-琼脂糖混合凝胶，进行单带释放电泳，比较观察三者的释放情况，特别是粒细胞的释放特点。

2.2.3 三层成分的单向电泳比较(多带释放) 将血浆、粒细胞、红细胞并排加入淀粉-琼脂糖混合凝胶，进行多带释放电泳，比较观察三者的释放情况，特别是粒细胞的释放特点。本次还加入全血标本，观察其有什么情况。

2.2.4 两层成分的双向电泳比较 将血浆和粒细胞并排加入淀粉-琼脂糖混合凝胶，进行双向电泳，比较观察二者的释放情况，特别是粒细胞的释放特点。

3 结果

3.1 三层成分的单向电泳比较(单带释放)结果 见图59-1。

(1) 血浆层有如下成分：白蛋白(最下方)、$α_1$球蛋白(□)、$α_2$球蛋白(○)、$β_1$球蛋白(◇)、$β_2$球蛋白(△)、纤维蛋白原(△)、γ球蛋白(▽)。

(2) 胞粒细层有如下成分：①在白蛋白与$α_1$球蛋白之间(箭头← →所指处)出现一种蛋白质，暂称$α_0$球蛋白或快泳粒细胞蛋白；②还有一种蛋白质位于$α_2$球蛋白与$β_1$球蛋白之间，可能是血红蛋白(联苯胺阳性)，可能是 HP-HB；③$α_1$与$α_2$之间还有一个弱带(联苯胺阳性)可能是Hx-Heme；④相当于$β_2$球蛋白处有一弱带，联苯胺阴性；⑤原点残留纸条稍前移(与血浆纸条、红细胞纸条比较)，使纸条前面的胶出现凹陷，说明此处可能有一大分子蛋白质，暂称粒细胞慢泳蛋白；⑥这里没有与血浆纤维蛋白原对应的成分。

(3) 红细胞层有如下成分：①少量白蛋白(污染)；②血红蛋白A_3；③血红蛋白A_1；④血红蛋白A_2；⑤CA(被泄露的血红蛋白覆盖而显蓝黑色)；⑥定释血红蛋白(短箭头←所指处)；⑦原点残留(原点处口形表现，联苯胺阳性)。

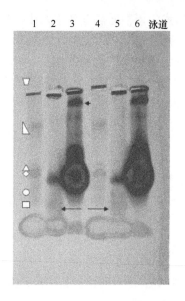
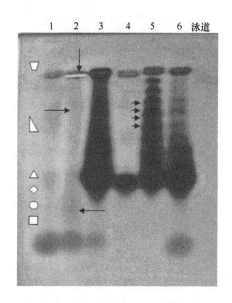

图 59-1 粒细胞内蛋白质的电泳释放(单带释放)
注释：单带释放(见短箭头←所指处)；泳道1=血浆，泳道2=粒细胞，泳道3=红细胞，泳道4=1、5=2、6=3

图 59-2 粒细胞内蛋白质的电泳释放(多带释放)
注释：多带释放(见4个短箭头→所指处)；泳道1=血浆，泳道2=粒细胞，泳道3=红细胞，泳道4=红细胞溶血液，泳道5=红细胞，泳道6=全血

3.2 三层成分的单向电泳比较(多带释放)结果　见图59-2。

(1) 血浆层：基本同图59-1，只是由于多带再释放使血浆各成分之间界限欠佳。

(2) 粒细胞层：基本同图59-1，①粒细胞特有的快泳蛋白位于 α_1 球蛋白与 α_2 球蛋白之间(长箭头←所指处)；②粒细胞慢泳蛋白位于纤维蛋白原稍后(长箭头→所指处)；③再有就是原点处(长箭头↓所指处)，呈口形，上下都有联苯胺阳性成分。

(3) 红细胞层：有密集的多带(多个短箭头→所指处)，原点也有大量血红蛋白残留。

(4) 全血层：基本同红细胞层，但多带明显减弱，来源于血浆成分对红细胞释放的影响，这就是所谓的"浆胞互作"(血浆与红细胞之间的相互作用)。全血里也看不到粒细胞快泳蛋白，这如何理解？

3.3 两层成分的双向电泳比较结果　见图59-3。与血浆相比，粒细胞有以下3个特点：①在 α_1 球蛋白的前方出现大量蛋白质(↑□处)；②在纤维蛋白原后边、原点前边，出现一些蛋白质(↑○处)；③在原点处及其稍前方出现大量蛋白质(↑◇处)，我们称之为"原点现象"；④在 β_1 球蛋白处也出现较多蛋白质(↑处)，可能是夹杂存在的血红蛋白 A_1。

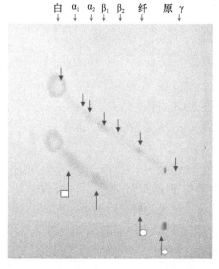

图 59-3 粒细胞内蛋白质的电泳释放(双向电泳)
注释：上层为血浆(有多个向下箭头↓者)，下层为粒细胞(有多个向上箭头↑者)。白=白蛋白，α_1=α_1球蛋白，α_2=α_2球蛋白，β_1=β_1球蛋白，β_2=β_2球蛋白，纤=纤维蛋白，原=原点，γ=γ球蛋白

4 讨论

粒细胞的特异蛋白质是什么？完整粒细胞内蛋白质的电泳释放研究有可能给出独特的答案。我们称完整细胞内蛋白质的电泳释放研究为"电泳释放蛋白质组学"，它与传统的蛋白质组学[1]不同之处就是使用完整细胞[2-8]，而不是破坏细胞后的蛋白质。电泳释放蛋白质组学的技术路线有三大步：①完整细胞内蛋白质的电泳释放；②对所释放蛋白质进行SDS-PAGE分析；③对SDS-PAGE分离出来的蛋白质进行质谱分析，从而明确粒细胞的特异蛋白质。

本文实验，相当于电泳释放蛋白质组学技术路线中的第一步。首先，可以对单向电泳时出现的集中特异带进行分析，再对双向电泳时出现的集中特异带进行分析，包括原点成分，寻找各组蛋白质之间相互关系。

参 考 文 献

[1] 姚峰. 慢性粒细胞白血病急变机制的比较蛋白质组学研究. 第一军医大学硕士学位论文, 2006.
[2] S.u Y, Shao G, Gao LJ, et al. RBC electrophoresis with discontinuous power supply-a newly established hemoglobin release test. Electrophoresis, 2009, 30: 3041-3043.
[3] Su Y, Gao LJ, Ma Q, et al. Interactions of hemoglobin in lived red blood cells measured by the electrophoesis release test. Electrophoresis, 2010, 31: 2913-2920.
[4] 秦文斌. 活体红细胞内血红蛋白的电泳释放. 中国科学：生命科学, 2011, 41(8): 597- 607.
[5] Su Y, Shen J, Gao LJ, et al. Molecular interaction of re-released proteins in electrophoresis of human erythrocytes. Electrophoresis, 2012, 33: 1042-1045.
[6] Kurien B T, Scofield R H. Protein Electrophoresis: Methods and Protocols, Methods in Molecular Biology, 2012, 869: 387-402.
[7] Su Y, Shen J, Gao LJ, et al. Molecular interactions of re-released proteins in electrophoresis of human erythrocytes. Electrophoresis, 2012, 33(9-10): 1402-1405.
[8] 秦文斌. 红细胞内血红蛋白的电泳释放——发现和研究. 北京：科学出版社, 2015.

第六十章　淋巴细胞内蛋白质电泳释放的初步研究

马宏杰[1]　高丽君[2#]　贾国荣[1#]　苏　燕[2]　周立社[2]　卢　燕[1]　李　喆[1]　李　静[1]
贺其图[1※]　秦文斌[2※]

(包头医学院　1. 第一附属医院　血液科；2. 血红蛋白研究室，包头，014010)

摘　要

目的：我们曾长期研究红细胞内血红蛋白的电泳释放，现在要扩大范围，进入其他细胞领域，本文是研究淋巴细胞内蛋白质的电泳释放。

方法：由急性淋巴细胞白血病患者血中分离出淋巴细胞成分，再用电泳释放技术观察其中蛋白质的释放情况。

结果：患者全血放置后分三层，上层为血浆，中层为淋巴细胞，下层为红细胞。淋巴细胞层的显微镜照相结果是以大细胞为主，核仁清楚，属于成人型淋巴细胞。单向电泳，比较淋巴细胞与血浆，可以看出淋巴细胞释放出来的成分比较集中。二次释放时又由原点释放出来一些蛋白质，它可能分子量比较大。淋巴细胞的双向电泳可以看出，第一向时释放出来成分基本上都在对角线上。第二向时由原点向上方释放出大量蛋白质。

结论：淋巴细胞内可能存在自己特有的蛋白质，其详细情况有待深入研究。

关键词　细胞成分电泳释放；淋巴细胞；原点成分；上升成分；电泳释放

1　前言

急性淋巴细胞白血病(ALL)是一种起源于淋巴细胞的 B 系或 T 系细胞在骨髓内异常增生的恶性肿瘤性疾病。异常增生的原始细胞可在骨髓聚集并抑制正常造血功能，同时也可侵及骨髓外的组织，如脑膜、淋巴结、性腺、肝等。我国曾进行过白血病发病情况调查，ALL 发病率约为 0.67/10 万。在油田、污染区发病率明显高于全国平均发病率。ALL 儿童期(0～9岁)为发病高峰，可占儿童白血病的 70%以上。ALL 在成人中占成人白血病的 20%左右。目前依据 ALL 不同的生物学特性制订相应的治疗方案已取得较好疗效,大约 80%的儿童和 30%的成人能够获得长期无病生存，并且有治愈的可能。吕艳琦[1]进行儿科急性淋巴细胞白血病细胞差异蛋白质组学分析，结果发现，ALL 患儿与健康儿童细胞内蛋白质表达存在明显差异，并鉴定出 8 种差异蛋白，其中谷胱甘肽转移酶 P 和抑制素显著增高，60S 酸性核糖核蛋白 PO 等明显减低。

当今的蛋白质组学研究，必须先破坏细胞释放出来蛋白质，然后对蛋白质进行各种实验

\# 并列第一作者

※通讯作者：贺其图，电子信箱：heqitu@136.com；秦文斌，电子信箱：qwb5991309@tom.com

分析，它的结果与完整细胞里边蛋白质的真实情况相差多少不得而知。完整细胞内各种蛋白质的存在状态非常复杂，其中包括多种形式的蛋白质相互作用，破坏细胞可能使这些相互作用减弱或消失。我们用红细胞做的实验证明了这一点：完整红细胞有血红蛋白 A_2 现象(HbA_2 与 HbA_1 之间的相互作用)，红细胞破坏后这种互作完全消失。本文作者的研究手段是不破坏细胞，将完整细胞直接加入凝胶电场进行电泳[2-7]，此时电泳释放出来的蛋白质应该与传统蛋白质组学的结果不大相同，也许能够查出这种完整细胞的特异蛋白质。详情参见正文。

2 材料及方法

2.1 标本来源 急性淋巴细胞白血病患者血液标本来自包头医学院第一附属医院血液科，患者，女，年龄23。

2.2 方法

2.2.1 分层试验 急性淋巴细胞白血病患者血液标本于室温放置过夜，次日观察淋巴细胞的沉降情况。

2.2.2 淋巴细胞层细胞的显微镜观察 取淋巴细胞层，显微镜照相，结果留图。

2.2.3 原点换位法单向电泳观察淋巴细胞释放情况 直接取淋巴细胞和血浆并排进行原点换位法单向电泳，比较它们的释放情况。开始时，泳道1=滤纸条含淋巴细胞，泳道2=滤纸条含淋巴细胞，泳道3=滤纸条含生理盐水，泳道4=滤纸条含血浆。开始通电，4小时后停电，泳道2与3滤纸条换位，继续电泳(相当于再释放)，直到血浆白蛋白接近终点，最后染氨基黑10B。

2.2.4 生理盐水洗过淋巴细胞的单向电泳 首先用生理盐水洗淋巴细胞(黏度增大)，然后与红细胞和全血并排进行多带单向电泳，比较观察淋巴细胞的释放情况。

2.2.5 生理盐水洗过淋巴细胞成分的双向电泳 将淋巴细胞用生理盐水洗后(黏度增大)，加在滤纸条上做双向电泳，观察第一向和第二向电泳释放情况。

2.2.6 含淋巴细胞滤纸片的电泳移动 将淋巴细胞用生理盐水洗后(黏度增大)，加在一较大圆滤纸片(直径=3mm)上，将滤纸片平放在较稀(琼脂糖为原来的1/2)的凝胶面上，再平压入胶，上边盖上塑料圆片(稍大于圆滤纸片，直径=4mm)，做双向电泳。不染色，直接在暗室观察结果，在黑色背景下照相留图。

3 结果

3.1 患者全血放置后分层情况 结果如图60-1。

由此图可以看出，患者全血放置后分三层：上层为血浆，中层为淋巴细胞，下层为红细胞。

3.2 淋巴层细胞的显微镜照相结果 见图60-2。

由此图可以看出，显微镜下的淋巴细胞，以大细胞为主，核仁清楚，属于成人型淋巴细胞。

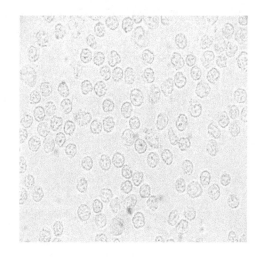

图 60-1 患者全血分层情况
注释：可分上中下三层，上层=血浆，中层=淋巴细胞，下层=红细胞

图 60-2 显微镜下的淋巴细胞
注释：淋巴细胞，以大细胞为主，核仁清楚

3.3 原点换位法单向电泳，观察淋巴细胞的释放情况 结果见图 60-3。

由此图可以看出，①淋巴细胞释放出来的比较集中成分(→所指处)，相当于血浆里的 α_2 球蛋白(○所指处)和 β_1 球蛋白(◇所指处)之间；②换位再释放时又由原点释放出来一些蛋白质(←所指处)，它的分子量可能比较大，与淋巴细胞内部结构结合比较牢固，初释放时出不来(泳道 2 的相应位置没东西)，再释放时才释放出来；③与泳道 4 原点相比，泳道 1、3 的原点滤纸片染色加深，我们称之为"原点现象"，也就是说，原点里还有东西没有释放出来，可能含有更大分子的蛋白质。

3.4 生理盐水洗过淋巴细胞的单向电泳，比较观察释放情况 结果见图 60-4。

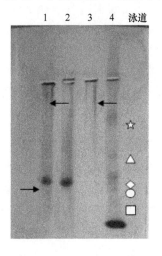
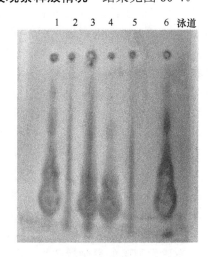

图 60-3 单向电泳，比较淋巴细胞与血浆
注释：淋巴细胞释放电泳，原点换位法[通电 4 小时后停电，泳道 2 与 3 滤纸条换位，继续电泳(相当于再释放)，直到血浆白蛋白接近终点，最后染氨基黑 10B]；开始时，泳道 1=滤纸条含淋巴细胞，2=滤纸条含淋巴细胞，3=滤纸条含生理盐水，4=滤纸条含血浆

图 60-4 单向电泳，比较红细胞、淋巴细胞和全血
注释：多带释放，半小时交替；泳道 1=红细胞，2=生理盐水洗过的淋巴细胞，3=全血，4=1，5=2，6=3

由此图可以看出，由生理盐水洗过的淋巴细胞释放出的蛋白质呈均匀的长带，与未洗的淋巴细胞不同(参见图 60-3)。看来，用生理盐水洗对淋巴细胞影响很大，可能生理盐水对淋巴细胞不是等渗，改变了细胞内蛋白质的存在状态，图 60-3 里的特异蛋白不见了。估计是没有释放出来，留在细胞内，电泳时留在原点。

3.5 生理盐水洗过的淋巴细胞双向电泳　结果见图 60-5。

由此图可以看出，第一向时，释放出来的成分基本都在对角线上，未见图 60-3 中的相应特异带。第二向时，由原点向上方释放出大量蛋白质(←所指处)、原点残留仍明显(↑所指处)，说明原点处还残留大量蛋白质。

3.6 生理盐水洗过的淋巴细胞与滤纸片一起电泳移动　见图 60-6。

由此图可以看出，通电后，圆滤纸片缓慢向阴极移动，两小时后圆滤纸片明显离开原点(↑所指处)，达到现在的位置(←所指处)。这说明，此蛋白质与滤纸片黏合牢固，而且具有明显的正电性、含有较多的正电荷，从而泳向阴极。

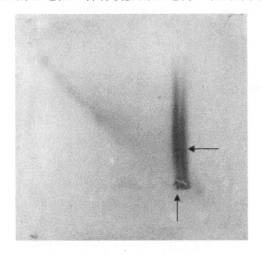

图 60-5　淋巴细胞的双向电泳
注释：双向电泳，第一向为普泳(相当于初释放)，第二向也是普泳(相对于第一向，它就是再释放)

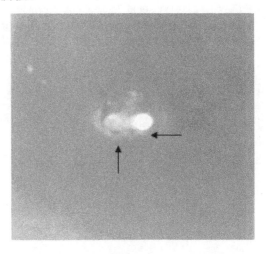

图 60-6　含淋巴细胞滤纸片的电泳移动
注释：含淋巴细胞滤纸片离开原点(↑所指处)退向阴极(←所指处)

4　讨论

淋巴细胞的特异蛋白质是什么？完整淋巴细胞内蛋白质的电泳释放研究有可能给出独特的答案。我们称完整细胞内蛋白质的电泳释放研究为"电泳释放蛋白质组学"，它与传统的蛋白质组学[1]不同之处就是使用完整细胞[2-7]、不是破坏细胞后的蛋白质。电泳释放蛋白质组学的技术路线有三大步：①完整细胞内蛋白质的电泳释放；②对所释放蛋白质进行 SDS-PAGE 分析；③对 SDS-PAGE 分离出来的蛋白质进行质谱分析，从而明确淋巴细胞的特异蛋白质。

本文实验，相当于电泳释放蛋白质组学的技术路线中的第一步。首先，可以对单向电泳时出现的集中带进行分析。再对双向电泳时出现的原点上升成分及后退成分进行研究，寻找各组蛋白质之间的相互关系。有趣的是，本实验中淋巴细胞经生理盐水处理前后变化很大，未用生理盐水处理的淋巴细胞在单向电泳时释放出来相当集中的特异蛋白区带，用生理盐水处理后标本的黏度加大、电泳时集中的蛋白带消失。而且，这种淋巴细胞能与滤纸片粘合，电泳时二者一起退向阴极，显示明显的正电性。现在看来，用生理盐水处理对淋巴细胞影响很大，可能是因为生理盐水对淋巴细胞不是等渗，改变了细胞内蛋白质的存在状态。电泳图60-3里的特异集中蛋白带，负电性明显(血浆 α_2 球蛋白和 β_1 球蛋白之间)，生理盐水处理后它在单向电泳中消失，有可能是它与淋巴细胞内其他正电性很强的蛋白质结合成大分子产物。双向电泳时的原点上升成分不集中，应当没包括上述特异蛋白，后者可能仍然留在原点。与圆滤纸片一起后退的成分里可能包括特异蛋白，详情有待进一步深入研究。

参 考 文 献

[1] 吕艳琦. 儿科急性淋巴细胞白血病细胞差异蛋白质组学分析. 郑州大学硕士学位论文, 2011.
[2] Su Y, Shao G, Gao LJ, et al. RBC electrophoresis with discontinuous power supply – a newly established hemoglobin release test. Electrophoresis, 2009, 30: 3041-3043.
[3] Su Y, Gao LJ, Ma Q, et al. Interactions of hemoglobin in lived red blood cells measured by the electrophoesis release test. Electrophoresis, 2010, 31: 2913-2920.
[4] 秦文斌. 活体红细胞内血红蛋白的电泳释放. 中国科学：生命科学, 2011, 41(8): 597- 607.
[5] Su Y, Shen J, Gao LJ, et al. Molecular interaction of re-released proteins in electrophoresis of human erythrocytes. Electrophoresis, 2012, 33: 1042-1045.
[6] Kurien B T, Scofield R H. Protein Electrophoresis: Methods and Protocols, Methods in Molecular Biology, 2012, 869: 387-402.
[7] 秦文斌. 红细胞内血红蛋白的电泳释放——发现和研究. 北京: 科学出版社, 2015.

第六十一章 胃癌细胞内蛋白质电泳释放的初步研究

张宏伟[1] 高丽君[3#] 闫斌[2#] 韩丽红[3] 郭春林[1] 秦文斌[3※]

(包头医学院 第一附属医院 1. 消化科 2. 普通外科，包头 014010；3. 包头医学院 血红蛋白研究室，包头 014010)

摘　要

目的： 我们曾长期研究红细胞内血红蛋白的电泳释放，现在要扩大范围，进入其他细胞领域观察非血红素蛋白的电泳释放，本文的研究对象是胃癌细胞，观察其中蛋白质的电泳释放。

方法： 取沉淀较少胃癌培养物(简称培养物Ⅰ)与培养基比较，进行双向电泳，观察二者有无差异。取沉淀较多的胃癌培养物(简称培养物Ⅱ)与培养基比较，进行双向电泳，观察二者有无差异。再将培养物Ⅱ双向电泳的原点结果放大，观察其细节。

结果： 培养物Ⅰ与培养基比较，结果没有明显差异。培养基与胃癌培养物Ⅱ结果出现明显差异，除了有原点上升成分外，还有一部分蛋白质离开原点、泳向阳极(前进成分)。

结论： 原点前移成分可能是胃癌细胞的主要特异蛋白质，具体内容有待深入研究。

关键词 胃癌细胞；电泳释放；双向电泳；原点成分；上升成分

1 前言

胃癌是我国常见的恶性肿瘤之一，在我国其发病率居各类肿瘤的首位。在胃的恶性肿瘤中，腺癌占95%，这也是最常见的消化道恶性肿瘤，在人类所有恶性肿瘤中名列前茅。早期胃癌多无症状或仅有轻微症状。当临床症状明显时，病变多已属晚期。因此，要十分警惕胃癌的早期症状，及时就医，以免延误诊治。胃癌细胞的蛋白质组学研究很多[1-4]，目的是寻找胃癌细胞特异的蛋白质。王平[1]用双向电泳发现22种与胃癌相关的蛋白质。周欣[2]用蛋白质组学研究证明，胃癌分化是一种多因素参与的复杂过程，糖酵解酶类可能成为胃癌分化相关的肿瘤标志物。李兆星[3]发现，不同分化胃腺癌组织之间差异蛋白有8种，可能参与肿瘤的发生、侵袭和转移过程。刘羽[4]证明，不同分化的胃癌组织中蛋白表达存在明显差异，本研究从39个差异明显的蛋白点里鉴别出48种蛋白质，最后确定6种，其中Serpin B1在胃癌中的表达具有一定的特异性。

必须指出，当今的蛋白质组学研究，必须先破坏细胞，释放出蛋白质，然后对蛋白质进行各种实验分析，它的结果与完整细胞里边蛋白质的真实情况相差多少不得而知。完整细胞内各种蛋白质的存在状态非常复杂，其中包括多种形式的蛋白质相互作用，破坏细胞可能使这些相互作用减弱或消失。我们用红细胞做的实验证明了这一点：完整红细胞有血红蛋白

\# 并列第一作者
※ 通讯作者：秦文斌，电子信箱：qwb5991309@tom.com

A_2 现象(HbA_2 与 HbA_1 之间的相互作用)，红细胞破坏后这种互作完全消失[5-6]。

本文作者的研究手段是不破坏细胞，将完整细胞直接加入凝胶电场进行电泳[7-10]，此时电泳释放出来的蛋白质应该与传统蛋白质组学的结果不大相同，也许能够查出这种完整细胞的特异蛋白质。详情参见正文。

2 材料及方法

2.1 标本来源 胃癌细胞培养物及其培养基，都来自包头医学院第一附属医院消化科。

2.2 方法

2.2.1 双向电泳比较培养基与胃癌培养物Ⅰ 取沉淀较少的胃癌培养物(简称培养物Ⅰ)与培养基比较，进行双向电泳，观察二者有无差异。

2.2.2 双向电泳比较培养基与胃癌培养物Ⅱ 取沉淀较多的胃癌培养物(简称培养物Ⅱ)与培养基比较，进行双向电泳，观察二者有无差异。

2.2.3 培养物Ⅱ双向电泳的原点细节 将培养物Ⅱ双向电泳的原点结果放大，观察其细节。

3 结果

3.1 双向电泳比较培养基与胃癌细胞培养物Ⅰ 结果见图61-1。

由此图可以看出，培养基与胃癌细胞培养物Ⅰ结果没有明显差异。

3.2 双向电泳比较培养基与胃癌细胞培养物Ⅱ 结果见图61-2。

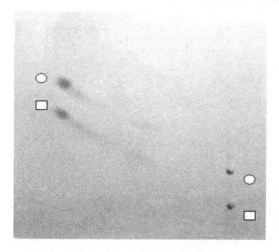 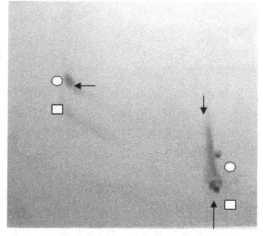

图61-1 双向电泳比较培养基与胃癌培养物Ⅰ
注释：上层(两个○之间)为细胞培养的培养基，下层(两个□之间)为细胞培养的胃癌培养物Ⅰ(含少量沉淀者)

图61-2 双向电泳比较培养基与胃癌培养物Ⅱ
注释：上层(两个○之间)为细胞培养的培养基，下层(两个□之间)为细胞培养的胃癌培养物Ⅱ(含多量沉淀)，↑所指处为加样的原点

由此图可以看出，培养基与胃癌细胞培养物Ⅱ结果出现明显差异。首先是培养基里的最前端成分(←所指处)在胃癌细胞培养物Ⅱ中消失，它可能是血清白蛋白，经蛋白酶消化成氨基酸，再合成大分子的胃癌细胞蛋白质。第一向电泳时，胃癌细胞留在原点(↑所指处)，第二向电泳时，释放出大量上升成分(↓所指处)。

3.3 培养物Ⅱ双向电泳的原点细节

结果见图61-3。

由此图可以看出，胃癌培养物Ⅱ在原点(□处)前方出现一种离开原点泳向阳极的蛋白成分(↑所指处)，第二向时再有一些蛋白成分离开原点和阳极成分而上升的一些蛋白质(↓所指处)。这说明，胃癌细胞的蛋白质可能至少有三大类：原点成分、阳极成分和上升成分。这些上升成分是第二向电泳时由原点上升的蛋白质，原来之可能是与前两类蛋白质相互作用而结合存在，第二向电泳时与之脱离而释放出来。现在看来，胃癌细胞的特异蛋白质可能存在于原点或其阳极成分中，作者推测，很可能存在于阳极成分中。

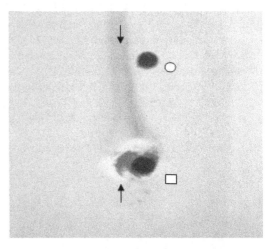

图61-3 培养物Ⅱ双向电泳的原点细节

注释：□所指处为培养物Ⅱ的原点，↑所指处为蛋白质离开原点、泳向阳极，↓所指处为第二向电泳时由原点上升的蛋白成分

4 讨论

胃癌细胞的特异蛋白质是什么？完整胃癌细胞内蛋白质的电泳释放研究有可能给出独特的答案。我们称完整细胞内蛋白质的电泳释放研究为"电泳释放蛋白质组学"，它与传统的蛋白质组学[1-4]不同之处就是使用完整细胞[7-10]而不是破坏细胞后的蛋白质。电泳释放蛋白质组学的技术路线有三大步：①完整细胞内蛋白质的电泳释放；②对所释放蛋白质进行SDS-PAGE分析；③对SDS-PAGE分离出来的蛋白质进行质谱分析，从而明确胃癌细胞的特异蛋白质。

本文实验，相当于电泳释放蛋白质组学的技术路线中的第一步。双向电泳中，第一向电泳时胃癌细胞培养物Ⅱ的释放成分都在对角线上，细看有蛋白质由原点前移(泳向阳极)，量较大。我们推测，这个前进成分可能包括胃癌细胞的特异蛋白质。原点成分和阳极成分在第二向电泳时都出现上升成分，这些上升成分是第二向电泳时由原点上升的蛋白质，原来它们可能是与前两类蛋白质相互作用而结合存在，第二向电泳时与它们脱离而释放出来。现在看来，胃癌细胞的特异蛋白质可能存在于原点或其阳极成分，很可能存在于阳极成分。作者推测，上升成分可能不是特异蛋白，但与其相互作用，维持其特异功能。上述一些推测是否合适有待进一步深入研究。

参 考 文 献

[1] 王平. 不同分化程度胃癌细胞差异蛋白质组学研究. 郑州大学硕士学位论文, 2006.
[2] 周欣. 应用蛋白质组学技术筛选和鉴定胃腺癌分化相关蛋白. 兰州大学硕士学位论文, 2009.
[3] 李兆星. 不同分化胃癌组织差异表达蛋白质的筛选和鉴定. 河北大学硕士学位论文, 2012.
[4] 刘羽. 应用蛋白质组学技术鉴定胃癌分化相关蛋白及 Serpin B1 表达与机制的研究. 河北大学博士学位论文, 2012.
[5] Su Y, Gao LJ, Ma Q, et al. Interactions of hemoglobin in lived red blood cells measured by the electrophoresis release test. Electrophoresis, 2010, 31: 2913-2920.

[6] 秦文斌. 活体红细胞内血红蛋白的电泳释放. 中国科学: 生命科学, 2011, 41(8): 597-607.
[7] Su Y, Shao G, Gao LJ, et al. RBC electrophoresis with discontinuous power supply – a newly established hemoglobin release test .Electrophoresis, 2009, 30: 3041-3043.
[8] Su Y, Shen J, Gao LJ, et al. Molecular interaction of re-released proteins in electrophoresis of human erythrocytes. Electrophoresis, 2012, 33: 1042-1045.
[9] Kurien B T, Scofield R H. Protein Electrophoresis: Methods and Protocols, Methods in Molecular Biology, 2012, 869: 387-402.
[10] 秦文斌. 红细胞内血红蛋白的电泳释放——发现和研究. 北京: 科学出版社, 2015.

第六十二章　小鼠胚胎成纤维细胞内蛋白质电泳释放的初步研究

姜树原[1]　高丽君[2#]　邵国[1※]　秦文斌[2※]

(包头医学院　1. 中心实验室；2. 血红蛋白研究室，包头　014010)

摘　要

目的： 我们曾长期研究红细胞内血红蛋白的电泳释放，现在要扩大范围，进入其他细胞领域、进入非血红素蛋白范畴。本文是研究小鼠胚胎成纤维细胞(NIH3T3)内蛋白质的电泳释放。
方法： 用小鼠成纤维细胞培养物直接进行双向电泳，观察其中蛋白质的释放情况。
结果： 第一向电泳时小鼠成纤维细胞培养物的释放成分都在对角线上，有较多原点后退成分。第二向电泳时在原点处出现较多上升成分。
结论： 原点后退成分可能是小鼠成纤维细胞的主要特异蛋白质，具体内容有待深入研究。

关键词　小鼠胚胎成纤维细胞 NIH3T3；双向电泳；原点成分；后退成分；上升成分

1　前言

小鼠胚胎成纤维细胞(NIH3T3)在科学研究中有着非常广泛的应用[1-6]，它的蛋白质组学研究[7]表明，由 NIH3T3 细胞分离出来的蛋白质，分成 SPA 处理组与对照组，两组双向电泳图谱中多数蛋白质的分布位置和表达量是一致的。软件初步分析显示，二者有近百个差异点，经过人工矫正减少为 31 个。从中筛选 10 个可信的差异蛋白点，最终结果有待进一步鉴定。

必须指出，当今的蛋白质组学研究，必须先破坏细胞，释放出蛋白质，然后对蛋白质进行各种实验分析，它的结果与完整细胞里边蛋白质的真实情况相差多少不得而知。完整细胞内各种蛋白质的存在状态非常复杂，其中包括多种形式的蛋白质相互作用，破坏细胞可能使这些相互作用减弱或消失。

我们用红细胞做的实验证明了这一点：完整红细胞有血红蛋白 A_2 现象(HbA_2 与 HbA_1 之间的相互作用)，红细胞破坏后这种互作完全消失[8-10]。

本文作者的研究手段是不破坏细胞，将完整细胞直接加入凝胶电场进行电泳[8-14]，此时电泳释放出来的蛋白质应该与传统蛋白质组学的结果不大相同，

也许能够查出这种完整细胞的特异蛋白质。详情参见正文。

\# 并列第一作者
※通讯作者：邵国，电子信箱：shootshao@163.com；秦文斌，电子信箱：qwb5991309@tom.com

2 材料及方法

2.1 标本来源 小鼠胚胎成纤维细胞(NIH3T3)培养物,来自包头医学院中心实验室,取一部分培养物做释放电泳,观察其释放特点。

2.2 方法

2.2.1 小鼠胚胎成纤维细胞培养物的双向电泳分析 用小鼠胚胎成纤维细胞培养物直接进行双向电泳,观察蛋白质释放情况。

2.2.2 培养物双向电泳原点的仔细观察 将培养物双向电泳的原点结果放大,观察其细节。

3 结果

3.1 小鼠胚胎成纤维细胞培养物的双向电泳分析 结果见图62-1。

由此图可以看出,第一向电泳时小鼠胚胎成纤维细胞培养物的释放成分都在对角线上,第二向电泳时在原点处出现较多上升成分。

3.2 小鼠成纤维细胞培养物双向电泳的原点细节 结果见图62-2。

由此图可以看出,第一向电泳时有原点成分和由原点后退成分,而且量较大。第二向电泳时又由原点释放出上升成分。

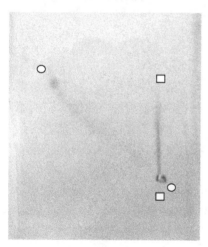

图62-1 小鼠胚胎成纤维细胞培养物的双向电泳分析

注释:两个○之间为第一向电泳结果,两个□之间为第二向电泳结果

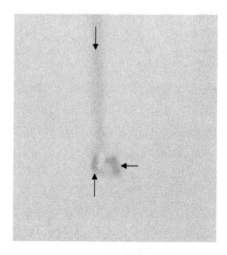

图62-2 小鼠胚胎成纤维细胞培养物双向电泳的原点观察

注释:↑所指处为原点成分,←所指处为后退成分,↓所指处为上升成分

4 讨论

小鼠胚胎成纤维细胞(NIH3T3)的特异蛋白质是什么?完整 NIH3T3 细胞内蛋白质的电泳释放研究有可能给出自己独特的答案。我们称完整细胞内蛋白质的电泳释放研究为"电泳释放蛋白质组学",它与传统蛋白质组学[7, 15]的不同之处就是使用完整细胞[9-15]而不是破坏细胞后的蛋白质。电泳释放蛋白质组学的技术路线有三大步:①完整细胞内蛋白质的电泳释放;

②对所释放蛋白质进行 SDS-PAGE 分析；③对 SDS-PAGE 分离出来的蛋白质进行质谱分析，从而明确小鼠胚胎成纤维细胞的特异蛋白质。

本文实验，相当于电泳释放蛋白质组学技术路线中的第一步。双向电泳中，第一向电泳时小鼠胚胎成纤维细胞培养物的释放成分都在对角线上，第二向电泳时在原点处出现较多上升成分。第一向电泳时有原点成分和由原点后退成分，第二向电泳时，原点成分上边出现上升成分，而后退成分就没有上移。

我们推测，这个后退成分可能包括 NIH3T3 的特异蛋白质。那些上升成分来自原点，与细胞结构结合比较牢固，所以第二次通电才被释放出来，由于上下拖拉很长，上边成分为分子量较小的蛋白质，下边成分是分子量较大的蛋白质。完整 NIH3T3 细胞内上升成分并不是来自后退成分，二者是什么关系，这是一个非常有趣的研究课题。

参 考 文 献

[1] 丁艳, 罗学刚, 申静, 等. Gankyrin 真核表达载体的构建及其在 NIH3T3 细胞中的表达. 中国药科大学学报, 2007, 38(2): 170-176.

[2] 张晓红, 唐圣松, 赵飞骏, 等. NIH3T3 细胞核内表达的 M-CSF 对细胞运动的影响. 解剖科学进展, 2007, 13(3): 239-234.

[3] 郑红, 汪思应, 杨晓明, 等. Tpr-Met 致 NIH3T3 细胞恶性转化的分子机制初探. 中国肿瘤, 2007, 16(9): 709-713.

[4] 王益民, 韦富康, 刘敏, 等. 成纤维细胞与创伤修复的研究进展. 中国修复重建外科杂志, 2000, 16(2): 126-128.

[5] 林霖, 傅欣, 张辛, 等. 骨形态发生蛋白 4 重组腺病毒的构建及其对 NIH3T3XB 细胞成骨分化作用的研究. 中国运动医学杂志, 2008, 27(3): 288-297.

[6] 韩焱福, 宋建星, 刘军, 等. 微囊化血管内皮细胞生长因子修饰 NIH3T3 细胞的制备及体外培养. 第二军医大学学报, 2008, 29(5): 485-494.

[7] 沈新明, 江培洲, 黄华, 等. 运用双向聚丙烯酰胺凝胶电泳和图像软件分析 TPA 处理 NIH3T3 细胞前后的蛋白质表达谱差异. 第一军医大学学报, 2004, 24(3): 310-313.

[8] Su Y, Gao LJ, Ma Q, et al. Interactions of hemoglobin in lived red blood cells measured by the electrophoresis release test. Electrophoresis, 2010, 31: 2913-2920.

[9] 秦文斌. 活体红细胞内血红蛋白的电泳释放. 中国科学: 生命科学, 2011, 41(8): 597-607.

[10] 秦文斌. 红细胞内血红蛋白的电泳释放——发现和研究. 北京: 科学出版社, 2015.

[11] Su Y, Shao G, Gao LJ, et al. RBC electrophoresis with discontinuous power supply – a newly established hemoglobin release test. Electrophoresis, 2009, 30: 3041-3043.

[12] Su Y, Shen J, Gao LJ, et al. Molecular interaction of re-released proteins in electrophoresis of human erythrocytes. Electrophoresis, 2012, 33: 1042-1045.

[13] Kurien B T, Scofield R H. Protein Electrophoresis: Methods and Protocols, Methods in Molecular Biology, 2012, 869: 387-402.

[14] Su Y, Shen J, Gao LJ, et al. Molecular interactions of re-released proteins in electrophoresis of human erythrocytes Electrophoresis, 2012, 33(9-10): 1402-1405.

[15] 邱宗荫, 尹一兵. 临床蛋白质组学. 北京: 科学出版社, 2008.

第六十三章 比较几种细胞的释放结果

1 结果比较

将第五十八~第六十二章的对应结果放在一起，比较它们的关系，见图63-1。

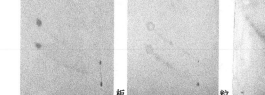

图63-1 比较几种细胞的释放结果
注释：板=血小板；粒=粒细胞；淋=淋巴细胞；胃=胃癌细胞；胚=NIH3T3

2 分类

2.1 按有无上升成分

(1) 原点上升型：淋巴细胞、胃癌细胞、NIH3T3。
(2) 非原点上升型：血小板、粒细胞。

2.2 按无原点沉淀

(1) 原点沉淀型：粒细胞、淋巴细胞、胃癌细胞、NIH3T3。
(2) 非原点沉淀型：血小板。

3 标本来源

血小板来自血库正常人，粒细胞来自慢性粒细胞白血病患者，淋巴细胞来自急性淋巴细胞白血病患者，胃癌细胞来自细胞培养，小鼠胚胎成纤维细胞来自细胞培养。

4 讨论

(1) 按有无原点沉淀来分类型可能比较合适，因为只有血小板来自正常人。
(2) 在这里，恶性疾病的双向电泳都有原点沉淀，这能总结为规律吗？仍需要观察和考验。
(3) 取各成分，做质谱分析，可能有助于弄清其中的关系。

第八篇 电泳释放蛋白质组学

前　　言

1 红细胞内血红蛋白的电泳释放内容很多，这里主要涉及初释放和再释放。

2 初释放，就是第一次通电后红细胞释放出来的血红蛋白，由于 HbA_2 位置特殊，又称为"血红蛋白 A_2 现象"。

3 再释放，就是第二次通电后红细胞释放出来的血红蛋白，此时初释放血红蛋白已经存在，在此基础上又出现了新的血红蛋白，它的位置相当于 HbA_1，第二次通电才能释放，推测它与红细胞膜的结合更为牢固。

4 取初释放血红蛋白进行蛋白质组学研究。

5 取再释放血红蛋白进行蛋白质组学研究。

6 当前蛋白质组学研究的核心技术就是双向凝胶电泳－质谱技术，即通过双向凝胶电泳将蛋白质分离，然后利用质谱对蛋白质进行鉴定。

第六十四章　红细胞初释放蛋白质组学

红细胞电泳释放蛋白质组学与普通的红细胞蛋白质组学有何不同？红细胞有普通蛋白质组学，是用红细胞溶血液做实验，发现许多斑点，但没有发现红细胞电泳释放时出现各种相互作用。红细胞初释放电泳蛋白质组学，是用完整红细胞电泳释放出来的血红蛋白做实验，初释放结果是红细胞内存在"HbA_1-HbA_2-Prx"式的相互作用，详见第八章内容。文中，除蛋白质组学内容外，还有一部分"交叉互作"的内容，可以删去不看。

第六十五章　红细胞再释放蛋白质组学

　　细胞电泳释放蛋白质组学与普通的红细胞蛋白质组学有何不同？红细胞有普通蛋白质组学，是用红细胞溶血液做实验，发现许多斑点，但没有发现红细胞电泳释放时出现各种相互作用。红细胞再释放电泳蛋白质组学，是用完整红细胞做再释放电泳释放出来的血红蛋白做实验。红细胞再释放蛋白质组学结果是红细胞内存在"HbA_1-CA_2=碳酸酐酶2"式的相互作用，详见第三十一章。

第六十六章 其他细胞电泳释放蛋白质组学展望

本章内容是展望，不是完成的蛋白质组学，是在现有电泳释放的基础上，提出蛋白质组学研究的切入点。

其他细胞有很多，这里主要是我们做过实验的几种细胞：血小板、粒细胞、淋巴细胞和胃癌细胞。每种细胞都做了释放实验，现在是在这些结果中寻找蛋白质组学研究的切入点。

第一节 血小板电泳释放蛋白质组学展望

1 血小板与其血浆的双向电泳

见图 66-1。

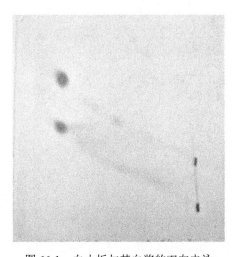

图 66-1 血小板与其血浆的双向电泳

注释：上层为血浆，下层为血小板。血浆方面，能看到白蛋白、$α_1$ 球蛋白、$α_2$ 球蛋白、β 球蛋白和纤维蛋白原，都在对角线上；原点没有上升成分。血小板方面，其在对角线上的成分与血浆不完全对应，而且出现多个离开对角线成分：白蛋白、α 球蛋白、纤维蛋白原。第一向电泳时紧跟白蛋白的成分，第二向电泳时随白蛋白水平上升。α 球蛋白、纤维蛋白原位置，也有类似情况。血小板成分的垂直线方面，原点二次释放明显，释放物顶部与白蛋白齐

2 蛋白质组学研究的切入点

图 66-2 中每个箭头(↓)所指处就是切入点。抠下这些脱离对角线的成分，进行蛋白质组学实验，就能发现其中的奥秘。我们的印象是，这里边可能有血小板里各种蛋白质之间的相

互作用。

上述印象的来源，是大鼠红细胞内血红蛋白之间的相互作用，详见图66-3。此图中有脱离对角线成分(箭头↓所指处)，后来发现大鼠血红蛋白之间的交叉互作[※]。根据这一点，我们推测血小板里的蛋白质也存在相互作用，其互作性质是否也属于交叉互作范畴，不得而知。

当然，血小板蛋白质的相互作用要比大鼠红细胞复杂得多，具体如何，有待进一步的科学实践。

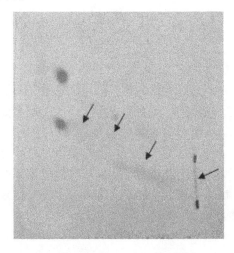

图66-2 血小板释放蛋白质组学的切入点
注释：箭头(↓)所指处就是切入点

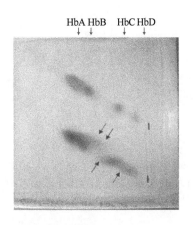

图66-3 大鼠红细胞及其溶血液的双向电泳
注释：上层为溶血液，下层为红细胞；考马斯亮蓝染色

第二节 粒细胞电泳释放蛋白质组学展望

1 粒细胞内蛋白质的电泳释放

见图66-4。

2 蛋白质组学研究的切入点

粒细胞的图66-4不同于血小板的图66-1。此粒细胞的图66-4里，粒细胞蛋白质与血浆蛋白有不同的部分，但是它们都在对角线上(箭头↑加□、↑加○、↑加◇处)，我们抠下这些成分，进行蛋白质组学研究，就能发现其中的奥秘。之前，我们推测血小板里可能存在多种蛋白质相互作用，本文的粒细胞如何？由于粒细

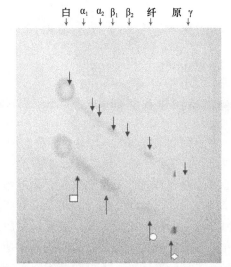

图66-4 粒细胞内蛋白质的电泳释放(双向电泳)
注释：上层为血浆(有多个向下箭头↓者)，下层为粒细胞(有多个向上箭头↑者)。白=白蛋白，$α_1$=$α_1$球蛋白，$α_2$=$α_2$球蛋白，$β_1$=$β_1$球蛋白，$β_2$=$β_2$球蛋白，纤=纤维蛋白，原=原点

[※] 参见秦文斌著《红细胞内血红蛋白的电泳释放——发现和研究》(科学出版社，2015)第101~105页

胞蛋白质没有脱离对角线，是否没有相互作用呢？这个问题不好解答。目前公认，各种细胞中普遍存在蛋白质相互作用，粒细胞不应例外。但是，粒细胞蛋白质没有脱离对角线，应当如何理解？是不同于交叉互作的其他互作吗？不得而知。期待未来的蛋白质组学研究结果能解决这个疑问。

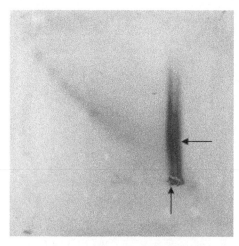

图 66-5 淋巴细胞的双向电泳
注释：双向电泳，第一向为普泳(相当于初释放)，第二向也是普泳(相对于第一向，它就是再释放)

第三节 淋巴细胞电泳释放蛋白质组学展望

蛋白质组学研究的切入点

由图 66-5 可以看到，淋巴细胞的结果又不同于血小板和粒细胞。其原点沉淀明显(箭头↑所指处)，由原点向上出现大量脱离对角线成分(箭头←所指处)，切入点就在这两个地方。我们抠下这些成分，进行蛋白质组学实验，就能发现其中的奥秘。箭头↑所指处是原点沉淀，它的蛋白质组学研究结果如何，不得而知。

第四节 胃癌细胞电泳释放蛋白质组学展望

1 双向电泳比较培养基与胃癌培养物Ⅱ

见图 66-6。

2 蛋白质组学研究的切入点

胃癌细胞的图 66-6 与淋巴细胞的图 66-5 比较相似。这里也是，原点沉淀明显(箭头↑所指处)，也是由原点向上出现大量脱离对角线成分(箭头↓所指处)。当然，切入点也就在这两个地方，我们抠下这些成分，进行蛋白质组学研究，就能发现其中的奥秘。箭头↓所指处的蛋白质组学研究结果，也可能是蛋白质之间的相互作用。原点沉淀处的蛋白质组学研究结果如何，不得而知。

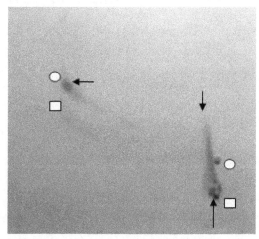

图 66-6 双向电泳比较培养基与胃癌培养物Ⅱ
注释：上层(两个○之间)为细胞培养的培养基，下层(两个□之间)为细胞培养的胃癌培养物Ⅱ(含多量沉淀)，↑所指处为加样的原点

第五节　NIH 3T3 电泳释放蛋白质组学展望

1　小鼠胚胎成纤维细胞 NIH3T3

所用样品由邵国老师教研室提供。由于只有细胞培养物，没有培养基，所以没有对照。双向电泳结果，见图 66-7。

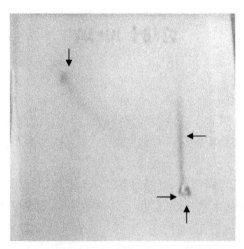

图 66-7　小鼠胚胎成纤维细胞培养物的双向电泳

注释：箭头↑所指处为原点(加样处)，箭头→所指处为"原点沉淀"，箭头↓所指处为培养物里可溶成分，箭头←所指处为"原点上升带"或"上升带"或"垂直线"

2　结果

第一向电泳时，NIH3T3 培养物中可溶成分泳出，不溶成分留在原点(沉淀)；第二向电泳时，一部分原点沉淀泳出，形成上升带。

3　蛋白质组学研究的切入点

图 66-7 与淋巴细胞的图 66-5 比较相似。这里也是原点沉淀明显(箭头→所指处)，也是由原点向上出现大量脱离对角线成分(箭头←所指处)。当然，切入点也就在这两个地方，我们抠下这些成分，进行蛋白质组学研究，就能发现其中的奥秘。

第九篇　制备电泳

前　言

　　1　各种电泳技术，几乎都涉及分离和制备功能。例如，用淀粉颗粒进行的淀粉板制备电泳，可以由红细胞溶血液分离和制备各种血红蛋白[1]，此法简单易行，主要用于分离制备，分析能力较差。

　　2　淀粉胶制备电泳可以分离制备各种蛋白质[2]，分析能力也很强，只是操作有一些复杂。

　　3　聚丙烯酰胺凝胶电泳也有制备型产品[3]，能够分离和制备多种蛋白质，但需专门设备，操作复杂，执行有一定难度。此产品不能用于分析，若想分析标本，要用专门的分析型电泳仪。

　　4　等电聚焦电泳，也有专门的制备型等电聚焦电泳装置[3]，它的性能及问题与制备型聚丙烯酰胺凝胶有类似之处。

　　5　我们的淀粉-琼脂糖混合凝胶电泳，操作比较简单，既能分析，又能制备。在制备方面，丽春红染色法功不可没。理由如下：丽春红染色时丽春红与蛋白质结合，此结合产物与凝胶的再结合不牢固，容易洗脱和分离制备。其他染色产物都与凝胶结合牢固，不容易洗脱。所以，我们能在丽春红染色后"抠胶"，来分离和制备各种蛋白质，产生我们自己独特的制备电泳方法。丽春红染色的这一特点，也是我们在实践中发现的，未见文献记载。将染色后的蛋白质与凝胶一起抠下来(我们称之为"抠胶"，下边常用这个词)，经"冻结-融化"等处理后，分离出来所要的蛋白质，从而达到制备目的。

　　6　抠胶取得的蛋白质，可以进行交叉互作。例如，由成人红细胞标本抠取 HbA_1、HbA_2、HbA_3，观察前二者之间的相互作用，再观察后二者之间的相互作用。结果是，HbA_2 与 HbA_1 之间有交叉互作，HbA_2 于 HbA_3 之间没有交叉互作。由脐带血抠出 HBF，再与成人 HbA_2 进行交叉互作，结果是存在交叉互作。

　　7　制备电泳的更深入的应用，是将抠得的血红蛋白与质谱分析连接起来，进入"电泳释放蛋白质组学"领域。例如，在 A_2 现象机制研究中，抠取单向电泳结果 HbA_1 与 HbA_2 之间部分，做质谱分析，结果发现，在血红蛋白 A_2 现象里，除了 HbA_2 与 HbA_1 相互作用外，还有 Prx(过氧化物还原酶)参加，即 HbA_1-HbA_2-Prx 形式存在。在再释放机制研究中，抠取再释放区带做质谱，发现再释放成分是 HbA_1 与碳酸酐酶 $2(CA_2)$ 的结合产物：HbA_1-CA_2。当然，再释放还有多种状态，有单带再释放和多带再释放，它们的抠胶情况也不一样。

参 考 文 献

[1] 曾溢滔，黄淑祯，陈美环. 血红蛋白分析技术 1 虹吸式淀粉板电泳. 上海医学，1979，2(2)：56-60.
[2] 蔡铎昌. 淀粉胶制备电泳. 西华师范大学学报(自然科学版)，1983，(1)：69-74.
[3] 何忠效，张树政. 电泳. 北京：科学出版社，1999：39-40.

第六十七章　制备电泳通则

(1) 根据需要，采取正常人、患者或动物的血液。
(2) 低速离心，弃上清液，留红细胞。
(3) 用生理盐水洗涤红细胞三次，低速离心，弃上清液，留下层红细胞。
(4) 向红细胞加蒸馏水、四氯化碳，置振荡器上，震荡30秒。
(5) 低速离心，取上清液(红细胞溶血液)，弃沉淀。
(6) 用此溶血液做普通电泳，丽春红染色，看到各种血红蛋白成分。
(7) 根据需要，连同所在的凝胶抠取所要的血红蛋白。
(8) 分别装入大EP管(1.5ml)，放入–20℃冰箱过夜。
(9) 到时取出，室温融化，高速离心，要上清液，备用(见以下研究)。
(10) 浓度太低者，需要浓缩，如HbA_2。

第六十八章　制备人 HbA_1 与 HbA_2
——用于交叉互作研究

1　正常人红细胞溶血液的单向电泳

取正常人血液，按上述通则操作，得到电泳图 68-1。

抠去 HbA_1 和 HbA_2 后，如图 68-2 所示。

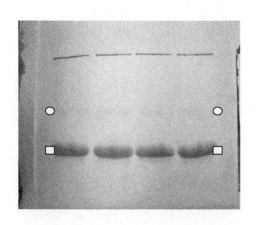

图 68-1　正常人红细胞溶血液的单向电泳
注释：两个□之间的区带为 HbA_1，两个○之间的区带为 HbA_2

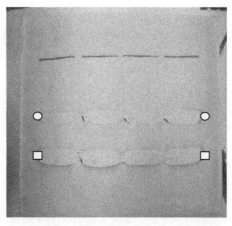

图 68-2　HbA_1 及 HbA_2 的抠胶情况
注释：两个□之间的位置为 HbA_1 抠去后情况，两个○之间的区带为 HbA_2 抠去后情况

2　按通则进行处理

3　进行"交叉互作电泳"

HbA_1 穿过 HbA_2，后者区带变形，如图 68-3 所示。

4　结果

证明 HbA_1 穿过 HbA_2 时，后者区带变形，说明二者发生了相互作用，我们把这种相互作用称之为"交叉互作"。说明我们制备的电泳方法有效。

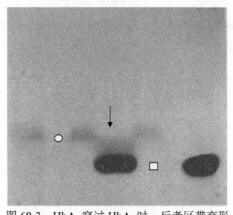

图 68-3　HbA_1 穿过 HbA_2 时，后者区带变形
注释：□两侧区带为 HbA_1，○两侧区带为 HbA_2，箭头↓所指处为 HbA_2 区带变形的地方

第六十九章 制备大鼠四种血红蛋白
——用于交叉互作研究

1 大鼠红细胞及其溶血液的双向电泳

取正常大鼠血液，按上述通则操作，得到如下电泳图 69-1。

结果：溶血液 4 个血红蛋白(染前红色)，含量 HbA＞bB＞bC＞bD，都在对角线上；红细胞有上述 4 种血红蛋白，又有脱离对角线成分，余从略。

讨论：抠取上述 4 种血红蛋白，直接做交叉互作实验。

抠胶后直接做交叉互作实验，见图 69-2。

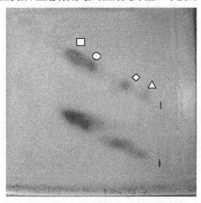

图 69-1 大鼠红细胞及其溶血液的双向电泳
注释：上层为溶血液，下层为红细胞，丽春红染色，再染考马斯亮蓝；□为 HbA，○为 HbB，◇为 HbC，△为 HbD

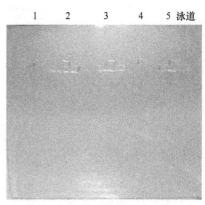

图 69-2 抠胶后直接做交叉互作实验
注释：泳道 1 为 HbB，2 为 HbC 穿过 HbB，3 为 HbD 穿过 HbB，4 为 HbC，5 为 HbC 穿过 HbD

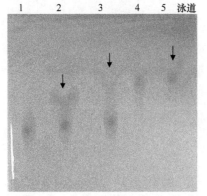

图 69-3 大鼠血红蛋白交叉互作
注释：泳道 1 为 HbB(对照)，2 HbB 穿过 HBC，后者区带变形(箭头↓所指处)，说明 HbB 与 HbC 发生交叉互作，3 为 HbB 穿过 HbD，后者区带变形(箭头↓所指处)，说明 HbB 与 HbD 发生交叉互作，4 为 HbC(对照)，5 为 HbC 穿过 HbD，后者区带变形(箭头↓所指处)，说明 HbC 与 HbD 发生交叉互作

2 结果

交叉互作结果，见图 69-3。还有大鼠其他血红蛋白之间的相互作用，此处省略。证明与人类血红蛋白一样，大鼠血红蛋白之间也存交叉互作。说明我们制备的电泳有效。

第七十章　制备 HbA$_2$ 与 HbA$_1$ 之间成分
——用于血红蛋白 A$_2$ 现象机制的研究

1　红细胞与其溶血液并排电泳

取正常人血液，按上述通则操作，得到如下电泳图 70-1。抠胶后情况如图 70-2 所示。

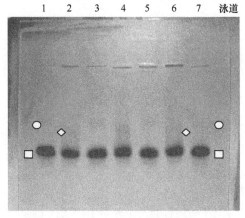

图 70-1　红细胞与其溶血液并排电泳

注释：泳道 1、3、5、7 为红细胞溶血液，泳道 2、4、6 为红细胞，两个 □ 之间为 HbA$_1$，两个 ○ 之间为 HbA$_2$，两个 ◇ 之间为 "红细胞 HbA$_2$"，准备抠取 HbA$_1$ 与 HbA$_2$ 之间部分，用于分析(对照)，准备抠取 HbA$_1$ 与 "红细胞 HbA$_2$" 之间部分，用于分析(标本)

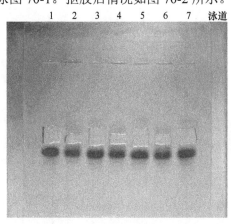

图 70-2　抠胶后情况

注释：泳道 1、3、5、7 为抠取 HbA$_1$ 与 HbA$_2$ 之间部分，用于分析(对照)，泳道 2、4、6 为抠取 HbA$_1$ 与 "红细胞 HbA$_2$" 之间部分，用于分析(标本)

2　按通则进行处理

3　2DE 和质谱分析

于 *Electrophoresis*(2010，31：2913-2920) 杂志发表，摘取内容如下所示。

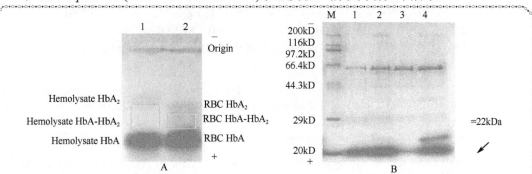

Figure 4　Proteins extracted from the starch-agarose mixed gel separated by SDS-PAGE. (A) The excised bands for HbA- HbA$_2$ and HbA$_2$ are indicated in this figure. Lane 1 contained hemolysate and lane 2 contained RBC. (B) The result of separation of the extracted proteins by 5%-12% SDS- PAGE. Lane M contained protein markers and lanes 1-4 contained hemolysate HbA$_2$, RBC HbA$_2$, hemolysate HbA-HbA$_2$ and RBC HbA- HbA$_2$, respectively.

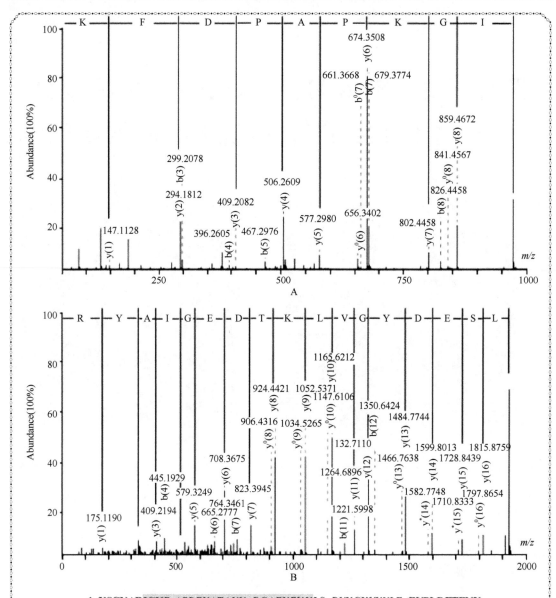

Figure 5. The nanoUPLC-ESI MS/MS spectra for the trypsin digested gel band. (A) The MS spectrum of a peptide with molecular weight 971.5440 and an amino acid sequence IGKPAPDFK is labeled. (B) The MS spectrum of a peptide with molecular weight 1927.9527 and an amino acid sequence LSEDY-GVLKTDEGIAYR is labeled. (C) The sequence coverage of thioredoxin peroxidase B is 77%.

a combined product of HbA$_2$ and HbA. However, if RBC HbA$_2$ = hemolysate HbA$_2$ + hemolysate HbA, the electrophoretic position of RBC HbA$_2$ should locate at the middle of hemolysate HbA$_2$ and hemolysate HbA. In fact, RBC HbA$_2$ was located near to hemolysate HbA$_2$. Thus, we speculated that there must be some other components (X) involving in this Hb complex. The electrophoresis results for hemolysate, RBCs, freeze-thawed RBCs and the supernatant of freeze-thawed RBCs indicated that integrity of the RBC membrane was necessary to mediate the interaction between HbA and HbA_2. When the RBC membrane was removed, the interaction was eliminated. If membrane integrity was damaged, the interaction would obviously be weakened.

To further explore the constitution of the Hb complex, proteins located between HbA and HbA_2 were extracted and separated by SDS-PAGE. The results showed a clear ≈22 kDa band in RBC HbA-HbA_2 but not in hemolysate HbA-HbA_2. LC/MS/MS detection of the digested ≈22 kDa band showed that this band was a peptide mixture mainly composed of TPx B, α-globin, δ-globin and β-globin. Traces of keratin appearing in the result (Supporting Information) was almost certainly due to contamination (Table 1).

Table 1 LC/MS/MS results for the ≈22 kDa band

NCBI accession no.	Name	Mass	Score	Queries matched
gi|9955007	Chain A, Thioredoxin Peroxidase B from red blood cells	21795	1218	R.SVDEALR.L R.GLFIIDGK.G K.TDEGIAYR.G R.IGKPAPDFK.A K.ATAVVDGAFK.E R.LSEDYGVLK.T R.QITVNDLPVGR.S R.QITVNDLPVGR.S R.GLFIIDGKGVLR.Q K.ATAVVDGAFKEVK.L K.EGGLGPLNIPLLADVTR.R K.EGGLGPLNIPLLADVTR.R R.KEGGLGPLNIPLLADVTR.R K.EGGLGPLNIPLLADVTRR.L K.EGGLGPLNIPLLADVTRR.L R.LSEDYGVLKTDEGIAYR.G R.LSEDYGVLKTDEGIAYR.G K.LGCEVLGVSVDSQFTHLAWINTPR.K + Carbamidomethyl (C) R.KLGCEVLGVSVDSQFTHLAWINTPR.K + Carbamidomethyl (C) R.KLGCEVLGVSVDSQFTHLAWINTPR.K + Carbamidomethyl (C) R.KLGCEVLGVSVDSQFTHLAWINTPR.K + Carbamidomethyl (C) R.LVQAFQYTDEHGEVCPAGWKPGSDTIKPNVDDSK.E + Carbamidomethyl (C) R.LVQAFQYTDEHGEVCPAGWKPGSDTIKPNVDDSK.E + Carbamidomethyl (C) R.LVQAFQYTDEHGEVCPAGWKPGSDTIKPNVDDSKEYFSK.H + Carbamidomethyl (C) R.LVQAFQYTDEHGEVCPAGWKPGSDTIKPNVDDSKEYFSK.H + Carbamidomethyl (C)
gi|4504351	Delta globin	16045	396	K.LHVDPENFR.L K.VNVDAVGGEALGR.L R.LLVVYPWTQR.F K.VVAGVANALAHKYH.- K.EFTPQMQAAYQK.V + Oxidation (M) K.GTFSQLSELHCDK.L + Carbamidomethyl (C) K.VLGAFSDGLAHLDNLK.G R.FFESFGDLSSPDAVMGNPK.V + Oxidation (M)
gi|161760892	Chain D, neutron structure analysis of deoxy human hemoglobin	15869	368	-.RHLTPEEK.S K.LHVDPENFR.L R.LLVVYPWTQR.F K.VNVDEVGGEALGR.L K.EFTPPVQAAYQK.V K.VVAGVANALAHKYH.- K.VLGAFSDGLAHLDNLK.G R.FFESFGDLSTPDAVMGNPK.V + Oxidation (M)
gi|47679341	Hemoglobin beta	11439	256	K.LHVDPENFR.- K.VNVDAVGGEALGR.L R.LLVVYPWTQR.F K.VLGAFSDGLAHLDNLK.G R.FFESFGDLSTPDAVMGNPK.V + Oxidation (M)

Continued

NCBI accession no.	Name	Mass	Score	Queries matched
gi|66473265	Homo sapiens beta globin chain	11480		K.LHVDPENFR.- R.LLVVYPWTKR.F K.VNVDEVGGEALGR.L K.VLGAFSDGLAHLDNLK.G R.FFESFGDLSTPDAV$\underline{M}$GNPK.V + Oxidation (M)
gi|229751	Chain A, Structure of hemoglobin in the deoxy quaternary state with ligand bound at the alpha hemes(a2)	15117	249	R.$\underline{M}$FLSFPTTK.T + Oxidation (M) K.LRVDPVNFK.L -.VLSPADKTNVK.A K.VGAHAGEYGAEALER.M K.TYFPHFDLSHGSAQVK.G K.TYFPHFDLSHGSAQVK.G K.VADALTNAVAHVDD$\underline{M}$PNALSALSDLHAHK.L + Oxidation (M)
gi|179409	Beta-globin	15870	247	K.LHVDPENFR.L K.EFTPPVKAAYQK.V K.VVAGVANALAHKYH.- K.VLGAFSDGLAHLDNLK.G
gi|4929993	Chain A, module-substituted chimera hemoglobin beta-alpha (F133v)	15780	240	K.LRVDPVNFK.L R.LLVVYPWTQR.F K.VNVDEVGGEALGR.L K.VLGAFSDGLAHLDNLK.G R.FFESFGDLSTPDAV$\underline{M}$GNPK.V + Oxidation (M)
gi|157838239	Chain A, hemoglobin thionville: an alpha-chain variant with a substitution of a glutamate for valine at na-1 and having an acetylated methionine nh2 terminus(a₂)	15278	226	R.$\underline{M}$FLSFPTTK.T + Oxidation (M) K.LRVDPVNFK.L K.VGAHAGEYGAEALER.M K.TYFPHFDLSHGSAQVK.G K.TYFPHFDLSHGSAQVK.G K.VADALTNAVAHVDDMPNALSALSDLHAHK.L + Oxidation (M)

TPx B has previously been called torin, calpromotin, thiol-specific antioxidant/protector protein, band-8, natural killer enhancing factor-B and is now named peroxiredoxin 2 (Prx 2)[24-25]. The human Prx 2 gene is located at 13q12, coding a 198 amino acid polypeptide with a molecular mass of 22 kDa[24]. There are 6 types of mammalian Prx isoforms (Prx 1-6)[26-28], of which Prx 2 is the third most abundant protein in RBC and has an important peroxidase activity that can protect RBCs against various oxidative stresses[24]. As we know, the RBC contains high levels of O_2 and Hbs and the continual auto-oxidation of Hbs produces many reactive oxygen species (ROS) such as O_2 and H_2O_2. Therefore, compared with other somatic cells, RBCs are exposed to a higher level of oxidative stress[13, 29]. The ROS in RBCs can damage proteins and membrane lipids but, as anucleate cells, RBCs are unable to synthesize new proteins to replace damaged proteins[26, 30]. Thus, RBCs must be well-equipped with many antioxidant proteins, which include catalase, glutathione peroxidase, and the emerging antioxidant enzyme, Prxs. For a long time, it was considered that catalase and glutathione peroxidase constituted the defense against ROS in RBC[14]. Recently, increasing attention has been given to the antioxidant role of Prxs in RBCs. Our experimental results show for the first time that in live RBC, Prx 2 binds with the globin chain of Hb to form a complex. This complex enables Prx 2 to be more effective in protecting Hb from oxidative stress. However, the manner by which Prx 2 interacts with these globins remains unknown. We believe some of the membrane proteins mediate the binding of this Hb complex because the HbA_2 phenomenon appeared only in RBC samples and not in hemolysate or the supernatant of freeze-thawed RBCs in which the RBC membrane was removed.

Furthermore, during our experiments, a unique method, ERT, was established for the study of protein-protein interaction in live cells. An increasing number of approaches, such as yeast two-hybrid systems, tandem affinity purification, protein chip, co-immunoprecipitation and glutathione-S-transferase pull-down methods, have been developed to detect protein-protein interactions[31]. However, functional protein-protein interactions are dynamic processes and many are maintained by non-covalent bonds. Therefore, the detection of interactions in live cells remains difficult. In addition, the loss of internal organelles greatly limits the use of interaction detection approaches such as TAP and GST pull-down in the RBC. With ERT, live RBCs were added directly onto the gel and the electric current perforated the membrane instantaneously. The protein complexes in live RBCs were thus released directly to the electric field and the different electrophoresis behaviors of RBC proteins could be directly compared with hemolysate, in which the protein-protein interactions would have been damaged during preparation, especially interactions mediated by membrane proteins. Thus, *in vivo* interactions in the RBC could be identified by ERT. Furthermore, this method can be used in other live cells. Evidence of *in vivo* protein-protein interactions may be found through finding differences in the electrophoretic behavior of living cells and corresponding cell lysates. However, the low resolving power of starch-agarose gels decreases the detection range and it is therefore mainly limited to the detection of interactions between high-abundance proteins. Interactions involving low-abundance proteins are difficult to detect but we believe that through the use of a high-resolution electrophoresis method, more information on such interactions may be found in the future.

This work was supported by grants from the Major Projects of Higher Education Scientific Research in the Inner Mongolia Autonomous Region (NJ09157) and the Key Science and Technology Research Project of the Ministry of Education. We also especially acknowledge all of the people who donated their blood samples for our research.

The authors have declared no conflict of interest.

References

[1] Kim, K. K. Kim, H. B., *World J. Gastroenterol.* 2009, *15*, 4518-4528.
[2] Bu, D., Zhao, Y., Cai, L., Xue, H., Zhu, X., Lu, H., Zhang, J., Sun, S., Ling, L., Zhang, N., Li, G., Chen, R., *Nucleic Acids Res.* 2003, *31*, 2443-2450.
[3] Hase, T., Tanaka, H., Suzuki, Y., Nakagawa S., Kitano H., *PLoS. Comput. Biol.* 2009, *5*, e1000550.
[4] Frishman, D., Albrecht, M., Blankenburg, H., Bork, P., Harrington, E. D., Hermjakob, H., Jensen, L. J., Juan, D. A., Lengauer, T., Pagel., P., Schachter, V., Valencia, A., in: Frishman, D., Valencia, A. (Eds) *Modern Genome Annotation*, Springer Wien New York, Vienna 2008, pp. 353-410.
[5] Börnke, F., in: Junker B. H., Schreiber F., *Analysis of Biological Networks*, Wiley-Interscience, New York 2008, pp. 207-232.
[6] Jung, S. H., Hyun, B., Jang, W. H., Hur, H. Y., Han D. S., *Bioinformatics* 2010, *26*, 379-391.
[7] Jung, S. H., Jang, W. H., Hur, H. Y., Hyun, B., Han D. S., *Genome Inform.* 2008, *21*, 77-88.
[8] Collura, V., Boissy, G., *Subcell Biochem*. 2007, *43*, 135-183.
[9] Perutz, M. F., Rossmann, M. G., Cullis, A. F., Muirhead, H., Will G., North, A. C., *Nature* 1960, *185*, 416-422.
[10] Perutz, M. F., *Brookhaven Symp. Biol.* 1960, *13*, 163-183.
[11] D'Alessandro, A., Righetti, P. G., Zolla. L., *J. Proteome Res.* 2010, *9*, 144-163.
[12] Goodman, S. R., Kurdia, A., Ammann, L., Kakhniashvili, D., Daescu O., *Exp. Biol. Med. (Maywood)* 2007, *232*, 1385-1408.

[13] Kakhniashvili, D. G., Bulla Jr., L. A., Goodman, S. R., *Mol. Cell Proteomics* 2004, *3*, 501-509.
[14] D'Amici, G. M., Rinalducci, S., Zolla, L., *J. Proteome Res.* 2007, *6*, 3242-3255.
[15] Alvarez-Llamas, G., de la Cuesta, F., Barderas, M. G., Darde, V. M., Zubiri, I., Caramelo, C., Vivanco, F., *Electrophoresis* 2009, *30*, 4089-4108.
[16] van Gestel, R. A., van Solinge, W. W., van der Toorn, H. W., Rijksen, G., Heck, A. J., van Wijk, R., Slijper, M., *J. Proteomics* 2010, *73*, 456-465.
[17] Zhang, Q., Tang, N., Schepmoes, A. A., Phillips, L. S., Smith, R. D., Metz, T. O., *J. Proteome Res.* 2008, *7*, 2025-2032.
[18] Eleuterio, E., Di Giuseppe, F., Sulpizio, M., di Giacomo, V., Rapino, M., Cataldi, A., Di Ilio, C., Angelucci, S., *Biochim. Biophys. Acta* 2008, *1784*, 611-620.
[19] Qin, W. B., Liang, Y. Z., *Chin. J. Biochem. Biophys.* 1981, *13*, 199-201.
[20] Qin, W. B., *Sheng Wu Hua Xue Yu Sheng Wu Wu Li Jin Zhan* 1991, *18*, 280-289.
[21] Su, Y., Shao, G., Gao, L., Zhou, L., Qin L, Qin W., *Electrophoresis* 2009, *30*, 3041-3043.
[22] Wang, H. X., Jin, B. F., Wang, J., He, K., Yang, S. C., Shen, B. F., Zhang, X. M., *Sheng Wu Hua Xue Yu Sheng Wu Wu Li Xue Bao* 2002, *34*, 630-634.
[23] Xia, Q., Wang, H. X., Wang, J., Liu, B. Y., Hu M. R., Zhang, X. M., Shen, B. F., *Zhongguo Yi Xue Ke Xue Yuan Xue Bao* 2004, *26*, 477-487.
[24] Schröder, E., Littlechild, J. A., Lebedev, A. A., Errington, N., Vagin, A. A., Isupov, M. N., *Structure* 2000, *8*, 605-615.
[25] Wood, Z. A., Schröder, E., Robin Harris, J., Poole, L. B., *Trends Biochem. Sci.* 2003, *28*, 32-40.
[26] Stuhlmeier, K. M., Kao, J. J., Wallbrandt, P., Lindberg, M., Hammarström, B., Broell, H., Paigen, B., *Eur. J. Biochem.* 2003, *270*, 334-341.
[27] Yang, K. S., Kang, S. W., Woo, H. A., Hwang, S. C., Chae, H. Z., Kim, K., Rhee, S. G., *J. Biol. Chem.* 2002, *277*, 38029-38036.
[28] Manta, B., Hugo, M., Ortiz, C., Ferrer-Sueta, G., Trujillo M., Denicola A., *Arch. Biochem. Biophys.* 2009, *484*, 146-154.
[29] Johnson, R. M., Goyette, G. Jr., Ravindranath, Y., Ho, Y. S., *Free Radic. Biol. Med.* 2005, *39*, 1407-1417.
[30] Halliwell, B., Gutteridge, J. M., *Free Radicals in Biology and Medicine*, Oxford University Press Inc., New York 1998.
[31] Drewes, G., Bouwmeester, T., *Curr. Opin. Cell Biol.* 2003, *15*, 199-205.

4 结论

血红蛋白 A_2 现象的机制是 HbA_2 与 HbA_1 相互作用，还有 Prx (过氧化物还原酶)参加，形成三联体：HbA_1-HbA_2-Prx。说明我们制备的电泳有效。

第七十一章 制备定时再释放区带
——用于再释放机制的研究

1 红细胞与其溶血液并排电泳(定时 15 分钟)

取正常人血液,按上述通则操作,定时 15 分钟,得到如下电泳图 71-1。抠胶后情况如图 71-2 所示。

取正常人血液,按上述通则操作,定时 30 分钟,得到如下电泳图 71-3。抠胶后情况如图 71-4 所示。

取正常人血液,按上述通则操作,定时 60 分钟,得到如下电泳图 71-5。抠胶后情况如图 71-6 所示。

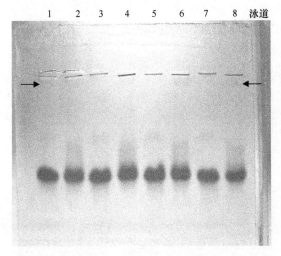

图 71-1 红细胞与其溶血液并排电泳(定时 15 分钟)
注释:泳道 1、3、5、7 为红细胞溶血液,泳道 2、4、6、8 为红细胞,两个箭头→←之间的区带为定释带,溶血液泳道没有区带,准备抠取定释带,用于分析(标本),准备抠取溶血液的对应部位用于分析(对照)

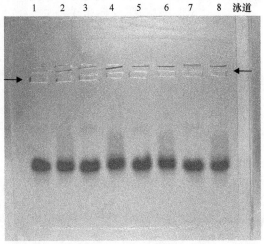

图 71-2 定时 15 分钟抠胶后情况
注释:两个箭头→←所指处为抠胶的地方,泳道 1、3、5、7,抠取相当于定释带的部位,用于分析(对照),泳道 2、4、6、8,抠取定释带的部位,用于分析(标本)

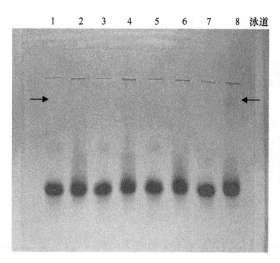

图 71-3　红细胞与其溶血液并排电泳(定时 30 分钟)

注释：泳道 1、3、5、7 为红细胞溶血液，泳道 2、4、6、8 为红细胞，两个箭头→←之间的区带为定释带，溶血液泳道没有区带，准备抠取定释带，用于分析(标本)，准备抠取溶血液的对应部位用于分析(对照)

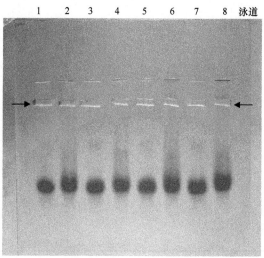

图 71-4　定时 30 分钟抠胶后情况

注释：两个箭头→←所指处为抠胶的地方，泳道 1、3、5、7，抠取相当于定释带的部位，用于分析(对照)，泳道 2、4、6、8，抠取定释带的部位，用于分析(标本)

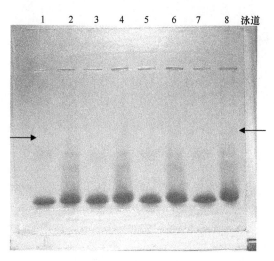

图 71-5　红细胞与其溶血液并排电泳(定时 60 分钟)

注释：泳道 1、3、5、7 为红细胞溶血液，泳道 2、4、6、8 为红细胞，两个箭头→←之间的区带为定释带，溶血液泳道没有区带，准备抠取定释带，用于分析(标本)，准备抠取溶血液的对应部位用于分析(对照)

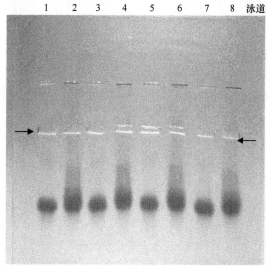

图 71-6　定时 60 分钟抠胶后情况

注释：两个箭头→←所指处为抠胶的地方，泳道 1、3、5、7，抠取相当于定释带的部位，用于分析(对照)，泳道 2、4、6、8，抠取定释带的部位，用于分析(标本)

2　按通则进行处理

3 2DE 和质谱分析

2DE 和质谱分析的研究于 *Electrophoresis*(2012，33：1402-1405)杂志发表，摘取部分如下所示。

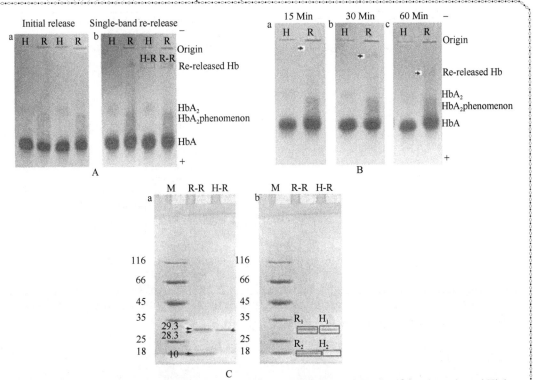

Fig. 1 Electrophoresis release test and the protein component isolation from single-band re-released Hb by 5%–12% SDS-PAGE. (A) Electrophoresis release test on starch-agarose mixed gel——(a) is the initial release electrophoresis；(b) is the single-band rerelease electrophoresis. H represents hemolysate sample，which is used as a control；R represents erythrocyte sample. R-R represents the re-released Hb band from erythrocytes；H-R represents the corresponding band from hemolysate. (B) The different location of re-released Hb band from erythrocytes at different resumed electrophoresis time after interruption——(a) the resumed electrophoresis time is 15 min；(b) the resumed electrophoresis time is 30 min；(c) the resumed electrophoresis time is 60 min. (C) The protein separation result of the single-band re-released Hb using 5%–12% SDS-PAGE——(a) the SDS-PAGE result of the R-R and H-R samples；(b) is the same as (a), only labels the name of the bands for MS detection. M represents protein marker；R_1 and R_2 represent the ～29 kDa and ～16 kDa bands of R-R sample respectively；H_1 and H_2 represent those corresponding bands of H-R sample，respectively.

Table 1 Proteins identified from three bands by MALDI-TOF MS

ID no.	NCBI accession no.	Name	Mass	P*I*	Score	Series matched	Coverage
H_1	GI：4557395	Carbonic anhydrase2, *Homo saplliens*	29 285	6.87	95	KYDPSLKPLSVSYDQATSLR.I RLNNGHAFNVEFDDSQDKAVLKG KGGPLDGTYR.L KYAAELNLVNWNTK.Y KAVOQPDGLAVLGIFLK.V KGKSADFTNFDPRG KSADFTNFDPRG	35%
R_1	GI：4557395	Carbonic anhydrase2, *Homo saplliens*	29 285	6.87	114	KHNGPEHWHKDFPIAKG KYDPSLKPLSVSYOQATSLRI RILNNGHAFNVEFD DSODKA KGGPLDGTYRL RUOFHFHWGSLDGOGSEHTVDKK KKYAAELHLVHWNTKY KAVQQPDGLAVLGIFLKV KSADFTNFDPRG RKLNFNGEGEPEELMVDNWRPAOPLK.N KLNFNG EGEPEELMVDNWRPAOPLK.N	56%

Continued

ID no.	NCBI accession no.	Name	Mass	PI	Score	Series matched	Coverage
	GI: 4502517	Carbonic anhydrase1, *Homo saplliens*	28 909	6.59	85	KTSETKHDTSLKPISVSYNPATAKE KHDTSLKPISVSYNPATAKE KEIINVGHSFHVNFEONDNRS KGPFSDSYRL RLFQFHFHWGSTNEHGSEHTV DGVKY KYSAELHVAHWNSAKY KESISVSSEQLAQFRS RSLLSNVEGONAVPMQHNNRPRP- TOPLKG	49%
R$_2$	GI: 4504349	HBB-HUMAN	16 102	6.75	156	KSAVTALWGKV KVNVOEVGGEALGRL RLLWYPWTORF RFFESFGOLSTPOAVMGNPKV KKVLGAFSOGLAMLDNLKG KVLGAFSOGLAHLDNK.G KGTFATLSELHCDKL KLHVDPENFRL RLLGNVLVCVLAHHFGKE KEFTPPVOAAYOKV KVVAGVANALAHKYH	89%

In the future, further information will be provided by using this method, and this will open up new ways to explore the mystery of erythrocytes.

4 结论

红细胞再释放机制是 HbA$_1$ 与 CA$_2$(碳酸酐酶 2)相互作用，形成二联体 HbA$_1$-CA$_2$。本实验中我们看到在红细胞再释放里看到碳酸酐酶。众所周知，在红细胞运氧过程中，血红蛋白与碳酸酐酶是互相配合的。再释放研究告诉我们，在红细胞里二者可以结合存在。说明我们制备的电泳有效。

第七十二章 制备电泳拾零

我们实验室还做过一些制备电泳,在本章简单介绍一下。

1 HbA$_2$ 与 HbA$_1$ 之间抠胶、定释

单向定点再释放电泳后,抠 HBA$_2$ 与 HBA$_1$ 之间和定释带两处的胶,结果见图 72-1。

2 多带释放,多处抠胶

单向多带再释放电泳后,按电泳的前后位置,抠再释放出来的多条带胶,结果见图 72-2。

3 双向双释放,多角度抠胶

双向双释放电泳后,多角度抠胶,结果见图 72-3。

4 于 HbA$_3$ 前、HbA$_2$ 后、CA 后抠胶

单向初释放电泳后,抠 HbA$_3$ 前、HbA$_2$ 后、CA 后的胶,结果见图 72-4。

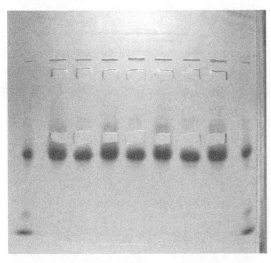

图 72-1 抠 HbA$_2$ 与 HbA$_1$ 之间和定释带两处胶

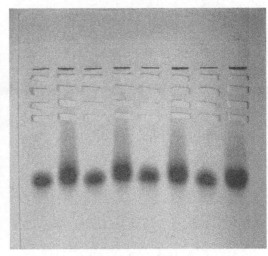

图 72-2 抠多带再释放后出来的多条带胶

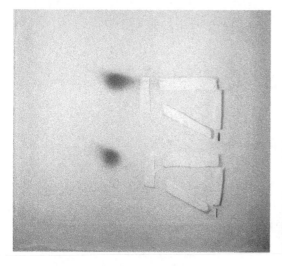

图 72-3　双向双释放，多角度抠胶　　　　图 72-4　于 HbA$_3$ 前、HbA$_2$ 后、CA 后抠胶

5　抠胶 HbA$_3$、HbA$_1$、HbA$_2$、CA

单向初释放电泳后，抠 HbA$_3$、HbA$_1$、HbA$_2$、CA 处的胶，结果见图 72-5。

6　草鱼血红蛋白抠胶

草鱼血红蛋白单向初释放电泳后抠胶，结果见图 72-6 中的右侧 2 个泳道。

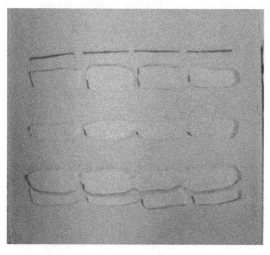
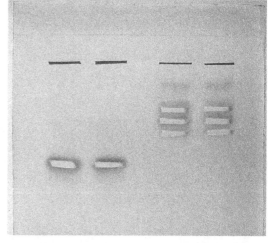

图 72-5　抠 HbA$_3$、HbA$_1$、HbA$_2$ 及 CA　　　图 72-6　草鱼血红蛋白抠胶，右侧 2 个泳道为草鱼血红蛋白

第十篇　无氧条件下红细胞内血红蛋白的电泳释放

前　言

1　以前的实验都是在空气中进行，也就是在有氧的环境里进行。

2　本篇实验是在缓冲液里加入偏重亚硫酸钠，凝胶内部的环境是还原的、无氧的，此时的电泳结果会怎么样？这就是本篇的研究内容。做这个实验的最初想法来自对梅花鹿镰状细胞的研究。

3　我们发现地中海贫血患者红细胞内血红蛋白的电泳释放异常，想再研究镰状细胞贫血患者的情况，但在国内找不到这种患者。此时想到梅花鹿的红细胞也是镰状，可以看看梅花鹿镰状红细胞的血红蛋白释放情况。但是，常规电泳未能检出镰状细胞的特点，在淀琼电泳中加入偏重亚硫酸钠后效果才显现出来。

4　无氧条件下红细胞内血红蛋白的电泳释放，结果如下所示：①无氧条件下血红蛋白A_2现象消失，红细胞内 HbA_2 与 HbA_1 的相互作用消失；②无氧条件下血红蛋白之间的交叉互作消失，HbA_1 穿过 HbA_2 时，后者的区带不变形；③无氧条件下再释放现象消失，红细胞内 HbA_1 与 CA_2 的相互作用消失。

第七十三章 无氧条件下的血红蛋白电泳释放
——偏重亚硫酸钠淀琼胶电泳

秦良谊[1,3#] 高丽君[1#] 苏 燕[1#] 高雅琼[1] 裴娟慧[2,4] 秦文斌[1※]

(包头医学院 1. 血红蛋白研究室;2. 病理生理教研室 包头 014010;3. 杨思医院检验科,上海 200126;4. 航天总医院,北京 100026)

摘 要

在梅花鹿镰状红细胞释放血红蛋白的研究中,我们发现偏重亚硫酸钠-淀粉-琼脂糖混合凝胶电泳显示出独特作用。此时,全血或红细胞标本都处于还原状态或无氧状态,其中血红蛋白的电泳行为与众不同。本文中我们比较研究了四组血液标本:分别来自正常人、HbE 病患者、梅花鹿、大鼠。结果表明,偏重亚硫酸钠淀琼胶电泳表现出一系列独特之处:①无氧条件下梅花鹿红细胞内血红蛋白的电泳释放增强;②无氧条件下大鼠红细胞内血红蛋白之间的相互作用明显减弱;③正常人的血红蛋白 A_2 现象反常;④HbE 病患者的血红蛋白释放试验结果更清晰,有利于鉴别某些异常血红蛋白。可以认为,偏重亚硫酸钠淀琼胶电泳技术成为血红蛋白电泳释放的又一新分支——无氧条件下的血红蛋白释放试验。

关键词 偏重亚硫酸钠淀琼胶电泳;无氧条件,血红蛋白 A_2 现象;红细胞;大鼠;梅花鹿

1 前言

偏重亚硫酸钠处理,能够促进血红蛋白 S 病患者的红细胞镰化[1],我们把它加入制胶缓冲液,用于梅花鹿的红细胞,发现它能改善梅花鹿红细胞内血红蛋白的电泳释放[2]。偏重亚硫酸钠处理,在无氧条件下对人类红细胞如何、对各种疾病鉴定效果如何,更是一无所知。本文就进入这个领域进行研究,本书的中心内容也是如此。

2 材料和方法

2.1 血液标本 大鼠为 SD 雄性大鼠,体重 250~300g,由北京维通利华实验动物技术有限公司提供。梅花鹿来自上海市动物园。正常人为本室科研人员(健康体检正常)。患者来自包头医学院第一附属医院和广西医科大学医学科学实验中心。

\# 并列第一作者
※ 通讯作者:秦文斌,电子邮箱:qwb5991309@tom.com

2.2 实验操作
2.2.1 缓冲液
(1) 普通淀琼胶电泳的制胶缓冲液，同文献[2]，为 TEB 缓冲液 pH 8.6。
(2) 偏重亚硫酸钠淀琼胶制胶缓冲液：在上述 TEB 缓冲液中加入 0.1%偏重亚硫酸钠[2]。
(3) 电泳槽中的缓冲液，同文献[2]，为硼酸盐，pH 9.0。

2.2.2 制胶
取 17cm×17cm 的玻璃板一块，称马铃薯淀粉 1.7mg，琼脂糖 0.24mg，加于装有 90ml 上述制胶缓冲液(普通缓冲液或含偏重亚硫酸钠的缓冲液)的三角烧瓶内，摇匀后，于沸水锅内煮 8 分钟，边煮边摇。取出后晾至 50℃左右平铺于玻璃板上。待完全凝固即可使用。

2.2.3 加样
将样品加入滤纸条后插入凝胶，加样后凝胶在槽中放置 2 小时，让还原剂充分作用，然后再电泳(普通淀琼胶与含偏重亚硫酸钠的淀琼胶同样处理)。

2.2.4 电泳
电泳条件参见文献[2]。

3 结果

3.1 患者与鹿血红细胞的单向释放电泳 见图 73-1。

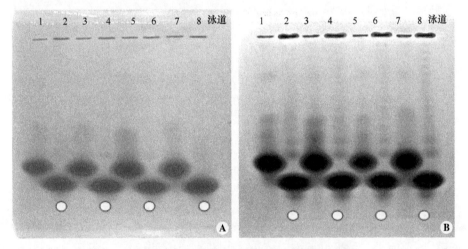

图 73-1 患者与鹿血红细胞的单向释放电泳

注释：患者与鹿血红细胞标本，A 为普通电泳，B 为偏重亚硫酸钠电泳。
泳道由左向右 1、3、5、7 为患者血红细胞，2、4、6、8 为鹿血红细胞标本(见〇处)

由图 73-1A、图 73-1B 可以看出，与普通电泳结果相比，偏重亚硫酸钠淀琼胶电泳时鹿血红细胞多带释放明显增强。这说明梅花鹿的镰状红细胞在无氧条件下才能释放出来多个血红蛋白区带。

3.2 大鼠红细胞的双向对角线电泳 见图 73-2。

由图 73-2A、图 73-2B 可以看出，普通淀琼胶电泳时大鼠红细胞及其基质出现明显的脱离对角线成分，而偏重亚硫酸钠淀琼胶电泳时没有明显的脱离对角线成分。这说明，大鼠红细胞内血红蛋白的相互作用需要有氧条件，无氧条件时互作消失。

3.3 正常人血红蛋白 A_2 现象的双向对角线电泳 见图 73-3。

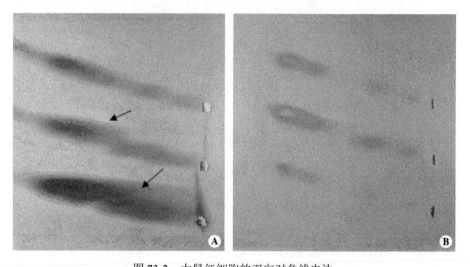

图 73-2　大鼠红细胞的双向对角线电泳

注释：标本为大鼠红细胞。A 为普通电泳，B 为偏重亚硫酸钠电泳。上层为溶血液，中层为红细胞，下层为红细胞基质

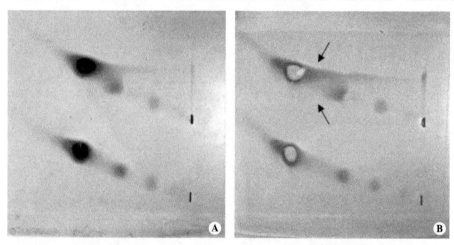

图 73-3　正常人血红蛋白 A_2 现象的双向对角线电泳

注释：A 为普通淀琼胶电泳，B 为偏重亚硫酸钠淀琼胶电泳。上层为红细胞，下层为红细胞溶血液

　　由图 73-3A、图 73-3B 可以看出，与普通淀琼胶电泳结果相比，偏重亚硫酸钠淀琼胶电泳时血红蛋白 A_2 现象也出现差异(参见 ↓、↑ 处)。这说明，无氧条件对血红蛋白 A_2 现象也有影响，而且其电泳图像特殊，含义如何尚待解读。初步认为是它将 HbA_2 上方的 HbA_1 与后来拖尾出来的 HbA_1 区分开(参见上边的 ↓ 处)。看来，血红蛋白 A_2 现象的内容还很丰富，需要今后深入研究。此时，无氧条件下的 CA 不少于有氧者！

3.4　本地患者红细胞的单向释放电泳　见图 73-4。

　　由图 73-4A、图 73-4B 可以看出，与普通淀琼胶电泳结果相比，偏重亚硫酸钠淀琼胶电泳时各种区带都明显清晰，血三系减少患者的标本的 HbA_2 现象异常。这说明，无氧条件下的电泳释放有其独特之处。

3.5　广西患者全血的单向释放电泳　见图 73-5。

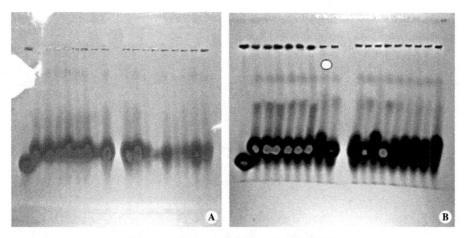

图 73-4 本地患者红细胞为单向释放电泳

注释：A 为普通电泳，B 为偏重亚硫酸钠电泳。共 18 个标本，第一泳道为鹿血红细胞，其他为同一批 17 个患者陈旧血的红细胞标本，其中第 8 泳道(见○处)为血液科血三系减少患者的标本

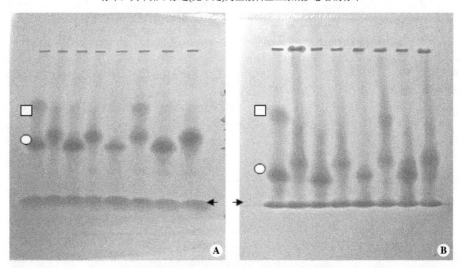

图 73-5 广西患者全血的单向释放电泳

注释：A 为普通淀琼胶电泳，B 为偏重亚硫酸钠淀琼胶电泳。同一批来自广西的 8 个全血标本，其中第一泳道的标本含 HbA(见○处)和 HbE(见□处)

由图 73-5A、图 73-5B 可以看出，与普通淀琼胶电泳结果相比，偏重亚硫酸钠淀琼胶电泳时血浆白蛋白的电泳速度明显变慢(见→处)，HbE(见□处)与 HbA(见○处)的距离变远，各泳道拖尾明显。这说明，无氧条件下有利于从电泳位置上区别 HbE 与 HbA_2。

4 讨论

众所周知，现在的各种电泳都是在空气(有氧环境)中进行，迄今为止我们的 HRT 也是如此。早期发现的血红蛋白 A_2 现象[3-6]、后来发现的再释放[7,8]，都是有氧环境下的电泳结果，不知道、也没想到无氧条件下情况如何。梅花鹿镰状细胞 HRT 研究促使创建偏重亚硫酸钠淀琼胶电泳，使红细胞成分的释放在还原条件下进行，从而出现一系列新情况：梅花鹿的镰状红细胞在无氧条件下才能释放出来多个血红蛋白区带；大鼠红细胞内血红蛋白的相互作用需要有氧条件，无氧时互作消失；无氧条件使血红蛋白 A_2 现象也受到影响，而且其电泳图像与有氧条件者相似，又不全同，含义如何尚待解读，初步认为是它将 HbA_2 上方的 HbA_1 与后来拖尾出来的 HbA_1 区分开，说明血红蛋白 A_2 现象的内容还有未解之处，需要今后深入研究；陈旧人血在无氧条件下区带相对更集中，这也是一个有趣的现象。总之，偏重亚硫酸钠淀琼胶电泳使我们进入一个新的研究领域，从电泳释放角度可以认为，它成为 HRT 的又一新分支——无氧条件下的血红蛋白电泳释放。当然，本文只是初步结果，刚刚开始，继续深入研究一定能给我们对红细胞内部情况的了解带来更多新知识。

参 考 文 献

[1] Harvey A. Itano and linus pauling a rapid diagnostic test for sickle cell anemia. Blood, 1949, 4: 66-68.
[2] 秦文斌. 红细胞内血红蛋白的电泳释放——发现和研究. 北京：科学出版社，2015.
[3] 秦文斌, 梁友珍. 血红蛋白 A_2 现象 I. 此现象的发现及其初步应用. 生物化学与生物物理学报, 1981, 13(2): 199-205.
[4] 秦文斌. 红细胞外血红蛋白 A 与血红蛋白 A_2 之间的相互作用. 生物化学杂志, 1991, 7(5): 577-587.
[5] 秦文斌. "红细胞 HbA_2" 为 HbA_2 与 HbA 的结合产物. 生物化学与生物物理进展, 1991, 18(4): 280-288.
[6] Su Y, Gao LJ, Ma Q, et al. Interactions of hemoglobin in lived red blood cells measured by the electrophoresis release test. Electrophoresis, 2010, 31: 2913-2920.
[7] 秦文斌, 高丽君, 苏燕, 等. 血红蛋白释放试验与轻型 β-地中海贫血. 包头医学院学报, 2007, 23(6): 259-263.
[8] Su Y, Shao G, Gao LJ, et al. RBC electrophoresis with discontinuous power supply – a newly established hemoglobin release test. Electrophoresis, 2009, 30: 3041-3043.

第七十四章　比较有氧和无氧电泳结果的三种方法

1 常规比较方法

做两次电泳,一次有氧,一次无氧。电泳后分别染色,比较两者结果。

2 一槽内两胶板法

一个电泳槽里放两个胶板,一个是偏重亚硫酸钠胶板,一个是普通胶板。电泳后分别染色,比较两者结果。

3 一胶板内两种标本法

一个胶板,两种标本,一个是标本里加入偏重亚硫酸钠;另一个是标本里不加偏重亚硫酸钠。在一个电泳槽里电泳,电泳后染色,比较两者结果。

第七十五章 无氧条件下正常分娩者红细胞内血红蛋白的电泳释放
——将偏重亚硫酸钠加入标本 无氧与有氧同在一胶板

高丽君[1] 宝勿仁必力格[1,2] 秦文斌[1※]

(1 包头医学院 血红蛋白研究室,包头 014010;2 包钢集团第三职工医院 检验科,包头 014010)

摘 要

以前的无氧实验,都是制胶缓冲液里含偏重亚硫酸钠。本文不同,用普通缓冲液制胶,无氧标本是将偏重亚硫酸钠加入血液或红细胞里,有氧标本不加此物质。然后,在同一胶板做二层双向电泳,丽春红-联苯胺染色,观察结果。本文检测的是正常分娩者的红细胞标本。结果表明,普通红细胞都有HbA_2现象,即有HbA_2与HbA_1之间的相互作用。但是,含偏重亚硫酸钠的红细胞,都没有HbA_2现象,即HbA_2与HbA_1之间没有发生相互作用。

关键词 正常分娩者;一个胶板;两种标本;红细胞加入偏重亚硫酸钠;二层双向电泳

1 前言

偏重亚硫酸钠处理,能够促进血红蛋白S病患者的红细胞镰化,我们把它加入制胶缓冲液,用于梅花鹿的红细胞,发现它能改善梅花鹿红细胞内血红蛋白的电泳释放。在此基础上我们建立起无氧条件下的血红蛋白释放实验(参见以上各章)。

本文与以上各章不同,用普通缓冲液制胶,无氧标本是将偏重亚硫酸钠加入血液或红细胞里,有氧标本不加此物质。然后,同一胶板,做二层双向电泳,丽春红-联苯胺染色,观察结果。详见正文。

2 材料与方法

2.1 材料 血液标本来自包钢集团第三职工医院检验科4号正常分娩者,女,26岁。

2.2 方法 正常分娩者红细胞的双向双层电泳。

(1) 处理:洗RBC方法。取200μl全血于1ml EP管中,加生理盐水至1ml。上下混合后,低速离心3分钟,弃上清液留下层沉淀。再加生理盐水至1ml,上下混合后低速离心3分钟,弃上清液留沉淀。重复洗3遍。下层的沉淀即为加样用的RBC。

※通讯作者:秦文斌,电子邮箱:qwb5991309@tom.com

(2) 制备：上层标本制备是 20μlRBC +2μl 饱和偏重亚硫酸钠。下层标本制备是 20μl RBC+2μl 生理盐水。上样 10μl。

(3) 电泳条件：定退普。普通电泳 2.5 小时后，停电 15 分钟再泳 30 分钟。倒极再泳 15 分钟。将胶板转向 90°，再普泳 1.5 小时。取下胶板。

(4) 染色：先染丽春红过夜，再染联苯胺。

3 结果

3.1 正常分娩者红细胞的双向双层电泳

丽春红染色结果见图 75-1。

(1) 结果：下层 A_2 现象正常，HbA_2 上边无现象。CA 在对角线上，数量较多。HbA_1 有下拉，无定释，无后退。

上层 A_2 现象不正常，消失了。CA 在对角线上，但数量明显减少。HbA_1 有下拉成分，无定释，无后退。

(2) 讨论：定释和后退都无，可能与标本陈旧有关。A_2 现象变化属于一板规律，无氧时 A_2 现象消失，A_2-A_1 互作断裂。无氧时 CA 减少，这种情况与乳腺癌类似，是癌症特有的吗？还是各种疾病都有呢？换一个非癌症标本观察，会有何现象？

3.2 正常分娩者红细胞的双向双层电泳

联苯胺染色结果见图 75-2。

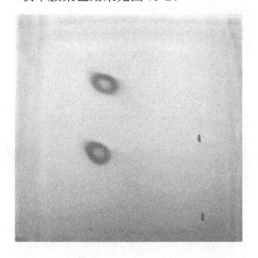 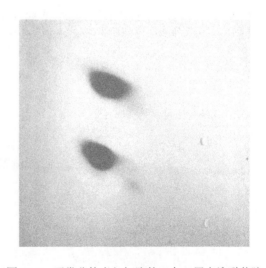

图 75-1　正常分娩者偏重红细胞双向电泳实验
注释：上层为偏重红细胞；下层无偏红细胞

图 75-2　正常分娩者红细胞的双向双层电泳联苯胺染色
注释：上层为红细胞含偏重亚硫酸钠；下层为红细胞不含偏重亚硫酸钠

(1) 结果：下层有 A_2 现象，CA 明显，原点处无特殊成分。上层 A_2 现象消失，CA 不明显，有减少现象。原点处有红色成分，类似沉淀物，怀疑为 CA 成分。

(2) 讨论：偏重亚硫酸钠使红细胞的 A_2 现象消失，无氧条件下红细胞的 A_2 现象消失。是由于偏重亚硫酸钠影响 CA 吗？有待继续观察研究。

4 讨论

偏重亚硫酸钠处理,能够促进血红蛋白 S 病患者的红细胞镰化[1],我们把它加入制胶缓冲液,用于梅花鹿的红细胞,发现它能改善梅花鹿红细胞内血红蛋白的电泳释放[2]。在此基础上我们建立起无氧条件下的血红蛋白释放实验(参见以上各章)。

现在是比较无氧和有氧条件下血红蛋白的释放情况,可以是两个胶板(有氧和无氧)互相比较,也可以一个胶板内两个标本(有氧和无氧)互相比较,本章属于后者。具体操作是用普通缓冲液制胶,无氧标本是将偏重亚硫酸钠加入血液或红细胞里,有氧标本不加此物质。然后,同一个胶板做二层双向电泳,丽春红-联苯胺染色,观察结果。

本文检测的是正常分娩者红细胞。结果表明,普通正常分娩者红细胞都有 HbA_2 现象,即有 HbA_2 与 HbA_1 之间的相互作用。但是,含偏重亚硫酸钠的正常分娩者红细胞,都没有 HbA_2 现象,即 HbA_2 与 HbA_1 之间没有发生相互作用。

由此看来,无氧和有氧条件对红细胞内部影响很大,有氧时红细胞内 HbA_2 与 HbA_1 结合存在,溶血处理时二者才彼此分开。无氧条件下红细胞内 HbA_1 与 HbA_2 并未结合,已经分开。无氧条件相当于有氧时的溶血状态,可以想象高原缺氧对人体的影响,而心肌梗死和脑梗死时严重缺氧的后果更为可怕。

参 考 文 献

[1] Itano HA, Pauling L. A rapid test for sickle cell anemia. Blood, 1949, 4: 66-68.
[2] 秦文斌. 红细胞内血红蛋白的电泳释放——发现和研究. 北京: 科学出版社, 2015.

第七十六章 无氧条件下脑出血患者红细胞内血红蛋白的电泳释放
——将偏重亚硫酸钠加入标本 无氧与有氧同在一胶板

高丽君[1] 宝勿仁必力格[1,2] 秦文斌[1※]

(1. 包头医学院 血红蛋白研究室，包头 014010；2. 包钢集团第三职工医院 检验科，包头 014010)

摘 要

以前的无氧实验，都是制胶缓冲液里含偏重亚硫酸钠。本文不同，用普通缓冲液制胶，无氧标本是将偏重亚硫酸钠加入血液或红细胞里，有氧标本不加此物质。然后，在同一胶板做二层双向电泳，丽春红-联苯胺染色，观察结果。本文检测的是脑出血患者的红细胞标本。结果表明，普通红细胞都有 HbA_2 现象，即有 HbA_2 与 HbA_1 之间的相互作用。但是，含偏重亚硫酸钠的红细胞，都没有 HbA_2 现象，即 HbA_2 与 HbA_1 之间没有发生相互作用。

关键词 脑出血；一个胶板；两种标本；红细胞加入偏重亚硫酸钠；二层双向电泳

1 前言

偏重亚硫酸钠处理，能够促进血红蛋白 S 病患者的红细胞镰化，我们把它加入制胶缓冲液，用于梅花鹿的红细胞，发现它能改善梅花鹿红细胞内血红蛋白的电泳释放。在此基础上我们建立起无氧条件下的血红蛋白释放实验(参见以上各章)。本文与以上各章不同，用普通缓冲液制胶，无氧标本是将偏重亚硫酸钠加入血液或红细胞里，有氧标本不加此物质。然后，同一胶板，做二层双向电泳，丽春红-联苯胺染色，观察结果。详见正文。

2 材料与方法

2.1 材料 血液标本来自包钢集团第三职工医院检验科 2 号脑出血患者，女，61 岁。

2.2 方法 脑出血患者红细胞的双向双层电泳。

(1) 处理：洗 RBC 方法。取 200μl 全血于 1μl EP 管中，加生理盐水至 1ml，上下混合后，低速离心 3 分钟，弃上清液留下层沉淀。再加生理盐水至 1ml，上下混合后低速离心 3 分钟，弃上清液留沉淀。重复洗 3 遍。下层的沉淀即为加样用的 RBC。

(2) 制备：上层标本制备是 20μl RBC+2μl 饱和偏重亚硫酸钠。下层标本制备是 20μl

※通讯作者：秦文斌，电子邮箱：qwb5991309@tom.com

RBC+2μl 生理盐水。上样 10μl。

(3) 电泳条件：定退普。普通电泳 2.5 小时后，停电 15 分钟再泳 30 分钟。倒极再泳 15 分钟。将胶板转向 90°，再普泳 1.5 小时。取下胶板。

(4) 染色：先丽春红染色，过夜拍照留图，再烤干后染联苯胺。

3 结果

3.1 脑出血患者红细胞的双向双层电泳

丽春红染色结果见图 76-1。

(1) 结果：下层 A_2 现象正常，HbA_2 上边无现象。CA 在对角线上，数量较多。HbA_1 有下拉，无定释，无后退。上层 A_2 现象不正常，消失了。CA 在对角线上，但数量明显减少。HbA_1 有下拉成分，无定释，无后退。

(2) 讨论：定释和后退都无，可能与标本陈旧有关。A_2 现象变化属于一板规律，无氧时 A_2 现象消失；A_2-A_1 互作断裂。无氧时 CA 减少，这种情况与乳腺癌类似，是癌症特有的吗？还是各病都有呢？换一个非癌症标本观察，会有何现象？

3.2 脑出血患者红细胞的双向双层电泳

联苯胺染色结果见图 76-2。

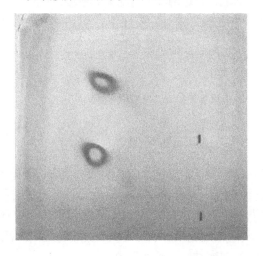 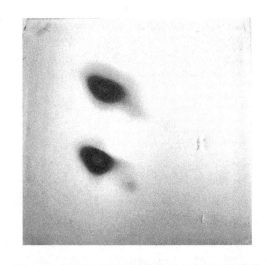

图 76-1 脑出血患者偏入红细胞双向电泳实验
注释：上层为偏入红细胞；下层为无偏红细胞

图 76-2 脑出血患者红细胞的双向双层电泳联苯胺染色
注释：上层为红细胞含偏重亚硫酸钠；下层为红细胞不含偏重亚硫酸钠

(1) 结果：下层有 A_2 现象，CA 明显，原点处无特殊成分。上层 A_2 现象消失，CA 不明显，有减少现象。原点处有红色成分，类似沉淀，疑为 CA 成分。

(2) 讨论：偏重亚硫酸钠使红细胞的 A_2 现象消失，无氧条件下红细胞的 A_2 现象消失。是由于偏重亚硫酸钠影响 CA 吗？有待继续观察研究。

4 讨论

偏重亚硫酸钠处理，能够促进血红蛋白 S 病患者的红细胞镰化[1]，我们把它加入制胶缓冲液，用于梅花鹿的红细胞，发现它能改善梅花鹿红细胞内血红蛋白的电泳释放[2]。在此基础上我们建立起无氧条件下的血红蛋白释放实验(参见以上各章)。

现在是比较无氧和有氧条件下血红蛋白的释放情况，可以是两个胶板(有氧和无氧)互相比较，也可以一个胶板内两个标本(有氧和无氧)互相比较，本章属于后者。具体操作是用普通缓冲液制胶，无氧标本是将偏重亚硫酸钠加入血液或红细胞里，有氧标本不加此物质。然后，有一个胶板做二层双向电泳，丽春红-联苯胺染色，观察结果。

本文检测的是脑出血患者红细胞。结果表明，脑出血患者红细胞都有 HbA_2 现象，即有 HbA_2 与 HbA_1 之间的相互作用。但是，含偏重亚硫酸钠的脑出血患者红细胞，都没有 HbA_2 现象，即 HbA_2 与 HbA_1 之间没有发生相互作用。

由此看来，无氧和有氧条件对红细胞内部影响很大，有氧时红细胞内 HbA_2 与 HbA_1 结合存在，溶血处理时二者才彼此分开。无氧条件下红细胞内 HbA_1 与 HbA_2 并未结合，已经分开。无氧条件相当于有氧时的溶血状态，可以想象高原缺氧对人体的影响，而心肌梗死和脑梗死时严重缺氧的后果更为可怕。

参 考 文 献

[1] Itano HA, Pauling L. A rapid test for sickle cell anemia. Blood, 1949, 4: 66-68.
[2] 秦文斌. 红细胞内血红蛋白的电泳释放——发现和研究. 北京: 科学出版社, 2015.

第七十七章 无氧条件下乳腺癌患者红细胞内血红蛋白的电泳释放
——将偏重亚硫酸钠加入标本 无氧与有氧同在一胶板

高丽君[1] 宝勿仁必力格[1,2] 秦文斌[1※]

(1. 包头医学院 血红蛋白研究室,包头 014010; 2. 包钢集团第三职工医院 检验科,包头 014010)

摘 要

以前的无氧实验,都是制胶缓冲液里含偏重亚硫酸钠。本文不同,是用普通缓冲液制胶,无氧标本是将偏重亚硫酸钠加入血液或红细胞里,有氧标本不加此物质。然后,同一胶板做二层双向电泳,丽春红-联苯胺染色,观察结果。本文检测的是乳腺癌患者的红细胞标本。结果表明,普通红细胞都有 HbA_2 现象,即有 HbA_2 与 HbA_1 之间的相互作用。但是,含偏重亚硫酸钠的红细胞,都没有 HbA_2 现象,即 HbA_2 与 HbA_1 之间没有发生相互作用。

关键词 乳腺癌;一个胶板;两种标本;红细胞加入偏重亚硫酸钠;二层双向电泳

1 前言

偏重亚硫酸钠处理,能够促进血红蛋白 S 病患者的红细胞镰化,我们把它加入制胶缓冲液,用于梅花鹿的红细胞,发现它能改善梅花鹿红细胞内血红蛋白的电泳释放。在此基础上我们建立起无氧条件下的血红蛋白释放实验(参见以上各章)。

本文与以上各章不同,用普通缓冲液制胶,无氧标本是将偏重亚硫酸钠加入血液或红细胞里,有氧标本不加此物质。然后,同一胶板做二层双向电泳,丽春红-联苯胺染色,观察结果。详见正文。

2 材料与方法

2.1 材料 血液标本来自包钢集团第三职工医院检验科 8 号乳腺癌血液标本,女,59岁。

2.2 方法 乳腺癌患者红细胞的双向双层电泳。

(1) 处理:洗 RBC 方法。取 200μl 全血于 1μl EP 管中,加生理盐水至 1ml,上下混合后,低速离心 3 分钟,弃上清液留下层沉淀。再加生理盐水至 1ml。再上下混合后低速离心 3 分

※通讯作者:秦文斌,电子邮箱:qwb5991309@tom.com

钟，弃上清液留沉淀。重复洗 3 遍。下层的沉淀即为加样用的 RBC。上层标本制备是 20μl RBC+2μl 饱和偏重亚硫酸钠。下层标本制备是 20μl RBC +2μl 生理盐水。上样 10μl。

(2) 电泳条件：定退普。普通电泳 2.5 小时后，停电 15 分钟再泳 30 分钟。倒极再泳 15 分钟。将胶板转向 90°，再普泳 1.5 小时。取下胶板。

(3) 染色：先丽春红染色过夜，拍照留图，再烤干后染联苯胺。

3 结果

3.1 乳腺癌患者红细胞的双向双层电泳丽春红染色结果　见图 77-1。

(1) 结果：下层有 A_2 现象，上层 A_2 现象消失。

(2) 讨论：偏重亚硫酸钠使红细胞的 A_2 现象消失，无氧条件下红细胞的 A_2 现象消失，等待联苯胺染色。

3.2 乳腺癌患者红细胞的双向双层电泳联苯胺染色结果　见图 77-2。

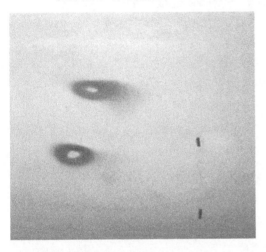

图 77-1　乳腺癌患者红细胞的双向双层电泳　丽春红染色
注释：上层为红细胞含偏重亚硫酸钠；下层为红细胞不含偏重亚硫酸钠

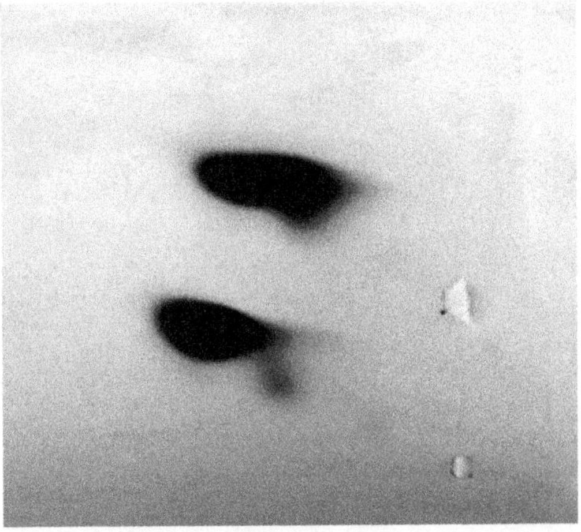

图 77-2　乳腺癌患者红细胞的双向双层电泳　联苯胺染色
注释：上层为红细胞含偏重亚硫酸钠；下层为红细胞不含偏重亚硫酸钠

(1) 结果：下层有 A_2 现象，CA 明显，原点处无成分。上层 A_2 现象消失，CA 不明显，有减少现象。原点处有红色成分，类似沉淀物，怀疑为 CA 成分。

(2) 讨论：偏重亚硫酸钠使红细胞的 A_2 现象消失，无氧条件下红细胞的 A_2 现象消失。是由于偏重亚硫酸钠影响 CA 的原因吗？有待继续观察。

4 讨论

偏重亚硫酸钠处理，能够促进血红蛋白 S 病患者的红细胞镰化[1]，我们把它加入制胶缓冲液，用于梅花鹿的红细胞，发现它能改善梅花鹿红细胞内血红蛋白的电泳释放[2]。在此基础上我们建立起无氧条件下的血红蛋白释放实验(参见以上各章)。

现在是比较无氧和有氧条件下血红蛋白的释放情况，可以是两个胶板(有氧和无氧)互相比较，也可以一个胶板内两个标本(有氧和无氧)互相比较，本章属于后者。具体操作是用普通缓冲液制胶，无氧标本是将偏重亚硫酸钠加入血液或红细胞里，有氧标本不加此物质。然后，同一个胶板做二层双向电泳，丽春红-联苯胺染色，观察结果。

本文检测的是乳腺癌患者红细胞。结果表明，普通乳腺癌患者红细胞都有 HbA_2 现象，即有 HbA_2 与 HbA_1 之间的相互作用。但是，含偏重亚硫酸钠的乳腺癌患者红细胞都没有 HbA_2 现象，即 HbA_2 与 HbA_1 之间没有发生相互作用。

由此看来，无氧和有氧条件对红细胞内部影响很大，有氧时红细胞内 HbA_2 与 HbA_1 结合存在，溶血处理时二者才彼此分开。无氧条件下红细胞内 HbA_1 与 HbA_2 并未结合，已经分开。无氧条件相当于有氧时的溶血状态，可以想象高原缺氧对人体的影响，心肌梗死和脑梗死时严重缺氧的后果更为可怕。

参 考 文 献

[1] Itano HA, Pauling L. A rapid test for sickle cell anemia. Blood, 1949, 4: 66-68.
[2] 秦文斌. 红细胞内血红蛋白的电泳释放——发现和研究. 北京: 科学出版社, 2015.

第七十八章 无氧条件下胃癌患者红细胞内血红蛋白的电泳释放
——将偏重亚硫酸钠加入标本 无氧与有氧同在一胶板

高丽君[1] 宝勿仁必力格[1,2] 秦文斌[1※]

(1. 包头医学院 血红蛋白研究室，包头 014010；2. 包钢集团第三职工医院 检验科，包头 014010)

摘 要

以前的无氧实验，都是制胶缓冲液里含偏重亚硫酸钠。本文不同，用普通缓冲液制胶，无氧标本是将偏重亚硫酸钠加入血液或红细胞里，有氧标本不加此物质。然后，同一胶板做二层双向电泳，丽春红-联苯胺染色，观察结果。本文检测的是胃癌患者的红细胞标本。结果表明，普通红细胞都有 HbA_2 现象，即有 HbA_2 与 HbA_1 之间的相互作用。但是，含偏重亚硫酸钠的红细胞，都没有 HbA_2 现象，即 HbA_2 与 HbA_1 之间没有发生相互作用。

关键词 胃癌；一个胶板；两种标本；红细胞加入偏重亚硫酸钠；二层双向电泳

1 前言

偏重亚硫酸钠处理，能够促进血红蛋白 S 病患者的红细胞镰化，我们把它加入制胶缓冲液，用于梅花鹿的红细胞，发现它能改善梅花鹿红细胞内血红蛋白的电泳释放。在此基础上我们建立起无氧条件下的血红蛋白释放实验(参见以上各章)。

本文与以上各章不同，用普通缓冲液制胶，无氧标本是将偏重亚硫酸钠加入血液或红细胞里，有氧标本不加此物质。然后，同一胶板做二层双向电泳，丽春红-联苯胺染色，观察结果。详见正文。

2 材料与方法

2.1 材料 血液标本来自包钢集团第三职工医院检验科 9 号胃癌血液标本。男，75 岁。
2.2 方法 胃癌患者红细胞的双向双层电泳。

(1) 处理：洗 RBC 方法。取 200μl 全血于 1ml EP 管中，加生理盐水至 1ml，上下混合后，低速离心 3 分钟，弃上清液留下层沉淀。再加生理盐水至 1ml 再上下混合后低速离心 3 分钟，弃上清液留沉淀。重复洗 3 遍。下层的沉淀即为加样用的 RBC。

※通讯作者：秦文斌，电子邮箱：qwb5991309@tom.com

(2) 制备：上层标本制备是 20μl RBC+2μl 饱和偏重亚硫酸钠。下层标本制备是 20μl RBC+2μl 生理盐水。上样 10μl。

(3) 电泳条件：定退普。普通电泳 2.5 小时后，停电 15 分钟再泳 30 分钟。倒极再泳 15 分钟。将胶板转向 90°，再普泳 1.5 小时。取下胶板。

(4) 染色：先丽春红染色，过夜拍照留图，再烤干后染联苯胺。

3 结果

3.1 胃癌患者红细胞的双向双层电泳丽春红染色结果 见图 78-1。

(1) 结果：下层 A_2 现象正常，HbA_2 上边没有成分；CA 在对角线上，数量较多；HbA_1 有下拉；无定释，无后退。上层 A_2 现象不正常，消失了；CA 在对角线上，但数量明显减少；HbA_1 有下拉成分；无定释，无后退。

(2) 讨论：定释和后退都无，可能与标本陈旧有关。A_2 现象变化属于一板规律，无氧时 A_2 现象消失，A_2-A_1 互作断裂。无氧时 CA 减少，这种现象与乳腺癌类似，是癌症特有的，还是各种疾病都有呢？换一个非癌标本观察会何不同？

3.2 胃癌患者红细胞的双向双层电泳联苯胺染色结果 见图 78-2。

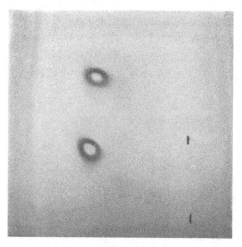 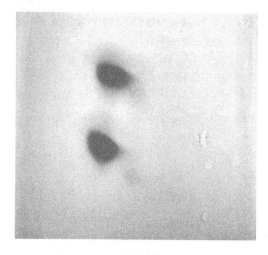

图 78-1 胃癌患者偏重红细胞双向电泳实验
注释：上层为偏重红细胞，下层为无偏重红细胞

图 78-2 胃癌患者红细胞的双向双层电泳 联苯胺染色
注释：上层为红细胞含偏重亚硫酸钠，下层为红细胞不含偏重亚硫酸钠

结果：下层有 A_2 现象，CA 明显，原点处无处分。上层 A_2 现象消失，CA 不明显，有减少现象。原点处有红色成分，类似沉淀，疑为 CA 成分。

讨论：偏重亚硫酸钠使红细胞的 A_2 现象消失，无氧条件下红细胞的 A_2 现象消失，是由于偏重亚硫酸钠影响 CA 吗？有待继续观察研究。

4 讨论

偏重亚硫酸钠处理，能够促进血红蛋白 S 病患者的红细胞镰化[1]，我们把它加入制胶缓冲液，用于梅花鹿的红细胞，发现它能改善梅花鹿红细胞内血红蛋白的电泳释放[2]。在此基础上我们建立起无氧条件下的血红蛋白释放实验(参见以上各章)。

现在是比较无氧和有氧条件下血红蛋白的释放情况,可以是两个胶板(有氧和无氧)互相比较,也可以一个胶板内两个标本(有氧和无氧)互相比较,本章属于后者。具体操作是用普通缓冲液制胶,无氧标本是将偏重亚硫酸钠加入血液或红细胞里,有氧标本不加此物质。然后,用一个胶板做二层双向电泳,丽春红-联苯胺染色,观察结果。

本文检测的是胃癌患者红细胞。结果表明,普通胃癌患者红细胞都有 HbA_2 现象,即有 HbA_2 与 HbA_1 之间的相互作用。但是,含偏重亚硫酸钠的胃癌患者红细胞都没有 HbA_2 现象,即 HbA_2 与 HbA_1 之间没有发生相互作用。

由此看来,无氧和有氧条件对红细胞内部影响很大,有氧时红细胞内 HbA_2 与 HbA_1 结合存在,溶血处理时二者才彼此分开。无氧条件下红细胞内 HbA_1 与 HbA_2 并未结合,已经分开。无氧条件相当于有氧时的溶血状态,可以想象高原缺氧对人体的影响,心肌梗死和脑梗死时严重缺氧的后果更为可怕。

需要充分的氧气,需要新鲜的空气,在这里我们看到红细胞内部的一些变化。

参 考 文 献

[1] Itano HA, Pauling L. A rapid test for sickle cell anemia. Blood, 1949, 4: 66-68.
[2] 秦文斌. 红细胞内血红蛋白的电泳释放——发现和研究. 北京: 科学出版社, 2015.

第七十九章 四种情况(正常分娩、脑出血、胃癌、乳腺癌)的比较和分析
——将偏重亚硫酸钠加入标本 无氧与有氧同在一胶板

这四种情况的详细内容参见前边第七十五～第七十八章。这里是将它们放在一起进行比较，从中寻找规律。

1 综合结果

见图79-1。

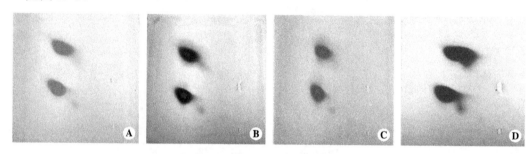

图 79-1 四种情况的综合结果

注释：4图都是红细胞的双向二层电泳结果，A为正常分娩、B为脑出血、C为乳腺癌、D为胃癌；上层都是偏重亚硫酸钠红细胞；下层都是无偏重亚硫酸钠红细胞

结果：上层都有 A_2 现象，即都有 HbA_2 与 HbA_2 之间的相互作用；下层都是 A_2 现象消失，即都没有 HbA_2 与 HbA_1 之间的相互作用。

2 讨论

现在看来，无氧条件时红细胞内部发生了很大变化，最明显的就是 HbA_2 现象消失。A_2 现象[1-4]是红细胞内 HbA_2 与 HbA_1 之间的相互作用，在红细胞内二者是结合存在的。这种结合只有在严重贫血时才遭到破坏。在这里，我们看到无氧条件下这种结合也遭到破坏，可见这种条件能够影响红细胞的内部结构，作用深刻。

HbA_2 现象是发生在红细胞内部的事情，后来我们发现在红细胞外 HbA_2 与 HbA_1 可以发生交叉互作[3-14]。现在的问题是，无氧条件下 HbA_2 现象消失，此条件下的血红蛋白交叉互作如何？也消失吗？下一章里我们就来讨论这个问题。

参 考 文 献

[1] 秦文斌, 梁友珍. 血红蛋白 A_2 现象 Ⅰ. 此现象的发现及其初步应用. 生物化学与生物物理学报, 1981, 13(2): 199-205.
[2] 秦文斌. 血红蛋白的 A_2 现象发生机制的研究"红细胞 HbA_2"为 HbA_2 与 HbA 的结合产物. 生物化学与生物物理进展, 1991, 18(4): 280-288.
[3] 秦文斌. 红细胞外血红蛋白 A 与血红蛋白 A_2 之间的相互作用. 生物化学杂志, 1991, 7(5): 577-587.
[4] Su Y, Gao LJ, Ma Q, et al. Interactions of hemoglobin in lived red blood cells measured by the electrophoresis release test. Humana Press, 2012, 869(17): 387-402.
[5] 秦文斌. 红细胞内血红蛋白的电泳释放——发现和研究. 北京: 科学出版社, 2015.
[6] 韩丽红, 闫斌, 高丽君, 等. 血红蛋白 F 与血红蛋白 A_2 之间的交叉互作//秦文斌著. 红细胞内血红蛋白的电泳释放——发现和研究. 北京: 科学出版社, 2015: 89-97.
[7] 韩丽红, 苏燕, 高丽君, 等. 血红蛋白 A_3 与血红蛋白 A_2 之间没有交叉互作//秦文斌著. 红细胞内血红蛋白的电泳释放——发现和研究. 北京: 科学出版社, 2015: 98-100.
[8] 裴娟慧, 高丽君, 秦文斌, 等. 活体大鼠红细胞内血红蛋白之间的交叉互作//秦文斌著. 红细胞内血红蛋白的电泳释放——发现和研究. 北京: 科学出版社, 2015: 101-105.
[9] 邵国, 苏燕, 王占黎, 等. 脊椎动物血红蛋白的分子进化——交叉电泳在血红蛋白研究中的应用//秦文斌著. 红细胞内血红蛋白的电泳释放——发现和研究. 北京: 科学出版社, 2015: 106-115.
[10] 苏燕, 邵国, 睢天林, 等. 几种鱼类血红蛋白不与人血红蛋白相互作用及其进化意义. 包头医学院学报, 2000, 23(2): 452-454.
[11] 于慧, 王占黎, 秦文斌, 等. 美国锦龟血红蛋白与人血红蛋白相互作用的生物信息学和计算机模拟研究//秦文斌著. 红细胞内血红蛋白的电泳释放——发现与研究. 北京: 科学出版社, 2015: 121-127.
[12] Yu H, Wang ZL, Qin WB. et al. Structuralbasis for the specific interaction of chicken hemoglobin with bromphenolblue: a computational analysis. Molecular Physics, 2010, 108(2): 215-220.
[13] 王程, 隋春红, 秦文斌, 等. 生物信息学方法推测脊椎动物血红蛋白相互作用机制//秦文斌著. 红细胞内血红蛋白的电泳释放——发现和研究. 北京: 科学出版社, 2015: 136-142.
[14] 高丽君, 苏燕, 秦文斌. 血红蛋白的聚丙烯酰胺凝胶交叉电泳//秦文斌著. 红细胞内血红蛋白的电泳释放——发现和研究. 北京: 科学出版社, 2015: 143-148.

第八十章　无氧条件下血红蛋白之间不能交叉互作
——一个电泳槽里两种胶板

高丽君　　宝勿仁必力格　秦文斌

(包头医学院　血红蛋白研究室，包头　014010)

摘　要

前边的实验证明，无氧条件下红细胞内的血蛋白 A_2 现象消失，说明此时 HbA_2 与 HbA_1 中间的相互作用消失。过去我们发现，红细胞内有 A_2 现象时，红细胞外 HbA_2 能与 HbA_1 发生交叉互作。现在是无氧条件下 A_2 现象消失，那么交叉互作如何变化？结果是，有氧条件下 HbA_1 穿过溶血液的 HbA_2 时出现"V"形改变，无氧条件下，HbA_1 穿过溶血液的 HbA_2 时未见"V"形改变，这说明，无氧条件下 HbA_2 不能与 HbA_1 发生交叉互作。

有氧条件下，血红蛋白形成携带氧的氧合血红蛋白，无氧条件下，血红蛋白为不带氧的还原血红蛋白，二者结构不同，功能也各异。所以，前者能够交叉互作，后者不能交叉互作。

关键词　无氧条件；有氧条件；血红蛋白 A_2 现象；HbA_2 与 HbA_1 之间的交叉互作

1　前言

无氧条件下红细胞内血红蛋白的电泳释放，是一个新课题，过去没有人研究过。之前的研究，发现无氧条件下红细胞内血红蛋白 A_2 现象消失。过去，我们发现红细胞内血红蛋白 A_2 现象[1-4]后，又发现红细胞外血红蛋白之间的交叉互作[3-8]。现在发现无氧条件下红细胞内血红蛋白 A_2 现象消失，推测红细胞外血红蛋白交叉互作也可能消失，下面就是对这个问题进行的研究和相应结果。

2　材料和方法

2.1　材料

标本来源　包钢集团第三职工医院 14 号正常产妇的全血。

2.2　方法

2.2.1　红细胞溶血液的制备　取全血，离心弃血浆，留红细胞。生理盐水洗 RBC，用 CCl_4 制备溶血液。

2.2.2　电泳纯 HbA_1 的制备　用溶血液做单向电泳，不染色，抠出含 HbA_1 凝胶部分，冻化、离心，收集含 HbA_1 上清，备用。参见图 80-1。

2.2.3　无氧胶制备　胶内加 1%偏重亚硫酸钠(偏重亚硫酸钠 1g+制胶用的 TBE10ml)，

大胶板加 0.9ml。

2.2.4 交叉互作-穿过实验 电泳过程中,让纯的 HbA_1 穿过溶血液。

3 结果

3.1 丽春红染色结果 见图 80-2。

由图 80-2 可以初步看出,有氧条件下有血红蛋白交叉互作,无氧条件下无血红蛋白交叉互作。

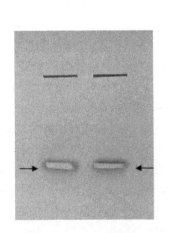

图 80-1 溶血液单向电泳 抠取含 HbA_1 的凝胶

注释:箭头所指处含 HbA_1 的凝胶已经抠出

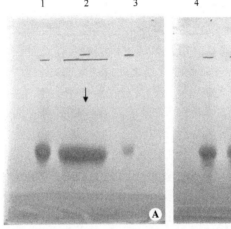

图 80-2 有氧和无氧条件下的交叉电泳 丽春红染色结果

注释:A 图为有氧条件下的交叉电泳,B 图为无氧条件下的交叉电泳;泳道 1、4 为溶血液对照,2、5 为 HbA_1 穿过溶血液,3、6 为 HbA_1 对照

结果:泳道 2 穿过处好像有一点"V"状(箭头所指处),泳道 5 穿过处看不见"V"状。

3.2 再染联苯胺后结果 见图 80-3。

由图 80-3 可以明确看出,有氧条件下有血红蛋白交叉互作,无氧条件下无血红蛋白交叉互作。

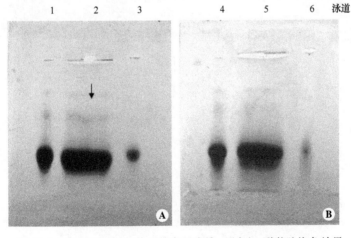

图 80-3 有氧和无氧条件下的交叉电泳 丽春红-联苯胺染色结果

注释:A 图为有氧条件下的交叉电泳,B 图为无氧条件下的交叉电泳;泳道 1、4 为溶血液对照,2、5 为 HbA_1 穿过溶血液,3、6 为 HbA_1 对照

结果:泳道 2 穿过处可见"V"形区带(箭头所指处),泳道 5 穿过处未出现"V"形区带。

4 讨论

无氧条件下红细胞内血红蛋白的电泳释放，是一个新课题，过去没有人研究过。以前各章的研究中，发现无氧条件下红细胞内血红蛋白 A_2 现象消失。以前，我们发现红细胞内血红蛋白 A_2 现象[1-4]后，又发现红细胞外血红蛋白之间的交叉互作[3-8]。现在发现无氧条件下红细胞内血红蛋白 A_2 现象消失，推测红细胞外血红蛋白交叉互作也可能消失，所以进行了本章的实验研究。实验结果是，有氧条件下 HbA_1 穿过溶血液的 HbA_2 时出现"V"形改变，无氧条件下，HbA_1 穿过溶血液的 HbA_2 时未见"V"形改变，这说明，无氧条件下 HbA_2 不能与 HbA_1 发生交叉互作。

有氧条件下，血红蛋白形成携带氧的氧合血红蛋白，无氧条件下，血红蛋白为不带氧的无氧血红蛋白，二者结构不同，功能也各异。所以，前者能够交叉互作，后者不能交叉互作。血红蛋白有 4 个亚基(2 条 α 链和 2 条 β 链)，每个亚基中含有 1 个血红素辅基。血红素 Fe 原子的第六配价键可以与不同的分子结合：有氧存在时，能够与氧结合形成氧合血红蛋白 (HbO_2)，无氧时为无氧血红蛋白 (Hb)。4 个亚基是通过盐桥(键)及氢键作用连接起来的。由于多个盐桥的存在，使整个血红蛋白分子的结构绷得相当紧密，不易与氧分子结合。但当氧与血红蛋白分子中的 1 个亚基的血红素的铁(Fe^{2+})结合后，产生别构作用，其四级结构将发生相当剧烈的变化，导致亚基间的盐桥断裂，从而使原来结合紧密的血红蛋白分子变得松散，易与氧结合。由此可见，血红蛋白有两种可以互变的构象：与 O_2 亲和力低不易与 O_2 结合的紧密型(T 型)和与 O_2 亲和力高容易与 O_2 结合的松弛型(R 型)(参见图 80-4)。

我们所创建的无氧条件下红细胞内血红蛋白的电泳释放，相当于红细胞处于静脉血中状态，也就是红细胞处于组织侧的状态。与此对应，在有氧条件下红细胞处于动脉血中状态，也就是红细胞处于肺侧的状态。无氧条件下红细胞内血红蛋白处于 T 型，HbA_2 不能与 HbA_1 结合，二者也无法进行交叉互作。与此相反，有氧条件下红细胞内血红蛋白处于 R 型，HbA_2 能与 HbA_1 结合，二者可以进行交叉互作。本文实验将血红蛋白立体构型与血红蛋白 A_2 现象联系起来，深化了人们对红细胞内血红蛋白存在状态的认识。

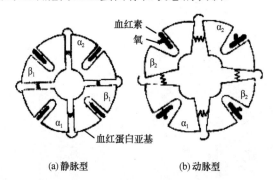

图 80-4 血红蛋白亚基结构的两种构型示意

注释：(a)静脉型=T 型(紧密型)=无氧血红蛋白，(b)动脉型=R 型(松弛型)=氧合血红蛋白；一个血红蛋白有 4 个珠蛋白链，$α_1$、$α_2$、$β_1$、$β_2$，还 4 个血红素，氧合时再加上 4 个 O_2，由于 O_2 的进入，血红蛋白分子由 T 型变成 R 型

参 考 文 献

[1] 秦文斌，梁友珍. 血红蛋白 A_2 现象Ⅰ. 此现象的发现及其初步应用. 生物化学与生物物理学报, 1981, 13(2): 199-205.

[2] 秦文斌. 血红蛋白的 A_2 现象发生机制的研究"红细胞 HbA_2"为 HbA_2 与 HbA 的结合产物. 生物化学与生物物理进展, 1991, 18(4): 280-288.
[3] 秦文斌. 红细胞外血红蛋白 A 与血红蛋白 A_2 之间的相互作用. 生物化学杂志, 1991, 7(5): 577-587.
[4] Su Y, Gao LJ, Ma Q, et al. Interactions of hemoglobin in lived red blood cells measured by the electrophoresis release test. Electrophoresis, 2010, 31: 2913-2920.
[5] Yu H；Wang ZL, Qin WB, et al. Structural basis for the specific interaction of chicken haemoglobin with bromophenol blue: a computational analysis. Molecular Physics, 2010, 108(2): 215-220.
[6] 秦文斌. 活体红细胞内血红蛋白的电泳释放. 中国科学生命科学, 2011, 41(8): 597-607.
[7] 秦文斌. 红细胞内血红蛋白的电泳释放——发现和研究. 北京: 科学出版社, 2015.
[8] 韩丽红, 闫斌, 高丽君, 等. 血红蛋白 F 与血红蛋白 A_2 之间的交叉互作. 中国医药指南, 2015, (5): 60-63.

第八十一章 有氧和无氧条件下糖尿病患者的红细胞指纹图

高丽君[#] 宝勿仁必力格[#] 高雅琼 韩丽红 苏 燕 周立社 秦文斌[※]

(包头医学院 血红蛋白研究室,包头 014010)

摘 要

目的: 创建糖尿病患者红细胞指纹图,比较研究有氧与无氧条件的差异。

方法: 取10支试管,第1管只加红细胞20μl,不加蒸馏水,第2管加红细胞18μl,蒸馏水2μl,第3管加红细胞16μl,蒸馏水4μl,以此类推,第10管加红细胞2μl,蒸馏水18μl。用这10种标本做双向十层电泳,第一向时加入前进和后退两种再释放,第二向为普泳。同时准备两个胶板,一个有氧条件,一个无氧条件。最后都是丽春红-联苯胺复染后观察结果。

结果: 从10个前进区带和10个后退区带可以看出有氧与无氧条件的差异。

结论: 无氧条件对糖尿病患者红细胞指纹图有一定影响。

关键词 糖尿病;红细胞;指纹图;有氧条件;无氧条件

1 前言

为了提高血红蛋白释放试验的分辨率,我们创建了相应的指纹图谱技术,在全血指纹图方面,首先研究了ABO血型的指纹图[1],本文是研究糖尿病患者红细胞指纹图,同时比较有氧和无氧条件的影响。这些都没人研究过,所以没有更原始的文献资料。

2 材料及方法

2.1 材料 糖尿病患者血液标本来自包钢第三职工医院检验科。

2.2 方法 基本同文献[1,2],具体如下所示。

2.2.1 标本处理 取10支试管,第1管只加红细胞,不加蒸馏水,第2~9管里所加蒸馏水由少到多、所加红细胞由多到少,构成一个连续的、完整的等低渗条件,具体操作参见表81-1。

表81-1 全血的等低渗全程处理

管号	1	2	3	4	5	6	7	8	9	10
红细胞(μl)	20	18	16	14	12	10	8	6	4	2
蒸馏水(μl)	0	2	4	6	8	10	12	14	16	18

注释:第1管为等渗,用原来的红细胞(含等量生理盐水),没有加蒸馏水;第2~10管为低渗,水量逐渐增加,红细胞相应减少;第10管中红细胞占10%,蒸馏水占90%。

[#] 并列第一作者
[※] 通讯作者:秦文斌,电子邮箱:qwb5991309@tom.com

2.2.2 双向电泳 用以上一系列标本直接做双向电泳。

(1) 第一向电泳：先普泳，再加入前进再释放和后退再释放。普泳=电势梯度 6V/cm，泳 2 小时 15 分钟左右，停电 15 分钟。前进再释放=电势梯度 6V/cm，再通电半小时。后退再释放=电势梯度 6V/cm，倒极再通电 15 分钟。

(2) 第二向电泳：普泳=电势梯度 6V/cm，倒极转向再泳 1 小时 15 分钟左右。

2.2.3 染色

(1) 先染丽春红：将凝胶板直接放入丽春红染液中过夜，取出、照相，再晾干或烤干。

(2) 再染联苯胺：将凝胶板直接放入联苯胺染液中，加 3% H_2O_2 直到血红蛋白变成蓝黑颜色，转入漂洗液(5%乙酸、1%甘油)换洗两次，每次 5 分钟，取出晾干。

2.2.4 结果保存 晾干凝胶与玻璃板结合，可长期保存。

3 结果

3.1 糖尿病患者红细胞指纹图(有氧条件) 见图 81-1。

由图 81-1 可以看出，前进带有 7~8 泳道，后退带也如此，只是前进带较强，后退带较弱。

3.2 糖尿病患者红细胞指纹图(无氧条件) 见图 81-2。

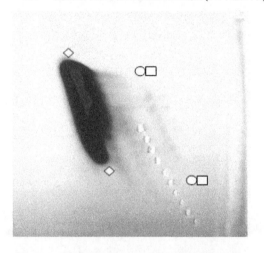

图 81-1 糖尿病患者红细胞指纹图(有氧条件)
注释：两个○之间为前进带，两个□之间为后退带，两个◇之间为血红蛋白 A_1

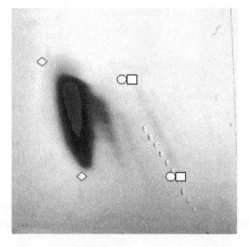

图 81-2 糖尿病患者红细胞指纹图(无氧条件)
注释：两个○之间为前进带，两个□之间为后退带，两个◇之间为血红蛋白 A_1

4 讨论

指纹图至少有两大类，全血指纹图和红细胞指纹图，文献[1, 2]涉及的都是全血指纹图，本文则是红细胞指纹图。现在看来，红细胞指纹图比较单纯，没有血浆成分的影响，全血指纹图相对复杂，任何血浆因素的变化都能影响它的结果，血型差异是典型的例子[3]。

本文所研究的红细胞指纹图，在有氧与无氧条件下的变化，不太明显，但是我们注意到，电泳过程中，10 个泳道血红蛋白的泳动情况各不相同，无氧条件下这些泳道出现"拐弯现象"，很有规律，其他标本也类似，所以在下一章里专门探讨。

由图 81-2 可以看出，前进带 1~10 泳道基本都有，后退带很弱，看不清。

参 考 文 献

[1] 秦文斌. 红细胞内血红蛋白的电泳释放——发现和研究. 北京: 科学出版社, 2015.
[2] 乔姝, 高丽君, 宝勿仁毕力格, 等. ABO 血型的双向全程释放电泳图谱. 包头医学院学报, 2016, 32(7): 5-7.

第八十二章 无氧胶中血红蛋白区带泳动时的拐弯现象

1 拐弯现象

在比较糖尿病红细胞指纹图电泳过程中，发现无氧胶血红蛋白区带泳动时出现拐弯现象，详见图82-1。由此图可以看出，电泳一开始，血红蛋白就有拐弯现象，开始时血红蛋白是 HbO_2，泳入无氧胶后，脱氧变成 Hb，它的泳道加快，出现拐弯。

在比较糖尿病红细胞指纹图电泳过程中，发现有氧胶血红蛋白泳动没有拐弯现象，详见图82-2。

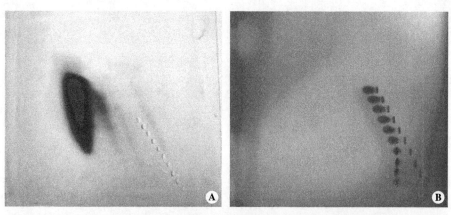

图82-1　糖尿病红细胞指纹图电泳过程中发现无氧胶拐弯现象
注释：A图为无氧条件下糖尿病红细胞指纹图结果，B图为其血红蛋白泳动过程中出现拐弯现象，未染色，在电泳槽中直接照相

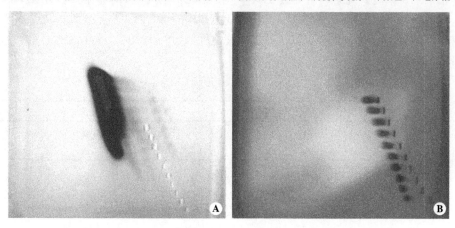

图82-2　糖尿病红细胞指纹图电泳过程中发现有氧胶无拐弯现象
注释：A图为有氧条件下糖尿病红细胞指纹图结果，B图为其血红蛋白泳动过程中没有拐弯现象，未染色，在电泳槽中直接照相

由此图可以看出，血红蛋白泳出后没有出现拐弯现象。原因是原来血红蛋白就是HbO_2，进入有氧胶后还是HbO_2，泳动速度不变，所以没有拐弯现象。

2 拐弯现象的连续观察

由图82-3可以看出，上边的5个图显示拐弯现象，下边的5个图没有拐弯现象。拐弯现象由最下边泳道(第10泳道)开始，逐步上移，直到最上泳道(第1泳道)，最后拐弯消失，又直行下去。

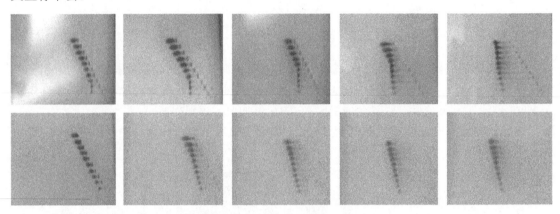

图82-3 无氧胶拐弯现象全程

注释：上边的5个图是无氧胶内红细胞指纹图电泳过程中出现的连续图像，无氧胶内10个泳道Hb(注意，不是HbO_2)的泳动过程，有拐弯现象！下边的5个图是有氧胶内红细胞指纹图电泳过程中出现的连续图像，有氧胶内10个泳道HbO_2的泳动过程，没有拐弯现象

3 拐弯现象的机制探讨

由表82-1可以看出，红细胞指纹图实验中10个管内红细胞的分布情况。电泳加样时，最上边泳道相当于第1管内容，其次泳道是第2管内容，以此类推，最下边泳道是第10管内容。第10管内容是红细胞：蒸馏水=2∶18，此时红细胞几乎完全溶血，相当于将游离血红蛋白加在第10泳道。开始时应当是HbO_2，在有氧胶里这个HbO_2不变，在无氧胶里，这个HbO_2变成Hb，这个Hb的电泳行为与HbO_2不同，所以出现了上边那样的"拐弯现象"。现在看来，与HbO_2相比，Hb显示一定的"负电性"，增加了向阳极泳动的力量，从而出现了拐弯现象。

表82-1 红细胞指纹图实验1～10另管的处理

管号	1	2	3	4	5	6	7	8	9	10
红细胞(μl)	20	18	16	14	12	10	8	6	4	2
蒸馏水(μl)	0	2	4	6	8	10	12	14	16	18

注释：第1管为等渗(没加蒸馏水)，第2管开始低渗……第10管严重低渗。

第八十三章 无氧释放总结

(1) 我们在"有氧条件下红细胞内血红蛋白的电泳释放"里，主要有以下三个发现。

1) 发现"初释放"，即"血红蛋白 A_2 现象"，主要内容是红细胞内 HbA_2 与 HbA_1 等相互作用，结合存在。

2) 发现交叉互作，首先发现 HbA_2 与 HbA_1 交叉互作，后来发现 HbF 也能与 HbA_2 交叉互作，还发现脊椎动物血红蛋白有的能交叉互作，有的不能，规律是羊膜动物血红蛋白能交叉互作，非羊膜动物血红蛋白则不能，这就是脊椎动物血红蛋白的分子进化。

3) 发现"再释放"，主要内容是红细胞内 HbA_1 与 CA_1(碳酸酐酶 1)相互作用，结合存在。

(2) 在"无氧条件下红细胞内血红蛋白的电泳释放"里，上述三个现象发生变化，具体如下所示。

1) 无氧条件下，初释放现象即 A_2 现象消失。

2) 无氧条件下，交叉互作现象消失。

3) 无氧条件下，再释放现象消失。

(3) 在人体的运氧过程中，肺部和末梢组织是两个终端。

1) 血液循环里的红细胞到了肺部，此时红细胞里的血红蛋白处于氧合状态：$Hb(O_2)_4$。以上三个现象都存在这种情况。

2) 到了末梢组织，此时红细胞里的血红蛋白处于还原状态，变成 Hb(没有氧)。以上三个现象也发生变化。

(4) 运氧过程中血红蛋白各种状态的动态平衡见图 83-1。

(5) 运氧过程中红细胞内血红蛋白存在状态的动态平衡见图 83-2。

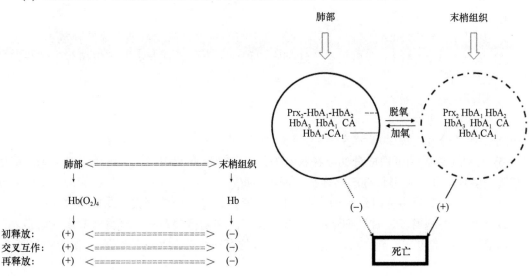

图 83-1 运氧过程中血红蛋白各状态的动态平衡示意   图 83-2 运氧过程中红细胞内血红蛋白存在状态的动态平衡示意

第十一篇 红细胞内蛋白成分的存在状态及其电泳释放图解

前　言

红细胞内各种蛋白质的存在状态如何？以下是我们的研究成果。红细胞内有 HbA_3、HbA_1、HbA_2、CA 等，它们的存在状态错综复杂，分述如下。

(1) HbA_3：主要是游离存在于红细胞内。

(2) HbA_1：情况较为复杂。

1) 游离存在量较大。

2) 结合存在的是少量。结合状态的还分两种，疏松结合和紧密结合。

A. 疏松结合的，一次通电就能释放出来，我们称之为"初释放"。

B. 紧密结合的，第二次通电才能释放出来，我们称之为"再释放"。

(3) HbA_2：它比较特殊，没有游离存在的，都在疏松结合的初释放里边，这里还有 Prx_2。

(4) CA：有两种情况。

1) 大部分游离存在。

2) 少部分 CA_1 与 HbA_1 结合，再共同结合于红细胞膜，第二次通电才能释放出来。

(5) 以上所说的，都是有氧状态下红细胞内蛋白成分的存在状态，无氧条件下情况发生以下一系列变化。

1) 初释放现象瓦解。

2) 再释放现象消失。

3) 所有成分都变成游离状态。

(6) 红细胞内各种蛋白成分的存在状态与红细胞的形态密切相关，这里主要讨论"球形红细胞增多症"和"靶形红细胞增多症"两种情况。

1) 球形红细胞增多症时红细胞内：①HbA_3 消失。②再释放消失。③初释放存在。

2) 靶形红细胞增多症时红细胞：①HbA_3 存在。②初释放存在。③再释放增强增多。

本篇用图解的方式，介绍红细胞内蛋白成分的存在状态及其电泳释放出来后的情况。

第八十四章　红细胞内的游离蛋白成分及其电泳释放

红细胞内的游离蛋白成分，主要有 HbA_3、HbA_1、CA 等，它们大量存在于红细胞内，未与红细胞膜结合，通电后很容易释放出来(图 84-1，图 84-2)。

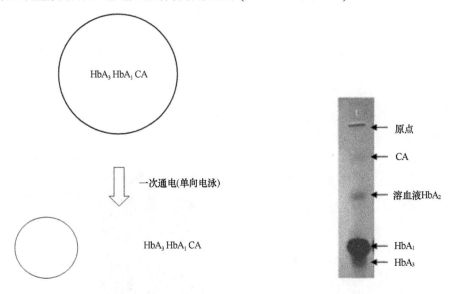

图 84-1　红细胞通电后游离蛋白成分释放示意　　图 84-2　红细胞内的游离蛋白成分及其电泳释放

注释：溶血液 HbA_2 是"红细胞 HbA_2"分解后的产物

第八十五章 红细胞内血红蛋白 A_2 等的初释放现象

红细胞内，HbA_2 与 HbA_1、Prx_2 互相结合存在，它们再与红细胞膜疏松结合，电泳时也比较容易释放(一次通电就能释放出来)，而且刚释放出来时它们还在互相结合，再通电时才能彼此分开(图 85-1～图 85-3)。

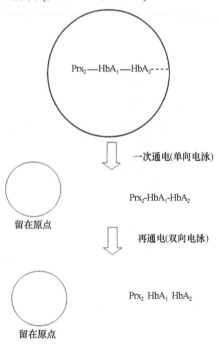

图 85-1 红细胞内成分释放示意

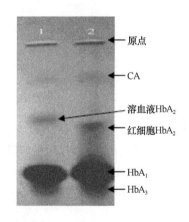

图 85-2 红细胞单向电泳的 HbA_2 现象

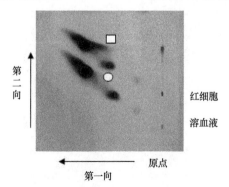

图 85-3 红细胞双向电泳的 HbA_2 现象

注释：第二向红细胞里的 HbA_2○ 与 HbA_1□ 分开

第八十六章 红细胞内血红蛋白 A_1 等的再释放现象

红细胞内 HbA_1 与 CA_1 互相结合(-)存在，它们再与红细胞膜紧密(—)结合，第一次通电时纹丝不动，第二次通电才能释放出来(我们称之为"再释放")，而且刚释放出来时它们还是互相结合存在，再通电时才能彼此分开(图 86-1，图 86-2)。

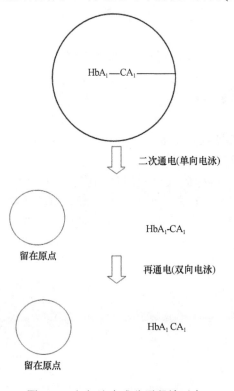

图 86-1 红细胞内成分再释放示意

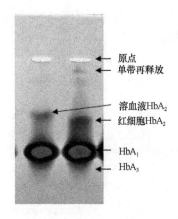

图 86-2 红细胞内成分再释放现象

第八十七章 红细胞内初释放现象与再释放现象同时存在状态及其电泳释放

在现实生活中，正常红细胞里初释放现象与再释放现象是同时存在的，而且游离蛋白成分也是同时存在的，这里专门讨论两个释放同时存在的情况。

初释放的参与者是 HbA_2 与 HbA_1、Prx_2，它们互相结合存在，再与红细胞膜疏松结合，电泳时也比较容易释放(一次通电就能释放出来)，而且刚释放出来时它们还在互相结合，再通电时才能彼此分开。

再释放的参与者是 HbA_1 与 CA_1，它们互相结合存在，再与红细胞膜紧密结合，第一次通电时纹丝不动，第二次通电才能释放出来(我们称之为"再释放")，而且刚释放出来时它们还是互相结合存在，再通电时才能彼此分开(图87-1)。

两种释放同时存在，会不会相互影响呢？

没有通电之前，二者同时存在，一个与红细胞疏松结合，一个牢固结合，互不影响。第一次通电时，初释放成分脱离红细胞，进入红细胞以外的凝胶，此时再释放成分纹丝不动，仍与红细胞膜牢固结合。第二次通电时，再释放成分脱离红细胞膜，进入红细胞以外的凝胶，此时再释放成分尚未解离，仍互相结合存在。但是，此时凝胶里的初释放成分，开始解离成为 HbA_2、HbA_1 和 Prx_2。如果再通一次电，凝胶里的再释放成分就开始解离成为 HbA_1 和 CA_1。电泳结果见图87-2。

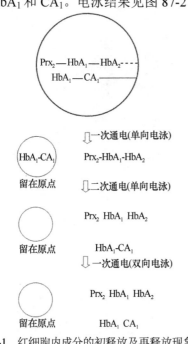

图87-1 红细胞内成分的初释放及再释放现象示意

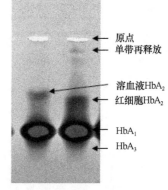

图87-2 红细胞内的初释放现象和再释放现象

第八十八章 红细胞内全部成分的存在状态及其电泳释放

红细胞内全部成分,指的是两种释放加上游离成分。游离蛋白成分不与红细胞膜结合,第一次通电就释放到红细胞外的凝胶中,而且还是游离存在(图 88-1,图 88-2)。初释放成分和再释放成分的情况,同第八十七章里的描述,这里不再赘述。

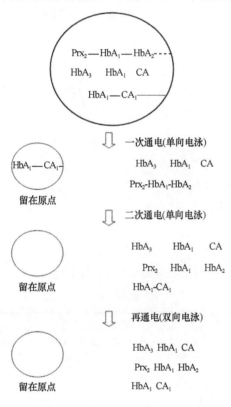

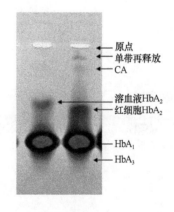

图 88-1 红细胞内全部成分的电泳释放示意　　图 88-2 红细胞内全部成分的电泳释放现象

第八十九章 无氧条件下红细胞内蛋白成分的存在状态及其电泳释放

前边各章内容都属于"有氧条件下"的情况，本章内容则是"无氧条件下"红细胞内蛋白成分的存在状态及其电泳释放。无氧条件下，情况变化很大，除了原来的游离蛋白成分，初释放蛋白成分也与红细胞膜脱离，变成游离成分。同样，再释放蛋白成分也与红细胞膜脱离，变成游离成分(图89-1～图89-3)。也就是说，红细胞内所有蛋白成分都游离存在，一次通电后全都释放到红细胞外的凝胶中，再通电，仍是这样。

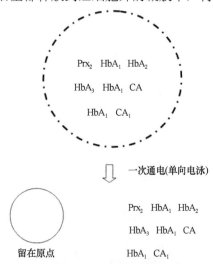

图89-1 无氧条件下红细胞内蛋白成分电泳释放示意图
注释：无氧条件下释放的特点是初释放现象消失，再释放现象也消失

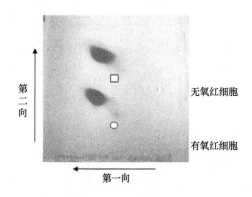

图89-2 红细胞有氧及无氧条件下的初释放
注释：上层结果中无 HbA_2(□)，说明无氧时没有初释放；下层结果中有 HbA_2(○)，说明有氧时有初释放

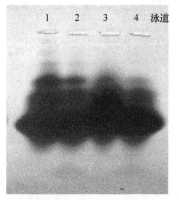

图89-3 红细胞有氧及无氧条件下的多带再释放
注释：泳道1来自有氧红细胞，有多带再释放；泳道3来自无氧红细胞，无多带再释放

第九十章 球形红细胞增多症时红细胞内蛋白成分的存在状态及其电泳释放

红细胞内各种蛋白成分的存在状态与红细胞的形态密切相关,这里有球形红细胞增多症和靶形红细胞增多症两种情况,本章主要讨论球形红细胞增多症。

球形红细胞增多症时红细胞内蛋白成分的分布情况是:①游离蛋白成分中 HbA_3 消失,这一点是球形红细胞增多症特有的;②再释放消失;③初释放存在(图90-1~图90-3)。

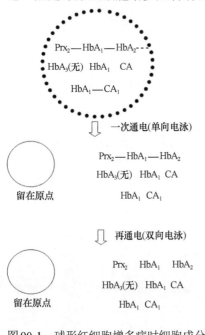

图90-1 球形红细胞增多症时细胞成分的电泳释放示意

注释:球形红细胞增多症的特点是没有 HbA_3、没有再释放、有初释放

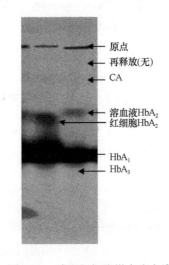

图90-2 球形红细胞增多症电泳情况

注释:①有初释放;②没有再释放;③ HbA_3 消失

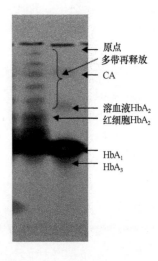

图90-3 对照(靶形红细胞增多症)电泳情况

注释:①初释放和 HbA_3 存在;②再释放增强、增多

第九十一章　靶形红细胞增多症时红细胞内蛋白成分的存在状态及其电泳释放

红细胞内各种蛋白成分的存在状态与红细胞的形态密切相关,这里有球形红细胞增多症和靶形红细胞增多症两种情况,本章主要讨论靶形红细胞增多症。

靶形红细胞增多症时红细胞内蛋白成分的分布情况是:①游离蛋白成分中 HbA_3 存在;②初释放存在;③再释放增强增多(图 91-1~图 91-3)。

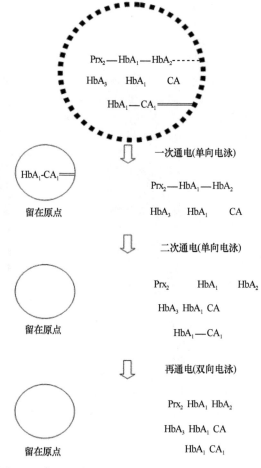

图 91-1　靶形红细胞增多症时细胞成分的电泳释放示意
注释:靶形红细胞增多症的特点是①初释放存在;②HbA_3 存在;③再释放增强、增多

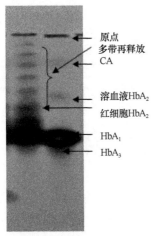 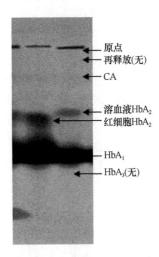

图 91-2　靶形红细胞增多症电泳情况　　　图 91-3　对照(球形红细胞增多症)电泳情况

注释：①初释放和 HbA_3 存在；②再释放增强、增多　　注释：①有初释放；②没有再释放；③HbA_3 消失

附 录

附录一 名 词 注 释[※][#]

淀粉-琼脂糖凝胶释放电泳专用名词

◎ **保温释放**

注释：将标本37℃保温一定时间后再做释放电泳。

◎ **初释放**

注释：将红细胞或含红细胞的全血加在淀粉-琼脂糖凝胶上，通电一次(不要停电)，就有血红蛋白释放出来，此时称为"初释放"。再参见"再释放"。

◎ **垂直线**

注释：双向淀粉-琼脂糖凝胶电泳后，由原点向上方释放的成分，呈直线状称其为"垂直线"，见附图1-1，附图1-2箭头←所指处。

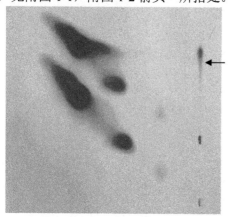 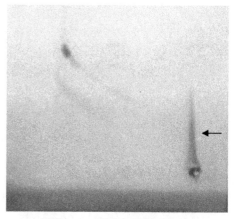

附图1-1 A_2现象双向电泳　　　　　附图1-2 胃癌细胞双向电泳

◎ **单带再释放**

注释：将红细胞或含红细胞的全血加在淀粉-琼脂糖凝胶上，通电后再停电-通电一次，此时又有一个血红蛋白区带放出来，此时称为"单带再释放"或简称"单带释放"。再参见"初释放""再释放"。参见附图1-3和附图1-4。

◎ **单向释放**

注释：用单向淀粉-琼脂糖凝胶电泳做"初释放"或"再释放"，都属于单向释放。与此对应的是"双向释放"。

◎ **等渗释放**

注释：标本不稀释(全血直接用，红细胞含等量生理盐水)，直接做释放试验。

※ "释放电泳"属于原创研究，所用词汇几乎都是前人未有，他人不易理解，故设此"名词注释"，供参考。

\# "名词注释"按汉语拼音排序。

◎ 低渗释放

注释：标本加一定量蒸馏水(全血直接加蒸馏水，红细胞不加生理盐水、加蒸馏水)，做释放试验。

◎ 等低渗全程释放

注释：同时有等渗和一系列低渗，简称"全程"，参见附表1-1。

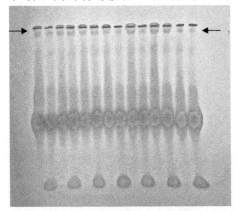

附图1-3 高血压病患者的单向单带再释放
注释：两个箭头→←之间为"单带再释放"

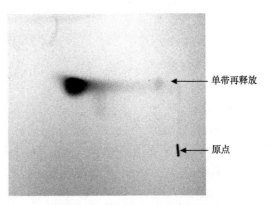

附图1-4 双向电泳中的单带再释放
注释：第一向电泳时"断电-通电"一次

附表1-1 标本(血液或红细胞)的等低渗全程处理

管号	1	2	3	4	5	6	7	8	9	10
标本(μl)	10	9	8	7	6	5	4	3	2	1
蒸馏水(μl)	0	1	2	3	4	5	6	7	8	9

注释：第1管为等渗，原标本，没有蒸馏水。2～10管蒸馏水量逐渐增加，标本相应减少，第10管中标本占10%，蒸馏水占90%。然后用它做释放实验，"初释放"或"再释放"，可以单带释放，也可以多带释放，参见附图1-5和附图1-6。

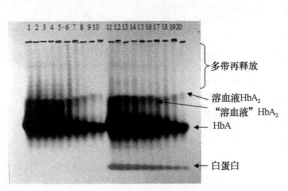

附图1-5 等低渗全程再释放电泳图
注释：这里是多带再释放，也称梯带或竹节带再释放。标本来自同一个，左侧10个泳道为其红细胞，右侧10个泳道为其全血。红细胞泳道号与上表的管号一致

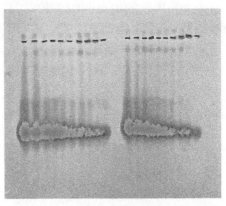

附图1-6 肝硬化患者红细胞全程电泳结果
注释：都是红细胞，左边为正常人标本，右边为肝硬化标本，再释放带不同

◎ 等低渗全程电泳

注释：参见"等低渗全程释放"。

◎ 等渗血红蛋白释放试验

注释：参见"等渗释放"。

◎ 低渗血红蛋白释放试验

注释：参见"低渗释放"。

◎ 第二次释放

注释：参见"再释放"。

◎ 第一次释放

注释：参见"初释放"。

◎ 电泳释放

注释：通常，红细胞释放血红蛋白，是通过裂解方法破坏红细胞放出其中的血红蛋白。"电泳释放"与之不同，不破坏红细胞，将完整红细胞(活红细胞)加在淀粉-琼脂糖凝胶中，通过电泳释放出来其中的血红蛋白。其他细胞内成分，也可以电泳释放。

◎ 电泳释放试验

注释：参见"电泳释放"。

◎ 电泳释放蛋白质组学

注释：参见"电泳释放"。通过电泳释放出细胞内的蛋白质，再用传统的蛋白质组学方法进行分析，研究和发现细胞内部蛋白质的一些奥秘(结构与功能、相互作用等)。这种想法和做法是我们提出了的。

◎ 电泳释放指纹图

注释：参见下边的"指纹图"。

◎ 淀粉-琼脂糖凝胶电泳

注释：将淀粉与琼脂糖混合制备凝胶，进行电泳分析。这种电泳方法的分辨率不太高，国内外几乎无人使用，我们则长期使用，已经超过四十年。这里的所有研究成果(早期的血红蛋白研究，后来的红细胞内血红蛋白的电泳释放、红细胞外的交叉互作等)都起源于这种电泳手段。我们的体会是，淀粉-琼脂糖凝胶硬度不强，有一定的黏度和流动性，可能有利于电泳释放的研究。

◎ 淀粉-琼脂糖凝胶双向电泳

注释：它在"血红蛋白 A_2 现象"的机制研究中做过贡献，参见下边的"双向电泳""对角线电泳"。

◎ 定点释放

注释：参见"电泳释放"和"再释放"，通过调节电泳时间使再释放出来的区带到达预定"地点"或位置。简称"定释"。

◎ 定时释放

注释：与"定点释放"是一回事，这里强调的是时间。

◎ 定释

注释：=定时释放或定点释放。

◎ 对角线

注释：做血红蛋白双向电泳，由原点到血红蛋白 A_1 之间画一条线，这就是"对角线"。标本为红细胞溶血液时，各成分(血红蛋白 A_1、血红蛋白 A_2、碳酸酐酶 CA)都在对角线上；标本为红细胞时，有血红蛋白成分离开对角线，首先是血红蛋白 A_2，它位于对角线的下方；还有血红蛋白 A_1 位于离开对角线的上方，与血红蛋白 A_2 上下呼应，参见附图 1-7。这说明，第一向电泳时由红细胞释放出来的血红蛋白 A_2 与血红蛋白 A_1 是结合存在的，第二向电泳时彼此分开，血红蛋白 A_2 泳速慢，下降到对角线下方(图中○处)，血红蛋白 A_1 泳速快，上升到对角线上方(图中□处)。

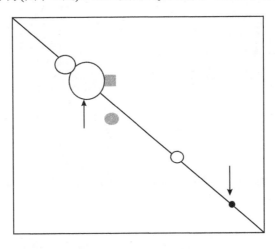

附图 1-7　对角线电泳模式

注释：由右下角的原点(↓所指处)到左上角的血红蛋白 A_1(↑所指处)，中间这条线就是"对角线"

◎　**对角线电泳**

注释：参见"对角线"。做双向电泳，观察有无成分离开对角线。对角线电泳是双向电泳的一种特例。目前国内外的做法都是将蛋白质样品加在凝胶的一端，第一次电泳后进行某种特殊处理(如利用非还原/还原方法处理二硫键)，再将凝胶转 90°进行第二次电泳，观察各电泳成分与对角线的关系，此双向电泳即为对角线电泳。如果两次电泳间的特殊处理对蛋白质无影响，则电泳图谱中的蛋白点都在对角线上；如果该特殊处理有影响，则电泳图谱中的蛋白点偏离对角线。根据特殊处理的性质，可获得变化蛋白点的相关重要信息。这些"对角线电泳"都来自双向聚丙烯酰胺凝胶电泳，并在二硫键鉴定、特别是在蛋白互作方面得到广泛应用。

这里介绍我们的双向对角线电泳，它的支持体是淀粉-琼脂糖混合凝胶。我们于 1991 年建立此方法，用于解决血红蛋白 A_2 现象的机制问题。后来用于血红蛋白再释放现象的机制研究。本书对角线电泳的特点是，样品为完整的红细胞，两向电泳之间不做任何处理，就能发现有无互作。

◎　**多带释放**

注释：参见"多带再释放"。

◎　**多带再释放**

注释：电泳过程中多次"断电—再通电"，又释放出来多个新的血红蛋白区带，故称"多带再释放"。参见附图 1-8～附图 1-10。

◎　**拐弯现象**

注释：无氧及有氧条件下电泳释放时出现的现象，见附图 1-11～附图 1-13。

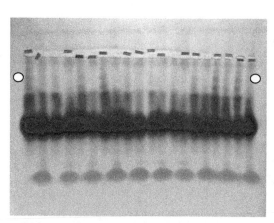

附图 1-8 糖尿病患者红细胞的多带再释放增强
注释：注意两个○之间的多带部分

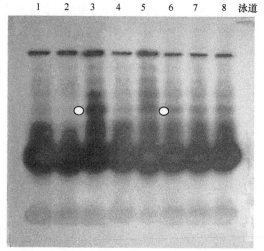

附图 1-9 β-地贫患者全血中红细胞的多带再释放增强
注释：泳道3为母亲，泳道5为女儿，注意两个○处的多带增强

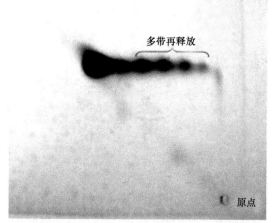

附图 1-10 α地贫患者红细胞的多带再释放增强
注释：第一向电泳时"断电-通电"多次这是双向电泳，多带再释放在对角线的右上方

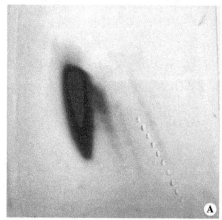 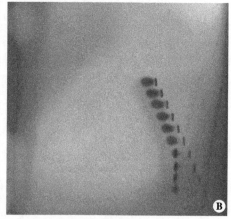

附图 1-11 糖尿病红细胞指纹图电泳过程中发现无氧胶拐弯现象
注释：A图为无氧条件下糖尿病红细胞指纹图结果，B图为其血红蛋白泳动过程中出现拐弯现象，未染色，在电泳槽中直接照相

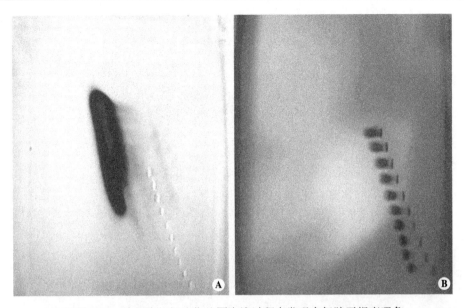

附图 1-12　糖尿病红细胞指纹图电泳过程中发现有氧胶无拐弯现象

注释：A 图为有氧条件下糖尿病红细胞指纹图结果，B 图为其血红蛋白泳动过程中没有拐弯现象，未染色，在电泳槽中直接照相

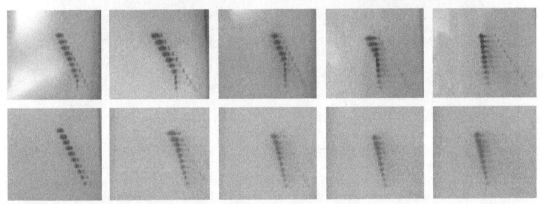

附图 1-13　无氧胶拐弯现象全程

注释：第一行的 5 个图是无氧胶内红细胞指纹图电泳过程中出现的连续图像，无氧胶内 10 个泳道 Hb(注意，不是 HbO_2)的泳动过程，有拐弯现象；第二行的 5 个图是有氧胶内红细胞指纹图电泳过程中出现的连续图像，有氧胶内 10 个泳道 HbO_2 的泳动过程，没有拐弯现象

◎ 后退电泳

注释：通常的血红蛋白电泳，各成分(血红蛋白 A_3、血红蛋白 A_1、血红蛋白 A_2、碳酸酐酶 CA)都是由负极泳向正极，我们称之为"前进电泳"。"后退电泳"与之相反，是前进到一定程度后调换电极使其"后退"，看原点是否还有血红蛋白释放(后退释放)出来，具体操作是停电，倒极(正负极对调)，有一些标本又由原点退出血红蛋白。目前，最突出的例子就是肝内胆管癌，它的后退非常明显，参见附图 1-14。

◎ 混合互作

注释：相互作用，与交叉互作对应，称为"混合互作"，这种互作比较多见，特别是各种物质与红细胞的相互作用，如附图 1-15 和附图 1-16。

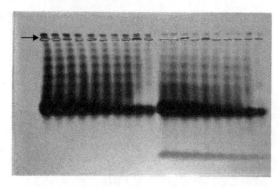

附图1-14 肝内胆管癌患者红细胞后退带明显

注释：全程双释放，左侧红细胞，右侧全血，"→"所指之处为"后退带"

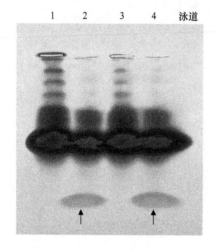 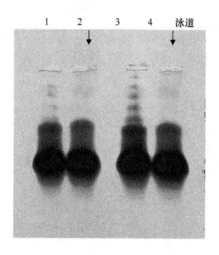

附图1-15 白蛋白对红细胞多带释放的影响电泳图

注释：泳道1、3为红细胞，泳道2、4为红细胞加白蛋白(见图中箭头↑所指处)。红细胞存在多带再释放，红细胞加入白蛋白后，多带再释放明显减弱。这就是白蛋白与红细胞的相互作用(混合互作)

附图1-16 丙种球蛋白对红细胞多带释放的影响电泳图

注释：泳道1、3为红细胞，泳道2、4为红细胞加丙种球蛋白(见图中箭头↓所指处)。红细胞存在多带再释放，红细胞加入丙种球蛋白后，多带再释放明显减弱。这就是丙种球蛋白与红细胞的相互作用(混合互作)

◎ 基质

注释：红细胞用CCl_4处理后离心，在下部CCl_4与上层溶血液之间的沉淀层即为"红细胞基质"(红细胞膜成分)。同样处理全血，得到的是"全血红细胞基质"，简称"全血基质"。不同疾病基质情况可有差别，参见"全血多组分电泳"及附图1-17。

◎ 交叉电泳

注释：交叉电泳由来已久，支持体是滤纸、琼脂或琼脂糖，标本是血浆蛋白。我们使用的支持体是淀粉-琼脂糖混合凝胶，标本主要是红细胞溶血液。用此技术，我们发现血红蛋白的"交叉互作"。参见附图1-18和附图1-19。

◎ 交叉互作

注释：一种血红蛋白穿过另一种血红蛋白，后者区带变形，这就是发生了"交叉互作"。参见"交叉电泳"。迄今为止，交叉互作只见于血红蛋白，不是普遍存在的互作方式，而与此对应的"混合互作"好像更为多见。参见"混合互作"。

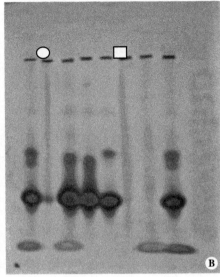

附图 1-17 不同疾病的基质情况

注释：A 图为肝豆状核变性，B 图为自身免疫性疾病，泳道 2○为全血基质，泳道 6□为红细胞基质；由上往下看，自身免疫性疾病患者两种基质都明显增多

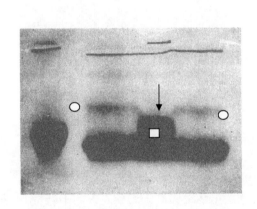

附图 1-18 血红蛋白 A_2 与血红蛋白 A_1 交叉互作

注释：箭头↓所指之处为 HbA_1(□)穿过 HbA_2(○—○)，后者变形

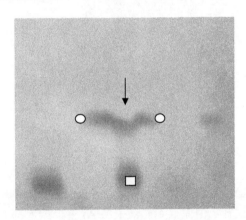

附图 1-19 血红蛋白 A_2 与中华鳖血红蛋白交叉互作

注释：箭头↓所指之处为中华鳖 Hb(□)穿过 HbA_2(○—○)，后者变形

◎ **交叉互作消失**

注释：无氧条件下做释放实验，交叉互作消失。参见"无氧条件下红细胞内血红蛋白的电泳释放"。

◎ **接力式初释放**

注释：简言之，就像运动会中"接力赛"那样。具体是用全血做初释放型(通电一次)单向电泳，其中红细胞内有高铁血红蛋白，就能释放出高铁血红素，传递给血浆中的白蛋白，后者接过高铁血红素，生成高铁血红素白蛋白(methemalbumin, MHA)，通常白蛋白无色，MHA 显暗红色。如果红细胞内没有高铁血红素，就不发生上述过程。以前，测 MHA 都是用血浆或血清，有或无 MHA，比较明确。过去无人用全血做此实验，不知道这方面的情况。现在看来，血浆或血清有 MHA，全血也有 MHA，这是一致的，说明有较严重的溶血。我们发现，有时血浆或血清没有 MHA，而全血出现 MHA，这是一个新问题，我们认为这是将要出现较严重溶血的状态，是预测"溶血风险"的有效指标。参见附图 1-20。

◎ **两个互作**

注释：一个"交叉互作"，一个"非交叉互作"（"混合互作"）。

◎ **两个释放**

注释：一个"初释放"、一个"再释放"。再参见"双释放"。

◎ **前进电泳**

注释："前进电泳"是针对"后退电泳"(参见前边此项)来说的。原因是，常规电泳("前进电泳")后有时原点处留有红色，甚至单带再释放或多带再释放后原点处仍有红色，后退一下会怎样？结果是前进、后退都可有再释放带，参见附图 1-21。

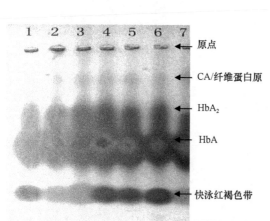

附图 1-20　全血中血红素接力式初释放
注释：看泳道 4、5、6 最下方的三个深色区带，即为接力释放者

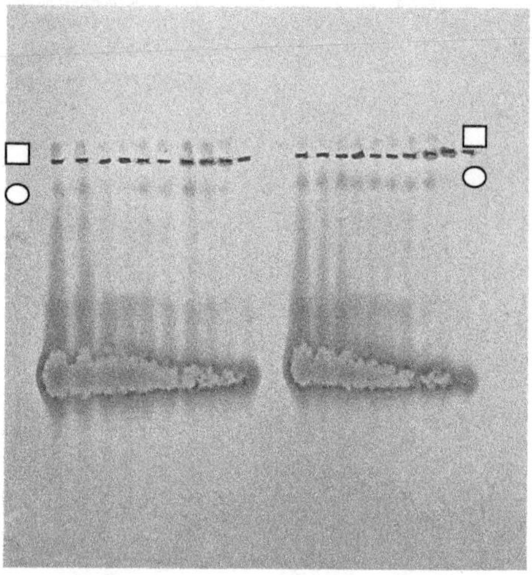

附图 1-21　肝硬化患者红细胞全程电泳结果
注释：两个〇之间的是"前进释放"，两个口之间的是"后退释放"

◎ **全程拖尾**

注释：由原点到血红蛋白 A_2 整个泳道都有拖尾。好像此时红细胞膜"关闭不严"，一直有血红蛋白"泄露"。

◎ **全血电泳**

注释：用全血直接做电泳。以前，有人做过血浆蛋白电泳、血清蛋白电泳、红细胞溶血液电泳(血红蛋白电泳)，但没人用全血做过电泳。我们是由红细胞释放电泳开始，延伸到全血，从中发现不少有趣的问题，有时用途比红细胞还大。红细胞电泳里，我们发现了"血红蛋白 A_2 现象"，全血电泳里，除了"血红蛋白 A_2 现象"外，还有"纤维蛋白原现象"，参见附图 1-22。

◎ **全血多组分电泳**

注释：拿来一份全血，用淀粉-琼脂糖凝胶电泳分析其中各个组分，全血、全血溶血液、红细胞、红细胞溶血液、基质(红细胞用 CCl_4 处理后的沉淀物)和血浆。将这些组分在同一胶板上电泳，就是"全血多组分电泳"。可以做初释放、再试放、定释或梯带；能看到血浆中有无 M 蛋白、MHA(高铁血红素白蛋白)，能看到溶血液中有无异常血红蛋白，血红蛋白 A_3、

血红蛋白 A_2、血红蛋白 F 是否明显增加，碳酸酐酶 CA 是否明显减少；从红细胞和全血，可以看到血红蛋白病等。这种电泳也是别人没有做过的。首先看上边"基质"处的两个图。再看附图 1-23：真性红细胞增多症。

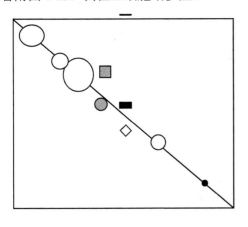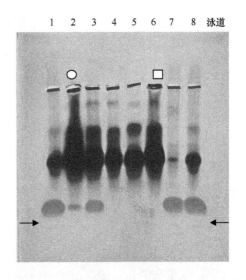

附图 1-22　全血的对角线电泳

注释：附图 1-22 是全血的双向电泳，这里既有"血红蛋白 A_2 现象"，又有"纤维蛋白原现象"。①上边的■(HbA_1)、下边的●(HbA_2)，它们都脱离对角线，这就是血红蛋白 A_2 现象。②上边的"长条■"(纤维蛋白原)、下边的◇(目前尚不明确为何物，推测它的存在)，它们都脱离对角线，这就是纤维蛋白原现象。③血红蛋白 A_2 现象的机制，基本明确。④纤维蛋白原现象的机制，尚不清楚。⑤这里还有"浆球互作"(血浆成分与血球成分之间的相互作用)。⑥全血电泳的奥秘还有很多，有待进一步发掘

附图 1-23　真性红细胞增多症的全血多组分

注释：泳道 1=全血溶血液，2=全血基质，3=全血，4=红细胞，5=红细胞溶血液，6=下边基质，7=浆液，8=1；此时全血和血浆出现前白蛋白(箭头→←处)，两种基质都特强(○、□处)

◎　全血红细胞释放

注释：全血中红细胞成分的电泳释放，不是游离红细胞成分的电泳释放。由于全血里有血浆，它会影响到红细胞，所以效果不同。参见"全血电泳"。

◎　全血溶血液

注释：全血用 CCl_4 处理，其中红细胞溶血，即为全血溶血液。全血冻化处理、用氯仿处理也可以，也是全血溶血液，效果有差别：冻化最温和，氯仿的变性作用最强。

◎　双释放

注释：指"游离红细胞的再释放"和"全血中红细胞的再释放"。同是红细胞，由于全血中有血浆，释放效果差异较大，形成一个大课题："环境因素(或血浆因素)对红细胞成分电泳再释放的影响"，内容丰富，潜力较大，理论上涉及血浆成分与红细胞之间的相互作用(混合互作)，临床上涉及多种疾病的诊断和鉴别诊断。参见"全血电泳"，再看附图 1-24。

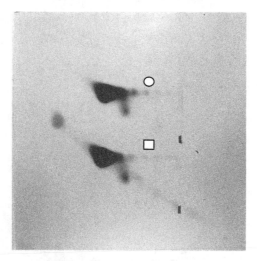

附图 1-24 双释放的双向电泳

注释：上层为红细胞，下层为全血；全血中红细胞的梯带释放(□下边)弱于红细胞(○下边)。说明此例全血中血浆成分对红细胞的多带再释放有抑制作用

◎ 双向电泳

注释：这里指的是"淀粉-琼脂糖混合胶双向电泳"，比等电聚焦与聚丙烯酰胺的双向电泳简单，但有自己的特色。用普通的电泳槽、一般缓冲液就能完成，可以将红细胞等细胞加到胶中，进行"血红蛋白释放试验 HRT"，还可以进行断电-再通电完成"再试放"。此法可做单层、双层、三层及更多层电泳，进行各种比较。参见附图 1-25～附图 1-29。

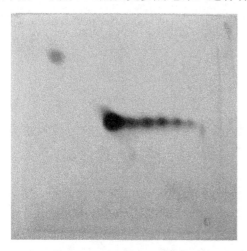

附图 1-25 单层

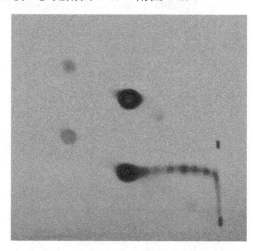

附图 1-26 双层

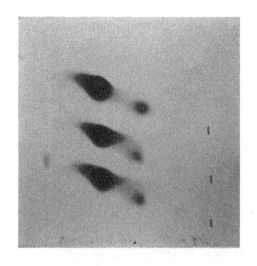

附图 1-27　三层

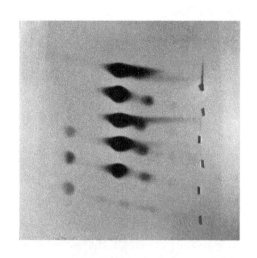

附图 1-28　六层

附图 1-29　十层

注释：这是拿全血全程的 10 个标本做双向电泳，也就是我们所说的"指纹图"。

◎　双向单层(一层)电泳

注释：一般双向电泳，只加 1 个标本。参见附图 1-25。

◎　双向双层(二层)电泳

注释：双向电泳，加 2 个标本。参见附图 1-26。

◎　双向三层电泳

注释：双向电泳，加 3 个标本。参见附图 1-27。

◎　双向六层电泳

注释：双向电泳，加 6 个标本。参见附图 1-28。

◎　双向多层电泳

注释：双向电泳，加多个标本。参见附图 1-26～附图 1-28。

◎　梯带

注释：相当于"多带"，像梯子脚蹬那样多，参见"多带再释放"，参见附图 1-25。

◎ 梯带释放

注释：电泳过程中多次切断电源，释放出多个区带，像"梯子氧"。参见"梯带"及"多带释放"。

◎ 无氧条件下红细胞内血红蛋白的电泳释放

注释：淀琼胶中加入偏重亚硫酸钠，造成还原环境、无氧条件。在此条件下，做红细胞内血红蛋白的电泳释放研究，发现血红蛋白 A_2 现象消失、交叉互作消失，与有氧条件下的实验结果形成鲜明对比。

◎ 细胞成分电泳释放

注释：将细胞加入凝胶后电泳，观察其中成分的释放情况。此时，不限于红细胞，任何细胞成分都可以电泳释放。我们做过血小板等 ERC。

◎ 纤维蛋白原现象

注释：做全血的双向电泳时，纤维蛋白原脱离对角线，这个现象就是"纤维蛋白原现象"。参见附图 1-30 和附图 1-31。

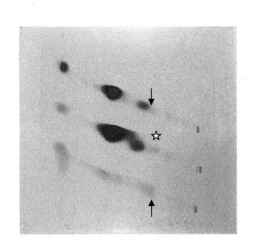

附图 1-30 双向三层电泳 比较全血溶血液、全血和血浆中的纤维蛋白原 FG

注释：血液标本来自包头医学院第一附属医院肾内科；双向三层电泳，上层为全血溶血液，中层为全血，下层为血浆；箭头↓所指的红色区带为全血溶血液里的 FG，它在对角线上；箭头↑所指的红色区带为血浆里的 FG，它在对角线上；☆下边的红色区带为全血中的 FG，它横过来，脱离了对角线。☆下边全血里的纤维蛋白原脱离对角线，这就是"纤维蛋白原现象"。临床标本中，大多数显示此现象，也有 FG 不脱离对角线者

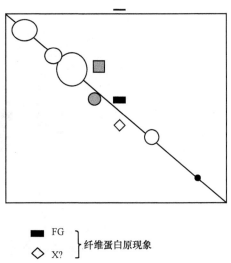

附图 1-31 全血的对角线电泳

注释：全血双向电泳中，能看到 FG，它脱离了对角线，"X"看不见(蛋白染色无效)，这些是作者的推测，有待进一步证实

◎ 血红蛋白释放试验

注释：将红细胞或含红细胞的全血加入淀粉-琼脂糖混合凝胶，通电后观察血红蛋白由红细胞释放出来的情况，这就是"血红蛋白释放试验 HRT"。HRT 分两大类：①"初释放"；②"再释放"，详见相应项目。

◎ 血红蛋白释放指纹图

注释：参见"指纹图"项下。

◎ 血红蛋白 A_2 现象

注释：将红细胞及其溶血液加在凝胶上并排电泳。此时来自红细胞的"血红蛋白 A_2"靠前（阳极侧），来自溶血液的血红蛋白 A_2 靠后（阴极侧），我们将这一对比电泳结果，称之为血红蛋白 A_2 现象。这是"血红蛋白 A_2 现象"的核心内容，其他细节还有：红细胞的血红蛋白 A_2 与血红蛋白 A_1 之间界限不太清楚，红细胞血红蛋白 A_2 与原点之间有轻度血红蛋白拖泄，红细胞的原点处可有少量血红蛋白残留。参见附图 1-32。

以上说的都是单向电泳，此现象的双向电泳更具特色，而且有助于对机制的理解。参见附图 1-33。此双向电泳图的下层为溶血液，上层为红细胞。溶血液的各成分（血红蛋白 A_3、血红蛋白 A_1、血红蛋白 A_2、碳酸酐酶 CA）都在对角线上；红细胞的各成分则不同：血红蛋白 A_3、血红蛋白 A_1、碳酸酐酶 CA 在对角线上；脱离对角线的成分有二：①血红蛋白 A_2 则在对角线下方（○处）；②在对角线上方与血红蛋白 A_2 上下对应处还有"血红蛋白 A_1"（□处），二者上下对应，它们中间有拖泄。这说明，红细胞内血红蛋白 A_2 与血红蛋白 A_1 是结合存在的，当第二向电泳时二者分开，形成图中上下对应的局面。

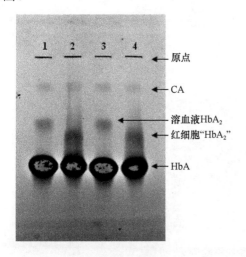 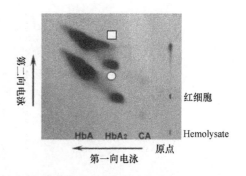

附图 1-32　血红蛋白 A_2 现象的单向电泳图
注释：注意溶血液 HbA_2 与红细胞 HbA_2 的位置关系；泳道由左向右，1、3 为溶血液，2、4 为红细胞

附图 1-33　血红蛋白 A_2 现象的双向电泳图
注释：来自溶血液的 HbA_2 与 HbA_1 在同一对角线上；来自红细胞的 HbA_2（见○处）在对角线下方，来自红细胞的 HbA_1（见□处）在对角线上方，上下呼应

◎ 血红蛋白 A_2 现象消失

注释：无氧条件下做释放实验，血红蛋白 A_2 现象消失。参见"无氧条件下红细胞内血红蛋白的电泳释放"。

◎ 原点残留

注释：用红细胞或含红细胞的全血做淀粉-琼脂糖混合凝胶电泳，可看到有多量血红蛋白释放出来。但是，电泳结束后原点处有时仍显红色，说明这里还有血红蛋白残留。原点残留血红蛋白与已经释放出来的血红蛋白有何区别？它能否再释放出来？为此，我们研究了再释放。参见"再释放"。不同疾病原点残留也不一样，参见附图 1-34。

◎ 原点泄漏

注释：通常情况下，电泳过程中要泳出若干集中的区带，整个泳道的背景是清晰的。用红细胞做释放试验过程中，有时由原点到前方(阳极侧)出现不集中的红色成分(由原点连续泄漏出来的血红蛋白)，称之为"原点泄漏"。这说明，红细胞膜有问题，有"漏洞"，不是一次泳出或分批泳出，而是一点一点地、拖拖拉拉地由原点释放出来。这种情况在正常人的红细胞也能见到(参见"血红蛋白 A_2 现象")，不同疾病原点泄漏也不一样，参见附图1-35。

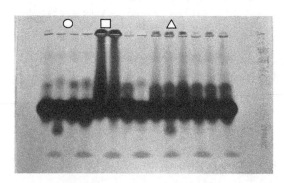

附图1-34　原点残留的不同情况

注释：□处原点残留最多，△处原点残留次之；○处没有原点残留

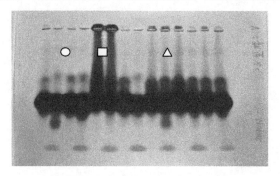

附图1-35　原点泄漏的不同情况

注释：8份癌症标本，有的无泄漏(○处)，有的有泄漏(△处)，泄漏最强的是肝内胆管癌标本(□处)

◎ 再释放

注释：将红细胞及含红细胞的全血加到淀粉-琼脂糖混合凝胶中，进行"血红蛋白释放试验 HRT"时，通电一次(不要停电)，就有血红蛋白释放出来，此时称为"初释放"。如果在此过程中人为地进行断电-再通电，就会有新的血红蛋白由原点释放出来，我们称之为"再释放"。"再释放"又分"单带再释放"和"多带再释放"。 一次断电-再通电，再释放出一条区带，称为"单带再释放"；多次断电-再通电，再释放出多条区带，称为"多带再释放"。参见附图1-36~附图1-38。

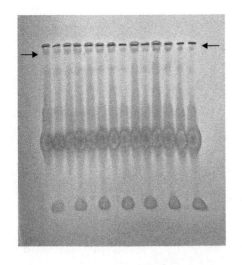

附图1-36 单向电泳中的单带再释放
注释：两个箭头(→←)之间的区带都是单带再释放

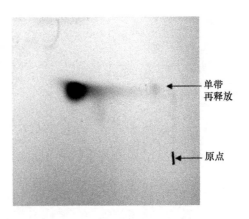

附图1-37 双向电泳中的单带再释放
注释：箭头(←)处为单带再释放，第一向电泳时"断电-通电"一次

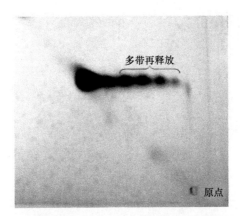

附图1-38 双向电泳中的多带再释放
注释：图中已标明多带再释放的位置，第一向电泳时"断电-通电"多次

◎ 指纹图

注释：指纹图或指纹图谱，文献里早有很多此类词汇，如蛋白质鉴定指纹图谱、芯片多肽指纹图谱、临床蛋白质指纹图谱、小灵芝抗肿瘤指纹图谱、DNA 指纹图谱、遗传指纹图谱、中药指纹图谱、黄酮类和酚类指纹图谱等。这里是我们提出的"电泳释放指纹图"或"血红蛋白电泳释放指纹图"。以下是我们指纹图的几个例子。附图1-39来自正常人标本，附图1-40来自地中海贫血患者，附图1-41来自肝硬化患者。现在看来，电泳释放指纹图有明显的鉴别意义，研究范围还在扩大中。

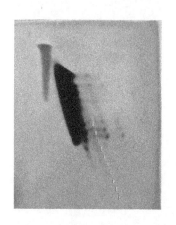 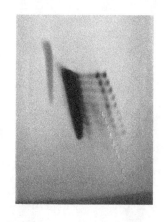 

附图 1-39　正常人血红蛋白电泳释放指纹图　　附图 1-40　地中海贫血患者血红蛋白电泳释放指纹图　　附图 1-41　肝硬化患者血红蛋白电泳释放指纹图

附录二 红细胞释放电泳(溶血液电泳)历年发表的文章

用红细胞做释放电泳，是由我们开始创建的，之前都是用红细胞溶血液做电泳，下面所列是我们在溶血液电泳方面的文章。

1964年
秦文斌，赵宗诚，吴永华. 几种哺乳动物血红蛋白的种间杂交. 生物化学与生物物理学报，1964，(4)：289.

1965年
赵宗诚，秦文斌，吴永华. 不同遗传类型羊血红蛋白与犬、兔血红蛋白的杂交. 生物化学与生物物理学报，1965，(5)：27。
秦文斌，赵宗诚，吴永华. 改进淀粉胶薄层电泳法及其对血红蛋白不均一性的分辨能力 生物化学与生物物理学报，1965，(5)：278.

1966年
赵宗诚，秦文斌，吴永华. Zhao ZC, Qin WB, Wu YH. Interspecific Hybridization of several mammalian hemoglobin. Scientia Sinica (中国科学)1966，(15)：360.

1974年
秦文斌，潘立民，黄静仁，等. 血红蛋白G. 新医学，1974，5(2)：66.

1975年
秦文斌. 血红蛋白H病. 内蒙古医学院学报，1975，(2)：96.
秦文斌. α-地中海贫血—血红蛋白H病. 包头科技，1975，(3-4)：12.
秦文斌. 血红蛋白分子杂交. 生物化学与生物物理进展，1975，2(3)：42，56-59.
秦文斌. 国内外异常血红蛋白研究近况. 内蒙古医学院学报，1975，(2)：137.
秦文斌. 氯仿-碘酒试验(CIT)对肾病综合征的诊断意义. 中华医学杂志，1975，55(1)：74.
潘立民，秦文斌. 白血病前期——一过性巨球蛋白血症. 新医学，1975，6(2)：71.
秦文斌. 用马铃薯淀粉直接作淀粉胶电泳. 内蒙古医学院学报，1975，(2)：5.

1976年
秦文斌，岳秀兰，陈其明，等. 脂蛋白分析用琼脂电泳. 中华医学杂志，1976，56(8)：

506.

1977 年

秦文斌, 王凤岐. 糖尿病血红蛋白. 包头医药, 1977, (1): 1.

秦文斌. 糖尿病与血红蛋白 A_1. 中华医学杂志, 1977, (57): 674.

1978 年

秦文斌. 麝香草酚絮状试验(TFT)反常与单株免疫球蛋白血症. 新医学, 1978, 9(5): 605.

秦文斌. 血红蛋白病的分子病理学. 生物科学动态, 1978, (2): 6.

秦文斌. 在我国发现的异常血红蛋白. 生物化学与生物物理进展, 1978, (5): 24.

秦文斌, 陈其明, 岳秀兰, 等. 血红蛋白 Constant Spring 及 CS-型血红蛋白 H 病. 中华医学杂志, 1978, (58): 399.

秦文斌, 李成林, 崔丽霞, 等. 包头地区汉族中发现的快泳异常血红蛋白——血红蛋白 J-包头. 生物化学与生物物理学报, 1978, 10(4): 333.

秦文斌, 崔丽霞, 陈其明, 等. 草原蒙古族牧民中发现两种 D 型异常血红蛋白. 遗传学报, 1978, 5(4): 263.

1979 年

秦文斌, 陈其明, 岳秀兰, 等. Qin WB, Chen QM, Yue XL, et al. Hemoglobin Constant Spring and CS-Type Hemoglobin H Disease. Chinese Medical Journal. 1979, 92(11): 787.

秦文斌, 王凤岐. Qin WB, Wang FQ. Diabets Mellitus and Hemoglobin A_3. Chinese Medical Journal, 1979, 92(9): 639.

秦文斌. 异常血红蛋白的分子遗传学. 国外医学(分子生物学分册), 1979, (1): 152.

秦文斌. 建国三十年来我国血红蛋白病临床研究进展. 输血与血液学杂志, 1979, (3): 6.

1980 年

秦文斌. 血红蛋白结构与功能——分子呼吸. 国外医学(分子生物学分册), 1980, (3): 109.

1981 年

秦文斌, 梁友珍, 陈其明, 等. 仿沉淀试验检查不稳定血红蛋白——证明血红蛋白 Constant Spring 是一种不稳定变异物. 中华血液学杂志, 1981, (2): 124.

秦文斌, 梁友珍, 王凤岐, 等. 用纸上电泳法测定红细胞碳酸酐酶及其在甲亢诊断中应用的初步探讨. 内蒙古医学院学报, 1981, 1(2): 105.

秦文斌. 在内蒙古自治区发现的异常血红蛋白. 内蒙古医学杂志, 1981, 1(1): 1.

秦文斌, 睢天林, 王海龙, 等. 在东乌珠穆沁旗蒙族牧民中发现的异常血红蛋白——血红蛋白 D-东乌. 内蒙古医学杂志, 1981, 1(4): 217.

秦文斌, 刘建, 崔珊娜. 内蒙古西苏旗牧区蒙古族中异常血红蛋白的调查. 西苏科技, 1981, (1): 6.

秦文斌, 刘建, 崔珊娜. 蒙古族牧民的 HbA_2 及抗碱 Hb 正常值. 西苏科技, 1981, (1): 8.

1983 年

秦文斌，睢天林，王海龙，等. 在蒙古族中首次发现血红蛋白 D-Los Angeles[β121(GH4)Glu→Gln]. 生物化学与生物物理学报，1983，15(6)：537.

1984 年

秦文斌. 血红蛋白病.北京：人民卫生出版社，1984.

秦文斌. 血红蛋白结构分析专辑. 包头医学院学报，1984，1(2)：1.

秦文斌，睢天林，闫秀兰，等. 在我国汉族中首次发现血红蛋白 J-Lome[β59(E3)Lys→Asn]. 包头医学院学报，1984，1(2)：1.

秦文斌，睢天林，闫秀兰，等. 在内蒙古自治区汉族中发现血红蛋白 E[β26(E8)Glu→Lys]. 包头医学院学报，1984，1(2)：26.

睢天林，秦文斌，岳秀兰，等. 血红蛋白 D-东乌=血红蛋白 D- Los Angeles[β121(GH4)Glu→Gln]. 包头医学院学报，1984，1(2)：20.

睢天林，秦文斌，闫秀兰，等. 在阿拉善盟蒙古族中遇到一例血红蛋白 E[β26(B8)Glu→Lys]. 包头医学院学报，1984，1(2)：32.

岳秀兰，秦文斌. 在包头地区发现一例血红蛋白 G-Taipei[β22(B4)Glu→lYS]. 包头医学院学报，1984，1(2)：42.

岳秀兰，闫秀兰，睢天林，等. 在包头地区发现一例血红蛋白 G-Coushatta[β22(B4)Glu→Ala]. 包头医学院学报，1984，1(2)：36.

周立社，秦文斌. 在包头地区发现一例血红蛋白 G-Chinese[α30(B11)Glu→Gln]. 包头医学院学报，1984，1(2)：50.

1985 年

秦文斌，睢天林，闫秀兰，等. 血红蛋白 G-呼和浩特=HbG-Taichung[α74(EF3)Asp→His]. 新医学，1985，16(1)：30.

秦文斌，睢天林，岳秀兰，等. 血红蛋白 Constant Spring. 遗传学报，1985，12(5)：395.

秦文斌，睢天林，岳秀兰，等. Qin WB，Ju TL，Yue XL，et al. Hemoglobin Constant Spring in China. Hemoglobin，1985，9：69.

睢天林，秦文斌，闫秀兰，等. 在我国汉族中发现一例血红蛋白 J-Lome[β59(E3)Lys→Asn]. 遗传与疾病，1985，2(4)：209.

睢天林，秦文斌，岳秀兰，等. 在我区遇到的一例苯酮尿症. 包头医学院学报，1985，2(2)：1.

岳秀兰，闫秀兰，睢天林，等. 一例遗传性胎儿血红蛋白持续增多症(HPFH)及其结构分析. 包头医学院学报，1985，2(2)：14.

1986 年

秦文斌，睢天林，岳秀兰，等. Qin WB，Ju TL，Yue XL，et al. Structure and Characteristics of Hb Constant Spring 7th International Congress of Human Genetics. Berlin German，1986，Part II：426.

睢天林，岳秀兰，秦文斌. Ju TL，Yue XL，Qin WB. Combining teaching experiments with scientific research. Symp Biochem Educat, 1986: 42.

岳秀兰, 闫秀兰, 睢天林, 等. 遗传性胎儿血红蛋白持续增多症(HPPH)一例报告. 中华血液学杂志, 1986, 7(12): 740.

1987 年

秦文斌. Qin WB. Hemoglobin Constant Spring and CS-type α-Thalassemia. Hemoglobin, 1987, (11): 588.

岳秀兰, 闫秀兰, 睢天林, 等. Yue XL, Yan XL, Ju TL, et al. Hereditary persistence of fetal hemoglobin, report of one case. International Meeting on Biochemistry. 1987, Abstracts: 18.

1988 年

睢天林, 岳秀兰, 王珂, 等. 氨基酸分析技术诊断苯酮尿症, Ⅰ微晶薄膜层析分析. 包头医学院学报, 1988, 5(1-2): 1.

闫秀兰, 睢天林, 岳秀兰, 等. 血红蛋白 Q-Thailand 的化学结构分析. 内蒙古医学杂志, 1988, 8(2): 68.

岳秀兰, 王海龙, 秦良宜, 等. 呼和浩特 9212 人群异常血红蛋白调查报告. 包头医学院学报, 1988, 5(1-2): 4.

1990 年

岳秀兰, 闫秀兰, 王海龙, 等. 在内蒙古地区发现一例血红蛋白 G-San Jose. 内蒙古医学杂志, 1990, (10): 1.

再次申明, 以上这些文章, 都来自溶血液电泳实验, 与完整红细胞释放无关。参加全国科学大会的文章, 就是发表于 1964～1966 年的文章。

附录三 红细胞释放电泳

1 红细胞初释放电泳简介

1.1 先做单向电泳，比较红细胞与溶血液的电泳差异 实验中发现，由红细胞释放出来的HbA_2(红细胞HbA_2)与溶血液HbA_2的电泳位置不同。红细胞HbA_2靠前(阳极侧)，溶血液HbA_2靠后(阴极侧)。这是怎么回事呢？目前还不清楚，暂时称为"血红蛋白A_2现象"。详情参见第四篇。

1.2 再做双向电泳，比较红细胞与溶血液的电泳差异 实验中发现，溶血液中各成分都在对角线上，而红细胞中各成分则不同，有的位于对角线上，有的则脱离对角线。脱离对角线成分有二，线上为HbA_1，线下为HbA_2。这说明什么问题呢？由实验推测，红细胞HbA_2可能是HbA_2与HbA_1的结合产物，在双向电泳的第二向电泳时彼此分开，快泳的HbA_1上升、慢泳的HbA_2下降，都脱离了对角线。在上、下两种血红蛋白之间，存在拖泄成分。详情参见第三篇。

1.3 质谱分析，弄清"血红蛋白A_2现象"的机制 取上、下两种血红蛋白之间部位，抠胶，做质谱分析，发现HbA_2与HbA_1互作，还有Prx(过氧化物还原酶)参加；复合物为HbA_1-HbA_2-Prx。详情参见第三篇。

1.4 "血红蛋白A_2现象"发生机制示意图 由下图可以看出，溶血液里，两个"小人"(HbA_2与HbA_1)没有"牵手"，游离存在。红细胞里，三个"小人"(HbA_1、HbA_2和Prx)"牵手"，形成复合物(附图3-1)。

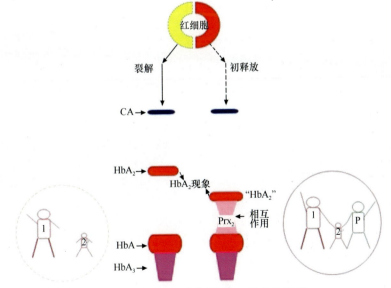

附图3-1 "血红蛋白A_2现象"示意

注释：图片来自"血红蛋白A_2现象"的发现和研究——活体红细胞内血红蛋白间的相互作用(1981~2017，秦文斌，包头医学院血红蛋白研究室)

2 红细胞再释放电泳简介

2.1 先做单向多带再释放电泳，比较红细胞与溶血液的电泳差异　电泳过程中，多次停电—再通电此时发现，由红细胞释放出来多个血红蛋白区带，溶血液则没有多带释放。这是怎么回事呢？目前尚不清楚，暂时称为"再释放现象"。双向再释放结果显示，再释放出来的成分很可能是 HbA_1；质谱分析结果，证明是 HbA_1-CA_1。详情参见第四篇。

2.2 再做双向多带再释放电泳，比较红细胞与溶血液的电泳差异　加样：上层为红细胞，下层为溶血液；第一向，多次停电—再通电，第二向，普泳。此时发现，由红细胞释放出来多个血红蛋白区带，与 HbA_1 平行，溶血液则没有多带释放。双向再释放结果显示，再释放出来的成分很可能是 HbA_1；质谱分析结果，证明是 HbA_1-CA_1。详情参见第四篇。

2.3 质谱分析，弄清"再释放现象"的机制　取以上实验中多带释放区带，抠胶，做质谱分析，发现 HbA_1 与 CA_1 互作，复合物为 HbA_1-CA_1。详情参见第四篇。

2.4 "单带再释放现象"发生机制　由附图 3-2 可以看出：红细胞里，两个"小人"（HbA_1-CA_1）牵手，形成复合物；H "小人"为 HbA_1，C "小人"为 CA_1。

2.5 "多带再释放现象"发生机制　由附图 3-3 可以看出：红细胞里，多个"双小人"（HbA_1-CA_1）"牵手"，每次停电—再通电释放出来一组"双小人"。

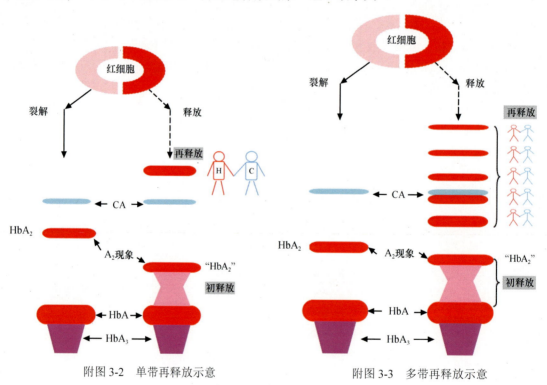

附图 3-2　单带再释放示意　　　　附图 3-3　多带再释放示意

附录四　电泳成分染色法

1　前言

多数电泳技术，都有电泳成分的染色问题。淀粉-琼脂糖凝胶电泳，涉及的染色法有丽春红染色法、氨基黑 10 B 染色法、考马斯亮蓝染色法、丽春红-联苯胺复染法。

1.1　丽春红染色法　此法不是常用的染色法，也不是较灵敏的染色法。我们把它与联苯胺染色法结合起来，先染丽春红，再染联苯胺，使血红蛋白显蓝色、非血红素蛋白显红色，对比鲜明，从而得到广泛应用。在长期使用过程中，我们又发现了它的另一优点。前边制备电泳中提到"抠胶"问题，抠胶操作都是用丽春红染色，然后冻结-融化处理，很容易将血红蛋白分离出来，从而达到制备目的。

1.2　氨基黑 10B 染色法　此法常用，特别是对血浆蛋白。对于血红蛋白，它不如联苯胺染色法。

1.3　考马斯亮蓝染色法　此法灵敏度较高。对于血浆蛋白，优于氨基黑 10 B 染色法。对于血红蛋白，可能稍逊于联苯胺染色法，但也有其优势。例如，在红细胞释放电泳结果中，在 HbA_3 的前方(阳极侧)出现一些细小成分，而联苯胺染色看不到这些东西。

1.4　联苯胺染色法　它是靠过氧化物酶活性，专门染血红蛋白或血红素蛋白，对于血红蛋白来说，其他染色法都不如此法。但是，对于血浆蛋白，本法是无能为力的。

1.5　丽春红-联苯胺联合染色法　血液标本，先染丽春红，再染联苯胺，出现红蓝搭配(血红蛋白显蓝色，非血红蛋白显红色)的鲜艳颜色，看起来很舒服。这是我们的原创技术，国外约稿中这种染色结果也特别醒目。

2　丽春红染色法

2.1　染色操作

2.1.1　丽春红染色液配制　先加 200ml 漂洗液于 1L 的容量瓶中，称取 1.0g 丽春红于该容量瓶中。加漂洗液至满刻度。

2.1.2　漂洗液的配制　先加 200ml 蒸馏水于 1L 的容量瓶中。用量筒量取 50ml 冰醋酸，20ml 丙三醇(甘油)于该容量瓶中。加蒸馏水至满刻度。

2.1.3　丽春红染色　将事先配制好的丽春红染液倒于 20cm×30cm 的白瓷盘中，再把电泳好的胶板平放进去，让染液没过胶面。约染 3 小时，或过夜。到时取出胶板，照片留图(附图 4-1)。

2.1.4　拍照留图　数码相机拍照。

2.1.5　胶板保存　室温自然干燥或在自制烤箱上烤干。

2.2　染色结果举例

2.2.1　巨球蛋白血症患者血清　双向电泳结果见附图 4-1。

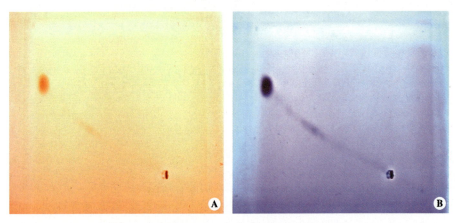

附图 4-1　巨球蛋白血症患者血清的双向电泳结果
注释：A=丽春红染色，B=氨基黑 10B 染色

结果：丽春红染色不如氨基黑 10B 染色清楚。

2.2.2　巨球蛋白血症患者血清单向向电泳结果　加与不加 DTT(二硫苏糖醇)，见附图 4-2。

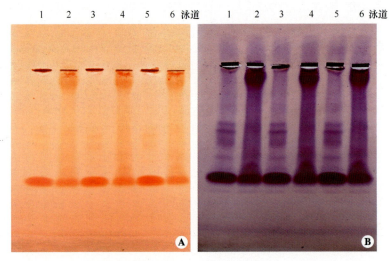

附图 4-2　巨球蛋白血症患者血清的单向电泳结果，加与不加 DTT
注释：A=丽春红染色，B=考马斯亮蓝染色；单数泳道为患者血清未加 DTT，双数泳道为患者血清加 DTT

结果：丽春红染色不如考马斯亮蓝染色清楚。

2.2.3　本书作者的红细胞指纹图　见附图 4-3。

结果：丽春红染色不如丽春红-联苯胺染色清楚。

2.2.4　本书作者的全血指纹图　见附图 4-4。

结果：丽春红染色不如丽春红-联苯胺染色清楚。

2.2.5　补充说明

(1) 丽春红染色法对蛋白质的显色不是最灵敏的，我们选择它的原因是它的"红色"。众所周知，联苯胺染色法对血红蛋白非常灵敏，血红蛋白被染成蓝色。如果把丽春红染色法与联苯胺染色法结合起来，血红蛋白显蓝色、其他蛋白显红色，电泳结果出来，一看就知道有无这两类蛋白质，它们的分布情况如何。详见下边的"丽春红-联苯胺染色法"。

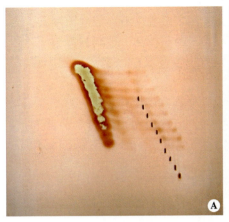 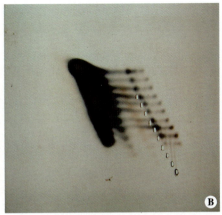

附图 4-3　本书作者的红细胞指纹图
注释：A=丽春红染色，B=丽春红-联苯胺染色

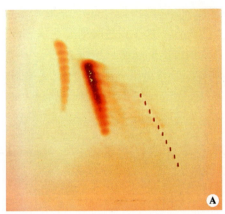 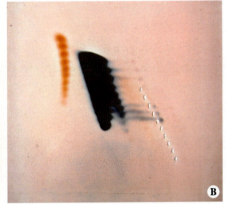

附图 4-4　本书作者的全血指纹图
注释：A=丽春红染色，B=丽春红-联苯胺染色

(2) 丽春红染色法的另一个优点或特点是有利于"制备电泳"：丽春红染色时丽春红与蛋白质结合，此结合产物与凝胶的再结合不牢固，容易洗脱和分离制备。其他染色产物都与凝胶结合牢固，不容易洗脱。所以，我们能在丽春红染色后"抠胶"，来分离和制备各种蛋白质，产生我们自己独特的制备电泳方法，详见第九篇制备电泳。

丽春红染色的这一特点，也是我们在实践中发现的，未见文献记载。

3　氨基黑 10B 染色法

3.1　染色操作

3.1.1　0.5%的氨基黑 10 B 染色液的配制　先加 200ml 漂洗液于 1L 的三角烧瓶中，称取 2.5g 联苯胺于该三角烧瓶中。再加漂洗液至 500ml，充分混合，待氨基黑 10B 完全溶解备用。

3.1.2　漂洗液的配制　先加 200ml 蒸馏水于 1L 的容量瓶中。用量筒量取 50ml 冰醋酸，20ml 丙三醇(甘油)于该容量瓶中。加蒸馏水至满刻度。

3.1.3　染色法　将事先配制好的氨基黑 10B 染色液倒于 20cm×30cm 的白瓷盘中，再把刚电泳好的胶板平放进去，让染液没过胶面。染 24 小时或更长时间，待取出胶板时再用清

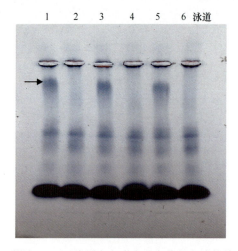

附图 4-5 巨球蛋白血症患者血浆与血清的单向比较电泳结果

注释：泳道 1、3、5=血浆，2、4、6=血清

水洗去胶面多余的染液，照片留图。

3.1.4 拍照留图　数码相机拍照。

3.1.5 胶板保存　室温自然干燥或置自制烤箱上烤干。

3.2 染色结果举例

3.2.1 巨球蛋白血症患者血浆和血清的单向电泳结果　氨基黑 10B 染色见附图 4-5。

结果：血浆有纤维蛋白原(箭头→所指处)，但未见巨球蛋白。巨球蛋白留在原点吗？血清没有纤维蛋白原，也未见巨球蛋白。巨球蛋白留在原点吗？

讨论：巨球蛋白留在原点。

3.2.2 巨球蛋白血症患者血清的单向电泳结果　比较氨基黑 10B 和考马斯亮蓝染色(附图 4-6)。

结果：两种染色法的结果差不多，考马斯亮蓝染色法的结果更清晰一些。

讨论：氨基黑 10B 染色法也可以，常用于血浆、血清蛋白电泳。

4 考马斯亮蓝染色法

4.1 染色操作

4.1.1 染色液的配制　2.5g 的考马斯亮蓝 G-250，450ml 30%冰醋酸，100ml 甲醇，加水至 1L。

4.1.2 染色法　将事先配制好的考马斯亮蓝的染液倒于 20cm×30cm 的白瓷盘中，再把染好丽春红并已烤干的胶板平放进去，让染液没过胶面。染 24 小时或更长时间，待取出胶板时再用清水洗去胶面多余的染液，照片留图。

4.1.3 拍照留图　数码相机拍照。

4.1.4 胶板保存　室温自然干燥。

4.2 染色结果举例

4.2.1 血红蛋白 A_2 现象的单向电泳结果，考马斯亮蓝染色　见附图 4-7。

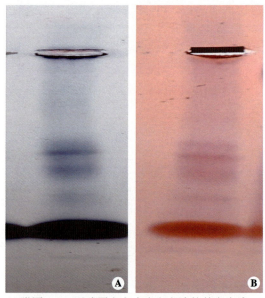

附图 4-6 巨球蛋白血症患者血清的单向电泳

注释：A=氨基黑 10B 染色，B=考马斯亮蓝染色

(1) 结果：与丽春红-联苯胺染色结果相比，考马斯亮蓝染色结果中多出以下几点：①多出快泳成分，两个□之间；②多出慢泳成分，两个○之间；③多出中泳成分，两个箭头→←之间。

(2) 讨论：慢泳成分可能是 CA_2 快泳成分为何种物质？尚不明确。中泳成分更复杂，红细胞者靠后，溶血液者靠前，为何种物质？尚不明确。

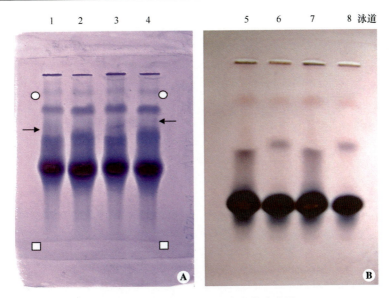

附图 4-7 血红蛋白 A_2 现象的单向电泳

注释：A=考马斯亮蓝染色，B=丽春红-联苯胺染色；泳道 1、3、5、7=红细胞，2、4、6、8=溶血液

(3) 结论：考马斯亮蓝染色法的灵敏度较高，上边是与丽春红-联苯胺染色法比较，若与其他染色法比较更显优势。

4.2.2 血红蛋白 A_2 现象的双向电泳结果，考马斯亮蓝染色 见附图 4-8。

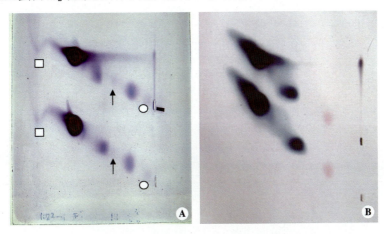

附图 4-8 血红蛋白 A_2 现象的双向电泳

注释：A=考马斯亮蓝染色，B=丽春红-联苯胺染色；上层=红细胞，下层=红细胞溶血液

(1) 结果：与丽春红-联苯胺染色结果相比，考马斯亮蓝染色结果中多出以下几点：①多出快泳成分，两个□之间；②多出慢泳成分，两个○之间；③多出中泳成分，箭头↑所指处。

(2) 讨论：慢泳成分可能是 CA_2 快泳成分为何种物质？尚不明确。中泳成分为何种物质？尚不明确。

(3) 结论：考马斯亮蓝染色法的灵敏度较高。上边是与丽春红-联苯胺染色法比较，若与其他染色法比较更显优势。必须指出，丽春红-联苯胺染色法的特点是能区分血红蛋白(蓝色)与非血红蛋白(红色)，考马斯亮蓝染色法没有这个功能。

5 丽春红-联苯胺染色法

5.1 染色操作

5.1.1 染色液的配制

(1) 丽春红染色液：先加 200ml 漂洗液于 1L 的容量瓶中，称取 1.0g 丽春红于该容量瓶中。加漂洗液至满刻度。

(2) 联苯胺染色液：先加 200ml 漂洗液于 1L 的三角烧瓶中，称取 0.5g 联苯胺于该三角烧瓶中。再加漂洗液至 500ml，于电磁炉上的水浴锅内加温至 75℃约 0.5 小时，待联苯胺完全溶解。

(3) 漂洗液的配制：先加 200ml 蒸馏水于 1L 的容量瓶中。再用量筒量取 50ml 冰醋酸，20ml 丙三醇(甘油)于该容量瓶中。加蒸馏水至满刻度。

5.1.2 染色法

(1) 丽春红染色：将事先配制好的丽春红染液倒于 20cm×30cm 的白瓷盘中，再把电泳好的胶板平放进去，让染液没过胶面。约染 3 小时，或过夜。到时取出胶板。

(2) 联苯胺染色：染色前现配染液。在白瓷盘内先加入硝普钠约 2g，再倒入刚配好的联苯胺液 500ml，再加入过氧化氢 1.4ml 混匀后。把已晾干或烤干的胶板放入染液中，约 20 分钟。边染边晃动瓷盘以便染色更均匀。

5.1.3 拍照留图　数码相机拍照。

5.1.4 胶板保存　室温自然干燥。

5.2 染色结果举例

5.2.1 前言
本书中，很多实验采用的都是"丽春红-联苯胺染色"，可达 99%。这也是我们的原创之一，国内外文献里，有丽春红染色和联苯胺染色，没有把二者结合起来的"丽春红-联苯胺染色法"。丽春红染色法对蛋白质的显色不是最灵敏的，我们选择它的原因是因为它的"红色"。

众所周知，联苯胺染色法对血红蛋白非常灵敏，血红蛋白被染成蓝色。如果把二者结合起来，血红蛋白显蓝色、其他蛋白显红色，电泳结果出来，一看就知道有无这两类蛋白质，它们的分布情况如何。最突出的例子是白蛋白，丽春红染色为红色。高铁血红素白蛋白(MHA)，联苯胺染色为蓝色。用全血、血浆或血清做电泳，然后用丽春红-联苯胺染色，观察白蛋白那个地方，若是红色，则证明正常；若是蓝色，证明有溶血疾病。

5.2.2 血红蛋白 A_2 现象的单向电泳　见附图 4-9。

(1) 结果：溶血液的 HbA_2(两个○所指处)与红细胞的"HbA_2"(两个□所指处)电泳位置不同。

(2) 讨论：这就是"血红蛋白 A_2 现象"单向电泳的标准图，从染色法来看，上边红色区带是碳酸酐酶，溶血液与红细胞的 CA 电泳位置相同。

5.2.3 血红蛋白 A_2 现象的双向电泳　见附图 4-10。

(1) 结果：下层各成分都在对角线上，上层有的成分在对角线上，有的不在对角线上(脱离对角线)，脱离对角线的成分是 HbA_2(○所指处)和 HbA_1(□所指处)，上下对应，中间有拖泄。

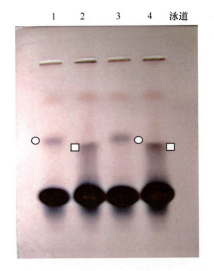
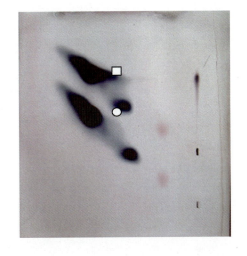

附图 4-9　血红蛋白 A_2 现象的单向电泳
注释：泳道 1、3=红细胞溶血液，泳道 2、4=红细胞；红色区带=CA(碳酸酐酶)，蓝色区带=血红蛋白

附图 4-10　血红蛋白 A_2 现象的双向电泳
注释：上层=红细胞，下层=红细胞溶血液；红色区带=CA(碳酸酐酶)，蓝色区带=血红蛋白

(2) 讨论：这就是"血红蛋白 A_2 现象"双向电泳的标准图。这个实验证明，"红细胞 HbA_2" = HbA_2-HbA_1，脱离对角线成存在相互作用关系，从染色法来看，红色区带是碳酸酐酶，溶血液与红细胞的 CA 都在对角线上。

5.2.4　多带再释放※的单向电泳　见附图 4-11。

(1) 结果：泳道 4、6 的区带特别多，即多带再释放。

(2) 讨论：从染色法来看，红色区带是白蛋白，黑色区带是血红蛋白，白蛋白没问题，血红蛋白有特殊情况。从疾病角度来看，轻型 β-地中海贫血患者全血出现多带再释放。遗传关系是母亲传给女儿，联苯胺染色效果良好。

5.2.5　多带再释放的双向电泳　见附图 4-12。

(1) 结果：看到"毛毛虫样"的多带再释放。

(2) 讨论：从染色法来看，红色区带是白蛋白和碳酸酐酶，蓝色区带是血红蛋白。从疾病角度来看，α-地中海贫血患者全血也出现多带再释放。联苯胺染色效果良好。

5.2.6　全血多组分电泳※　见附图 4-13。

(1) 结果：与其他人群相比，泳道 2、6 的基质明显增多。

(2) 讨论：此时作者 82 岁，基质增多的原因不明。联苯胺染色效果明显。

5.2.7　全程电泳※　见附图 4-14。

(1) 结果：红细胞全程中，多带释放增强、后退释放增强；全血全程中，多带释放和后退释放也都比较强。

(2) 讨论：这是全程电泳中典型的全面增强病例，实际上，它是来自肝内胆管癌导致的重度"梗阻性黄疸"。联苯胺染色效果良好。

※ 多带再释放：释放电泳里，第二次通电造成释放，称为"再释放"；再释放时，放出一个区带的，称为"单带再释放"；反复多次通电，放出多个区带的，称为"多带再释放"，这里是多带再释放。

※全血多组分电泳：是分离全血中的各种成分，然后做单向电泳，比较各种人群之间的差异。这里有 7 项：全血溶血液、全血基质、全血、红细胞、红细胞溶血液、红细胞基质、血浆。

※全程电泳，也称"等低渗全程电泳"，即由等渗到低渗的全部过程，而且还有红细胞全程和全血全程，一次完成。

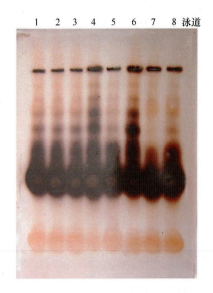

附图 4-11 多带再释放的单向电泳

注释：这是 8 份全血标本，泳道 4、6 是轻型 β-地中海贫血标本，4 是女儿，6 是母亲，5 是父亲，其他人都是正常人。最下边一排红色区带是白蛋白，上边的黑色区带是血红蛋白

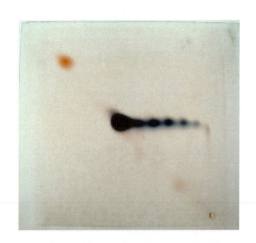

附图 4-12 多带再释放的双向电泳

注释：这是一个 α-地中海贫血患者全血的多带再释放双向电泳结果；左上方的红色区带是白蛋白，右下方的红色区带是碳酸酐酶，中间横行蓝色多带是血红蛋白

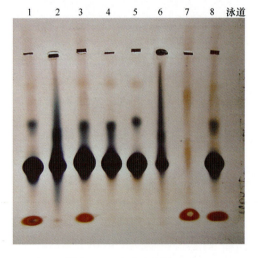

附图 4-13 全血多组分电泳

注释：这是本书作者的全血多组分电泳结果。泳道 1=全血溶血液，2=全血基质，3=全血，4=红细胞，5=红细胞溶血液，6=红细胞基质，7=血浆，8=泳道 1

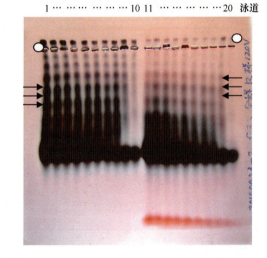

附图 4-14 肝内胆管癌的全程电泳

注释：1～10=红细胞全程，11～20=全血全程，右下方多个红色区带=白蛋白，多个箭头→←之间=多带释放，两个○之间=后退带(后退释放)

5.2.8 指纹图电泳※　见附图 4-15。

(1) 结果：前进释放全面增强，后退释放也全面增强。

(2) 讨论：这是 α-地中海贫血患者的全血指纹图，其特点就是两种释放(前进和后退)都增强，而且是全面增强。联苯胺染色效果良好。

※指纹图电泳：可以理解为全程的双向电泳，全程是等低渗的单向电泳，如果把它改成双向电泳，就是这里所说的"指纹图"。全程里，一次完成红细胞和全血的两部分电泳，指纹图不行，只能分开，成为"红细胞指纹图"和"全血指纹图"。附图 4-15 是全血指纹图。

5.2.9 全血高铁血红素白蛋白单向电泳　见附图 4-16。

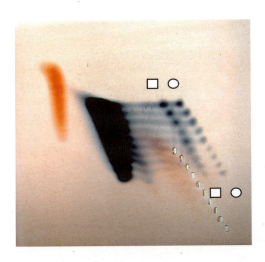

附图 4-15　α-地中海贫血患者的全血指纹图
注释：左上方的红带是白蛋白，右下方的红带是碳酸酐酶和纤维蛋白原；两个□之间的是前进带(前进释放带)，两个○之间的是后退带(后退释放带)

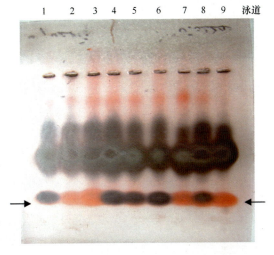

附图 4-16　全血高铁血红素白蛋白(MHA)单向电泳
注释：两个箭头→←之间为血浆白蛋白部分，其中红色区带为正常的白蛋白，黑色区带为 MHA

(1) 结果：有 MHA 的标本存在溶血。

(2) 讨论：做全血的单向电泳，通过染色法，能够发现有无溶血问题。

5.2.10　全血高铁血红素白蛋白双向电泳　见附图 4-17。

(1) 结果：有 MHA 的标本存在溶血。

(2) 讨论：做全血的双向电泳，通过染色法，能够发现有无溶血问题。

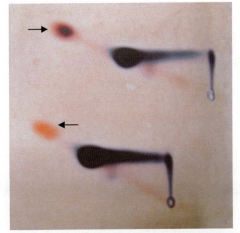

附图 4-17　全血高铁血红素白蛋白(MHA)双向电泳
注释：上下两层，为两个全血标本，箭头←所指红色区带为白蛋白，箭头→所指处红黑区带为白蛋白里含 MHA

附录五　电泳图画[※]

Ⅰ　花(附图 5-1)

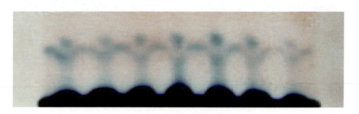

附图 5-1　花
注释：血红蛋白 A_2 与 A_1 之间的交叉互作，单向电泳

Ⅱ　鸟(附图 5-2)

Ⅲ　虫(附图 5-3)

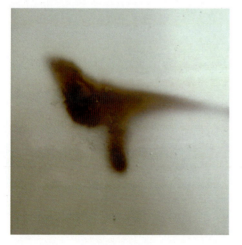

附图 5-2　鸟
注释：干燥综合征患者红细胞的双向电泳

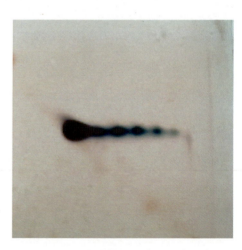

附图 5-3　虫
注释：α-地中海贫血患者红细胞的双向电泳

Ⅳ　鸭子戏水(附图 5-4)

Ⅴ　游泳比赛(附图 5-5)

[※] 图片来自我们的释放电泳和交叉电泳。

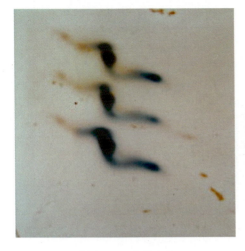

附图 5-4 鸭子戏水

注释：乌龟红细胞的双向三层电泳

附图 5-5 游泳比赛

注释：乌龟血液的双向三层电泳

Ⅵ 双排键电子琴(附图 5-6)

Ⅶ 高楼夜景(附图 5-7)

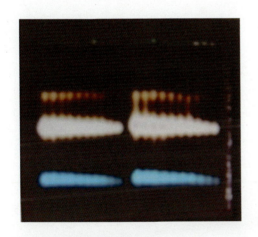

附图 5-6 双排键电子琴

注释：球形红细胞增多症患者血液的全程电泳

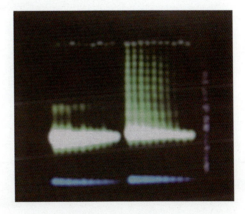

附图 5-7 高楼夜景

注释：正常人与 α-地中海贫血患者血液的全程电泳

Ⅷ 瀑布(附图 5-8)

Ⅸ 水中月影(附图 5-9)

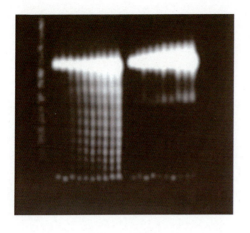

附图 5-8　瀑布
注释：α-地中海贫血患者血液的全程电泳

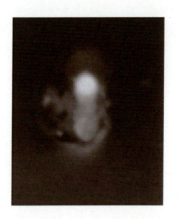

附图 5-9　水中月影
注释：将淋巴细胞白血病患者的淋巴细胞放在圆滤纸片上电泳，有蛋白质释放到淀琼凝胶中

X　夜空流星(附图 5-10)

XI　航天飞机/宇宙飞船(附图 5-11)

附图 5-10　夜空流星
注释：鸡血与人血的双向双层电泳

附图 5-11　航天飞机/宇宙飞船
注释："血红蛋白 A_2 现象"的双向双层电泳

XII　火箭发射/反导系统(附图 5-12)

XIII　飞船降落/太空舱返回(附图 5-13)

附图 5-12　火箭发射/反导系统

注释：淋巴细胞白血病患者淋巴细胞的双向电泳

附图 5-13　飞船降落/太空舱返回

注释：乳糜胸患者血液双向电泳

XIV　漫山云雾(附图 5-14)

附图 5-14　漫山云雾

注释：血红蛋白 A_2 与 A_1 的交叉互作，双向电泳

XV　惊涛骇浪(附图 5-15)

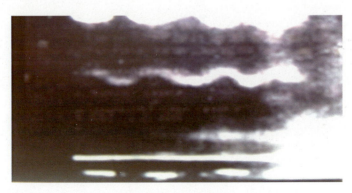

附图 5-15　惊涛骇浪

注释：血红蛋白 A_2 与 A_1 的交叉互作，单向电泳